PRODUCTIVITY IMPROVEMENT IN CONSTRUCTION

McGraw-Hill Series in Construction Engineering and Project Management

Consulting Editor

Raymond E. Levitt *(Stanford University)*

Barrie and Paulson: Professional Construction Management
Jervis and Levin: Construction Law: Principles and Practice
Koerner: Construction and Geotechnical Methods in Foundation Engineering
Levitt and Samelson: Construction Safety Management
Oglesby, Parker, and Howell: Productivity Improvement in Construction
Peurifoy and Ledbetter: Construction Planning, Equipment, and Methods
Peurifoy and Oberlender: Estimating Construction Costs
Shuttleworth: Mechanical and Electrical Systems for Construction

Also available from McGraw-Hill:

Schaum's Outline Series in Civil Engineering

Each outline includes basic theory, definitions, and hundreds of solved problems and supplementary problems with answers. The current list includes:

Advanced Structural Analysis
Basic Equations of Engineering
Descriptive Geometry
Dynamic Structural Analysis
Engineering Mechanics, 4th ed.
Fluid Dynamics
Fluid Mechanics and Hydraulics
Introduction to Engineering Calculations
Introductory Surveying
Reinforced Concrete Design, 2d ed.
Space Structural Analysis
Statics and Strength of Materials
Strength of Materials, 2d ed.
Structural Analysis
Theoretical Mechanics

Available at your college bookstore

PRODUCTIVITY IMPROVEMENT IN CONSTRUCTION

Clarkson H. Oglesby
Stanford University

Henry W. Parker
Stanford University

Gregory A. Howell
University of New Mexico

McGraw-Hill Publishing Company

New York   St. Louis   San Francisco   Auckland   Bogotá   Caracas
Hamburg   Lisbon   London   Madrid   Mexico   Milan
Montreal   New Delhi   Oklahoma City   Paris   San Juan
São Paulo   Singapore   Sydney   Tokyo   Toronto

This book was set in Times Roman by the College Composition Unit
in cooperation with Ruttle-Shaw & Wetherill, Inc.
The editors were B. J. Clark and John M. Morriss;
the production supervisor was Louise Karam.; the designer was Rafael Hernandez.
Project supervision was done by The Total Book.
R. R. Donnelley & Sons Company was printer and binder.

PRODUCTIVITY IMPROVEMENT IN CONSTRUCTION

3 4 5 6 7 8 9 0 D O C D O C   9 3 2 1

ISBN   0-07-047802-3

Library of Congress Calaloging-in-Publication Data

Oglesby, Clarkson Hill (date)
  Productivity improvement in construction.

  Based on: Methods improvement for construction
managers / Henry W. Parker, Clarkson H. Oglesby (date)
  Bibliography: p.
  Includes index.
  1. Engineering—Management. I. Parker, Henry W.
II. Howell, Gregory A. III. Parker, Henry W. Methods
improvement for construction managers. IV. Title.
TA190.035 1989   624'.068   88–9346
ISBN 0–07–047802–3
ISBN 0–07–047803–1 (solutions manual)

ABOUT THE AUTHORS

CLARKSON H. OGLESBY is Silas Palmer Professor of Civil Engineering, Emeritus, Stanford University. He was a regular faculty member for 31 years, until 1974, and has continued his active participation in teaching and research since then. Before coming to Stanford during World War II, he was employed in construction for 12 years. During the years that Professor Oglesby has been associated with the Construction Engineering and Management Program at Stanford, over 1000 graduate engineers have entered the industry and teaching positions in many parts of the world. He has served as a technical advisor on programs in construction education in Chile, Australia, South Africa, and at many universities in the United States. He is currently a member of the Academic Advisory Council of the Construction Industry Institute.

Many of the concepts in this book originated in courses that Professor Oglesby taught in the areas of productivity improvement and the "people side" of construction management. In addition to this pioneering work in education for construction managers, he is known for his research and writing on highways and for his book *Highway Engineering,* 4th ed. (John Wiley, 1982). He is coauthor, with Henry W. Parker, of *Methods Improvement for Construction Managers* (McGraw-Hill, 1972).

Professor Oglesby holds an A.B. degree in Engineering and the Degree of Engineer in Civil Engineering, both from Stanford University. He is an Honorary Member of ASCE. In 1964 he received the Golden Beaver Award for Outstanding Service to the Construction Industry. He has received Outstanding Educator awards from the American Road and Transportation Builders Association and the National Society of Professional Engineers as well as two Transportation Research Board awards for technical papers in highway economics and planning. In 1965–66 he was a Fulbright Professor at Imperial College, London. He is a Registered Civil Engineer in California and a member of Phi Beta Kappa, Tau Beta Pi, and Sigma Xi.

HENRY W. PARKER is Professor Emeritus of Civil Engineering at Stanford University, where he taught for over 20 years in the graduate Construction

Engineering and Management Program. Before going to Stanford, he was employed in the heavy construction industry in management and engineering for 15 years.

During his years at Stanford, Professor Parker taught courses in productivity improvement. His research on this topic contributed greatly to the development and use of timelapse equipment as a tool for recording and improving construction operations. He also developed strategies to assist construction managers to initiate productivity-improvement programs. He directed and participated in research in construction safety and lectured on this topic to many industry groups. A most significant finding was to link supervisors' attitudes and actions to construction accidents and the disruptions and costs that they bring. Professor Parker coauthored, with Oglesby, *Methods Improvement for Construction Managers* (McGraw-Hill, 1972).

Parker's construction experience covers work in the United States, Canada, and South America as a superintendent, engineer, and assistant project manager and has involved work on canals, railroads, highways, tunnels, and dams, as well as industrial projects. He was employed by Winston Bros. Co. of Minneapolis and its partners from 1947–1962, when he left to join the faculty at Stanford. Professor Parker's consulting has taken him to Venezuela and Canada as well as through much of the United States. He has lectured and consulted for numerous trade organizations and contractors, and has been an arbitrator and expert witness in productivity and safety matters. He was a member of the Underground Construction Research Council for 9 years.

Professor Parker is a graduate of Dartmouth College and the Thayer School of Engineering. He is a Registered Civil Engineer in California and New Hampshire, a Fellow of the American Society of Civil Engineers, and a member of Sigma Xi, Phi Beta Kappa, and Tau Beta Pi.

GREGORY A. HOWELL is associate professor of civil engineering at the University of New Mexico, Albuquerque and a consultant who specializes in the organization and management of construction projects. Before he organized the consulting firm, Howell Associates, he worked as an engineer in both heavy and general building construction, and he managed Timelapse, Inc., a firm which manufactured special timelapse camera and projection equipment.

Professor Howell's consulting activities for the past 15 years have taken him to all parts of North America and to South America and Africa. He has worked on power plants, petrochemical facilities, commercial and industrial buildings, and highways, for both owners and contractors. In addition he has been retained as an expert witness on productivity-related disputes. His clients include many major U.S. contractors and owners.

Howell received his B.S. and M.S. degrees in Civil Engineering from Stanford University. He is a Registered Civil Engineer in New Jersey and California and a Member of the American Society of Civil Engineers. He has authored a number of articles and conducted many seminars dealing with construction productivity. He served in the Civil Engineer Corps of the USNR with the Seabees in Vietnam and Thailand and as an Aide to the Commander, South West Division, Naval Facilities Engineering Command.

CONTENTS

PREFACE

This book is a refocused and greatly expanded version of an earlier work titled *Methods Improvement for Construction Managers,* by Henry W. Parker and Clarkson H. Oglesby (McGraw-Hill, 1972). That book was written because of the authors' conviction that changing management techniques and operating procedures could improve on-site productivity. Today that conviction is stronger than ever. It has been reinforced by the original authors' more recent experiences augmented by the third author's background. These, in sum, present about 100 years of combined experience in observing, teaching, managing, consulting, and doing research in construction productivity improvement. Because the need to improve productivity is even more urgent than it was 16 years ago, we and many of our colleagues feel that this book can fill a large void in the construction literature.

We wish to make clear at the outset that improvements in productivity and safety go hand in hand. Accident prevention is always an important element in productivity improvement; every mention of productivity improvement implies accident reduction as well.

In the late 1970s a turning point was reached when the long-neglected question of productivity in construction was raised insistently and loudly. It came with the general realization that construction productivity had fallen dramatically—some say by about 3 percent per year—in each of the previous ten years. One influential group, the Business Roundtable, representing some of the largest buyers of construction, became so concerned that it sponsored and carried through the "Construction Industry Cost Effectiveness Study." This study, completed in early 1983, resulted in 23 hard-hitting reports, each prepared by executives and specialists from buyers of construction, contractors, and construction educators. These reports outlined many contributing factors leading to cost increases and proposed solutions to correct them.

Following the study's conclusion, the Construction Industry Institute, made up of representatives of buyers, contractors, and university specialists, and based at the University of Texas at Austin, was created to implement the Business Roundtable's recommendations by carrying out research and disseminating its findings on many of the construction-related problem areas identified in the earlier study. Many other agencies and companies and their management, as well as academics, have become involved in similar activities sponsored by industry, universities, and governmental agencies such as the National Science Foundation and branches of the armed services. All have targeted some of their construction research on factors related to construction productivity. Our book draws on the findings of this research as well as improvements that have been made in industry practices, both to refocus and augment the approaches outlined in the earlier book.

The basic approaches presented here have been developed over a period of some 35 years, during which two of the authors and other faculty members have taught courses dealing with productivity-improvement techniques to students in the graduate Construction Engineering and Management Program in the Civil Engineering Department at Stanford University. These experiences have been enhanced by the hands-on experience of all three authors in managing construction work, consulting, and conducting workshops for groups representing owners, designers, contractors, and, at times, labor unions.

The difference in approach between this and the earlier book comes from a realization by the authors that they had failed to target the key audience, namely, construction managers at all levels. Rather, it was naively assumed that presenting the techniques for improving productivity in books, university classrooms, or at workshops would assure their ready adoption by the construction industry. In reality, these adoptions seldom occurred. Since that earlier time it has become clear to us that any efforts to improve on-site productivity must first gain the attention and enthusiastic support of managers, particularly those at the top. Only if they take the concepts seriously will any sustained productivity-improvement efforts be undertaken on site. For this reason, our new book emphasizes the savings in hard dollars that can be realized by improving productivity. The focus is directed at the mental sets that higher-level managers must develop and the actions that they must take if productivity-improvement efforts are to be successful.

This book has been written and organized to be useful to several audiences, ranging from students preparing to work in the industry to those charged with management responsibilities at all levels. These include:

• *Students in undergraduate engineering, architectural, or construction-management programs:* For this group, the amount of time available in the curriculum to study construction operations in the field is limited; therefore it is important to make that experience meaningful. This can be done by employing the sections of the book which focus attention on techniques for observation, analysis, and problem solving which can improve productivity and safety in on-site construction operations.

• *Students in graduate-level construction programs:* These students have time to dig in more deeply and learn the specifics of productivity and safety improvement in on-site operations. In addition, the book introduces the overall organizational approaches and management strategies required to make productivity improvement happen, and supports these with in-depth presentations on how to work with people, on safety, and on the more advanced, computer-based techniques for productivity improvement that are now being developed.

• *Field personnel who desire or are charged with the responsibility for applying the concepts underlying productivity and safety improvement on their projects:* This group may encompass any or all among the following: project managers, engineers, planners, line supervisors, including superintendents and foremen, and planning and methods-improvement specialists. The book serves each of these groups in different ways. Higher-level management and planners will be more concerned with developing the strategies, mechanisms, and commitment to set productivity-improvement approaches in motion and make them work. Those actually directing the work will want to know the procedures and techniques that must be introduced to improve their performance.

• *Organized oversight, productivity improvement, or worker teams estab-lished by owners or contractors:* As with field personnel, the book addresses a variety of activities that may concern and be the focal point of such groups.

• *Individuals looking for ways to improve their performance through study and application of selected aspects of productivity improvement:* Again, particular sections of the book will provide a focus by indicating approaches that have been found to be effective in developing productivity-improvement ideas and then getting people to implement these findings.

• *Groups, possibly organized by companies of trade associations, to study or implement efforts in specific areas:* The book will provide direction and examples for those interested in organizing and implementing productivity-improvement programs, applying specific techniques, dealing with people as workers and human beings, and developing strategies to improve safety.

To better fit the needs of these diverse audiences, the subject matter of this book has been organized as follows:

Chapter 1: offers a prescription that must be followed if efforts to improve productivity are to be successful.

Chapter 2: provides background on the construction industry, its cultures, how it is organized and operated, the parties and people involved, and how all these and other factors influence efforts to improve productivity.

Chapter 3: proposes policies, procedures, structures, and reward systems that will create a climate favorable to improving productivity and safety, and, with them, timeliness and quality.

Chapter 4: views on-site construction as a system involving activities that take place off-site, on-site, and at the work face, and examines the role of people in this system.

Chapters 5 through 8: thrust the reader into the approaches and techniques which apply most directly to on-site productivity improvement and the approaches that have been found to be successful.

Chapters 9 through 11: provide concepts and specific data on the roles that people play in productivity improvement as organisms, as human beings that react to management's strategies, and as planners developing and implementing productivity-improvement programs.

Chapter 12: deals with safety and environmental health in the following ways: it proposes positive approaches to accident prevention, offers data about accidents and the out-of-pocket costs of unsafe practices, points out that carelessness can lead to accidents or be injurious to health, describes current approaches that can be applied to achieve on-site accident reduction, and outlines the accident-reduction activities of various agencies aimed at decreasing accidents and thereby lowering their costs in both human and money terms.

Chapter 13: briefly introduces some of the newer and sometimes more complex approaches to productivity improvement, particularly those that are being made available as computers become working tools both off and on site.

Appendices A through D: supply useful data to support and reinforce the concepts and practices given in the text.

The authors have made strong efforts to provide references for those who wish to explore specific topics in greater detail. As a practical matter, only a small number from those available were cited: these were chosen, regardless of their age, as the best on each specific topic. The references generally were chosen from sources available in large public or university libraries or because they can otherwise be obtained easily.

In its entirety, this book addresses many of the key issues in on-site construction productivity and safety. It focuses primary attention upon on-site situations, both away from and at the work face. After all, the site is where the action is; and the site is where, through the actions of managers or workers, productivity improvements finally take place.

Clarkson H. Oglesby

Henry W. Parker

Gregory A. Howell

ACKNOWLEDGMENTS

This book summarizes under one cover what the authors have learned for themselves and from others during their combined experience totaling about 100 years. During that time, they have profited from the input of colleagues on the Stanford faculty and some 1000 graduate students in the Construction Engineering and Management Program at Stanford University, who have provided challenge, support, and leg work. The knowledge and experience of faculty at other schools and a host of individuals in the construction industry have also been drawn upon, and their contributions are gratefully acknowledged. Inputs from these sources have been utilized, in part, by employing each author's published reports, and in such cases they have been cited as references. Many others have contributed greatly by passing on their experiences informally or in unpublished form.

Efforts to improve construction productivity and safety reported here follow four general tracks, in each of which a number of people have made outstanding contributions. These tracks are: (1) techniques for recording and analyzing data; (2) the "people side," involving individuals as organisms and human beings, and as organized in groups to make productivity improvement a reality; (3) accident prevention as an economic, human, and organizational problem; and (4) advanced approaches stemming primarily from computer applications.

Specific recognition in each of these areas is given here to those whose work most influenced the authors, either through their writings or personal contacts. The names of many others, not listed below, appear in the references.

In the recording and analyzing data track: J. W. Fondahl, W. L. Jones, and Henning Jacobsen, who in the 1960s developed and applied timelapse to construction operations; and, more recently, B. C. Paulson and Lloyd Waugh and students working with them, who brought videotaping as an advancement over timelapse photography into classroom and professional use. A number of individuals, including Glenn Ballard, J. D. Borcherding, G. J. Lemna, D. F. Rogge, and R. L. Tucker, have employed questionnaires effectively, and H. R. Thomas Jr. and others have made work sampling a more effective tool. The examples of productivity-improvement studies in Chapter 8 and Appendix C come directly from the authors' experience or the work of Stanford students.

In the human-relations and organizational tracks: The early teaching of the late C. L. Thorpe at Stanford and the research of J. D. Borcherding focused the findings of behavioral scientists on the roles workers play in improving construction productivity. More recently J. D. Borcherding, A. Laufer, W. F. Maloney, and J. Fillen, among others, have published data on further advances. Work dealing with "organizing to make productivity possible" has involved, among others, J. T. O'Connell, V. E. Sanvido, C. B. Tatum, J. A. Vanegas, and J. M. Williams, as well as the authors.

In the accident prevention field: Early research by L. D. deStwolinski, followed later by studies by Harry Eckstein, Jimmie Hinze, R. E. Levitt, M. R. Robinson, and N. M. Samelson; and research on structural and other failures, coupled with the effects of sharply rising insurance costs, has focused serious management thinking and actions on the economic and human aspects of accident prevention.

In the "computer-based, advanced approaches" field: Pioneering equipment simulation and economic studies by Paul Tiecholz, James Douglas, and Axel Gaarslev, and modeling by Ronald Woodhead and D. W. Halpin set the stage for intensive activities in these fields. More recent efforts on equipment economics, made by Glenn Sears and M. C. Vorster, and on simulation by a host of researchers, among them B. C. Paulson and D. W. Halpin and their associates, are acknowledged.

Studies in automation and robotics have involved, among others, Paulson, A. Wiszawski, D. A. Sangrey, and Japanese companies such as Shimizu.

In applications of "expert systems" to construction: research and applications began very recently. Involved, among others, are: M. R. McGartland, C. T. Hendrickson, J. E. Dickman, T. A. Kruppenbacker, R. E. Levitt, and I. D. Tommelein.

Glenn Ballard, Mike Casten, and Jerry Talley, construction consultants, have made substantial contributions both through their professional activities and by participating in brainstorming sessions to develop concepts and ideas which have now been proved by field testing.

Specific individuals, with their company affiliations who, to the authors' knowledge, successfully applied the techniques outlined in this book to actual projects included the following: Jack Basler, Pacific Telesis; Al Burkhart, Hensel-Phelps; Jim Carroll, Morrison-Knudsen; David Grubb, Swinerton and Walberg; Warren Jackman, Contra Costa Building and Construction Trades; Ted Kennedy, B., E., & K; Jim Manning, Pacific Gas and Electric; Eric Miller, Sentinel Group; Fred Moore, Hawaiian Dredging and Construction; Peter Nosler, Rudolph and Sletten; Howard Peek, Brown and Root; Jerrold Poole, CMGM; and Tom Rowe, Bechtel.

Other companies have realized substantial benefits by employing productivity improvement techniques on some of their projects. These include Dillingham Corporation of Canada, Exxon Research and Engineering, Florida Power and Light, Fluor, Fruin-Colnon, Granite-Ball-Groves, Guy F. Atkinson, Hewlett Packard, Ingenieros Civiles Asociados S. A., Jaynes Inc., Kalman Floor Co., Lagoven S. A., LTA Construction, Nello L. Teer Co., Scott Co., Sohio Construction Co., and Williams and Burrows.

In many of the applications noted above, management personnel and consultants alike acknowledged that the key to successful implementation was the active and willing participation of foremen and crews.

Preparing the manuscript for submission to the publisher was an arduous task. It was made easier by Ardis Oglesby, whose critical review of all the manuscript and page proofs was indispensable; Glenn Katz, who advised on computer applications; Jane Kiefer, who transferred earlier writings to the computer; Karen Hansen, whose critical evaluation of the manuscript was most helpful; and Jorge Vanegas, who organized it for submission to the publisher in both printed and computer-disk form.

McGraw-Hill and the authors would like to thank Carroll Dunn, The Business Roundtable Construction Committee; Jerald Rounds, Arizona State University; Glenn Sears, University of New Mexico; and Richard Tucker, University of Texas.

Finally, all three authors greatly appreciate the support and forbearance of their wives during the long gestation and writing phases involved in producing the finished book.

CHAPTER 1

PRODUCTIVITY IN ON-SITE CONSTRUCTION—THE STATE OF THE ART AND A PRESCRIPTION FOR IMPROVING IT

CONSTRUCTION INDUSTRY PERFORMANCE— THE STATE OF THE ART

The construction industry builds for industry, business, individuals, and governmental agencies. All about us are plants, buildings, roads, housing, systems to supply water and dispose of wastes, and many other facilities that are required to keep our modern society viable. In the United States, construction is nearly its largest industry. In 1986 the construction industry employed about 4.4 million people; in all its phases, expenditures were $389 billion dollars, or 9 percent of the gross national product. Table 1-1 shows how these total dollar expenditures are distributed among the common general classifications.

Observers close to the construction industry see several portents of change ahead. The first is that in some segments, the domestic volume of construction, stated in uninflated dollars or as a percentage of the gross national product, has been shrinking. Some predict that it will continue to do so because of regulation; lower expenditures for plants, buildings, housing, and other facilities by private interests; decreased government spending; and, possibly, high interest rates. In addition, foreign contractors, many of whom have strong engineering, management, and research capabilities, are increasingly challenging American firms for work overseas as well as at home. For example, Japanese, German, and French firms are acquiring partial ownership in U. S. construction firms and, in a few instances, establishing their own companies.

Some among the many buyers of construction have become increasingly concerned with ever-increasing costs, high accident rates, late completions,

1

TABLE 1-1
VOLUME OF NEW CONSTRUCTION IN 1986 IN THE UNITED STATES,
BY MAJOR CATEGORIES*

Categories	$ in millions	Percent of total
Total—all construction, new and improvement	388,817	100.0
Private owners		
Residential buildings, new or improvement	187,148	48.1
Nonresidential—buildings, plants, etc.		
Industrial	13,747	3.6
Office	28,591	7.4
Other commercial	28,170	7.2
Other: hotels, institutional, miscellaneous	20,663	5.3
Farm—nonresidential	2,046	0.4
Public utilities, privately held	33,948	8.8
All other private construction	2,276	0.6
Total—all private owners	316,589	81.4
Public agencies		
Buildings	23,494	6.0
Highways and streets	23,359	6.0
Military facilities	3,919	1.0
Conservation and development	4,668	1.2
Sewer systems	8,105	2.1
Water supply	3,370	0.9
Other public	5,313	1.4
Total—all public construction	72,228	18.6

*Source: Construction Review, U.S. Department of Commerce, July–August, 1987.

and poor quality. They have realized that the problem cannot be blamed entirely on the construction trade unions and their restrictive work rules or featherbedding or on worker intransigence, lack of skills, and unwillingness to work, as has often been done in the past. Rather, they recognize that there are a myriad of other causes which can lead to ineptness in the way construction projects are conceived, planned, and administered by owners, designed by engineers or architects, and managed and executed by contractors.

In the late 1970s, members of the Business Roundtable, an organization of the presidents of some 200 of the largest corporations in the United States, decided that something should be done about construction performance in all its aspects. Their enterprises make large investments in fixed facilities, primarily industrial and process plants and commercial buildings, and rising costs were becoming a major problem. From this concern came the Construction Industry Cost Effectiveness Project (CICE), which involved fact-finding studies on pertinent issues by 23 task forces made up of knowledgeable people from Roundtable companies, design firms, contractors, and universities. The reports of each of these groups and a summary have been widely distributed.

The Business Roundtable cost effectiveness study found deficiencies in almost all aspects of the construction process, ranging from planning and design through the construction process itself. Among them are those outlined in Report A-6, titled "Modern Management Systems" (November 1982), which deals with present construction management practices. It states that

> The construction industry has been criticized, to a large extent justifiably, for its slow acceptance and use of modern management methods to plan and execute projects. Many people both inside and outside the industry view this as a primary cause of serious delays in schedules and large cost overruns that have plagued the industry in recent years. Yet there is no lack of modern cost-effective management systems that can provide project managers with all the controls they need.

The report faults owners as well as contractors, stating that

> Many owners do not seem to be aware of the economic payoff from the appropriate use of modern management systems, and therefore are unwilling to incur the costs of operating the systems on their construction projects.

This and other Roundtable reports support two propositions regarding industry performance. These are that (1) owners, designers, constructors, field supervisors, and craftsmen should all be involved in efforts to improve performance by developing project goals and working toward their fulfillment, and (2) systems and techniques are available or will be developed by means of which construction managers can be more effective. The reports also strongly imply that all too often, construction managers are not aware that modern management systems for developing project goals and carrying them out are available; and if they are aware of them, they do not insist that the systems be used.

After the Business Roundtable studies had defined and clarified many of the key issues in construction performance, the next step was to undertake research on what could be done to improve it and to circulate the results of the findings. This is being done in part through the Construction Industry Institute (CII), created in 1983, and based at The University of Texas at Austin. Its activities are directed and financed by a group of owners, consultants, and contractors; the research it sponsors is carried on primarily at universities. (References to the publications of both CICE and CII appear often in this book.) In addition, governmental research organizations such as the National Science Foundation and branches of the U. S. military are showing substantially more interest in the construction industry's performance and have become increasingly active in research to find ways to improve it. Furthermore, universities with programs in construction management are becoming involved in research and in introducing the findings into their curricula.

Concerns about performance are not confined to construction alone; they pervade all American industry. For example, Ronald Schmitt, senior vice-president and chief scientist of General Electric, is quoted in *Engineering Education News*, April 1987, as saying that "the challenge for engineering education today is to retain our creative strength while improving our execution." In the context of this book, execution means "construction."

PERFORMANCE VS. PRODUCTIVITY

The word "performance" involves all aspects of the construction process. This book is primarily concerned with "productivity," one facet of performance, with attention focused primarily on on-site activities or others which directly affect it. The industry faces many other problems, but with the exception of safety, the other topics are not addressed here.

It is important at the outset to make a distinction between performance and productivity. Performance as applied to on-site or associated activities is a broad, inclusive term, encompassing four main elements, namely, productivity, safety, timeliness, and quality. More specifically, these can be described as:

1 Productivity, which is measured primarily in terms of cost, with "satisfactory productivity" usually meaning "work accomplished at a fair price to the owner and with a reasonable profit for the contractor."

2 Safety, that the project is accident-free within reasonable limits.

3 Timeliness, interpreted both as "on schedule" and "everything is on hand when needed."

4 Quality, which means that the facility and all its elements meet the specified requirements and perform in a manner which satisfies the owner's needs. It does not mean "gold plating" or "better than needed to do the job."

It is clear that these four elements are not mutually exclusive; each impinges on the others. One can be sure that projects, if managed well by knowledgeable people who carry out the work properly the first time it is undertaken, would take care of them all. To make this point clear, consider the following ways in which the other three elements affect productivity:

• *Safety* is often seen primarily in terms of efforts to protect individuals from death and injury, thereby reducing human suffering and accident costs. But accident prevention has a direct impact on productivity as well. All unsafe acts and events, even those that do not result in injuries, disrupt the work in a variety of ways and divert the attention of management from its primary function, which is to get the job done.

• *Timeliness* has two elements. The first and large-scale one is that of completion of projects on time, so that owners have use as was anticipated when the projects were undertaken. The second and equally important one might be called "scheduling." It involves making sure that at the job level, all the elements necessary to carry through a particular task are available. It is obvious that this element of timeliness has important implications for productivity.

• *Quality* also has two dimensions. The first and overall one is that of the completed project functioning as the owner intended. The second concerns the many details involved in producing this result, for if the details that affect quality are not carried through properly the first time they are undertaken, there is a high probability that redoing will be called for; and reworking has devastating consequences for productivity, not only in time spent and cost incurred with redoing, but often more importantly, in its effect on morale and the willingness of job personnel to work productively.

The authors have focused this book primarily on the first two categories: productivity (cost effectiveness) and safety. They reason that (1) the details of these two key elements have been given less attention or sometimes have been ignored by practitioners, educators, and writers, and (2) the details of the techniques underlying timeliness (including scheduling) and quality and its control have widespread coverage in the literature and in educational programs, if not in industry practices. Even so, the impact on productivity of timeliness and quality have been addressed often in the succeeding chapters in this book.

PRODUCTIVITY DEFINED

Productivity has a variety of meanings. In nationally developed statistics it is commonly stated as constant in-place value divided by inputs, such as worker-hours. For the owner of an existing or contemplated plant or other property or equipment, it may be the cost per unit of output produced by the facility. For the contractor, a rough measure often is the amount or percentage which costs are below (or above) the payment received from the owner. Basically, all these approaches, and others that might be cited, attempt to measure the effectiveness with which management skills, workers, materials, equipment, tools, and working space are employed at or in support of work-face activities to produce a finished building, plant, structure, or other fixed facility at the lowest feasible cost. It is productivity in this sense that this book addresses and for which it proposes organizational arrangements, systems, procedures, and attention to people to bring about improvements.

When a construction project is first undertaken, it is usually the case that all the parties involved—owner, designer, suppliers, and constructor alike—attempt to do their parts correctly. But projects must progress through many complex steps involving these and possibly other parties and many individuals and are carried out under a variety of complications. Thus, changes in the original concepts and the details supporting them are almost bound to occur. Because of these certainties about construction, it follows that a book such as this must have two purposes if it is to deal with productivity objectively. The first is to present guidance and techniques aimed at "how to do a task right the first time." The second purpose is to recognize that there are bound to be imperfections in planning and carrying out the actual tasks as originally conceived; to propose ideas and approaches with which to find, analyze, and correct those imperfections on the job while it is under way; and to avoid them on tasks and projects that will follow. It is far too easy in a book such as this one to appear to be second-guessing the people who are out there fighting the battles. Such is not the intent here; rather it is to provide ways for thinking about and doing complex tasks right the first time or to improve on the original efforts if they must be repeated.

Readers, and particularly students anxious to get their teeth into techniques for improving productivity at the work face may be disturbed that this book does not plunge directly into those topics. The authors, relying on many years' experience in observing productivity improvement efforts in the field, have chosen to give attention in the early chapters to a broader but overriding issue, which is to develop project goals and an understanding of and commitment to them in the minds of all those involved in a particular activity. Important detail follows in succeeding chapters. This order has been adopted deliberately, since experience has shown that efforts to improve productivity will fail unless they rest on a carefully prepared foundation, developed in three successive ways as follows:

1 Goals set, agreed to, and accepted by all involved with the project or activity. Everyone at every level of the involved organization or organizations, but especially those at the top level, must be on board and firmly committed to the effort; otherwise it is headed for failure. Those at lower levels quickly sense higher-management's intentions and behave in accordance with them.

2 A plan of action, including the necessary organization and financing, to develop an approach, carry through the required procedures, and put the findings to work.

3 Knowledge by means of which the plan of action can be accomplished.

The rest of this chapter is concerned with goal setting, an activity that is necessary to make any enterprise succeed. It is written in the context of on-site construction, but the concepts have universal application. Following this, Chap. 2 describes the distinctive characteristics of construction which make the application of goal setting for it unique. Chapter 3 proposes a plan of action for accomplishing the goals outlined above, recognizing the constraints imposed by the industry's uniqueness. The remainder of the book provides a knowledge base incorporating both techniques and human concerns from which, given goals and a plan of action, productivity improvement and accident prevention can be implemented successfully.

DEVELOPING GOALS FOR CONSTRUCTION PROJECTS

An examination of how improvements in performance have come about in other industries and occasionally in construction shows one common starting point: establish in highest management and at all other levels in all the organizations involved a philosophy or set of goals for the enterprise, as proposed above in general terms. Once these goals are adopted, there must be a plan of action and knowledge to carry it out. Energy, time, and money must be committed. As will be demonstrated in this book, the rewards from doing so far outweigh the costs.

For on-site and supporting construction activities, which are the areas of special attention in this book, as for any other venture, goals must be adopted and carried through by management in all the parties in the enterprise. In many instances, owners should take the lead, since they have the most at stake in the long run. It is they who initiate projects, pay for them, and use them after completion. Among others with substantial stakes in the enterprise are designers, contractors, subcontractors, suppliers and their office and field staffs, and, where appropriate, craftsmen and their organizations.

In broad terms, a set of objectives or goals for on-site and supporting construction activities, if fulfilled, will be rewarding to and reflect credit on all the individuals involved and on the organizations to which they owe their loyalties. Among the common measures for these rewards are:

1 Costs or budgets must match or improve on those projected at the beginning.

2 Completion must be on schedule or ahead of it.

3 Quality must fit the anticipated service requirements.

4 The work must be largely free of accidents or health hazards.

5 When completed, the final product must serve the owner's purposes.

6 The outcome must leave the individuals involved with a feeling that they have participated in a worthwhile venture.

Given a carefully thought out set of objectives or goals as described above, two other steps must follow before they can be achieved. These are:

1 Develop a structure for implementing the goals which considers the operating, informational, and human aspects. Operational includes an organizational structure with assigned responsibilities at various levels in each affected organization. Informational is concerned with the generation, collection, and dissemination of appropriate and reliable data in a timely manner. Human involves the thinking through and establishment of a broad scheme for recognizing, measuring, and rewarding accomplishments or identifying and dealing with unacceptable performance.

2 See that the necessary effort, time, concern, training, and money are allocated so that the tasks called for by the planned structure can be accomplished. In doing so, it is essential to recognize that actions in any of the three areas (operational, informational, and human) will strongly affect the other two.

Anyone knowledgeable about construction knows that carrying through and implementing all the goals and procedures for all the parties is difficult. On most jobs such broad objectives and plans for carrying them out, if they are stated at all, become submerged and lost sight of in the day-to-day concerns. Perceived conflicts of interest and the feelings and egos of individuals also complicate their execution. But there are examples where some or all of the parties have knowingly or unknowingly adopted similar courses of action as targets. These jobs seem to go right. All parties become cooperators rather than adversaries and work to a common purpose.

Here, at the beginning of this book, it is important to underscore some of the constraints peculiar to construction which make it difficult to establish and implement a set of common objectives such as those outlined above. Included would be the following:

1 Construction operates differently from other industries (see Chap. 2 for details). For example, most construction projects are unique (one of a kind) and generally fast moving, so that organizations are not static but must be rebuilt again and again with different designs and designers, management, materials, equipment, and crews. Consequently, there is little repetition and few second chances to learn from earlier mistakes. Too often, common practice considers productivity (costs), safety, schedule, and quality as separate objectives, with responsibilities for them delegated to different people in each organization. Again, owner, designer, constructor, and craftsmen and their representatives have different basic objectives which affect their views of the relative importance of these factors and of their effects on their individual or organizational well being. Combining these factors—lack of repetition, accelerated scheduling, a variety of designs and viewpoints, and several parties—makes the task of setting objectives, much less carrying them out, particularly difficult in on-site construction.

2 The contractual structure under which projects are normally carried out is seldom conducive to cooperation among the parties or to joint efforts to improve productivity. Often there is no contractual relationship at all between some of the affected parties. Because of these conditions, the parties assume that there are boundaries they cannot step across; rather they tend to deal at arms length and even become adversaries, all looking out for their own interests when things go wrong. As is pointed out in Chap. 3, efforts to improve job relations and productivity require that the parties work together harmoniously either within or outside formal contracts to generate ideas and develop solutions.

3 The traditional hierarchical management structure within each organization, developed to get the work accomplished, blocks free discussion and exchanges of ideas. Because these efforts take time and seem to delay progress, they are seldom welcomed.

4 The usual attitude of construction people, managers and workers alike, is to get on with the job. This does not provide a climate, lead time, or a thoughtful, searching approach necessary to develop and carry out new or innovative ideas.

How, then, are these conditions being changed on projects where cooperation among the parties and productivity improvement efforts have been successful? It begins when those at the top of or others with influence in each affected organization jointly adopt a clear set of objectives, a structure, and an operating plan something like those outlined above and carry through on them in the following ways:

1 Develop an attitude (win-win) at all levels in each organization which makes cooperation rather than conflict (win-lose or lose-lose) essential.
2 Give efforts to develop cooperation a top priority among the many demands on time and energy.
3 Plan and agree jointly with other affected parties on a specific and continuing program of action to create and maintain a win-win attitude. This must be fitted to the situation, which will be different for each project.
4 Commit the time, money, and human resources required to implement the plan's successive stages. This involves setting up a formal or informal organization or other mechanism separate from those based on the usual contractual or management structures. Its purpose is to develop innovative approaches both to management and technical matters and to handle problems that cannot or should not be dealt with through contractual agreements or the traditional lines of authority. (Chapter 3 offers further discussion.)
5 Make these priorities clear to everyone at all levels in the affected organizations. Among the steps are:
   a Use every opportunity and a variety of strategies to get the message out.
   b Develop and implement a system for all the parties that rewards cooperation and improved performance in all the four areas, namely, productivity, safety, schedule, and quality. The reward system must somehow recognize contributions in these four areas by a single clearly defined and stated set of measures patterned to fit the individuals involved.
   c Insist on commitment to the plan. As a last resort, discipline or even get rid of those who can not or will not fit into the new approach.
6 Follow through by carefully monitoring progress and results. At the same time, do not expect spectacular accomplishments overnight. Changes in past ways of thinking, practices, and the attitudes that underlie them do not come easily or without stress and strife. Be patient.

The lists of objectives and requirements for implementing them demonstrate that any program to improve on-site construction has many facets. A full-blown program will call for efforts and talents in several areas. This book addresses a number of the techniques involved. But the primary effort is directed at improving productivity, with some attention to the specifics of safety. Detailed approaches to timeliness and quality, except as by-products of improved productivity, are left to others.

In numerous instances in the past, the most successful performance or productivity improvement programs have been instituted on distressed projects where management was desperate; cooperation rather than conflict and confrontation seemed the only way to avoid catastrophe. In most of these cases, a change in higher-level project management of either owner, designer, contractor, or all three was a first step. Only then the essential trust, cooperation, communication, and commitment at all levels in the organizations—which were needed to get the project back on track possible—were restored.

The specialists or consultants who work on such rescue missions will, of necessity, focus on specific areas where impacts can be made quickly, since it is too late to develop a full-fledged and carefully integrated program. In cases where their efforts have been influential in bailing the projects out or at least minimizing the damage, the usual approach is first to develop and implement procedures that will improve job morale and bring cooperation among the parties. Only then are efforts undertaken to improve productivity. The specialists have learned by experience that without good morale and cooperation, introducing and implementing a meaningful program is impossible. They also know that this is best accomplished when a project is first begun, when cooperation is easy to establish, rather than waiting for adversarial relationships and other trouble to develop.

Even given these examples of the success that cooperation brings to productivity improvement under adverse conditions, the sad fact is that the lessons learned seem not to have affected the attitudes and activities of higher-level management in either the company involved or many others in the construction industry. All too often, new projects are started in ways that seem to ignore the lessons learned from earlier troubles. The parties seem to forget or ignore the proven, sensible, and far-better approach of instituting a program of cooperation at the beginning of a project and modifying it as necessary at every significant change point thereafter.

The authors are fully aware that to institute a comprehensive productivity-improvement program is costly, since it involves staffing, training, and making a commitment of resources from the beginning. Then it must be carried forward as the project continues. But many such programs have paid off. A few cases where this approach has been executed successfully are described later in this book. In at least one instance, not only owner, designer, contractor, and subcontractors but union representatives as well were involved. But the sad fact is that, in general, even the concept of joint objective or goal setting is largely unknown in the construction industry, much less a formal, fully integrated approach to productivity improvement.

SUMMARY

Many construction people who have come up following traditional practices will scoff at the notion that on-site construction will be done more cheaply, safely, and quickly and to required quality by instituting the practices for developing cooperation and teamwork set out here at every level and in every phase of a project. Their approach is to start with the assumption that the other parties are adversaries, not cooperators, and that their goals and those of the others cannot be reconciled. To them construction is a win-lose game, with everyone for himself, as is attested to by the large number of books, articles, workshops, and lawsuits devoted to claims.

In contrast to that attitude, this book proposes that the parties to construction projects set formal goals and establish the procedures necessary to develop cooperation among them and the individuals that are involved. The authors are not alone in advocating this approach. One indicator comes from the landmark conclusions of the Construction Industry Cost Effectiveness Study sponsored by the Business Roundtable, an effort now being carried forward and implemented through the Construction Industry Institute. It strongly emphasizes the necessity for cooperation accompanied by strong attention to education and training.

To summarize, the step-by-step approach outlined here involves developing and implementing practices aimed at improving construction productivity, and recognizing their implications in organizational, communication, and human terms. This in turn calls for a commitment of time, effort, talent, and money. It is important also to recognize that the principles apply to almost any construction enterprise, large or small, private or public. In many cases, different and largely untried approaches will be required. But the evidence is strong that the effort will be worthwhile.

CHAPTER 2

PRODUCTIVITY—THE CLIMATE FOR IT IN ON-SITE CONSTRUCTION

INTRODUCTION

This chapter examines several characteristics of the construction industry and its management makeup and practices that impinge directly on productivity. It outlines some of the difficulties facing construction managers that divert attention from the broader objectives, goals, systems, and procedures that are necessary for a focus on productivity, which, in turn, keeps them from utilizing the abilities and creativeness of all the parties involved in the construction process.

Readers who know the construction industry well may find little that is new in the earlier sections of this chapter. But the later portions will provide a less conventional way of looking at the industry's peculiarities, and also will clarify the issues that can make efforts to improve productivity very difficult.

Topics covered in this chapter include:

- Characteristics of the industry—its parties and the relationships among them
  - The way in which projects are commonly organized
  - Staffing, managing, and communication
  - The sources and backgrounds from which on-site managers come
  - Reasons why on-site management is a complex task
  - Performance measures and reward systems and how they affect efforts to improve productivity

Overall, the chapter aims to set the stage on which efforts to improve productivity are carried out.

Readers should view the climate outlined here as a constraint to be overcome rather than as an insurmountable obstacle, if the aim is to improve performance in general and productivity in particular.

CHARACTERISTICS OF THE CONSTRUCTION INDUSTRY

The Parties in Construction

In thinking about productivity in construction it is helpful first to look at the parties involved and the roles they play. In simple terms, construction is an enterprise involving four major groups or parties, all of whose actions can influence productivity. These are:

1 *Owners,* who conceive and sometimes modify projects, usually designate the sites; arrange for design, financing, and construction; and operate the completed facility.

2 *Designers* (usually engineers or architects), who convert the owner's conceptions into specific and detailed directions through drawings and specifications. They may operate separately or as part of a design-construct team.

3 *Constructors* (contractors and subcontractors), who manage the efforts necessary to turn the designers' directions (drawings and specifications along with contract documents) into completed structures or plants. Their function is to provide leadership and managerial staff; assemble a work force; issue instructions; furnish a work plan or method; provide materials, equipment, and tools at or available to accessible work faces.

4 *The labor force* (particularly foreman and craftsmen), that through its skills and efforts, working individually or in crews directed by foremen, transform the directions depicted in plans and specifications into reality. Following methods developed by themselves or management, the workers put together the ingredients (materials, information, equipment, tools, and working space) supplied them through the efforts of the constructor's management, to give completed tasks which combined become projects.

The efforts of all four of these parties are essential if an owner's project is to become a reality. But final completion depends on craftsmen or crews performing individual tasks at work faces. Furthermore, the cost, quality, and timeliness of their outputs depend not only on their skills and desires to work but also on the performance of the three other parties who control the ingredients necessary for productivity at the work face.

To date, most of the efforts to maintain excellent or to improve poor productivity as well as to place the blame for failure has been directed at activities and craftsmen at the work face. Much still remains to be done in this area, and this book addresses those issues. But many of the high costs, errors, and delays have been wrongly attributed to work-force inefficiency or malingering. Actually, these deficiencies may have been brought about by actions, lack of actions, or errors of the other three parties; their failure to perform has made it impossible for craftsmen or crews to be productive.

The Knowledge and Skills Demanded by Construction

Each construction project, as defined by the owner, calls on the knowledge and skills of specialized engineers and/or architects to produce the designs, drawings, and contract documents to describe completely the work to be accomplished. Then, having provided the site and financing, the owner usually contracts with a constructor to convert these documents into reality. In turn, the constructor assembles the necessary materials and supplies, equipment, and skilled personnel in office and field to complete the project.

Because the need for construction to serve a wide variety and class of owners varies greatly in function and magnitude, the industry over time has developed many specialized skills in design, management, and craftsmanship. These may be brought to bear on a given project under a variety of arrangements and forms of contract. Always, some among the many specialized skills are needed to do the design and carry through construction and start-up. The following incomplete listing of these skills and where they are found offers one way to show the construction industry's complexity.

- *In the owner's organization:* Personnel able to award and administer contracts, pay the bills, and monitor costs, timeliness, and quality. Under certain circumstances the owner may employ specialists in design and construction management or even undertake the construction.
- *In the design or design-and-build firms (or the owner's organization):* Engineers and architects and possibly other specialists with the skills necessary to carry through the design and to make at least preliminary schedules and cost estimates.
- *In the constructor's home or business office or offices:* Personnel able to price the work, purchase major materials and supplies; procure, manage, and supply construction equipment; attend to and monitor financial and business matters of all sorts, and, to a greater or lesser degree, supervise field operations.
- *In the constructor's or owner and/or builder's operation on site:* A management and often an engineering team with talents appropriate for the project.
- *Craftsmen and craft supervisors with the skill needed to carry through the actual construction:* Depending on the nature of the work, this could include some or all of the following: carpenters, laborers, ironworkers (structural and reinforcing), equipment operators, surveyors, concrete finishers, pipe fitters, welders, sheet-metal installers, electricians, air-conditioning specialists, bricklayers, stone masons, painters, and glaziers.

In sum, every construction project, large or small, involves a number of parties and the skills that each can bring.

How the Construction Industry Operates

Construction is a highly diversified industry operating in a complex society. For this reason, the notion that improvements in its productivity will come easily does not make sense. A few statistics about the industry are given here to indicate how great that complexity is.

In 1986, yearly expenditures for construction in all its various forms totaled about $389 billion or 9 percent of the total national income of the United States. As listed in Department of Commerce statistics for that year, construction income was $182 billion. This can be compared with all manufacturing of durable goods, $411 billion; manufacture of nondurable goods, $274 billion; wholesale trade, $210 billion; and retail trade, $301 billion. Thus, there is substantial support for the claim that construction is among our largest industries. This probably comes as a surprise to most people, since the automobile industry is so much in the news. However, the expenditures for new automobiles (in 1985) totaled only $87 billion.

Construction is far different from, for example, the motor-vehicle industry, which is dominated in the United States by three large manufacturers and their sales outlets, with one firm, General Motors, accounting for more than half the total. In contrast, construction is carried out by about 450,000 contractors ranging from giants with annual volumes in the range of $5 billion to individuals operating alone or with only a few employees. Medium- and small-sized firms predominate; in fact, the average number of employees is about 10; 75 percent of the firms have four or fewer employees. The largest firm holds only 2 percent of the total market; and the top 400, less than one-third. Individual construction projects also vary widely in scope; the largest fall in the $1 billion range, involve several thousand workers and a management hierarchy half-a-dozen levels deep, and last for several years. The smallest projects involve only a few workers, have no hierarchy at all, and are completed in a few days or weeks. In another dimension, construction can be separated into about 25 quite distinct specialties. These, in terms of finished product, are commonly separated into five categories: home building, commercial building, industrial, highway and heavy, and marine. Only the exceptional and usually very large firm operates in more than one or two of these areas.

Construction is a custom rather than a routine, repetitive business. To a great extent each project in each of its categories is designed and built to serve a special need. Although specific design skills, such as engineering and architecture, and crafts, such as carpentry, pipe fitting, welding, or equipment operation, are needed over and over again, the outputs differ in size, configuration, and complexity. Furthermore, in most cases, after design is completed, the specialized workers and materials, tools, and equipment needed for each finished element are brought together at each project site. Also, of necessity, craftsmen shift from project to project and often from employer to employer.[1] By contrast, in manufacturing, the workers with the needed tools and equipment stay at one place, and assemblies and materials flow past their locations. Employment is at one site for a substantial period of time. As a further complication, many construction operations of necessity are carried on out of doors, so that managers and workers must cope with the delays, inconveniences, and discomforts which rain or snow and temperature extremes can bring.

Construction labor, which numbers about 4.4 million, both union or nonunion, also tends to be stratified, primarily along craft lines. For example, there are 15 AFL-CIO craft unions and these encompass tradesmen offering some 60 different skills. On the contractor side, one can list at least 12 active national associations or trade groups, and many of these will have one or more affiliates in each state. It is no wonder that these groups seem to bring a babel of voices and points of view to issues that affect the industry.

On-site construction differs from manufacturing and many other enterprises in another very important regard: its end product is highly visible to all involved, be they project or construction managers, engineers, superintendents, foremen, or crafts workers. To be in a smoothly running construction project can be highly rewarding and satisfying if management makes this possible. On the other hand, frustration and dissatisfaction become high and incentive is lost when for any one of a variety of reasons people's efforts produce few tangible results, when craftsmen feel impelled to look busy while doing nothing productive, or when they must tear out and redo earlier work. Unfortunately, as discussed in more detail in Chap. 10, frustration and dissatisfaction are too often the result. A principal aim of this book is to demonstrate proven ways by means of which construction will be productive and rewarding for all involved in it.

This thumbnail sketch of the construction industry and how it can reward people or turn them off points up a paradox that must be faced. All those involved in construction—owners, designers, large or small contractors operating in a variety of fields, employer and employee representatives, or skilled or unskilled workers—find rewards and satisfactions when construction projects are carried out efficiently, safely, on time, and with suitable quality. But the diffusion of construction activities among so many parties and enterprises in so many diverse segments, each of which is pursuing other more specific and tangible goals, makes a focus on productivity extremely difficult and leaves little time for a concentrated effort to improve present-day practices or to transfer this knowledge to subsequent projects. It is only human for those involved to

---

[1] One argument advanced for craft unions is that they provide a pool of craftsmen from which individual contractors can draw when needed. Even so, the evidence is clear that many workers, whether union or nonunion, are continually employed for periods of several years by a sole contractor who operates in a single area.

blame others rather than themselves for a lack of progress. Owners and contractors claim that the workers are at fault; workers and their representatives point to management's inability to make productive work possible. What is needed is a good-faith effort from all involved to lay aside their differences. Learning about the highly productive methods now in use and developing means for improving less-satisfactory ones may then be possible. This book, particularly in Chaps. 4 through 8, and 13, addresses these topics.

How Individual Projects Are Organized

Figure 2-1 is a typical on-site organization chart for a relatively large construction project being carried out by a single contractor. It is, of course, one of many that might be employed. The chart outlines the formal relationships among the various management positions: that is, it shows the paths set up by management through which authority and information are supposed to flow down and information is expected to flow up through the organization. For example, craft foremen receive instructions from and report back to craft superintendents. The same relationship applies between craft superintendents and the project superintendent, and the project superintendent and the contractor's project manager. Figure 2-1 also shows, at the top, an owner's project manager who deals with each (or possibly several) contractor's project managers on construction matters but who also supervises or coordinates design, the owner's inspection and management groups, if they exist, and possible government inspection as well.

Figure 2-1 clearly depicts the formal channels for authority and communication. Flow is downward from the contractor's construction manager to general superintendent to craft superintendent to the craft foremen who actually supervise the work. This hierarchy is commonly referred to as the *line*. Reporting flows upward through this same chain. Figure 2-1 also shows a construction or project engineer and other engineers who report to him as well as managers in several other areas. Those carrying out these functions are commonly referred to as *staff*. They do not directly supervise construction. Rather, their activities are intended to facilitate the construction process.

Management theorists label organizations such as those represented by Fig. 2-1 as classic or bureaucratic. Among their characteristics are a unity of command, a clear delegation of authority and responsibility, and enough levels in the hierarchy to permit supervisors at each level to control the next level down (called span of control). It is also described as impersonal, meaning that the boxes are drawn and then individuals are put in them.

With the bureaucratic approach, the formal communication system follows the lines of authority and control. For example, job schedules developed by the planning and scheduling engineer pass through and are approved by (see Fig. 2-1) the construction or project engineer, the project superintendent, and the craft superintendent before reaching the craft foreman. Disagreements would be resolved by the construction project manager. Disputes between

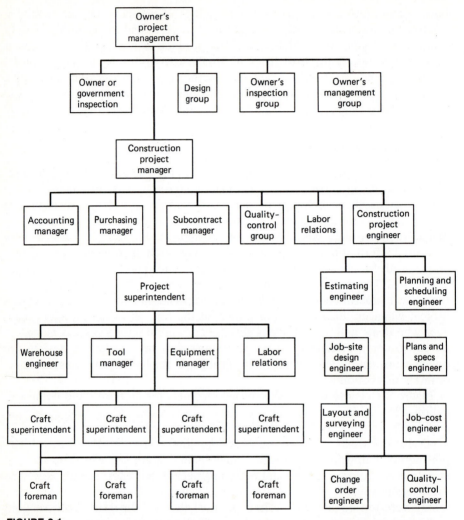

FIGURE 2-1
An example of the organizational and management structure for a large construction project.
(For other examples, see books such as D. S. Barrie and B. C. Paulson, Jr., *Professional Construction Management, McGraw-Hill,* New York, 1984.)

craft foremen or superintendents and one or more engineers, if formalized, would likewise involve the entire chain, with final settlement at the construction project manager's level.

Asserting that communication and dispute settlement on construction projects follow the circuitous routes outlined above would be ridiculous. In any harmonious organization, the responsible parties deal directly with one another. Even so, on projects with deep hierarchies and particularly when a project is over budget and behind schedule or has quality problems, the parties involved tend to follow the bureaucratic channels as a means of protecting themselves and their group. Consultants involved with projects that are in trouble almost always find very little communication outside these channels.

For small construction projects, on-site management might include only the line organization from project superintendent down, and even these functions might be consolidated. Other functions, such as accounting, purchasing, and dealing with subcontractors might be handled in the contractor's central office. Still others, such as engineering, might involve a single individual or not be needed at all. In fact, although a well-managed small project may on paper have an organizational plan, the parties often operate as though it did not exist and ignore the supposed lines of authority and accountability. On the other hand, for super projects the line and staff organization shown in Fig. 2-1 would probably be expanded greatly with, for example, separate project superintendents for major elements such as civil, mechanical, piping, and electrical work. It should be clear that on such projects, management can become more formal, which can make the flexibility needed to react to changed circumstances hard to achieve.

An organization chart such as Fig. 2-1 implies that the management relationships on projects remain static. This is not nor should it ever be the case. Responsible managers recognize that the management structure should change as a project goes through its various phases. These phases usually are as follows:

Phase 1: Inception and start-up. This includes planning and scheduling individual operations, and determining the needs of labor and staff, and arranging for procurement of equipment, materials, and other essential elements.

Phase 2: Production. Emphasis is on substantially completing the major elements. Often individual operations can be carried out independently, possibly in separate locations, and largely free from interference from others.

Phase 3: Integration. Tying together the major elements to produce a completed whole.

Phase 4: Start-up and operation. Depending on contract arrangements, either the constructor or the owner will test out and run the facility initially until it is fully operational, and possibly for a period thereafter.

On large industrial projects on restricted sites, the uniqueness of each of these phases is easily recognizable. There are other kinds of construction such as buildings, housing, and highway projects where more than one phase may be going on at the same time and the distinctions among them are less apparent. In some instances the owner may be involved in phase 1; an example could be ordering long lead-time materials or machines. Without question, however, this phasing exists to some degree on every project, even the simplest ones.

Planning for the various phases will demand a different combination of management and craft skills, and possibly a modified management structure to carry each phase through effectively. A simple example is a rural highway project where successive managers and crews do clearing, structures, grading, and paving separately and in a prearranged sequence. In building and housing, subcontracting provides a useful arrangement for handling some or all of phases 1, 2, and possibly 3. In contrast, on the large industrial, chemical, and power projects on a single site, differences in management and craft requirements generate very difficult problems for on-site management.

On large projects in particular, a "people" problem can arise in a bureaucratic organization because of differences in backgrounds and career paths between line and staff. As discussed in more detail later, most individuals in the line, often including the project manager, are older and have reached their positions following long experience in the skilled trades. On the other hand, staff personnel, particularly engineers or graduates of construction-management programs, are younger and college educated. Where such differences exist, establishing rapport and informal, friendly personal relationships is often more difficult.

The intent of this brief discussion of the bureaucratic organizations under which construction projects supposedly operate is to highlight one of the many obstacles that stand in the way of programs set up to improve productivity. All too often these organizational and people problems go unrecognized. Forums, parallel organizations, work teams, or similar arrangements, discussed in Chap. 3, are proposed as a way to mitigate this problem.[2]

THE FLOW OF PEOPLE, INFORMATION, IDEAS, AND RUMOR ON CONSTRUCTION SITES[3]

Figure 2-2 is an attempt to diagram the complex ways in which people and information (written, verbal, and otherwise communicated) flow either directly or as feedback on construction projects. Whereas Fig. 2-1 shows positions and lines of authority and responsibility, Fig. 2-2 gives details of these flows and how they operate.

The heaviest solid or dashed lines in Fig. 2-2 show the flow of workers and foremen (and sometimes superintendents) to the work site where they complete construction tasks. The workers who form the crews come from a labor pool, are in some manner directed to the work site, and are hired by the foreman and assigned to a crew.[4]

---

[2] For a detailed survey and analysis of the organizational structures of 12 large power projects, see C. B. Tatum, *Technical Report 279,* Department of Civil Engineering, Stanford University, Stanford, Calif., November 1983. See also H. R. Thomas et al., *Journal of Construction Engineering and Management,* ASCE, vol. 109, no. 4, December 1983, pp. 406–422., C. B. Tatum and R. B. Fawsett, *ibid,* vol. 112, no. 1, Mar. 1986, pp. 49–61, C. B. Tatum, *ibid,* vol. 112, no. 2, June 1986, pp. 259–272, and H. R. Thomas and A. C. Bluedorn, *ibid,* vol. 112, no. 3, Sept. 1986, pp. 358–369.

[3] For an excellent discussion of this topic see A. D. Russell and E. Triasse, *Journal of the Construction Division,* ASCE, vol. 108, no. C03, Sept. 1982, pp. 419–437. This paper focuses on management information systems in building construction.

[4] On union projects workers are assigned to a project through a hiring hall. On nonunion jobs, the contractor may hire on site or use some other local or area arrangement. Obvious as this people flow may seem, it is shown here to make clear the importance of a work force. For without

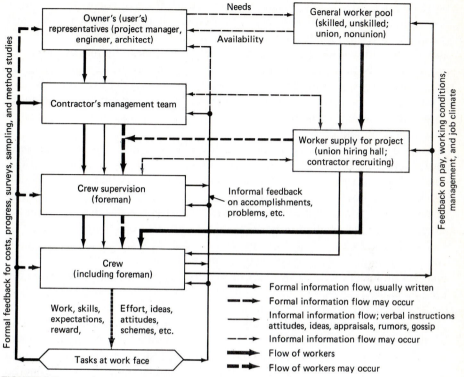

FIGURE 2-2
Parties involved and the flow of people, information, ideas, and speculations before, during, and after construction crews perform designated tasks.

The information flows diagrammed in Fig. 2-2 demonstrate that communications of any kind involve a very complex process and that they will be strongly influenced and possibly biased by those who generate and transmit them. If the messages are seen by the senders as being to their benefit, they will be complete and straightforward. If, on the other hand, the consequences of generating information are threatening or make an individual look inept or stupid, the message will be incomplete and garbled. Message receivers will apply the same mental processes in making their interpretations. The point to be recognized at the outset of this discussion of job communications of any kind is that they are, at best, incomplete, inexact, and often biased.[5]

As shown in Fig. 2-2, communications on a project are both written and oral. The middle-weight solid and dashed lines in Fig. 2-2 show two flows of

foremen and crews, tasks would not be accomplished, regardless of management's efforts. Furthermore, the diagram emphasizes the fact that workers and crews generate much of the informal feedback that flows around and beyond the project.

[5] Communication and its counterpart, listening, are discussed in far more detail in Chap. 10.

written or recorded communications. Downward flows involve such items as plans, specifications, and a variety of written instructions. Upward flows carry feedback of several kinds about the accomplished task. It can be seen by examining the downward lines that each level in the hierarchy makes inputs appropriate to its function. The detail given depends on the complexity of the project and how its management operates. Of particular concern for this book is the character and completeness of the written communications between the contractor's management team and the foreman and, to a lesser degree, between foreman and crew.

The upward middle-weight line near the left margin of Fig. 2-2 represents formal feedback on costs, progress, and other matters. Its accuracy and fairness and the parties with whom management shares it (dotted horizontal double lines) and how it is collected and used are among the most important influences on efforts to improve productivity. This topic is discussed later in this chapter under the heading of performance measures and reward systems.[6]

The thin lines in Fig. 2-2 represent informal information flows. They are commonly thought of as verbal instruction and feedback, but are far more than that. As indicated, a large segment of the downward informal flows shown in Fig. 2-2 consists of verbal instructions to augment or serve as a substitute for written directions or plans. These suffer from the fact that they may be incomplete, inexact, and subject to misunderstanding or distortion, and leave no dependable record. On the other hand, the give and take that should accompany verbal instructions can make them more effective than written ones and can be an excellent way to develop ideas, clear up misunderstandings, and generate enthusiasm for the task to be done. This sort of exchange is important in productivity improvement.

Figure 2-2 shows two ways that informal feedback flows upward from the task. The first is through the contractor's organization and possibly to the owner's representatives. It can cover a variety of matters, including the problems that were encountered and the means for improving current operations. The second feedback is from the crew and possibly the foreman to the worker-supply pool concerning such factors as job conditions and working relationships.

Informal communications can have a marked effect on productivity. As already indicated, they are a primary channel for developing or filling in schedules, procedures, and many other factual details needed to carry out tasks and for getting feedback on what has happened. However, informal communications carry many other messages, covering such things as expectations, attitudes, and judgments. Senders often have purposefully made their messages incomplete or have filtered out unpleasant details. The communications may reflect personal or group animosities. Moreover, what the sender thought was

---

[6] For the results of a survey showing the subject matter that is commonly dealt with in construction organizations at the various levels, see K. A. Tenah, *Journal of Construction Engineering and Management*, ASCE, vol 110, no. 1, Mar. 1984, pp. 101–118.

transmitted is filtered again by the receivers. Among the best indicators of how projects are going is to explore and analyze the quantity, quality, and intents of this informal information flow, either by careful listening or, in some cases, by interviews or questionnaires. These topics are explored in more detail in Chaps. 7 and 10.

SOURCES OF ON-SITE CONSTRUCTION MANAGERS

To be effective, managers of on-site construction at all levels ranging from foreman to project manager must have knowledge and skills in areas such as (1) knowing the construction process itself and how it is accomplished; (2) employing techniques for finding better approaches to this process; (3) being able to plan, schedule, and monitor all aspects of the work; and (4) getting the work accomplished by being adept at instructing, directing, and motivating people. Traditionally, construction managers have developed these capabilities the hard way, starting as craftsmen and learning as they worked over the years until they advanced into management positions. More recently, another route has developed, in which individuals begin with a college education aimed at providing a superficial acquaintance with construction but developing knowledge in the other areas listed above. This is followed by assignments which provide a detailed knowledge of construction itself. Successful managers have usually followed one or the other of these career paths. Each is discussed briefly below.

The Traditional Path to On-Site Construction Management

Construction and its technology and management developed from crafts such as masons, carpenters, and laborers and continues to be craft oriented. Stone masons and laborers built the great pyramids, temples, and palaces of the ancient world. Skilled stone masons, along with carpenters and laborers to serve them, constructed the magnificent cathedrals and castles of the Middle Ages. Later, building was controlled by specialists such as carpenters, brick masons, and roofers, among others. As construction has expanded into railroad, road, and dam building, mass housing, high-rise structures, and industrial plants, new crafts with other skills have developed, such as specialists in operating equipment; installing machines, vessels, and fixtures; and in joining, controlling, and powering them. As in the past, then, construction relies on workers skilled in a variety of crafts.

It is entirely logical that on-site construction is not only carried out but is also directed by craftsmen, whether they came up through the unions, often beginning as apprentices, or entered the work force in other ways. Today, a majority of those who supervise on-site construction as well as the owners of many construction firms were originally in the crafts. Also, it is usual to leave the control of on-site work largely to craft superintendents or foremen who have had long experience and have worked up to their present positions over a period of time. Most of them lack training in planning or management, except that which has been gained on the job. There are, of course, exceptions. For example, increasing numbers have taken or are taking foreman or superintendent training programs either conducted by employer associations such as the Associated General Contractors or given in-house by individual owners or contractors. But experience is the main criterion.

This discussion is in no way intended to denigrate the abilities of managers who have come up through the crafts. Their talents are an essential element in successful field construction management. But craft orientation coupled with a long period of doing it the way they learned and know may create a mental set that hampers certain basic ways of thinking needed to improve productivity, such as a tendency (1) to accept orders from above without question or challenge; (2) to assume that the traditional ways of carrying out operations are the most efficient; (3) to follow the conventional practice of leaving actual accomplishment of tasks to foremen and craftsmen from the appropriate trade without examining their approaches first; (4) to fail to challenge themselves as well as foremen and workers to look for creative and less costly methods for carrying out their work; and (5) because of an orientation toward work-face activities, to fail to recognize that away-from-the-work-face situations may primarily control productivity and therefore give them critical attention.

These deficiencies, attributed here to managers from the crafts but common to many others, can be corrected with management education and practice, encouragement from higher management, and sufficient time to find and carry out the new approaches. Unfortunately, seldom are any of these corrective measures taken.

The Less-Traditional Path to On-Site Construction Management

Since World War II, much of the ownership and home-office management of the larger construction firms has shifted toward professionals educated in such areas as engineering, construction, law, business, and accounting. In addition, some firms have moved away from traditional contracting and expanded their operations to include land development or construction management or design with the aim of serving the building, manufacturing, chemical, and power markets through design-construct or construction-management enterprises. In these large firms, most of their professionally educated people tend to be concentrated in central or field offices and primarily perform design, staff, or coordinating functions. After a time some of these people may move into on-site management where they must get hands-on experience in running work before they are qualified to take full charge.

Others among these professionally educated people go directly into site management, usually working under craft supervisors or having strong support from them and craftsmen as well. They have much to absorb about field practices, but, in many cases, do so quickly, since they have the mental tools and acuity to learn rapidly.

Some construction executives have proposed that college-educated people, particularly engineers and architects, should replace construction supervisors from the crafts for all levels above foreman. It is claimed that their education permits them to overcome the weaknesses of supervisors from the crafts. But this is not a feasible suggestion, first because there are not enough college-educated people available to make even a dent in the need, and second, no person familiar with both professional construction education and construction field practices sees this as a likely possibililty. The new graduates lack hands-on experience in craft techniques and in getting and keeping crews at work. And, although they are conditioned to accept the notions about and techniques underlying planning, scheduling, and improving work, few have practiced the techniques in the real world.

In other words, it is almost impossible for on-site construction managers, be they from the crafts or professionally educated, to have backgrounds that encompass both work-face experience in several crafts and also management education dealing with planning, scheduling, and handling people problems. Furthermore, because of the usual craft-structured organizational plan and many other pressures, if persons with such qualifications existed, there would be neither time nor opportunity to put all their talents to work. Improving productivity, then, requires pooling the skills and knowledge of the two groups. As discussed in more detail in Chap. 3, this can best be done by developing activities which will augment the line organization under which site management is carried out.

COMPLEXITIES IN THE CONSTRUCTION-MANAGEMENT SITUATION

It is common knowledge that in the real world of on-site construction, those in charge seldom get around to establishing and carrying out effective and continuing productivity-improvement programs. In looking for reasons, two obstacles that loom large are: first, managers in general are unaware that proven ways of going about and implementing productivity improvement programs are available. This book has been written to address the problem by demonstrating some proven ways. The second obstacle is that managers are too occupied with the complexities involved in getting the work done to think about, much less carry out, organized programs. The next few pages examine some of these complexities, to put them in perspective so that they can be approached as constraints to be worked around, rather than insurmountable barriers which make efforts toward productivity improvement impossible.

The General Management Climate

For more than 60 years it has been common practice to define a manager as one who plans, organizes, controls, coordinates, and leads. This description conjures up the image of everything in order and running smoothly. However, for most managers in every industry and at all levels the real-life situation is better described as chaotic, unplanned, and charged with improvisation. *Fortune* magazine, May, 1976, states that studies have shown that managers worked 55 to 60 hours a week; never got past the first item on a list of things to be done; spent 40 to 60 percent of their time in meetings, 90 percent of which were called by others; and averaged only 15 to 30 minutes between interruptions. This situation had changed little by 1985, when an average work week of 56.9 hours was reported. Although these facts are from industries other than construction, the general picture applies to it as well.

One observer of management work has explained its complexities by subdividing what managers do into the large variety of roles they must play, as follows: first, interpersonal (figurehead, leader, liaison); second, informational (monitor, disseminator, spokesperson); and third, decisional (entrepreneur, handler, resource allocator, negotiator). These terms suggest that the manager is expected to be all things to all people, both within and outside the organization.[7]

From this brief discussion of what managers do, it is easy to see why they pass up opportunities to improve productivity. They are overloaded with work and belabored by demands for their attention so that little time, energy, or capacity remain to attack new topics or acquire new techniques or skills. If this situation is to be rectified, management first must become aware that new and better ways exist and then provide resources, training, and time to make improvement possible.

How Construction Differs from Manufacturing

Those involved in productivity improvement in other industries often criticize construction managers for failure to apply the common techniques of planning and methods-and-time study to their problems. Although these charges are partially true, they do fail to recognize that on-site construction has unique problems. A few of these, covered in somewhat different form earlier, are:

1 Most projects or their individual work phases are of relatively short duration. One consequence is that management teams and possibly the work force must be assembled quickly and cannot be shaken out or restructured before the project or work phase is completed. Also, planning and tooling up for an operation can be done only once. Unfortunately, many owners and contractors carry this short-duration thinking on to longer projects. For example, the need for long-range planning may be overlooked, adjustments to management structure and staff may be foregone, and lessons learned from earlier mistakes may not be applied. In addition, management may take a here today, gone tomorrow attitude toward lower-level supervision and the labor force and neglect valuable selection and training procedures.

2 On-site work stations are not permanent. In a manufacturing plant, for

[7] See H. Mintzberg, *The Nature of Management Work,* Harper & Row, New York, 1973, and *Harvard Business Review,* July–Aug. 1975.

example, a given operation is assigned to and carried out in one place. In contrast, specialized construction crews progress from location to location. Often this must be done in sequence; for example, crews doing excavation, concrete forms, reinforcing steel, and concrete placing and finishing must finish before vessels or structures can be set in place. Often, the additions made by succeeding crews modify or further restrict the available working space.

3 The final product is usually of unique design and differs from work station to work station so that no fixed arrangement of equipment or aids such as jigs and fixtures is possible. Even so, given that many operations are highly repetitive—for example, a weld is a weld—standardization of procedures and special means for making the work go faster and more easily should not be overlooked.

4 Because construction is a preliminary step leading to a completed facility, the layout and arrangements may make access for construction difficult and permanent provisions for safety impossible.

5 Because construction often needs highly skilled craftsmen rather than unskilled workers, individual crews, whether union or nonunion, usually do specialized operations. This means that tasks which must be carried out in sequence have crew following crew. Unless schedules are done carefully and are accompanied by commitments to see that they are met, delays and cost overruns are almost inevitable.

6 Operations are commonly conducted out of doors and are subject to all the interruptions and variation in conditions and the other difficulties that rain, snow, heat, and cold can introduce.

7 Construction often involves large-scale, cumbersome, and heavy assemblies of vessels or other components that are difficult to handle and fasten in place.

8 The owner is deeply involved in the construction process while the purchaser of manufactured goods is not. Buyers of the usual manufactured products seldom have access to the plant where they are made, nor do they deal directly with factory managers. For example, most electric appliances are stock items which reach the buyer through wholesalers and retailers. Purchasers merely select a particular make and model. Even with semicustom-built items such as automobiles, buyers have almost no input into basic design details such as engine bore and stroke, drive-train characteristics, or the physical and chemical properties of steels, plastics, or paints. Rather, purchases are made from a limited number of machines on a take it or leave it basis. Rarely do buyers deal personally with anyone at any level in the manufacturer's management hierarchy or labor force.

Even when the contractor bids a lump sum on a supposedly complete set of plans and specifications, owner's representatives are involved since they stipulate how to adapt the design to site or other conditions. Also, they monitor the quality, dimensioning, and placement of materials. Before acceptance, there is a final review, which often brings a sizable "punch list" of supposed or actual deficiencies to be corrected. On fast-track or other projects where the overall project concept may not have been finalized or at least construction begins before design is completed, even more owner, designer, or other interventions are to be expected.[8]

The point being made here is that in manufacturing, the buyer does not enter the scene; in construction, the buyer is in it from beginning to end. Thus, construction managers have an added set of players with which to contend. Usually, these individuals or groups have different purposes, concerns, and reward systems. As a result, communication often becomes difficult, and clashes become commonplace. Adversarial relationships (see below) can easily develop, with each party attempting to frustrate the other or cover up its own mistakes.

This brief discussion about how construction differs from manufacturing is not intended to explain away the failure of construction managers to make jobs run better by paying careful attention to away-from-the-work-face matters, improving the job climate, and applying productivity-improvement techniques. Rather, the point is that such efforts must not be undertaken blindly, but must be modified to fit the particular conditions of each project and its specific management and labor problems.

Certain Kinds of Construction Projects Are Very Complex

The complexities of large projects—high-rise buildings, for example—have often been met by having general contractors or construction managers serve as the owner's representatives in planning, coordinating, and managing. Among other duties, they arrange to bring on site successively or as needed contractors who do site preparation, excavation, and building-frame erection. These are followed by specialty contractors for plumbing, mechanical and air conditioning; electrical; elevators; and finishing activities of all sorts. In addition to sequencing the operations, when work and storage space is restricted, they allocate and schedule the temporary use of exterior space, interior areas, elevators, and hoists. Finally, they see that final corrections to meet the owner's acceptance are coordinated and approved.

Complex industrial and power facilities call for all the coordination mentioned above for buildings. In addition, this coordination may begin earlier, at the conceptual and design phases. Modifications to adjust to changes in technology or owner or regulatory-agency requirements may add to the difficulties. It is understandable why coordinating and controlling such projects are difficult assignments indeed.[9]

---

[8] A parallel from manufacturing to fast-track construction is a contract to develop a prototype of a new military aircraft or other weapon where requirements and design are continually changing during production. Here, the tradition is that contracts are cost-reimbursable. In contrast, cost-reimbursable contracts for construction are suspect and neither owners nor contractors are comfortable with them. They are usually illegal for public works projects. One could ask, "Does this mental set against cost-reimbursable contracts in construction make sense?"

[9] See reference to C. B. Tatum, *op cit*. for added details.

In addition to the increasing complexity in projects themselves, site management is held accountable for many requirements unrelated to the project per se, among which are governmental regulations or labor agreements that require reporting on wages and fringe benefits, equal employment and business opportunities, and safety matters, to name only three. To a large degree these complexities and requirements have crept up on job-site management. And even with assistance on paperwork from home-office or job-based computers, they add another straw onto job-management's overloaded back.

Construction Projects Can Be Plagued by Adversarial Relationships

Construction projects provide a fertile ground for adversarial relationships to develop and grow both within and outside the contractor's organization. Management-labor differences, already alluded to, are an obvious area of conflict. Others can include:

1 Between-craft conflicts, not only at the work-face but at all levels below the top of the organization when operations are organized along craft lines as they usually are.

2 Line-management–engineering conflicts. The on-site engineering organization usually reports to the project manager (see Fig. 2-1). Its responsibilities commonly include laying out the site, handling any engineering design called for at the job site, making available the needed plans and specifications, and keeping track of changes and change orders. It may also be charged with responsibility for developing schedules and for inspection and quality control (see below).

At the root of many line-management–engineering conflicts are differences in objectives. Line management is primarily interested in and judged by its output as measured by the cost system (see below). Failures or delays in engineering activities can adversely affect these costs. On the other hand, engineering is commonly seen by higher management as a "cost center" and encouraged to spend as little as possible, which often means it is understaffed and cannot comply quickly and fully to line-management's requests or demands for service. Other functions often assigned to engineering such as inspection for quality may delay the work or require that it be redone. This also adversely affects progress and costs, which are line-management's main concern. But work done poorly or deficiencies uncorrected may reflect unfavorably on engineering, which will insist that its orders be carried out.

3 Line-management–service function conflicts. Every construction project needs a variety of services such as purchasing, warehousing, and tool and equipment supply and management. On small projects, provision of such services may be handled as a part-time assignment by an individual who carries out other functions for line management, engineering, or accounting. On large, complex projects, however, separate departments or organizations may be created to carry out these functions.

Just as with the line-management–engineering conflicts mentioned above, individuals carrying out service functions have different objectives and often are judged by different measures from those of line management. Almost always they are seen as cost centers rather than profit centers and are charged against that always suspect account, overhead. The pressure on individuals who manage these services is to keep costs down rather than to provide service. Also, objectives may differ. For example, purchasing may see its function as finding the best price for a given item even though a bargain may be unsuitable or poor quality and ultimately lead to an increase in field costs. Again, those in charge of warehousing and handling tools or equipment may take pride in keeping inventory low or imposing rigorous checkout procedures to prevent theft, even though such practices may cause delays and lead to higher costs to field operations.[10]

4 Conflicts between line management and the contractor's quality-control organization. In most privately financed and lately on some governmental projects, quality control has become a separate function in the on-site organization. It may be carried out through a quality-control manager or engineering, as noted above, and the reports are directed to top site management. Occasionally, the report is sent to the home office or to the owner, bypassing on-site management entirely.

From a human-behavior point of view, placing both production and quality-control responsibilities in the contractor's organization can lead directly to conflict. For even when line management is committed strongly to quality, it costs money and time to observe certain niceties that inhibit production, or to correct errors or omissions detected by quality-control personnel. And as stated earlier, money and time are the usual measures by which line management is judged. In contrast, performance of the quality-control personnel may be measured, or they may feel that they are measured, by their ability to detect errors. For reasons such as these, it can be argued that assigning quality control to the contractor is, in principle, wrong.[11]

In almost all instances, quality-control inspections have an adverse effect on productivity, either by introducing another step into the production process or by calling for rework or at least changes in procedures. As with engineering and services, quality control is usually seen primarily as a cost and individuals in charge of it may be rewarded for keeping expenses down. In other instances, it is rewarded for being a watchdog; seldom is it allowed to be an ally of line management.

5 Contractor-subcontractor conflicts. Building construction is usually under the control of a general contractor who holds a prime contract with the owner. The firm commonly does only a small portion of the work (often con-

---

[10] The discussion here is merely intended to highlight an area of conflict. Ways to improve these situations are discussed later.

[11] For added discussion see M. Isaak, *Journal of the Construction Division*, ASCE, vol. 108, no. CO-4, Dec. 1982, pp. 481–484.

crete forming and placing). The remainder, possibly 80 percent or at times even 100 percent of the total, is handled by specialty firms that have subcontracts with the general contractor for a portion of the work. On a major building there may be 20 or more subcontractors. In the heavy and highway segments of construction, general contractors usually do a larger portion of the work with their own forces and may subcontract only specialty items. For industrial and housing projects either of the patterns outlined above may be followed, depending on the circumstances. In any event, whenever there are subcontractors on a site, there is an opportunity for conflicts to develop. Every one of the parties is in business to make a profit, and the activities or practices of the general contractor or certain subcontractors may be costly to others. The list of potential conflicts is long; an example would be arranging access by different subcontractors to a particular work station at a specific time. The responsibility of the general contractor's site manager is to resolve such disputes and to keep the project moving—a function which demands much ability.

The most common contractor-subcontractor differences are over progress and final payments for the work accomplished. General contractors gain by delaying and by making smaller progress payments; on the other hand, subcontractors want prompt, large settlements. Other sources of argument revolve around the subcontractor's work schedule, payment for claimed extra work, quality, safety, and responsibility for such extra tasks as cleanup. Here again, the point is that differences in objectives lead to conflicts, and these can, and usually do, adversely affect all aspects of a project.

6 Owner-contractor disputes. The arguments over promptness and level of payments and quality of work, just mentioned, apply equally to owner-contractor relationships. Often there are also issues revolving around deficiencies or discrepancies in the plans, specifications, or site conditions different from those originally anticipated. Another group of disputes centers on time of completion. If a project is delayed, the owner may withhold payment, claiming damages under a liquidated damages or other section of the contract; in rebuttal, the contractor will claim that payment is due because the delays resulted from actions of the owner or changes in the plans or site conditions. Or the contractor may claim added payment for the costs of accelerating the work to complete the project on time. As with contractor-subcontractor relationships, the interests of the parties differ markedly. Unless there is good faith and a willingness to compromise by both parties, the differences will develop into adversarial relationships that take the attention of all involved away from getting the job done.

7 Contractor–construction-manager disputes. Owners sometimes employ construction managers to serve as their representatives on construction projects. This is an alternative to the general-contractor–subcontractor approach. Construction managers are commonly contractors, engineers, individuals, or firms which specialize in this activity. Often they advise the designers on constructibility, including costs, of alternative designs.

Under the construction-manager approach, there is no general contractor, and contracts for each major subdivision of the work are directly made between the owner and an appropriate construction firm. The construction manager advises on the terms and conditions of these contracts, serves as coordinator during construction, and handles the details of payment, inspection, and acceptance.

Contractor–construction-manager disputes follow the same lines as those described above under owner-contractor disputes, since construction managers stand, in essence, in the owner's shoes. But as separate parties with different interests, they may also develop differences with the designer or even the owner and these can also affect relationships on the work site.

8 Contractor–third-party disputes. Many individuals or agencies, representatives of designers, local or federal officials, special interests such as environmentalists, and even the courts, have a say in construction operations. Seldom are they helpful and adversarial relationships may often develop that must be resolved.

This brief recounting of a few adversarial relationships that can develop on construction projects is far from complete. It is intended to highlight another set of pressures on on-site managers that can consume precious time and mental energy. Furthermore, research has shown that when a controversy exists, the primary focus of attention is on winning or at least resolving it, so that many other important matters are postponed by all those involved.

These built-in characteristics of the construction process often foster conflicts that take management's attention away from concerns about productivity. It is for this reason that this book focuses so heavily on developing mutual respect, teamwork, and commitment from all the parties.

Performance Measures and Reward Systems for Construction Are Often Unsatisfactory

Almost all construction companies use formal cost accounting systems and progress reports to measure the performance of all levels of job management (see Fig. 2-2). All too often these are the only measures employed. In essence, they record the expenditures charged against each significant class or unit of work and the quantity of it completed to date or in a given time period. The sum of the expenditures divided by the quantity completed gives a unit cost. This is compared with an estimate or budget, and the difference, or ratio, between the two is taken as a measure of performance. Results are distributed to or circulated among higher-level (home-office) management and job management down to, and sometimes including, foremen. Appropriate summaries are also reported to the estimating department to provide information for bidding similar items on subsequent work.[12]

In general, the individual items in the cost report cover physical accomplishment at the work face which can be seen and measured. Examples include

---

[12] There are numerous books and articles on cost accounting for construction. For a discussion of their characteristics, see P. Tiecholz, *Journal of the Construction Division,* ASCE, vol. 100, no. 3, Sept. 1974, pp. 255–263, and *ibid.* vol. 100, no. 4, Dec. 1974, pp. 561–570.

such items as cubic yards of excavation, square feet of concrete forms, cubic yards of concrete placed, or lineal feet of pipe or conduit installed. They are often the pay items in unit-price contracts. In most cases, costs and accomplishments are reported by field supervision, although some contractors may have the reporting done by cost engineers or clerks.

At first glance, this form of cost control seems to offer construction managers a valuable tool for both assessing a going project and providing data for estimating the cost of future work. And it can fulfill this function on simple, straightforward operations that are not affected by away-from-the-work-face foul-ups and when almost all the costs and accomplishments are for a single crew doing a single task at the work face. However, for more complex situations such a cost system incorrectly used may deceive rather than inform management, and lead to conflicts, less-efficient operations, and strained relationships among project personnel.

All too many construction executives and field managers are wedded to their cost systems and resent anyone who criticizes them. Even so, those individuals will usually agree about some of the following difficulties that cost systems bring:

1 The supposedly firm data reported by the cost system drives out consideration of important factors or concepts that are not quantified in money terms. For example, as mentioned earlier, activities such as planning, engineering, purchasing, warehousing, and tools are commonly classed as cost centers or overhead, and outlays for them are clearly reported . On the other hand, the usual cost system cannot measure, and does not report, the losses occasioned if these services are carried out poorly and result in a substantial increase in overall costs. Often, these overhead items become the first targets when a project is in financial trouble. Actually, as will be demonstrated later, in many cases the pressure to economize on these items may be the principal reason that the project is losing money.

2 Because it focuses on unit costs for individual work-face items rather than on the costs generated by individual foremen and crews, the cost system may obscure the true magnitude of expenditures. For example, reporting forming costs of (say) $4 per square foot, particularly if within the estimate, does not attract management's attention. Such a number taken alone does not emphasize the facts that a foreman with a crew of ten, at $600 per week per worker has responsibility for $300,000 per year in raw labor costs and that the general foreman directing four such crews controls $1,200,000 annually. Neither does the $4 per square foot figure make the important point that if by training, work-improvement studies, or some other means, the effective work percentage is increased, the savings can be substantial. For example, if the effective-work percentage in the situations just described were increased from 30 to 40 percent, the gains would be $100,000 and $400,000 per year, respectively.

3 Reported costs for crews at the work face may not show their inefficiencies. For example, properly conducted work sampling or other observation might show that productive activities were underway only 20 percent of the time. To a casual observer, workers might appear to be busy, since they have long since learned to do so in order to keep themselves and their foremen out of trouble. But the real situation may well be that the workers cannot be productive even if they want to be, since some or all among plans, instructions, materials, tools, or equipment may be lacking. However, despite such inefficiencies, the unit cost may meet the estimate simply because it is derived from earlier projects which were also inefficient. Under this circumstance, the already overburdened manager logically assumes that all is well and devotes attention to other pressing problems.

4 The effects on productivity of support activities are not measured. Engineering activities, such as providing schedules and plans, expediting, handling tools and toolrooms, and transporting and lifting equipment are commonly classed as overhead, and management tries to keep their costs down. But, as indicated above, poor performance of these services can cause inefficiencies and delays at the work face—effects that are not reflected in the cost system.

5 Managers may be held accountable for costs that they do not control. Delays resulting from failures in support activities such as those just mentioned, from waiting for preceding crews to finish, from rework, or from having to cope with a congested work site may force a crew to be idle or inefficient, but costs are still charged against the operation. Again, costs of more than one crew may be lumped into a single cost item. In such instances, none of the managers involved will or should be expected to be accountable. Furthermore, adversarial, noncooperative attitudes will likely develop.

6 The cost system directs attention to items running over budget. There is a strong tendency to neglect items running under budget, although substantial savings still may be possible.

7 Where an item is running over budget, pressures develop to hurry, and thereby lower the quality of workmanship. Furthermore, things such as cleanup that can be left for succeeding crews or shifts will remain undone.

8 If the cost system is used as a basis for judging performance, and particularly if it is considered to be unfair, reporting may not be complete or honest. In order to avoid having poor records, managers will juggle costs among items, charging work against items that are underrunning the budget rather than against those that are overrunning it. Again, they may overload such items as cleanup if several crews are permitted to charge work against them. On the performance side, field supervision will often report more progress than is actually accomplished in order to cover up and to get out from under management pressures, even if only temporarily.

9 Given that the cost system is considered to be fair in that it reports the activities of a single foreman and crew, management too often puts the findings to the wrong use. "Management has a scapegoat!" "There is some specific person to blame for the cost overrun." Actually, the report only says that something is awry. Other causes might be that the estimate or budget is too low, or that plans, materials, tools, or equipment are lacking. As construction

safety research has proved so conclusively, when something goes wrong, the effective manager digs into the situation rather than blaming others. But this is not easy to do, given the environment in which such deficiencies occur.

Many of the difficulties just described strongly reflect two basic human tendencies that managers neglect at their peril. The first is to protect one's personal well-being, even if this involves covering up one's mistakes or shifting blame to others. The other is to please the boss or its counterpart, do not give the boss bad news. Good news is welcome; bad news makes the boss dislike and blame you.

Some may feel that this brief critique of construction cost systems is unfair and too harsh. Unfortunately, these systems have often let managers down. Instead of providing guidance, they have given poor, if not wrong, indications; and managers acting on them have made decisions leading to all sorts of difficulties.

Informal Feedback Systems

As discussed earlier, in addition to the formal progress and cost reporting system, feedback on job progress and problems is reported up through the management hierarchy in a variety of other ways. Foremen give oral reports to superintendents and they, in turn, pass the information on to construction managers, either informally or at scheduled times or meetings. Also, some form of written communication scheme may augment or be a substitute for word-of-mouth reporting. Again, information may be passed to and assembled by a project scheduler or coordinator, if there is one. And, as in any situation where groups of people are involved, there is the grapevine through which information flows outside established channels. As shown on Fig. 2-2, workers as well as management have their own feedback systems which, unfortunately are often their only but oftentimes erroneous source of information.

Communication in all the ways just mentioned is an integral part of any organized work situation. It generates information and impressions which may become a way of measuring and evaluating performance. And, as with the cost system, the affected individuals, in order to protect themselves and carry good news to the boss, feel pressures to color, distort, withhold, or even falsify the messages.[13]

Construction Management Has Not or Will Not Fully Use Labor's Talents

Construction management's attitude toward labor is paradoxical. On the one hand, it often leaves planning and carrying out work-face tasks to foremen

---

[13] The complexities of communication and means for improving it are discussed in more detail in Chap. 10.

who, in turn, must delegate it to craftsmen, each of whom manages his or her own efforts. On the other, it is prone to blame labor for low productivity and poor quality. Rarely have the talents of labor been enlisted to find the reasons for low output or to devise better, more efficient methods.

Management offers several explanations for its attitude toward labor. One is that the labor force is not permanent but comes and goes, a charge that is only partially true. Another is that workers, encouraged by their unions, often hold down productivity through slowdowns, placing quotas on output, work stoppages, featherbedding, jurisdictional disputes, and even sabotage to extort concessions from management. At times these charges may be true. But often such claims are employed by management at all levels to explain away its ineptitudes, such as failures to provide direction, materials, tools, equipment, or an accessible workplace. As will be shown in Chap. 10, in most instances workers gain satisfaction from productive work and are frustrated when it is not effective. How to tap labor's knowledge and desire to be productive and to teach it appropriate techniques for doing so is one of the most important topics covered in this book.

Overcoming Obstacles Brought On by Construction's Management System

This section has summarized the obstacles that are usually taken as givens by on-site construction managers. But it has been demonstrated that many of them can be mitigated or avoided entirely by employing sensible management approaches which include time management, team building, planning, and attention to people. Some of the techniques needed to make this happen are given later in this book. Then, without the necessity of devoting time to putting out the fires brought on by these difficulties, it should be possible to get to the important tasks associated with productivity improvement.

SUMMARY

This chapter has attempted to look at the organization; information flow; people, particularly management people; management climate; uniqueness of construction projects; and the adversarial relationships and performance measures that make construction operations different from other industries. Emphasis has been given to how each of these factors impinges on efforts to improve productivity. Chapters 1 and 2 have set the stage for a more detailed examination of the topics in the chapters which follow.

CHAPTER 3

PRODUCTIVITY IN CONSTRUCTION—WHAT MANAGEMENT CAN DO TO IMPROVE IT

INTRODUCTION

Chapter 1 discussed construction industry performance and the differences between performance and productivity, defined productivity, and developed the concept of project goals for all the involved parties and what they should be if productivity improvement is to be realized. Chapter 2 examined the situations unique to construction under which these goals must be achieved. This chapter proposes management approaches that should be adopted if the goal of improved productivity is to be achieved. Chapters which follow deal with techniques for improving productivity, assuming that management has made it possible for them to be followed by adopting and implementing appropriate goals and procedures.

This chapter deals with management approaches to improve productivity under the following headings:

- Investigations of productivity—what they do and do not tell
- What has been done to date to improve productivity
- Questions that should be asked about productivity in on-site construction
- Tracks which management must understand and follow if it is to start and continue productivity improvement
- How a productivity-improvement program is put together
- Summary

INVESTIGATIONS OF CONSTRUCTION PRODUCTIVITY— WHAT THEY DO AND DO NOT TELL[1]

As mentioned before, a reported decline in productivity in construction in the United States has been causing great concern. One commonly cited set of statistics to support this claim is from the American Productivity Center (J. W. Kendrich). Its study states that between 1968 and 1978 output per unit of construction labor, in constant dollars, fell 29 percent. In contrast, this ratio increased 36 percent from 1948 to 1958 and 34 percent from 1958 to 1968. Business Roundtable Report A-1 cites a 22 percent decline in construction productivity between 1970 and 1979. Again, The Summary Business Roundtable report cited a drop of 3 percent a year from 1973 to 1979 and 8 percent in the 1979–1980 year. These and other statistics were widely cited at that time, particularly by those who wished to place the blame for productivity losses on labor. Actually, Alfeld, in the Roundtable report, concludes that such claims can be neither supported nor refuted. He states that "since the quality of the available data is inadequate for precise analysis, economists have found the apparent productivity decline either impossible to explain or explain away." Alfeld makes the point that the important but unanswered question is, "How much of the national construction productivity decline, if it exists, can be ascribed to the involved parties, and what to do about it ?" He states that the answer will not be found in any of the Bureau of Labor Statistics (BLS) data or anyone else's. This can be largely attributed to the fact that the data are aggregated and therefore have very little meaning in a highly segmented and regionally different industry. For example, the productivity of small local contractors who employ the same labor force or the same kinds of projects may have remained constant or even improved, but the aggregated data, which includes large complicated projects, may unbalance the results. Furthermore, because of the adversarial climate which often exists between owners and contractors on the one hand and organized labor on the other, such data as exists may be employed in a misleading manner.

According to Alfeld, the characteristics of the large-scale industrial projects of particular concern to the Business Roundtable probably have made them especially susceptible to lower productivity. These characteristics are cited here, primarily to indicate the broad scope of the productivity problem and management's and labor's contribution to it. Some of Alfeld's conclusions may not be as applicable in the late 1980s, nor to smaller, less complex projects; others apply to them with equal force. But they provide room for thought. They are, in modified form:[2]

---

[1] Much of the discussion which follows is drawn from *A Report on the Availability and Interpretation of Construction Productivity Statistics,* September 1980, prepared by Louis E. Alfeld for the Business Roundtable. Its findings, along with others, are summarized in the Roundtable Report A-1 titled *Measuring Productivity in Construction,* September 1982. See also *Business Round Table Report A-2* titled "Construction Labor Motivation."

[2] The tremendous cost and time overruns in the construction of atomic power plants is a recent and classic example which illustrates Alfeld's views. The construction involved a complicated and highly technical product coupled with stringent safety requirements, which led to many design changes that in turn, brought delays and rework. In addition, projects were of mammoth, almost

1 Quality requirements imposed by owners or regulatory agencies are more stringent, so that projects have more regulation or interference. These often produce tighter and more frequent inspections and continual design changes during construction, which can, in turn, increase delays and create rework, thus lowering output.

2 Lower average labor skills are available at large projects or at remote sites, where any available person, skilled or not, is hired in order to fill manning requirements, and where dismissal for poor work rarely occurs. Increased open-shop construction may also be a factor in situations where it offers lower average skills or where experienced, trained union workers have other jobs or refuse to work open shop.

3 The traditional management systems, particularly for planning, scheduling, and controlling materials are unable to cope with increased scale and complexity and with difficult staffing problems.

4 The fixed-fee or cost-plus contracting practices sometimes employed on such projects place less pressure on contractors to hold down costs or to press for rapid completion. This practice may also place owners in the role of construction managers, an area in which they may not have suitable skills.

5 A higher rate of innovation caused by changing economic pressures (e.g., energy) produces new designs, techniques, and materials faster than project management and the labor force can learn to fabricate and install them efficiently.

6 The lower cost of labor relative to the more rapidly rising cost of such other inputs as financing (time) and capital equipment has sometimes resulted in accelerated schedules (fast tracking) which use more labor (at a lower average productivity) to finish projects earlier.

7 The work ethic among workers, particularly those younger and better-educated, may have been less strong or absent.

Alfeld points out that solutions to the first six problems rest largely with owners, designers, or constructors, not with labor. Most thoughtful managers would agree. Alfeld's seventh reason—loss of the work ethic—normally commands the most attention among managers, be they owners or contractors, who are looking for someone other than themselves to blame. Alfeld points out that this loss of the work ethic may well be the least important. Furthermore, as other research has shown, even the most highly motivated workers lose incentive when a job is so poorly managed that their contributions make little difference to the outcome. All too often, delays which prevent or restrict production at the work face teach craftsmen to pretend that they are busy. Also, changes or errors requiring rework soon kill the incentive to produce, for work done will only have to be ripped out and done again.

overwhelming, size and often on remote sites, so that staffing was with transient and often less-skilled management and labor.

The point of this brief and incomplete discussion of the factors affecting construction productivity is to make clear that many factors can play a part in either decreasing or improving productivity, all of which deserve attention.

WHAT IS AND WHAT IS NOT BEING DONE TO IMPROVE PRODUCTIVITY IN ON-SITE CONSTRUCTION

To say that constructors do not recognize that productivity means higher profits is absurd. To this end they have eagerly made many investments and adopted a wide variety of schemes where the return on investments or other payoffs were apparent and measurable. A few examples among many that have been adopted are new models and kinds of equipment and labor-saving devices such as special concrete forms and accessories, automated machinery, and many jigs and fixtures. On the other hand, constructors are less perceptive of and receptive to planning and other approaches to productivity improvement where the payoffs are difficult to quantify but the costs in management effort or special staff to implement them are real and highly visible.

There is undisputed evidence that the approaches described in this book will pay substantial if not huge dividends. An early example comes from Weldon McGlaun who was for many years manager of construction for Proctor & Gamble. As an owner, this company has taken an active role in the management practices of constructors who build plants for it on a cost-reimbursable basis. McGlaun has stated that "We believe the ratio of dollars saved to dollars invested in an effective planning program is in the ratio of 4 to 1 to 8 to 1." He subdivides these savings among planning, including methods studies, 50 percent; forecasting, 10 percent; scheduling, 15 percent; and cost control, 25 percent.

Another report of estimated savings, these from implementing CICE recommendations, has been given by C. D. Brown of DuPont, 1985 Chairman for the Construction Industry Institute. He reports in *Engineering News Record,* Aug. 14, 1986, p. 48, a 10-to-1 ratio between savings and expenditures. More specific are the results of a recent study of several large industrial projects, where carefully planned and executed productivity improvement programs combining a variety of approaches have been undertaken. (See Chaps. 5 through 11 for added detail about some of the procedures used.) A survey of several of them by Victor Sanvido[3] indicated ratios of estimated savings to expenditures in the range of 15 to 1. One can be certain that on many other projects, less elaborate and often informal programs for productivity improvement are being carried out using portions of the more elaborate approaches just mentioned. Seldom will these be reported, either because construction people are too busy to do so or because they do not wish to reveal a competitive advantage. Un-

---

[3] See *Technical Report 273,* Department of Civil Engineering, Stanford University, Stanford, California, Mar. 1983.

fortunately, also, measurement of savings through a contractor's cost system is almost impossible, so that even individuals who carry them out will not have supporting evidence.

Less specific support for the notion that savings result when time is devoted to careful planning and analysis comes in statements from site management, which reports that projects seem to run better and productivity is higher when starts by the work force are delayed inadvertently. Unfortunately, there are no quantitative measures of these savings: even worse, the wasted time of project supervision during this delay period is seen as a cost that should have been avoided or as a reason for filing claims, rather than as an activity that probably would save money.

These examples of savings are few and scattered and are largely unknown or ignored by the industry in general. For example, a survey reported in *Engineering News-Record,* Dec. 13, 1984, p. 58, indicates little industry interest in productivity improvement. It states that only 361 of 950 contractors responded at all to a questionnaire asking about applications of findings of the Business Roundtable Construction Industry Cost Effectiveness Study. Of those responding, only 5 percent reported that their companies had "acceptable" programs aimed at improving productivity. These percentages are probably high, since relatively few contractors know of the study and its recommendations. Another piece of evidence showing the lack of concern of the industry for productivity improvement is offered by David Arditi.[4] He reports on 1979 and 1983 surveys conducted by *Engineering News-Record* to which only 15 percent of the top 400 contractors replied. The respondents did indicate choices from a number of listed home-office and field operations where they thought productivity could be improved the most at headquarters and in the field. They also expressed an interest in helping to identify problems and evaluate results. But the answers to two questions were particularly revealing. To the question, "Would you contribute funds to support programs in productivity improvement" only 18 percent responded yes. And to the question, "Would you conduct programs aimed at improving productivity" only 26 percent responded yes. Clearly these responses indicate that although industry leaders favor someone else doing something about productivity-improvement, they are reluctant to undertake it in their own firms. It must be concluded that the notion that productivity improvement efforts pay off handsomely is not widely shared.

The attitudes just described for large contractors actually permeate the industry. A survey of small- and medium-sized contractors produced strikingly similar results.[5]

---

[4] See D. Arditi, *Journal of Construction Engineering and Management*, ASCE, Vol. 111, no. 1, Mar. 1985, pp. 1–14.

[5] See E. Koehn and S. B. Caplan, *Journal of Construction Engineering and Management*, ASCE, vol. 113, no. 2, June 1987, pp. 327–339.

QUESTIONS THAT SHOULD BE ASKED ABOUT PRODUCTIVITY IN ON-SITE CONSTRUCTION

Productivity can be defined as the ability to produce an abundance or richness of output. Productivity improvement, then, treats efforts to enhance present operations from an existing level of output to more nearly approach that desired abundance and richness. This book will demonstrate (1) techniques by means of which successful efforts to improve productivity can be organized and carried out, (2) how excellent productivity in construction has been obtained, and (3) how less excellent productivity can be improved.

It has already been stated that construction has been and is now being done successfully, if not as effectively as is possible. In the final analysis, this has been accomplished by craftsmen or crews at work faces utilizing methods, information, materials equipment, and tools usually stipulated and arranged for, or supplied by, others. It follows that to maintain high construction productivity where it exists or to enhance less productive operations, attention must be focused on what happens at or impinges on activities at the work face.

This book focuses on four topics, which can be summarized as follows:

1 How commitments to productivity improvement can be developed, implemented and maintained at all levels in construction organizations.

2 How all those who are responsible for on-site activities can manage them in a way that will support or enhance efficient on-site and work-face and related operations.

3 The approaches and techniques that are available through which productivity on site and at the work face can be improved.

4 How existing or new away-from or at-the-work-face practices which improve productivity can be recorded, made known, and put to work in individual firms and by the construction industry generally.

The first two topics on this list are the subject of the remainder of this chapter. The remainder of the book addresses the others.

The authors recognize that the many activities and the skills they demand, other than those treated here, are needed by owners, designers, constructors, managers, field supervisors, and craftsmen to carry out construction projects successfully. Numerous books, manuals, and other publications which treat many of these technical and management topics are available. "More Construction for the Money," January 1983, the *Summary Report* of the Construction Industry Cost Effectiveness Study of the Business Roundtable, is a starting place for exploring; and far more detail will be found in the 23 Roundtable reports on which the *Summary* is based. The aim of this work is to treat only the four questions just listed, which have to a large degree been neglected by educators and industry alike.

TRACKS WHICH MANAGEMENT MUST FOLLOW TO IMPROVE PRODUCTIVITY

In approaching the topic of productivity improvement, it is helpful to think of the factors which must be recognized, observed in detail in a given situation, and covered by an organizational and implementation plan. Here, these factors are referred to as tracks, divided into (1) culture, (2) management skills, (3) team building, (4) strategy-structure, and (5) reward system. Some of these topics have been addressed earlier but in a somewhat different context. They are discussed here one by one.[6]

Readers may wonder why the authors have introduced this rather abstract track concept into a book dealing with productivity improvement. Their answer is that plans which have not covered all of these tracks will have little chance of success. To illustrate: Consider a plan that has been devised that will overcome the cultural differences among participants, provide the necessary problem-solving skills, be staffed with a dedicated team, and have a strategy and procedures to get the scheme going and to maintain it. Suppose also, however, that it lacks a suitable reward system or carries unacceptable risks as seen through the eyes of the members of the originating and implementing groups. In this case, one can be almost certain that the effort will have little chance of success. This sad fact has been demonstrated over and over again where productivity-improvements schemes have been attempted. The same result can be expected if the implications of any other among the five tracks have been overlooked.

The paragraphs which follow briefly describe each of the tracks in a construction context.

The Culture Track

Each of the parties in construction—owner, designer, contractor (or construction manager), and labor—has a distinct culture and way of measuring the success or failure of a construction project. These will dictate the way they approach or participate in efforts to improve productivity. Although there are variations within each culture, some of the important distinctions among them are as follows:

The Owner Culture Owners or their representatives are managers of business or industrial enterprises or of facilities built for the public. In this role they conceive projects, arrange for their design and construction, pay for them, operate them, and finally shut them down. Their concerns about these activities and how efforts to improve productivity will affect them will dictate their behavior.

Project costs are a principal concern in the owner culture at four project stages: (1) when it is formulated, (2) when individuals involved are deciding whether or not to undertake it, (3) during and after construction, particularly as owner representatives are held accountable for cost increases during construction, and (4) as the facility is operated and maintained. Timeliness is weighed by owners in terms of the consequences in costs or other measures

---

[6] The concept of five tracks is adapted from R. H. Kilmann, *Beyond the Quick Fix*, Jossey-Bass, San Francisco, 1984.

when the completion is earlier or later than anticipated. Quality, as a measure of the project's performance during its service life, can carry heavy weight both in cost and other factors. Finally, all owners have some concerns about accidents, because they lead to liability or other losses. Some but unfortunately not all owners also place great importance on safety because they see it as a moral responsibility or because it affects their public image.

The Designer Culture   Productivity as a descriptor of cost, quality, timeliness, and safety affects the interests of designers, but somewhat differently than it affects those of owners. Designers have a stake in cost, because if the outlay that the owners must make to carry out their plans and specifications at the conceptual or preliminary design phases appears to be too high, the project may not get beyond one of these and the fees associated with detailed design and possibly construction supervision may not materialize. At the construction phase following detailed design, costs affect the designer in two opposing ways. Design omissions or errors may rebound and lead to claims or fee reductions as well as a loss in reputation, which may mean a loss of present clients or an inability to get new ones. On the other hand, if design fees are based on a percentage of construction costs, a very common arrangement, added costs mean added revenues, sometimes accompanied by a charge that the designer has gold-plated the project to increase fees. Difficulties with timeliness, quality, and, perhaps, safety during the construction phase may have adverse effects on the designer and bring claims or reflect on the individual's or firm's reputation and its prospects for repeat or new clients. In general, the construction manager culture has the same concerns as do designers, plus some of those discussed below under contractor culture.

The Contractor Culture   Without question, the primary motivator among contractors is profit; projects and the firm's overall success are judged as good or bad primarily on this basis, which is understandable, since survival of a firm and rewards to owners and employees are dependent on it. In addition, and to varying degrees depending on the character of the work and the contracts under which it is performed, there will be concerns about the firm's reputation for quality, timeliness, and safety and general standing in the construction community. Even these are indirect measures of profitability, inasmuch as they affect a firm's ability to get new work.

It is important to recognize that even though profit is the primary motivator in the contractor culture, its effects can be substantially different depending on the form and details of the contracts employed by owners and the climate they create. In the public sector, owner agencies are usually required by law to solicit competitive, fixed- or unit-price bids and award contracts to the lowest responsible bidder who submits a regular (complete and not obviously unbalanced) bid, with responsibility judged largely by financial solvency. Past records of shoddy workmanship, time overruns, safety violations, or sharp dealings regarding contract clauses or claims seldom can be considered, although, human nature being what it is, a repetition of past behaviors can be expected on new projects. This competitive-bid constraint does not apply to owners in the private sector, unless their management chooses to use it. But if such a choice is made, the competitive bid culture will emerge and the same behaviors described for the public sector can be expected. The difference is that private owners, in contrast to public ones, have a choice among contractors and contract types and clauses and can weigh their effects on the underlying attitudes and probable behaviors of the contractors they select.[7]

The Management-Skill (Problem-Solving) Track

Productivity improvement does not take place in a vacuum; it requires careful study and an ordered approach. Regardless of whether the effort is by an individual or a group, there must be a well-planned step-by-step attack on the problems being considered. In Kilmann's scheme (op. cit.), these steps are (1) sensing problems, (2) defining problems, (3) deriving solutions, (4) implementing solutions, and (5) evaluating outcomes. (For a somewhat different approach, see the section on ''Decision making'' in Chap. 10 of this book.)

The Team-Building Track

Teams set up to carry through an organized productivity-improvement effort must include representatives of all the affected parties if they are to accomplish their objectives. Who is involved depends on the particular situation, which can range from the on-site activities of a single crew to concerns involving owner, designer, contractor, and subcontractors, and possibly labor representatives.

The first step in instituting the team-building track must be a commitment from managers at the top of each of the affected parties. They must make clear to their subordinates that they believe the effort is both necessary and worthwhile and prove their sincerity by allocating the resources of money, time, and talent and by establishing a follow-up procedure. Only then should team members be designated. They should meet the following criteria, among others:

• The abilities that each can bring to the effort. Position in an organization's hierarchy should not be a factor. In the on-site efforts within the contractor's

---

[7] A detailed discussion of the complexities of the many possible contract forms for construction and the clauses in them that can affect the relationships between owners, designers, and contractors is outside the scope of this book. A few that can be the most troublesome are work-scope definition, the completeness of supporting documents, effects of and payment for design or construction changes, definition of what constitutes reimbursable costs, payment procedures, and requirements for cost and schedule reporting and control. For a recent and penetrating analysis of this topic, see ''Impact of Various Construction Types and Clauses on Project Performance'', *CII Publication 5-1, July 1986.*

organization, groups can well include individuals from high in the line, office personnel, engineers, and superintendents or foremen.

• A suitable temperament and willingness to be a team player. Kilmann proposes that a deliberate effort be made to avoid possible troublemakers or if they must be included, to provide group or individual counseling for them.

Regardless of the sorts of activities that the group is to undertake, it will probably be necessary to provide help and coaching on how to develop data, conduct effective meetings, and report and implement the group's findings. This is not an easy task and often it is advisable to bring in an outside consultant unless the effort is large enough to support an in-house specialist.

An important aspect of team building is a recognition that it takes time to establish rapport and trust among the members of a group. At first, it will seem that little is being accomplished. Both participants and higher management must be warned not to expect substantial results quickly. Observers of such groups report that the first productive result comes slowly, but those which follow come at an ever-increasing rate.

The Strategy-Structure Track

Once groups are established and their area or areas of activities designated, strategies must be devised and a structure developed to implement their findings. As discussed further in Chap. 4, both the strategy to be employed and the structure for implementation will fall into two general categories, (1) those which involve the system and will require organizational cooperation, restructuring, or a reassignment of responsibilities, or (2) those which involve an activity or activities that fall within the province of a single supervisor.

It is to be expected that the responsibility and initiative for implementing the strategies developed by the productivity-improvement teams and the structure to carry them out will fall on line management. The reasoning underlying this tactic is outlined later in this chapter.

The Reward-System Track

Productivity improvement efforts are carried out by people, and the strategies and structures that are proposed will be implemented by them. It follows that the rewards for and risks to each of the parties will largely control their behaviors. As already mentioned, experience has shown that schemes that do not recognize the rewards and penalties associated with them are seldom implemented.

These human behavior, or "reward-system," factors and ways of dealing with them are discussed at considerable length in Chaps. 5 and 10. In the simplest terms, they might be reduced to three: (1) what is done must be to the advantage of the individual and the group to whom loyalty is owed, (2) the end product must please the individual's boss, and (3) there must be "cover" for the individual if something goes wrong.

The rest of this chapter outlines methods for developing productivity-improvement programs. Underlying these approaches is a recognition that the tracks just outlined must always be considered.

PUTTING TOGETHER PRODUCTIVITY-IMPROVEMENT PROGRAMS

Contrasts between Productivity Improvement and Production Organizations and their Functions

Figure 2-1 shows the structure of the line organization for a typical construction job, and Fig. 2-2 illustrates how information flows in such organizations. In the broader context of a project as a whole, this line structure must be expanded to include owner and designer as well, as specified in the contractual relationships among the parties. These structures work well on the aspects of projects that are carefully and fully defined with complete plans and clear specifications and where there are no substantial changes in site conditions or hitches in the field operations. Some management scientists may propose other arrangements such as matrix management, but the authors of this book feel that for the usual construction project, where coordination and team play are essential, this form of management structure, modified to fit special circumstances, makes sense. It maintains the status quo. The duties and obligations of the parties are defined; everyone knows what is to be done, how it is to be done, and who is to do it.

It is when something goes wrong or changes are to be made, as is common in construction, that this traditional management arrangement gets into trouble. For example, there is usually no contractual or line relationship between the designer and the contractor, because the owner has separate agreements with these parties. It follows that there is no established mechanism which can force designer and contractor to communicate formally to make changes or to settle differences in a manner that is agreeable to both of them and the owner as well. Rather, the owner must be the intermediary. Agreements, if finally reached, involve the line organizations of all three parties. This procedure is cumbersome and time-consuming; and on fast-moving construction projects, it often comes into play too late. Given this situation, it is easy to see why it can become difficult to introduce changes that would be mutually beneficial to the owner as well as to the other directly affected parties.

Other instances in construction can be cited where productivity-improvement schemes must be foregone because the changes they require cannot be accomplished through the line organization. Included among them are situations where the owner or designer should communicate directly with subcontractors or materials suppliers. Examples can also arise within a contractor's organization where even though a contract is not involved, cooperation and communication among its various sections such as engineering and the line or purchasing and the field make innovation difficult.

The line organization is particularly ill-situated to take advantage of ideas regarding innovation or productivity improvement offered by lower-level managers or craftsmen. Its entire philosophy and method of operation is set up to do it the way you were told to do it. Thus, unless there is a strong counterpressure from higher-level management, the easy way is to follow usual or designated methods, even though better ones can be found and brought forward. This difficulty is made even worse in situations in which time pressures say get it done now.

The examples just cited illustrate why productivity improvement is an activity wherein top-down or contractual management approaches do not fit, since the aim of productivity-improvement efforts is to find and introduce change in the way things are being done. Furthermore, these changes may affect established practices such as the roles of the various parties, the concept or design of the project, or the way operations and individual tasks are to be managed or carried out. All these can challenge the authority of line managers and threaten their egos and security. It is easy to see why boat rockers, who are the "yeast" in productivity-improvement efforts, are considered to be mavericks and troublemakers by line management.

In contrast to the established line organization whose task is to carry out an established course of action, effective productivity-improvement groups have no specifically assigned tasks and little or no formal organization. Their goal is to make recommendations, not decisions. Their approach involves a group of individuals sitting at a round table trading ideas, with no one in charge and with all the parties, regardless of rank in the organization or organizations, having equal status. In sum, productivity-improvement groups must have a different charter than that of the line organization.

Before a problem-solving group undertakes its work, there must be a meeting of the minds in the group on certain key questions. Among which are:

- Do we want to be involved?
- Do we have clear objectives?
- Are these objectives the same for all of us, or can we reach a consensus and agree on them?
- Do we have the methods, resources, and technology to achieve our objectives?
- Do we have a system through which our objectives can be accomplished?

To be effective, problem-solving groups, must have an able chairperson, or possibly a facilitator from outside, who helps the group carry on meaningful discussions and prepare its recommendations. Among the functions of the chair or facilitator are, without taking charge, to (1) keep the discussion on track, (2) quiet those who wish to dominate, (3) encourage the quiet to bring their ideas forward, (4) help the group develop consensus, (5) assist in recording and putting together the ideas that are generated, and (6) be involved in developing a procedure for implementing the group's recommendations.

Each problem-solving group, to be effective, must have among its members' skills covering all the aspects of the problem being considered. In many instances, after discussion, a consensus on a reasonable solution will be reached. With other problems, no single solution seems to be appropriate. In such cases, a technique called brainstorming, intended to encourage creative thinking, is often employed. Brainstorming is a four-step process, which proceeds roughly as follows:

Step 1: All members of the group are encouraged to propose any solution that they can contrive, regardless of how far out or impractical it may seem. These are recorded on a blackboard or tear sheet where they can be seen by the group. During this step, a rule, to be strictly enforced by the group, is that criticisms or derogatory comments about any of the proposals will not be permitted.

Step 2: When no other proposals are forthcoming, members of the group are encouraged to propose changes in or modifications to the recorded proposals. As in step 1, no discussion or criticism is permitted.

Step 3: Proposals as submitted or amended are discussed critically, one at a time, and those found by consensus to be unrealistic are eliminated from the list. If at all possible, decisions by voting are to be avoided.

Step 4: The remaining schemes or combinations of them are discussed in detail and a proposed solution is developed, again by consensus. Where alternative solutions have merit, these are also developed fully for presentation to the decision-maker.

Solutions recommended by such problem-solving groups, however reached, may often involve changes that those farther down in the line organization are not able to make because, as indicated above, they cut across or ignore organizational or contractual barriers or propose changes in work assignments. In such cases, the recommendations must go to a level where action is possible.

Productivity-improvement teams (or individuals) may be involved at several different levels or stages in the project's life. Their activities can encompass groups involving:

1 All the parties to a project, both on- and off site. These groups are often referred to as oversight teams or parallel organizations.

2 Organized on-site groups representing the affected parties, commonly called work teams or quality circles.

3 Organized on-site crew productivity and quality efforts.

4 Consultants or in-house specialists acting as facilitators for groups or as troubleshooters for, or advisors to, management.

At any of these levels, it is important that there be strong support from those at the top of the line organization or organizations and also that the teams or

TABLE 3-1
A COMPARISON BETWEEN SOME OF THE IMPORTANT ORGANIZATIONAL AND
BEHAVIORAL CHARACTERISTICS OF TRADITIONAL MANAGEMENT PRACTICES AND
PRODUCTIVITY-IMPROVEMENT ACTIVITIES

Traditional Management approach	Productivity-Improvement approach
Organization characteristics	
Purpose is to get work done; emphasis is on output and practicability	Purpose is to find better solutions to organizational, technical, and people problems
Coordination and decision making follow chain of command	Position in hierarchy is unimportant; has no chain of command; does not make decisions
Power is based on position in hierarchy; leadership is at the top	Power and leadership are based not on position or rank but on knowledge and ability to pursuade
Functions are specialized; task assignments and division of labor are by skills	Functions may be specialized but may be diffuse, covering a spectrum of topics
Operations are task-oriented	Operations are change-oriented
Problems are solved by following precedent and explicit rules and procedures	Problem-finding and problem-solving techniques are employed to develop new approaches
Information and feedback limited to that made available through the system	Full access is given to all pertinent information and feedback
Human-behavior situation	
Rewards come through conformance and following the plan; penalties possible for those who do not follow the rules or who make waves	High rewards for innovation and for challenging usual practices; lower vulnerability; shielded from reprisals

individuals work at their tasks long enough to develop trust and rapport among team members and with the line organization.

Some of the most important contrasts between how the line organization and productivity-improvement activities must function to be effective are outlined in Table 3-1. A detailed examination of the comparisons given there is well worth the effort. Without bringing forward each item in detail, it can be seen that the traditional approach relies on top-down management to get a specified job done. In contrast, the productivity-improvement approach is aimed at instituting change. To make such efforts effective, there must be a hang-loose approach which accepts and encourages challenges to what the line is currently doing. This difference is illustrated in all the examples given below.[8]

---

[8] See Chap. 10 for additional discussion of this topic from a human relations and communications perspective.

Developing Acceptance of the Productivity-Improvement Concept

To get acceptance of productivity-improvement as a concept requires that those involved must understand certain principles that can be developed from the five track approach outlined above. From these will evolve ways of thinking that are essential if success in generating and implementing ideas is to be achieved. Among them are the following:

1 Trust and respect at all levels in the parties involved.

2 Agreement that all parties have somewhat different but legitimate aims which motivate them. Any proposed action which compromises these must be aired and agreed upon.

3 A mental set accepting that cooperation is better than controversy is in the best interests of all the parties, since it is the only way to develop win-win solutions. Furthermore, it must be accepted that adversarial approaches are detrimental, since they lead to win-lose situations, which means that the parties must be willing to forego fixed positions and reach compromises.

4 Recognition of the reality that line management will, at least at first, react negatively to productivity-improvement efforts if they appear to threaten the status quo and the line's authority, ego, or perquisites, or make it appear to be incompetent.

5 Getting the group to overcome certain ways of thinking and operating that may cause resistance to productivity-improvement proposals. Among these are:
- Looking for scapegoats by blaming others when something goes awry. Labor and management alike have each often labeled the other as villain in such situations.
- Overcoming the "I have always done it this way, so why change" mental attitude, or its counterpart of "This way of doing it is good enough. Don't bother me with new ideas." This attitude is particularly prevalent when those who are supposed to implement improvements are overloaded with other responsibilities.
- Impatience with the slow place at which the new approach catches on and shows results.

Any good construction manager could add to this list. Giving it here is intended to make clear that proposing changes and getting them carried out are two quite different matters.

Getting Productivity Improvement Started

As pointed out in Chap. 2, construction, as contrasted to manufacturing, is fast moving; projects and the tasks within them come and go rapidly and people are accustomed to quick action. But it takes time to introduce the changes in thinking discussed above. A program must go through a cycle involving (1) setting up a system, (2) developing positive attitudes and building a team which has a desire both to improve productivity and to work together, and (3) establishing the processes through which productivity improvements are found and put into effect.

Consultants or in-house specialists who have been involved in setting up effective productivity-improvement programs find that to gain acceptance and develop credibility they must demonstrate that they are on site to help job management, not to threaten it or make it look bad. Three among the techniques which they recommend are:

1 Start slowly. Work first on small problems which have a high probability of success. Suggested changes should not embarrass or offend the egos of job personnel.

2 Propose other changes that are highly visible, which can serve as attention-getters. These need not be productivity-related. For example, they might involve improving site appearance, access, parking, and sanitary facilities, or setting up a variety of programs to increase a feeling that management cares (see Chap. 11 for examples).

3 In choosing operations to investigate, select those that seem to be going well, for example, ones that are under budget and ahead of schedule. It is usually possible to work with job personnel on such operations to find and implement further improvement without being seen as a threat.

EXAMPLES OF PRODUCTIVITY-IMPROVEMENT APPROACHES

As indicated, a range of approaches to productivity improvement have been carried out successfully on construction projects. Seldom do they bloom fully overnight; rather they involve detailed study and take time to implement. Almost without exception, they have followed a progression of steps such as those outlined in Fig. 3-1. All start with top management deciding to proceed and providing the necessary resources; then an appropriate group is established, the needed technical support is made available, a problem (or problems) is tackled, recommendations are made to top management, top management decides whether to implement or reject the recommendation, and finally, and very important, top management explains its decision. Differences among the schemes to be discussed below in more detail, are (1) who initiates the effort, (2) study-group membership, and (3) procedures for implementing recommendations.

Oversight Teams (Parallel Organizations)

As indicated earlier, line organizations follow the management and behaviors listed on the left side of Table 3-1. Some may argue that with enlightened management, behaviors of people who operate under this formal structure will not have the undesirable consequences that can flow from the rigidities inherent in the system. However, anyone who has observed how management operates on projects will admit that in the hustle and bustle of getting the job done, fol-

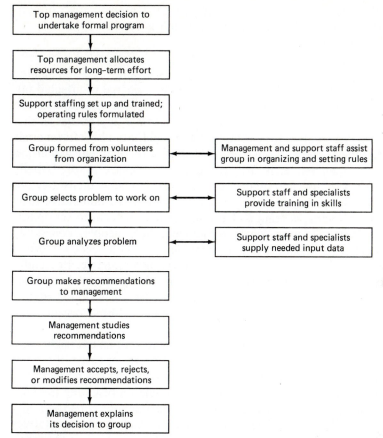

FIGURE 3-1
Flowchart showing procedures for forming and operating formal oversight or
work teams.

lowing these patterns sometimes leads to less-than-optimum job conditions. It
is almost certain that management behaviors on impacted jobs will follow the
characteristics listed on the left in Table 3-1, as individuals scurry around to
protect their firm's and their own personal interests and reputations and their
legal position under the contracts. Actions such as these can make a bad situation even worse.

Oversight teams, organized to have the characteristics shown on the
right side of Table 3-1, have increasingly been employed to overcome these
difficulties. They have been effective in finding solutions to problems in all
areas where group problem solving is helpful. It must be understood, however, that oversight teams are not substitutes for formal organizations.
Rather, they are proposed as an addition to them. Thus they demand time
from already busy personnel, and cost money. Those who use them feel

that the payoff exceeds the cost several times over. Unfortunately, in contrast to the investment in time and money, which is readily apparent, the payoff is hard to measure.

A first reaction when the concept of oversight teams is suggested is to claim that they exist informally on every well-run project. Certainly decision making on many important matters is done only after those in charge of affected operations meet, discuss the situation, and determine a course of action for job management to follow. The difficulty with such informal approaches is that rather than foreseeing trouble, they usually are brought into play after a crisis has developed. Often, preparations for such spur-of-the-moment gatherings may be lacking or incomplete, if such gatherings take place at all. Usually meetings are subject to time and other pressures, and key individuals may not be able to attend. Some may feel threatened and will be uncooperative. The benefit of having an oversight team is that it forces management to set up formal procedures that will bring together knowledgeable people, commonly those most affected by the final decisions, provide them with needed information, and give them time to study troublesome problems and recommend a course of action in time for the findings to be useful. This group neither adopts nor implements its recommendation, although it may be called on for advice. Rather, decision making is left to appropriate people in the formal organization or organizations.

Ideally, the owner of a project should be the party to start an oversight team beginning at the time the designer is first brought on board, or earlier if several parties in the owner's organization are involved in initiating the project. The team's membership will change over time, with people representing the contractor being added as soon as possible (see the discussion of constructibility in Chap. 5). It might be appropriate at a later date to include representatives from the subcontractors and labor unions. As people are added to the team, others might be dropped when the interests they represent are no longer affected by the team's actions.

It is possible, of course, for contractors rather than owners to set up oversight teams involving their own forces, subcontractors, and even representatives of owner, designer, and unions. It should be clear that no single way to implement oversight teams will fit the wide variety of construction projects. For example, on very large projects it might be wise to create a permanent steering committee composed of top managers in the affected organizations. It would set policies and operating rules and possibly select the topics to be studied. Then task forces composed of personnel knowledgeable in a particular area would be assembled to study a problem and develop recommendations and procedures for implementing changes. These would be submitted to the steering committee for review and implementation. On smaller projects, a less elaborate approach is probably sensible, but as already indicated, implementation of its recommendations would fall on line management.

Essential to the success of any oversight team is access to data. Sometimes this can be obtained from job records; in other instances it must be gathered either by team members or by knowledgeable staff made available by the management of the concerned parties. If for some reason any of these individuals withholds information, the effort is bound to fail, since it will create dissension and destroy trust.

A common practice in setting up oversight teams is to provide a facilitator to help the team get started and to coach them on how to communicate effectively, conduct successful meetings, and prepare data and conclusions for submission to line management. All too often, these techniques are unknown to construction people and skilled help can make their task go more easily.

Organized Productivity-Improvement Teams (Quality Circles)

There is nothing new about having people work together in teams on construction projects. By far the larger part of work-face activities involves crews of several persons. Planning and scheduling are or should be group activities at levels ranging from key project managers through superintendents and foremen. A team of specialists or supervisors from several levels is often brought together to tackle special problems involving, for example, materials, equipment, methods, or space. In all these instances, inputs are solicited, with decisions made after discussion of workable alternatives. Instructions may then be prepared so that the line will know how to carry through the recommendations. Increasingly, on well-managed projects, job managers or groups of specialists meet at regularly scheduled times and consider specific agendas. However, none of these activities meet the criteria for work teams or quality circles. Work teams involve deliberate, organized efforts by top company management to engage all levels of the work force in productivity studies to take advantage of group thinking and interchange in solving problems.

Applications of formal work teams (quality circles) are rare in construction today, although some instances can be cited. For example, there is a very large design-construct firm that has some 80 circles within or cutting across the fields of engineering, purchasing, warehousing, field supervision, and data processing. A number of other firms are using them in more modest numbers.

Formal attempts at group problem solving in industrial situations, under the name of quality circles, were first reported in about 1961 in Japan. At that time, Japanese products had the reputation of being shoddy and unreliable. But by combining statistical quality-control methods, first developed in the United States, with company-organized group efforts to improve quality, Japanese firms made great advances and were soon producing goods such as automobiles, cameras, and television and audio devices meeting the highest standards at competitive prices. In the western world, industries began to react to the competition offered by Japan and by the late 1960s had focused attention on what the Japanese were doing. By about 1973 several American companies were experimenting with quality circles. Today many firms have them in successful operation.

Several organizations have activities and/or publications and training programs in the quality-circle field. Among them are the International Association of Quality Circles, The American Society for Quality Circles, The Quality Circle Institute, and the American Management Institute. In addition, consultants who specialize in this area are available on a fee basis. Excellent written material and short courses are available from some of these sources.

Attempts by industries in the United States to install and maintain quality circles have not always been successful. Among the cited reasons are:

1 A failure to recognize that the relationships between management and labor are radically different in Japan and the United States. In Japan, management and labor have always operated as a team, with management offering employment for life and often a dual pay scale, one part of which is a salary and a second, a substantial annual or semiannual bonuses based on company profits. All these provisions have given the employee a vested interest in the company's success. In the United States, in contrast, employment is usually casual, with the employer relatively free to reduce the work force or dismiss individuals unless restrained by union or legislative restrictions or other conditions, and the employee is free to move from company to company. Except for managers, pay is commonly for hours or by the week or month worked at some predetermined rate. Bonuses at the worker level are almost unheard of. If they exist, management sets the conditions or bargains with the union over their terms. Under such conditions, the employees may feel that they have little stake in the company's profits or even its survival. Sometimes an adversarial situation exists, with a complete absence of trust or respect between the parties. It follows that whereas in Japan the work force would accept and cooperate in efforts to implement quality circles, in the United States workers may be indifferent to or resist them as another management ploy to increase profits by speeding up the work pace and ultimately reducing the size of the work force.

2 In many U.S. companies, top managers have missed the advantages offered by quality circles by failing to give them enthusiastic support and continued attention, and by not making the substantial financial investments needed if they are to succeed. Also, managers have failed to recognize that it takes time to overcome the animosities, prejudices, and fears of the workers; and until this happens, a program will not be accepted. In addition, training and implementation by a special staff are required before there are tangible results which management can implement.

3 Quality circles can appear to be a threat to middle managers and foremen, who fear they will no longer be bosses with authority and stature but only team members or on the outside looking in. Experience has shown that unless these fears and prejudices are overcome, attempts to develop quality circles will be difficult.

Figure 3-1, referred to earlier in the discussion of oversight teams, is a flow diagram outlining the steps employed in instituting and carrying through a successful and ongoing work-team (quality-circle) program in a large construction company. Although this scheme appears to be far too elaborate for the usual construction company or for all but the largest and longest individual projects, it can serve as a useful guide for less elaborate undertakings, since each step must be implemented if the effort is to be successful.

Examining Fig. 3-1, one sees the following:

1 Commitment to the plan and the allocation of resources to it for a considerable period of time. Companies, large or small, that do substantial amounts of work in a given area so that management and crews move from project to project fit this prescription, as do single large projects of long duration. But the approach may fit other situations as well; for example where a local contractor employs the same crew over and over again to do the same kinds of tasks. The key to a successful program is continuity at the team level.

2 Management agreement that the teams will operate on company time, usually for an hour or more a week. To do otherwise says loudly and clearly that management is expecting few results from the team's efforts and that the plan is nothing but a poor effort toward better employee relations.

3 Setting up the groups and development of rules for their operation require careful attention. Commonly the group is made up of six to ten volunteers from the work force who have been informed of the activity and its purposes and wish to be involved. No one should be coerced into participating, and it should be made clear that a person may withdraw at any time. Membership probably will be rotated among the interested parties, and groups that have become inactive or quit working may be abolished. Furthermore, it should be made clear that the only reward is in the satisfaction that results from making the work situation better by improving management-worker relationships and by helping the company to be more profitable. How the work team or its members participate in awards from the suggestion system, if one exists (see Chap. 11), should be made clear. However, this should be considered incidental to the main purpose of the group.

4 Staff support and management cooperation are essential. When the teams are formed, they will need help in knowing what is expected of them. When the groups are numerous, as with a well-developed program in a large industrial plant, full-time staff of one or more may be necessary. If less than full-time staff is involved, it must be made clear that this activity has top priority.

In most instances, members of work groups will not know the techniques involved in holding effective meetings, locating and solving problems, or analyzing operations. Consequently, training in these areas must be the first order of business and will provide the agenda for a number of the group's early meetings.

5 Given that the group has settled on a problem, it must make an analysis and arrive at one or more solutions. Sometimes brainstorming is a first step, and in some cases this may bring agreement on a possible solution. In other instances it may be advisable to bring a line manager or specialists from the staff to provide data or expertise or information from company records. Effectiveness and enthusiasm will be destroyed if management does not cooperate fully, even to the extent of disclosing what may be considered confidential information.

6 The rules for group operation must be made clear. Usually the group selects its own chairperson and secretary or recorder. The problem or problems to be considered will be developed by consensus. Sometimes the rules say that management will not have a voice in topic selection, although, in practice, as common interests develop, the group tends to focus on problems that are among management's concerns. It is important that topics be limited to items that affect productivity, quality, safety, or worker-management relations. It must be made absolutely clear that the work group is not intended as a vehicle through which members can air personal or other grievances. This is not its function, and attempts to deal with such matters get in the way of useful activities.

7 It is of utmost importance that the group understand that although its recommendations will be given serious consideration, the decision on implementing them rests with management, since factors other than the proposal standing alone must be considered. But management must realize that it cannot arbitrarily dismiss a proposal and that a reasonable explanation of its actions must be given. To do otherwise will destroy the group's morale and interest and management's credibility.

In summing up this discussion of work teams (quality circles) it should be repeated that they have been valuable to many U.S. industrial concerns and to the few contractors that have tried them. Not only do they produce excellent ideas, but they can be valuable in cutting across the rigid lines of authority existing in many companies, thereby permitting lower-level supervision and workers to make inputs from their skills and concerns to the benefit of company and workers alike.[9]

Organized Crew-Level Productivity-Improvement Efforts

It will be emphasized throughout this book that lower-level supervisors, skilled construction craftsmen, and even people classed as common labor have knowledge or ideas that can make their efforts more productive and safe. The question is how can these talents be developed and used, given the usual hectic rush to get the work done. Every good foreman does this informally as a

---

[9] For supplementary information on quality circles in Japan, a report on the practices of four contractors in the United States who are employing them, and a detailed bibliography, see B. A. Gilly, A. Touran and T. Asai, *Journal of Construction Engineering and Management*, ASCE, vol. 113, no. 3, Sept. 1987, pp. 427–439.

part of working out a detailed work plan for each task, but this is a hit-or-miss thing which can be improved upon.

Today, some constructors have undertaken efforts to tap the worker's reservoir of knowledge and interest by formalizing their approach to it. Companies are already required by OSHA regulations and company practices which preceded or flowed from them to conduct crew (tailgate) safety meetings (see discussion in Chap. 12). The idea is to encompass discussions of productivity and quality into these or, if appropriate, into added formally scheduled meetings. Some of these efforts have been highly successful, as have the earlier efforts on safety; others have been dismal failures. In any event, by setting aside time in the work week for crews to talk about productivity and quality says clearly that management feels it is important and pays off.

If they are to be effective even though relatively informal, these meetings must follow the general pattern outlined in Fig. 3-1 and discussed above under the heading of formal work teams. This means initiating by higher management, developing or providing skills in both analytical procedures and methods to make groups operate effectively, carrying on group discussion, formulating recommendations, and finally ensuring either adoption by management or, if not, an explanation as to why the recommendation was not used. It is essential that a written report on the group's activities, and recommendations and how its recommendations were handled go to higher management and be acknowledged by it. Without this feedback, any early enthusiasm can quickly dissipate.

Crew meetings to discuss productivity and safety can follow a variety of patterns, depending on the approach and the effort and advance training that management chooses to put into them. At one extreme could be informal brainstorming and analysis sessions wherein methods or other topics are discussed and, possibly, a recommendation to implement the suggestions made by the crew to the foreman. At the other would be carefully planned sessions wherein the crew would learn and apply some of the simpler techniques such as process charts or crew balancing. In such instances, the foreman or a crew member might be trained beforehand in the use of these techniques or an outside resource person brought in to assist.

It should be clear from this brief discussion that formal crew-level productivity-improvement activities do not come about easily. They require a conviction on the part of management that they are worth the money and effort, followed up by strong and sustained effort.

Getting and Implementing Ideas Generated by Individuals

Seasoned construction craftsmen not only have the skills needed to carry out their trades, but many of them have a knack for and ideas about doing their work efficiently. Furthermore, they have worked on other projects and have absorbed ideas from them. All this knowledge and talent is available if management creates a climate favorable to using it. In other industries and on some large construction projects, substantial efforts are made to tap this resource with formally organized suggestion systems. These are described in

Chap. 11. But even without providing such organized approaches and the recognition and awards they commonly carry, higher management can tap this resource. To do so it must make clear to supervision and workers alike that ideas about ways of improving productivity are welcome and will be given serious consideration. Instilling this concept is a difficult task which requires learning and honestly exercising skills in dealing with people. Some of the approaches to this problem are discussed in Chap. 10.

The Role of Staff Specialists or Consultants in Productivity-Improvement Efforts

From the discussion just concluded, it should be clear that to develop and operate productivity-improvement programs at any level calls for skills not always available in the contractor's home office or on the job sites. A number of very large contractors employ full-time specialists to plan and implement their productivity-improvement programs. It is very important that these individuals have the full support of management at the highest levels. Where companies have several large projects going, a productivity specialist will be on the staff of the company president or some other high official in the home office. In addition, a specialist will be assigned to each project. Usually this individual reports to the project manager on the job site rather than to the specialist in the home office so as to make clear to managers on the project that the specialist is on the project team and that the cooperation of all is expected.

There are other arrangements for using productivity-improvement specialists that have been very successful. For example, in one firm that does many high-rise buildings, the specialist reports directly to the company president. Usually he spends a month or two on a project while it is in its early stages. There he works with superintendents, foremen, and crews helping to develop management systems, schedules, and methods. After certain important and repetitive operations are begun, he will work with management and crew members to improve them. Also, he is on call at other times. Yet another approach is to use assignments in productivity improvement as a training ground for younger individuals that the company is bringing along. In this role they learn a great deal both about construction itself and the problems of getting along with others.

Engaging consultants skilled both in productivity improvement and facilitating group interaction is another way to get productivity-improvement programs started and running well while at the same time training in-house personnel. Often this is the logical approach, particularly in smaller companies. One of the serious difficulties with this plan is that the program may die once the consultant has moved on.

This brief discussion is intended to demonstrate that there are many approaches at several organizational levels that will lead to successfully developing and maintaining a productivity-improvement program or group of programs. But they do not come about merely by wishing them into place. They require management nurturing and persistence and a considerable amount of money.

OVERCOMING DIFFICULTIES THAT PRODUCTIVITY-IMPROVEMENT PROGRAMS CAN BRING

The concept of a formal productivity-improvement program is unknown to most managers and others involved in the construction process, whether they are employed by owner, designer, or constructor. These individuals are accustomed to operating through line organizations and under the relationships that follow from specific contracts. Thus, to them, establishing separate groups and activities which ignore lines of authority and contractual relationships seems ridiculous. Neither do they feel comfortable with operations where the standing and status of the participants are based on knowledge in a particular area rather than on position in the hierarchy. All in all, such arrangements can be seen as a real threat to an individual's or an organization's authority, position, self-esteem, and financial well-being. Even jobs may seem threatened.

The first step, then, in establishing a productivity-improvement program is to overcome these prejudices and mental sets. Steps that must be taken include the following:

- Carefully supply full information about what the program is, how it will be carried out, and who will administer it.
- Emphasize that the program is not intended to take over the functions of line management. Its assignment is to find and study productivity problems, suggest solutions, and report back to the line. The decisions about adopting and implementing them, possibly in modified form, rests with line management.
- Make clear that top management instituted the program and will provide the financial and administrative support, follow-up, and, if necessary, discipline, to make it succeed.
- Have a clearly defined and spelled-out reward system which recognizes the contributions of those in the program and the line's success in implementing it.
- Make haste slowly. The perceived threats from such a program do not go away overnight.
- Straighten up problems and grievances not necessarily related to productivity which busy management has neglected as trivial, but are not to those involved.
- Use care in selecting the first problems to tackle. It is better to choose operations that will be repeated and seem to be going well than those in trouble and about which those involved are sensitive. The former are nonthreatening; the latter are threatening.

It is essential in instituting a program to understand which parties in the line organizations are threatened by the productivity-improvement plan and the items selected for study and analysis. There is no fixed rule on this. However, it might be generalized that on the one hand, top management, which institutes the scheme, and the labor force, which is accustomed to accepting direction, are the least threatened. On the other hand, middle management, such as superintendents and foremen, may feel most threatened since, as they see it, their authority, status, expertise, and credibility may be challenged. Sometimes their jobs may seem to be on the line. And, to be honest, this may sometimes be the case. On well-run, smoothly flowing jobs there may be less need for so many levels of supervision or so many in each level.

In sum, productivity-improvement programs do not come into being nor can they run and be productive without planning, continual effort, and thinking about the plan and the affected people and their reactions. But it has been demonstrated that they can pay off.

SUMMARY

This chapter has examined the concept of productivity and how it might be measured, what has been done to date to improve it in the construction industry, and the ways of thinking that must be followed if productivity improvements are to be carried forward successfully. Finally, it describes approaches that can be employed in tackling productivity improvement at different stages and levels in the construction process.

The authors of this book maintain that without first establishing mechanisms such as those described here, efforts to develop productivity improvement programs that will pay off and continue will fail, or at least be fraught with difficulties.

CHAPTER 4

PRODUCTIVITY IMPROVEMENT IN ON-SITE CONSTRUCTION VIEWED AS A SYSTEM

INTRODUCTION

Chapters 1, 2, and 3 described the uniqueness of the construction industry and suggested ways of organizing effective productivity improvement programs. This chapter offers an overview of the system or group of interdependent processes and activities under which such programs must function. It summarizes the succeeding chapters, which are subject-specific. These are Chaps. 6 through 8 and 12 and 13, which treat the techniques of on-site productivity improvement and safety, and Chaps. 9 through 11, which consider the very important but less tangible human factors which must be considered in any successful effort to improve productivity.

With respect to productivity improvement, it is important to recognize that the ultimate purpose of construction is to build, and the building is done by craftsmen at work faces. In many respects, construction operates like a professional sports organization. With it, much goes on to support the players' efforts; and while the players cannot perform effectively without the support of coaches, trainers, playbooks, and administrative staff, only the players on the field actually win or lose. So it is with construction organizations: the work is done at the work face by individuals or foremen and crews, and the purpose of the rest of the organization, both off site and on site, is to support work-face activities. On-site construction, then, can be pictured as a system developed to build something. For it to operate efficiently requires that attention be given to all the elements, including those that occur off site, and on site but away from the work face as well as at the work face.

Figure 4-1 is a simple flow diagram showing the parties and the other elements which make inputs to the construction system for a simple straightforward project. This diagram clearly indicates the flow of activities which begins with an owner and finally leads to a foreman and crew at the work face delivering a product. In construction, this product is a completed task which is, in turn, one element of a completed project.

The topics summarized in this chapter and treated in more detail in later chapters are:

- Off-site or away-from-the-work-face support for work-face activities
- At-the-work-face approaches to productivity
- Humans as workers in the construction environment
- Human behavior as an element in generating and implementing commitment to productivity

FIGURE 4-1
Diagram showing the parties and their inputs to the construction process.

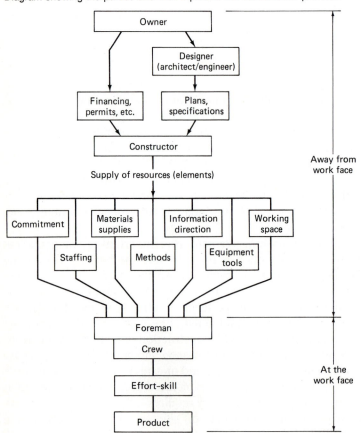

- Recording and disseminating knowledge about productivity improvement
- Steps in and possible approaches to productivity improvement

OFF-SITE AND AWAY-FROM-THE-WORK-FACE APPROACHES
TO PRODUCTIVITY IMPROVEMENT

To the uninitiated, and even to many experienced construction hands, the construction process seems simple. It combines operations for assembling resources such as those shown in the middle portion of Fig. 4-1 and for letting crews or individual craftsmen at the work face get on with the job. But an analysis of what is required to provide this support off site and on site but away from the work face demonstrates that the operation is complex indeed. And this support is essential regardless of the size or complexity of the operation or project.

On small projects that involve only two layers of management, superintendents and foremen have developed informal procedures to cover breakdowns in the support system. On such jobs, when difficulties arise, someone bird-dogs the problem and gets it solved. But on large projects with complicated designs, several levels of management, some of which are off site, coupled with specified materials or other items having long lead times, and special equipment and tools, these informal methods cannot correct system errors. Nor is improvising possible. Errors, omissions, or other failures away from the work face not only throw the entire operation out of kilter, but also bring obvious and highly visible delays, cost overruns, and possibly safety and quality problems. On smaller jobs, these foul-ups bring the same sorts of delays and costs, but the resulting crises are resolved on site and generally go undetected because of the insensitivity of the cost-reporting or other control systems.

By looking at the away-from-the-work-face activities that impinge on productivity at the work face, one can get what might be called a global perspective of the construction process. Figure 4-2 illustrates this global situation by cataloging many of these away-from-the-work-face factors. In this figure, inputs are divided into those involving planning on the one hand and resource supply on the other. Many of these impacts are generated by external forces far beyond the reach of those responsible for work-face activities. Among these may be off-site actions of governmental agencies, the owner, the designer, the constructor's home office, along with the on-site actions of those who supply resources and do planning. Figure 4-2 also shows, on the right side, that even after the task is accomplished at the work face, there will be resulting impacts. This diagram, which has work-face activities at its heart, clearly supports the argument offered earlier that activities at the work face are central to the construction process.

Another concept that can be deduced from Fig. 4-2 is that adverse consequences result from failures to perform work-face activities on schedule, effectively, properly, and safely. Negative effects will flow backward along the supply and planning chains as well as in the output direction. An interesting

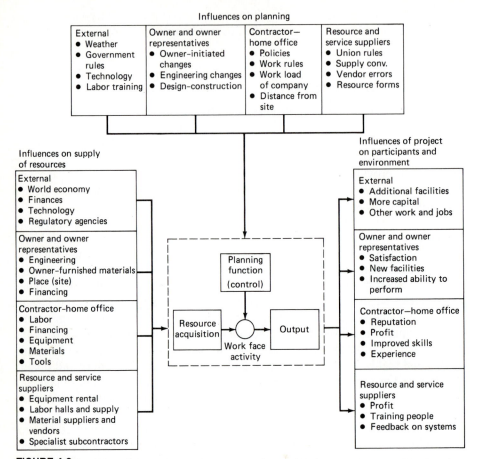

FIGURE 4-2
Away-from-the-work-face influences on the construction process. (*Adapted from V. E. Sanvido, Technical Report 282. Department of Civil Engineering, Stanford University, Stanford, Calif., June, 1984.*)

and enlightening exercise is to assume a failure to accomplish a given work-face task and, using Fig. 4-2, trace the repercussions.

Figure 4-3 shows that the construction process involves two kinds of management inputs: resource controls, shown on the left; and planning, shown on the right, which involves a work plan or process by which physical inputs will be changed to outputs. Both are essential. Also, Fig. 4-3 shows on the left that under some circumstances, on-site efforts may be necessary to bring off-site resources onto the site. Furthermore, as shown at the bottom of the figure, there is a feedback loop involving repair and maintenance of partially used resources which must be handled to avoid future difficulties.

Figure 4-4 is an expansion of Fig. 4-3 to show the situation for a large complex construction project. It depicts the many interrelationships involving

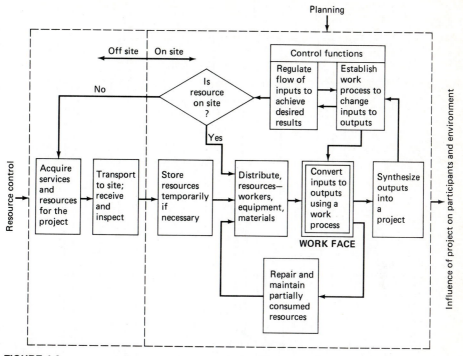

FIGURE 4-3
Diagram showing management activities that support work-face procedures. (*Adapted from V. E. Sanvido, Technical Report 282. Department of Civil Engineering, Stanford University, Stanford, Calif., June 1984.*)

management's resource control (staff functions) on the left of the diagram and operational control (line functions), down the center. Both these functions become more specific and detailed as they move from project to area to discipline, and, finally, on the operational side, to foreman and craftsmen. On the right side, Fig. 4-4 shows feedback on output flowing back to the various operational levels. The diagram also shows both inputs and feedback between engineering (staff) and the line down to the discipline superintendent level.

The activities diagrammed in Fig. 4-4, which seem very complex, focus on only a single activity of one crew. On a very large project, there could be hundreds of activities going on simultaneously, all of which require this same level of attention.

Table 4-1 has been developed in an attempt to depict the complexities of away-from-the-work-face decisions that must be made for a single work-face task. The table also can function as a checklist to see that all the bases are covered in a given situation. Along the top of the table are listed the inputs required at the work face, as shown earlier in Fig. 4-1. Listed vertically along the left side of the table are some of the important factors affecting some, and

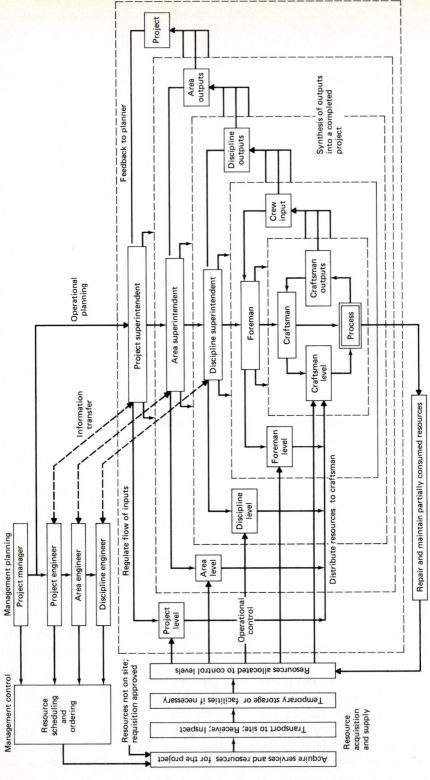

FIGURE 4-4
Interplay among authority, responsibility, and communications on every work-face task on a large construction project. (*Adapted from V. E. Sanvido, Technical Report 282, Department of Civil Engineering, Stanford University, Stanford, Calif., June 1984.*)

TABLE 4-1

AWAY-FROM-THE-WORK-FACE INPUTS AND THEIR TRAITS THAT AFFECT
PRODUCITIVITY AT THE WORK FACE

Factors that may affect each input	Clear of external constraints	Staffing	Instructions	Methods	Materials	Equipment	Tools	Suitable work space
Organizational planning and communication								
Source								
Special characteristics								
Control location								
Coordination requirements								
How moved or transported								
Indentification procedure								
Timing and its limitations								
Information-flow procedure								
How tied to work-face task								
People								
Responsible parties								
Experience and training required								
How tied to commitment								
How tied to accomplish-ment								
How tied to individual evaluation								

Required inputs

possibly all, of these inputs. To analyze a work-face activity completely, it is necessary to answer the questions raised by each input-factor combination, as defined by their intersections in the table. There may be no input-factor relationship for some specific work face operations. Also, the list of factors might change with different project or task situations. However, regardless of the differences in complexities among projects and their tasks, most of the activities called for at a given input-factor intersection will require consideration, to be followed in most cases by a decision and some action by some individual or group. Otherwise, adverse consequences can result, whether the operation is as simple as installing an electrical outlet or as complex as setting up and connecting a key vessel in a large industrial plant. The difference will be in the number of parties involved and the level of complexity of the individual inputs.

No attempt has been made here to fill in the blanks in Table 4-1 for individual input-factor relationships, because they will differ widely among tasks. Discussions later in this book deal with specifics about some of them. The reasons for introducing the topic now is to demonstrate the complexity of the away-from-the-work-face element and to reinforce the point made in Chap. 1 and treated in more detail later in this book; organization, planning, communication, and people problems all must be considered.

The concept and its implications illustrated by Table 4-1 cannot be overemphasized. Each of the relevant interfaces must be dealt with in some manner, either before or as the task is undertaken. If this is not done, then it becomes a choke point and the foreman or crew must search out the missing elements or call off the operation. As the entries made in each space in the table will show, individual elements may be controlled or supplied by different managers in various ways and may be obtained, stored, distributed, and accounted for by a variety of procedures. Furthermore, as the table indicates, most of the inputs can have important consequences for the people involved.

There is a specific reason for introducing the concepts underlying Table 4-1 so early in this book. If studied carefully, it will bring into sharp focus the functions conducted off site and away from the work face that impact productivity at the work face. It also demonstrates how much cooperation, coordination, skill, and knowledge are needed to perform them. Table 4-1 adapted to a project or a single activity can thus serve as a checklist for advance planning and scheduling, since it can direct attention to the individual inputs that should be covered in the work plan. For troubleshooting, it offers a means for zeroing in on how, when, and where things have gone awry. For this book, it establishes a format for approaching the several away-from-the-work-face problems to be discussed in detail in Chaps. 5 through 8. In them, each topic is addressed first by looking at how it can be handled well, and, second, by examining what can be done when problems arise.

AT-THE-WORK-FACE APPROACHES TO PRODUCTIVITY IMPROVEMENT

Enhancing productivity at the work face involves (1) getting people at all levels to learn and employ imaginative thinking patterns and techniques, and (2) creating a job climate in which concepts and ideas that are developed can be carried through. There is no right way for accomplishing these goals. At times, simply asking the six basic questions of what, why, when, where, how, and who at the right time and place is sufficient. Other situations involve questionnaires or interviews or the traditional observational, work-sampling, and methods-and-time-study techniques augmented with recording and analytical tools such as time-lapse photography, videorecorders, computers, or simulation. All levels of job personnel from craftsman to top management can be involved in some of these.

To a large degree, the lack of attention to possible productivity improvements at the work-face results because the planning and control of decisions about work-face activities are left to craft superintendents or foremen, or to the craftsmen themselves. Seldom do they, or higher management for that matter, know or think of using the techniques for productivity improvement. Field personnel usually are extremely busy, are confronted by many problems, and have neither the time nor the inclination to look back at tasks already accomplished. Numerous studies have proved conclusively that under such pressures, people seldom are receptive to new ideas. Furthermore, able as these people are in getting the work done, their tendency is to follow familiar and proven approaches and not to look for new ones. Other industries have learned that to improve productivity, management intervention is necessary; but with only a few exceptions, construction management generally does not follow this practice.

Thinking and acting to improve work-face productivity in construction-related activities began over 80 years ago with Frederick W. Taylor, the father of scientific management. He pointed out the fallacy of expecting improvements in work methods if management failed to address them or left them to even the best craftsmen or craft supervisors. Yet today, few construction managers or craftsmen have heard of Taylor or of Gilbreth's later studies of bricklaying and other construction tasks.

This book, in Chaps. 5 through 8 and in App. C, describes and gives examples of a variety of techniques for observing, discussing, recording, and analyzing construction operations to improve productivity at the work face. These chapters will show how a variety of techniques have led to substantial improvements in methods and significant cost reductions. Even where the results have been less dramatic than those shown here, it has been demonstrated that such careful analysis can further improve already efficient procedures.

HUMANS AS WORKERS IN THE CONSTRUCTION ENVIRONMENT

Construction often demands that workers do hard physical tasks outdoors in all kinds of weather. To maximize their productivity and minimize dangers to their safety and health, managers must understand workers' abilities and limitations under the conditions imposed by the site as well as the materials, supplies, and equipment employed. Among the factors that must be taken into account are the following:

- Short-term physical fatigue, as affected by the abilities of humans as machines, the energy demands of certain tasks, and techniques, including tool selection, to reduce effort or strain
- Daily and weekly fatigue
- Twenty-four-hour fatigue
- Long-term fatigue from occasional or scheduled overtime
- Mental and stress fatigue, and boredom
- The effects of heat, relative humidity, and cold on productivity

In planning for productive and safe work, factors such as these must be considered. They are merely listed here to make the reader aware that they will have important consequences. They are discussed in more detail in Chap. 9.

HUMAN RESOURCES AS AN ELEMENT IN GENERATING AND IMPLEMENTING COMMITMENTS TO PRODUCTIVITY IMPROVEMENT

All the elements required to implement productivity at the work face are carried out by or through people. Only by informing all affected individuals and getting them committed to carrying through in appropriate ways will the best-intended approaches to productivity improvement be accomplished. In essence, this means retaining people's already established positive attitudes and actions or changing negative ways of thinking and acting.

Behavioral scientists have been observing and analyzing management and worker practices and behaviors under various conditions for more than half a century and have developed several helpful ways of thinking about them. From their findings, which are discussed in more detail in Chap. 10, the evidence seems conclusive that a key ingredient is how managers assess and relate to, or are perceived as assessing and relating to, the people they direct. Either their actual or perceived gut feeling is that people are basically lazy and must be driven (McGregor's theory X) or that people are basically sincere, honest, and anxious to work (McGregor's theory Y).

Although they might operate differently, both theory X or theory Y managers can get people to work if those they direct need employment and can be coerced because their basic physiological or safety needs are threatened. Seldom in construction in the United States today, however, are construction managers or craftsmen in such dire straits. Consequently, managers with an actual or perceived theory X mental set are deprived of their most important strategy for getting people to work and therefore can seldom be effective with those whose cooperation is essential if a high level of productivity is to be reached and maintained. Theory Y managers do not face this handicap.

Some construction executives and managers seem to exhibit both theory Y and theory X mental sets, depending on the groups they are dealing with. On the one hand they hold to theory Y in dealing with their project management and support personnel such as engineers. On the other, they follow theory X behaviors in their approach to craftsmen and even foremen, particularly those from the unions. This behavior results from real or imagined situations on earlier projects, so that they assume that all workers are, by choice, lazy and unwilling to work productively. Whether correct or not, this thinking is self-defeating. With such attitudes, theory X managers forego any chance of utilizing the talents of those craftsmen who want to be cooperative and productive.

Table 4-2 attempts, on a single page, to summarize and contrast the effects of actual or perceived theory X and theory Y mental sets on the behavior to be expected from subordinates and on their willingness to make the commitments necessary for high productivity. As with most concepts employed to analyze human behavior, these groupings cannot be considered as applicable in all situations. Even so, to ignore them is dangerous.

Chapters 1 to 3 have emphasized the importance of overall job climate and the support of higher management in productivity improvement. A listing of some of the important project-level ideas, as summarized in Table 4-2 and enlarged upon in Chap. 10, are the following:

• To be effective, managers must be sensitive to people, believe in them, and provide an opportunity for and encourage them to be part of activities that can enhance productivity
• Management behavior can be either autocratic or participative, depending on the manager's style, the people being directed, and the situation
• Management must create and maintain conditions under which all those involved are willing to make commitments and be excited about being an important part of a winning team
• Neither managers nor fellow workers should be soft on or accepting of repeated failures to perform
• Excellent managers attract and keep excellent subordinates and craftsmen. Poor managers drive them away

Unquestionably, the attitudes and behaviors of top-level company and project managers are the most important of all in developing positive attitudes toward productivity. They set the tone that makes implementing productivity concepts and actions such as those described in this book effective. Smart subordinates quickly pick up cues about what will please the boss. For example, if carrying bad news means trouble for the carrier, the boss will never hear it. Again, if the top boss's attitude is that staff or craftsmen must be driven or treated as children, it will become the prevailing management style from top to bottom of the company or project. Observant people read these messages clearly and conform unless they quit or are fired first.

Productivity, in the sense used here, does not mean ill-conceived attempts to urge or coerce workers to physically overdo or accelerate the work pace beyond what is reasonable and safe in a given situation. Again, Taylor was

TABLE 4-2

VARIOUS ACTUAL OR PERCEIVED MANAGEMENT MENTAL SETS ABOUT AND ACTIONS TOWARD PEOPLE AND THEIR EFFECTS ON SUBORDINATES' CHOICE OF WORKPLACE, BEHAVIOR, AND WILLINGNESS TO MAKE COMMITMENTS*

Manager's and lower-level management's mental sets			
Despotic or insensitive		**Creative-sensitive**	
People are lazy and must be driven (theory X)†	Subordinates should be treated like children (theory X or Y)†	People are mature, basically good and honest, and anxious to work productively, be helpful, and cooperate with management (theory Y)†	
Management's behavior pattern			
Autocratic or feigned participative but actual decisions made without participation	Truly autocratic decisions made without participation	Autocratic but makes decisions after informal consultation when it is appropriate	Participative— discusses situations when appropriate and often accepts consensus reached by subordinates; assumes responsibility for decision, however made
Climate in which job personnel operates			
1. Unacceptable to people with drive and initiative; suitable for those with a to-hell-with-it attitude 2. Destroys initiative 3. Provides no upward communication, feedback or new ideas.	1. Comfortable for those wishing a passive boss-knows-best situation 2. Discourages initiative 3. Little upward communication or feedback or generation of new ideas	1. Management and workers are willing to make commitments; have a win-win attitude; find personal satisfactions in and feel a part of a winning team; will welcome challenge when expectations of them are realistic 2. Managers and work force expect high performance of others; failure after a fair trial is not acceptable 3. Managers at all levels will: (a) Encourage people to take initiative (b) Create channels for good communication and feedback (c) Enlist all in efforts to generate new approaches to work planning, scheduling, and methods, and to dealing with people	

*Underlying this tabulation is the notion that when lower-level management and craftsmen have relative freedom to move from job to job, they will shift to the ones where they are comfortable with the management style.
†See Chap. 10 for added discussion of theory X and theory Y leadership styles.

among the first to prove that overexertion and speedups are counterproductive. And more recent research has demonstrated conclusively that such practices greatly increase the potential for accidents. Yet the belief is all too prevalent today among construction managers that gains in productivity come mainly by demanding and getting a higher level of physical exertion from workers. The very term ''pusher'' used so commonly to define the foreman's role casts him and his supervisors as drivers of others. As will be demonstrated in this book, methods that improve productivity almost always reduce the physical demands on workers and, moreover, provide them and management satisfaction from a job well done. The slogan Work Smarter Not Harder expresses this idea well.

Two other concepts not included in Table 4-2 that are discussed in Chap. 10 are essential to productivity and deserve special mention. Unless management recognizes them, their efforts are almost certain to fail. These are:

- Commitment is required: motivation is not enough
- Accountability must flow upward as well as downward

The word "commitment" rather than "motivation" was employed deliberately in the preceding paragraphs to make an important distinction. Motivation means a desire to do well; commitment means a willingness to carry that desire into action. With respect to productivity, this difference is crucial. As pointed out in Table 4-2 and discussed in more detail in Chap. 10 there is strong evidence that construction people, workers and managers alike, prefer to be productive and get satisfaction from doing so. In most instances, then, motivation is not the problem; the difficulty is to turn motivation into commitment. Before this transition can occur, even the most highly motivated person first sizes up the situation and asks, "Can I carry through on what is proposed?" and "Will it be to my advantage to do so?" If the answer to both questions is yes, then a commitment is made; if it is no, the forthright person will refuse to make a commitment. Others may say that they will perform, knowing in advance that they will not, but for a variety of reasons choose not to say so. This difference between motivation and commitment applies to everyone on the job, regardless of position.

Ways of getting commitment to a project and its goals have been discussed in Chap. 1, so it is suggested that the reader review that chapter. But getting commitment to carry out specific operations or tasks is of equal importance and is discussed briefly here and in more detail in Chap. 10.

To understand the difference between motivation and commitment, it is helpful to look at a real-life construction situation. Consider, as an example, the willingness of foremen and crews doing piping to agree to complete tasks by a certain time. Before they agree, they will expect answers to questions about things away from the work face that usually are beyond their control, such as:

1 Are the drawings and other information ready, complete, and correct?

2 Are the vessels, valves, fittings, and other materials on site and accessible?

3 Have work methods been preplanned, or will they be left to foremen and crews? Are they safe?

4 Is the crane scheduled for this task and in good working order?

5 If needed, has proper and safe scaffolding been installed?

6 Are suitable tools and power to operate them available and in working order?

7 Is the workplace clear and accessible?

If all these answers are positive or if there is trust based on past experience that promises will be kept and that the time allocated for the tasks is realistic, then and only then will the foremen commit themselves and their crews to perform. To repeat: the point of all this is that motivation of foremen and their crews is not enough. Commitments will only be made after it is clear that the impediments to performance that are beyond the foreman's control have been cleared away and that expectations for output are attainable.

Many field construction managers and superintendents will challenge the proposition that all these preliminaries are required to get commitment from foremen and their crews. They have seen from personal experience that good construction people lacking one or more of these requirements will scurry around and where possible, make do and finish the task.[1] If not, they will improvise some other task so that their craftsmen are busy or at least look busy. However, these same supervisors know that before long the able foremen will refuse to commit themselves to meet estimated times or costs—if, of course, they haven't quit first.

The example given here is for foremen and crews, since they are at the end of the performance chain. But the principle applies in getting commitment at any level. Commitment will come only after the affected person or group knows that possible impediments over which they have no control will be eliminated. Furthermore, repeated failure to perform at any level will make it impossible to get commitment and performance by those at all lower levels.

An ingredient that must accompany commitment is upward accountability. Traditional management systems define accountability in terms of subordinates being responsible to their superiors, with managers at each level controlling, evaluating, rewarding, and, if necessary, disciplining those below them. But where productivity is concerned, accountability has to be reversed and go up rather than down the management chain. The primary need is for a mechanism by which those responsible for bringing the essential elements to the work face are given credit for their successes and held accountable for their failures. It is obvious that neither the usual management structure nor cost-reporting systems offers suitable measures for judging the performance of higher-level managers. Furthermore, the notion that project managers or engineers, superintendents, or general foremen relinquish authority and disciplinary powers to foremen seems bizarre indeed. But if one accepts (1) that activities at the work face control and all others serve them and (2) that commitment of those involved is required, then this upward accountability is essential. To illustrate from the example cited above, the engineer whose drawings are faulty, the superintendent who fails to provide working space or a crane, the general foreman who forgets the scaffolding, or the toolroom manager whose stock is deficient, may be responsible for poor productivity. Should not foremen be in a position to hold those above them accountable, rather than taking the blame when their unit costs are out of line?

---

[1] Field observations show that these procedures are common. On some projects, foremen are away from their crews as much as 60 percent of the time doing these catch up jobs.

In a small way, measures to establish this upward accountability are being developed. Among the procedures discussed later in this book are craftsmen and foremen surveys and questionnaires, as well as various observational, analytical, and work improvement techniques. But only a beginning has been made toward establishing accountability at levels higher than the foreman who is burdened today with recriminations for actions, inactions, or other failures beyond his control.

The preceding discussion of management mental sets, commitments, and accountability presumes that those affected clearly understand management's intentions and actions. All too often, this is not the case because these desires have not been communicated effectively and completely. However, without this clear communication, the best-planned schemes and commitment at the top will be ineffective.

Communication processes are complex. Their forms and how to make them effective are discussed in detail in Chap. 10. However, presenting a short list covering the topics, interests, and concerns that must be communicated in an appropriate way, if schemes to enhance productivity are to be carried through to completion seems in order. Incidentally, this list can be very useful as a check to be applied before starting any operation. The five items are:

1 Are the programs, approaches, and techniques to be employed clearly understood by all those involved?

2 Are those carrying out every element of the program physically able and knowledgeable enough to carry out the program?

3 Do all participants mentally accept the proposed scheme as one they can carry out?

4 Do those involved feel that carrying through the planned scheme will be consistent with the interests of the person, organization, or group to which they owe their loyalty?

5 Do those involved feel individually that carrying through the planned scheme will be compatible with their personal interests and well-being?

Some construction people will feel that the emphasis given here to generating and implementing commitment, upward accountability, and communication is unnecessary if not heretical. One manager put it this way, "All this is a lot of bunk—if you want something done, tell them to do it and take no back talk!" But the fact is that the attitudes and behaviors of the most effective construction managers at all levels, if examined carefully, will follow these concepts consciously or unconsciously.

STEPS IN AND APPROACHES TO PRODUCTIVITY IMPROVEMENT

Given that productivity enhancement is a general objective for a firm, project, or manager, a specific plan and progression for it must be developed. Only for a large project, and in a company with past experience and success, would it be possible to charge ahead with all or even most of the approaches and techniques that have been effective elsewhere. In most instances, a modest beginning is better, with at most a few approaches to demonstrate management's strong interest and support. These should be selected to show quick results. In the beginning, the undertaking must be explained fully to those affected; otherwise it will be interpreted as a threat and met with strong resistance.

In many instances, successful starts toward productivity improvement have resulted from a low-key approach. Since positive job climate and demonstrated management concern about people are necessary steps to obtaining commitment, these often should be given attention first. For example, talking to craftsmen or circulating questionnaires among them might disclose dissatisfactions such as too few, dirty, or poorly located toilets, difficulties or discrimination in parking arrangements, or long delays in entering and leaving the work site. At the same time, staff conversations or questionnaires may disclose problems in the management structure or conflicts between line and staff. Acknowledging and quickly correcting such obvious deficiencies will clearly demonstrate management's concern and willingness to act. Publicizing a company's concern about craftsmen's safety as a part of the preemployment routine and while on the job, a valuable effort in its own right (see Chap. 12), is another technique. The first reaction of some managers to actions such as these may be hostile, since they seem to be personal indictments. But on second thought, most managers usually recognize that these apparently peripheral things that they have been too busy to see are important. Even with such objections, these approaches have worked well as starters.

A variety of other ways of finding likely areas for productivity improvement, pinpointing specific problems, finding solutions and implementing the results are in use. Some of these shed light on problems away from the work face. Others have work face applications. Many, if not most of them, give insight into shortcomings in both areas. A listing of these approaches, with details given in later chapters, includes the following:

1 Data gathering
- Informal: looking and talking it over
- Using questionnaires and interviews
- Formal, short-time activity sampling
- Detailed project-wide work-sampling
- Mapping and recording present layouts and operating procedures
- Drawing site or work-station layouts
- Producing still photography
- Time-lapse filming
- Videotaping

2 Formal analysis tools
- Producing process flowcharts
- Using crew balance charts

3 Problem solution techniques
- Training in problem solution
- Individually examining data
- Brainstorming
- Creating crew or management analysis activities
- Maintaining quality circles

4 Implementation
- Creating suitable project climate through management efforts, management activities, project newsletters, awards, prizes, questionnaires
- Communicating effectively
- Removing obstacles to obtaining commitment
- Making needed change in procedures, techniques, materials, tools, and work situation quickly and decisively

Figure 4-5, which deserves careful study, asks questions and suggests some simple procedures for implementing each of these steps.

FIGURE 4-5
Suggested steps for achieving productivity improvement.

I. Record the job (list all details of the job as it is currently performed).

II. Analyze every detail
  A. Ask the six basic questions of each detail
    1. What is its purpose?
    2. Why do it this way?
    3. When is the best time to do it?
    4. Where is the best place to do it?
    5. How is the best way to do it?
    6. Who is the best qualified to do it?
  B. Evaluate the job layout; the tools, equipment, materials used; material flow; and safety

III. Devise a better method
  A. With an understanding of the desired objective, develop a better method, by answering the six basic questions:
    What ⎫   ⎧ Eliminate unnecessary detail
    Why ⎪   ⎪ Rearrange for better sequence
    When ⎬ → ⎨ Provide better tools, devices, jigs, materials
    Where⎪   ⎪ Simplify to make it easier, faster
    How ⎪   ⎪ Plan for safety
    Who ⎭   ⎩ Consult with others who have an inherent interest
  B. Write up a detailed version of the better method
  C. Generate a substitute solution

IV. Implement the method
  A. Sell the method
    1. To the boss
    2. To the foreman
    3. To the workmen
  B. Once approval is received, put the better method to work immediately
  C. Continue to use the new method—it may take a little time to learn all the details
  D. Give credit and praise where they are due

In assessing the system aspects of possible productivity improvements in a specific situation, a useful approach often is to determine (1) whether its roots are away from or at the work face and, (2) whether it requires changes to the system or is an isolated spot problem where for some reason a workable system has broken down or is performing poorly. Two simple examples of these distinctions are illustrated in Table 4-3.

With problems such as those illustrated in Table 4-3 identified and classified, one or more of several tools for analyzing and reporting that are listed above can be used to isolate the difficulties. However, as indicated, solving and implementing can involve quite different corrective measures and personnel. For example, the specific spot problems described in Table 4-3 might be corrected almost at once by suggestions or directions. On the other hand, the system problems may reach deep into the project organization and reflect how responsibility has been assigned and even affect status and rewards of various people. In either case, however, evidence that isolates either spot or system problems can have far deeper significance. It often may be a surface reflection of serious management or other deficiencies or conflicts among individuals within the organization.

Productivity improvements isolated by data-gathering will often seem to have simple, straightforward solutions. With solid management cooperation and insistence, these can be easily implemented. Even when such is the case, however, first solutions should be analyzed to see if further refinements are possible. A common recommendation by specialists in productivity improvement is always to develop at least a second solution after a first one has been tentatively identified.

Given the large number of approaches to productivity enhancement, choices about which approaches or techniques to use must be made. To carry out this decision-making process is not an easy, straightforward task on construction projects. Often it is done under mental overloads introduced by re-

TABLE 4-3
EXAMPLES ILLUSTRATING DIFFERENCES BETWEEN SYSTEM AND SPOT PROBLEMS

	Classification of problem	
Location of the problem	**System problem**	**Spot problem**
Away from the work face	Tool procedures create severe shortage of tools for all crafts	Simple special tool missing from toolroom or not working
At the work face	Person in charge of cranes used for concrete placement fails to assign one to task	Concrete placement slow because crane has been positioned so that truck access is difficult; costs are high because placement crew is too large

acting to crises or time constraints which make creative thinking almost impossible. In every situation, it is helpful to adopt a step-by-step approach which involves (1) assembling data, (2) finding problems, (3) developing choices, and (4) identifying the most suitable or acceptable solution. Furthermore, within this orderly decision making process there is a need to utilize recent knowledge from the behavioral sciences about (1) barriers to creative and critical thinking such as job pressures, distractions, and interruptions; (2) the limitations of the human mind as it organizes and processes information; and (3) linear or left-side vs. big-picture or right-side thinking. (See discussion in Chap. 10.)

Putting together these factors with a lack of time, force of habit, and a very natural resistance to change means that many decisions about productivity are less than the best ones—a very common difficulty that has been labeled with the coined word "satisficing," which means achieving the target in a way that is good enough but not the best.

Given that near-optimal decisions have been made on what and how to improve productivity in a given situation, the final step is to actually get it done. As emphasized earlier, production is accomplished by craftsmen or crews at the work face, which means that the plan must be communicated clearly to those who are to carry it out, accompanied by strong support by higher-level management. There are many ways to accomplish these aims, some of which are discussed in greater detail in the chapters which follow.

RECORDING AND DISSEMINATING KNOWLEDGE ABOUT PRODUCTIVITY-IMPROVEMENT TECHNIQUES

To enhance productivity at the work face, attention must be paid there as well as to the away-from-the-work-face activities on which work-face operations depend, and to the people involved, both managers and workers. If solutions are implemented, another concern should be to record and pass along the knowledge gained so that it is not lost and can be applied elsewhere on this or subsequent projects. In other words, why reinvent the wheel for each operation on every project or for each manager or crew. Unfortunately, the answer seems to be that with a few notable exceptions, such as those given in App. B, reinvention is a common practice in construction. The figures and tables given earlier in this chapter, which indicate the complexity of the construction process, serve to explain this difficulty. The first question to be asked is, "Which among the many advances on which operations and from which projects should be recorded and passed along in a formal process; and which should go unrecorded and their use left to a chance passing along by word of mouth as individuals recall them?" The second question, equally if not more important is, "If the information were made available, would field managers use it?"

In explaining why their companies do not have formal methods for recording and passing on effective techniques or ways of dealing with people, executives often claim that this communication happens informally. For example, it is stated that superintendents and foremen remember good practices and procedures and employ them on their next jobs. Again, project managers or superintendents may structure the organization and arrange materials supply and delivery, equipment or tool supply, or access to the work site using techniques that have worked well earlier. These practices may be effective on small jobs or in small companies where managers are long-time employees and know each other so that word of mouth or other informal ways of communicating will work. But observations on projects or in companies too large for these simple knowledge-transfer methods to apply indicate that detailed knowledge of better practices developed by one crew on a site may not reach or be applied by other crews on that site, much less on other sites. Neither are better management approaches or techniques for serving away-from-the-work-face activities practiced on other projects. The evidence seems clear that the construction industry in general and the individual companies within it have not seen the need to formally record and disseminate information on productive practices. Being pressed for time and being action-oriented, they have rather assumed that informal approaches to information transfer are sufficient. And, as mentioned earlier, the cost of setting up a formal system for reporting and disseminating information in a company is substantial and is obvious and measurable, while no hard measures exist for appraising the benefits. In an industry which operates with a short-range viewpoint so that dollars made on each project are used to measure success, little thought is given to retaining and reusing knowledge that will pay off in the long run.

Recording and disseminating information on the many techniques and procedures through which productivity has been improved on an industry-wide basis is an even more difficult problem. As was discussed in Chap. 2, construction is a diverse and fragmented industry. It involves many owners, designers, constructors, and craftsmen. Those in each of these groups have their own motivators, practices, talents, and knowledge. To date, with a few exceptions, there is no organized way for collecting and transferring detailed knowledge of better practices and procedures through the industry, particularly in the area of productivity. In fact, individual constructors may feel it is to their advantage to withhold knowledge that they have developed.

Chapter 5 of this book demonstrates techniques by which productivity problems can be attacked through planning and scheduling. Chapters 6 through 8 outline a variety of data-gathering, recording, and analysis procedures. In the short run, at least, individual firms, either owners or contractors, will be responsible for applying these approaches to their own situations, recording the most successful ones, and spreading the knowledge of them through their organizations.

SUMMARY

This chapter has briefly taken a systems look at on-site productivity improvement and has summarized the various elements that make up that system. The intention has been to set the stage for more extensive treatment of these elements in the chapters which follow. Their sequencing by topical areas is:

Chapters 5 through 8—Techniques
Chapters 9 through 11—The people side
Chapter 12—Safety
Chapter 13—Advanced approaches to productivity improvement that are with us or that are on the horizon

CHAPTER 5

FORMAL PREPLANNING FOR ON-SITE CONSTRUCTION

INTRODUCTION

In its broadest sense, preplanning for on-site construction provides the thinking, arranges for the necessary elements, establishes the requirements, and develops the operating rules for all that happens at the work face. It involves all the parties to a project and their people, except for the crews that carry out the intended operations at the work face. To recap from earlier discussions, on most projects these parties are: (1) owners, who outline and think through and define the requirements to be satisfied by the building, plant, or other facility; (2) designers, who portray the project in plans and specifications; and (3) constructors, who bring together the elements required and see the project through to completion. In some instances, owner, designer, or a separate party may take on some of the management functions of these three in a role referred to as "construction manager." All these parties must plan, with their efforts directed at the areas for which they have responsibility and done at levels of detail suited to their responsibilities. Often the parties may deal with the same topic, but from different viewpoints. For example, owner and designer will look at site layout primarily in terms of access to and operation of the completed facility, while the constructor is concerned with access for each set of tasks during the construction phase. Again, the owner and designer are concerned with the final placement, operation, and maintenance of individual machines, vessels, or other components of an in-place system, while the constructor is responsible for installation, with its problems of access, lifting, placement, and connection. Thus, preplans dealing with the same topic will have multiple focuses and will be developed to different levels of detail.

In comparison with planning in industrial situations, preplans for construction often suffer from fast decay brought on by unanticipated changes. These can initiate a chain of dislocations and delays, many of which could not have been foreseen at the time the preplan was developed. Those preparing preplans must recognize that all plans are subject to change and must provide flexibility and a means for quick adjustment. Otherwise reality disappears, and when this happens, those for whom the preplans were made ignore them and improvise to keep the work going.

Formal preplanning, done correctly, involves five steps or phases, which are:

1 Planning the planning process
2 Gathering information
3 Preparing the preplan
4 Disseminating the pertinent information in the plan to all affected parties
5 Evaluating the results of the planning efforts and its consequences

Experience has shown that success in preplanning requires that attention be given to all five items. However, in practice, items 1 and 5 are often neglected. At the start (item 1), the effort itself is often undertaken without clearly defining what the preplan is to accomplish and how the ends will be achieved. Then, after construction is completed, little attention is usually given to item 5, evaluation. Rather, planners rush on or are rushed on to new activities and do not have or take the opportunity to learn from the preplan's successes and failures.[1]

This chapter is concerned primarily with the on-site aspects of formal preplanning viewed through the eyes of the constructor. For that reason, it considers the viewpoints of owner, designer, or constructor's home office only insofar as their preplanning affects on-site activities.

The topics addressed here fall under the following main headings:

• What formal preplanning is
• Common constructor attitudes toward formal preplanning
• Whether or not formal preplanning pays
• How formal preplanning efforts should be staffed
• Tools employed in formal preplanning
• Examples of formal preplanning

FORMAL PREPLANNING

On-site construction involves turning ideas depicted on plans and in specifications into a completed structure or plant by assembling, combining, and erecting a number of parts and pieces. For this to be done effectively, safely, on time, and with suitable quality requires extensive planning off site and on site, including many individual work faces. Planning activities, done informally on

[1] For added discussion, see A. Laufer and R. L. Tucker, "Is Construction Planning Really Doing Its Job," in *Construction Management and Economics*, London, 1987.

most jobs, have always been a normal function of construction management. They still are today. From the time a project is conceived, some form of planning goes on almost continually. It first involves the owner, an architect or engineer, and sometimes a construction manager. After the project is formalized into plans, specifications, and contract documents, the constructor continues to plan by developing a schedule and ordering materials, equipment, and other items of importance in the overall execution of the project. Following this, on-site planning involves assembling the necessary elements at the various work faces as depicted earlier in Fig. 4-1 and 4-2 and finally carrying out the steps necessary to bring these elements and the workforce together to complete the job.

There are three levels of formal planning that should be a part of the constructor's execution of every job. These are:

1 An overall plan that provides a general outline of work: for example, which tasks will be done when and what elements are necessary to do them. The analytical tools employed may include Gannt (bar) or flowcharts which show sequences, including precedence or arrow diagrams. Very few constructors would consider starting a project without having done this kind of planning, at least informally.

2 Contingency plans to fall back on in case the original plan goes awry. These cover incidents that management may not have expected or wanted to occur. Common surprises in construction include changes in project scope, unanticipated site conditions, unusual weather, late deliveries of critical materials or equipment, work stoppages or labor shortages, and accidents. Because such happenings are common in construction, advance thinking to anticipate them is important. And yet, because they often are unexpected and bring crises which call for quick adjustments, planning for them is often done quickly and haphazardly. In such cases the old adage "haste makes waste" applies in full force.

3 Detailed planning for work execution at the task level. Such planning, if done formally and in an organized manner, normally is written and gives minute details about time and place and how people, skills, materials, tools, plans, equipment, and space are to be employed. This kind of careful written planning is too seldom done by constructors. Rather, they do the operation, whatever it is, the same way as the last time, assuming that the same external conditions apply. In general, they rely on informal, often verbal, communications, with the prethinking and execution left largely to the foreman, sometimes with help from the superintendent. This gets the work done, but often less effectively than if planning were given more attention. Many of the managers at these levels will tell you that they have too little time left from other duties to do detailed written planning. Commonly they would not do it, even if they had the time, because it is not the way they have learned to operate. And usually there is no pressure from higher-level management to change their past practices. Under these circumstances, formal planning will not take place. If detailed written advance planning is to be done, higher management must assign it a higher priority than other activities and either put pressure on field supervision or make someone else responsible.

WHAT IS INVOLVED IN PREPLANNING?

The premise of this book is that detailed advance planning (preplanning) under the three levels outlined above plays an important role in improving productivity and ensuring that a project is completed on time, safely, and under budget. As already discussed in Chap. 4, preplanning in essence involves setting down procedures in detail about who, what, why, how, when, and where; and it is done well in advance of the time when particular tasks are to be undertaken. This approach will replace the usual practice of many field supervisors, which can best be described as reactive rather than active. It is the process of cooperatively thinking through in advance the details of the task to be done; of anticipating interferences, shortages, and other pitfalls before execution is begun, of allowing time for field supervisors to actually manage their work rather than having to run around picking up the pieces or putting out fires.

Types of preplans range from those dealing with major blocks of work and the elements needed to carry them out to daily written instructions for a small crew. In their most common form, they involve sketches, drawings, and listings of materials, equipment, and tools and their locations along with written instructions which indicate how the images shown by the drawings are to be turned into a completed physical structure. Sketches, often isometric projections, may be needed to show relevant details not on the engineering drawings. Also included are information on the source and disposition locations of temporary materials as well as any modifications they may require. Pipe fabrication and metalwork details and excavation schematics also are common preplanning aids. Notes regarding inspection requirements or conflicts with other crews and careful instructions regarding safety precautions can be included. Some preplans also show beginning and ending times or anticipated durations of each activity. Several examples of preplans, ranging from simple to complex, are shown in App. B.

As indicated, preplanning is done early to be sure that drawings, tools, materials, and special equipment for an operation will be on hand, or to provide acceptable substitutes. This permits making alternative plans for an operation's execution or for postponing it. Furthermore, given this lead time, all involved have a chance to make inputs, discuss, and possibly modify the proposed plan of execution.

The timing of preplanning varies with the individual operation. For making important work assignments or for large complicated tasks requiring special materials, machinery, or equipment that must be designed, ordered, fabricated, and delivered, preplanning must be done as much as a year or more in advance of the need. At the other extreme, for a routine task taking a day or two that is to be assigned to a foreman and small labor crew, a week or two may be enough lead time. A rule of thumb suggests that preplans for crews doing routine work be ready about 10 days before the scheduled starting date.

Preplanning takes many forms, all of which have one goal—to make management more effective. But because the operations to be preplanned vary widely in scope, even on a single project, discussions of how to preplan sometimes turn into arguments about the need for preplanning, how much paperwork is required, and who will be responsible. These arguments are really not pertinent and can be divisive. As approached here, preplanning is defined broadly. It is an organized system of advance planning that produces detailed written instructions for a given task, be it assignment of responsibility for away-from-the-work-face activities, coordination among work assignments, or instructions at the crew or individual worker level. In every instance, it requires that an individual designated by management think through the details of an operation. (Table 4-1 illustrates the factors that a preplan might address.) After consultation with other concerned parties, that person then prepares a written set of instructions.

It is worth repeating that preplanning is not a new concept. It has always been done, customarily in someone's head. But there are tremendous advantages to writing it down. For as with scheduling or cost-reporting and tracking, preplanning provides a way of organizing data and, in turn, formalizing the thought processes that managers always have used. By putting the plan down on paper, it is then in a form that can be readily communicated to all those concerned with the particular task.

COMMON ATTITUDES OF CONSTRUCTION MANAGERS TOWARD PREPLANNING

Construction managers often react to suggestions that they undertake preplanning on a formal basis with contentions such as (1) there are already too many tasks for supervisory people to do effectively or (2) to preplan will mean hiring more staff. Both statements are usually true. Field construction management is often overloaded, but with much of its time spent correcting yesterday's mistakes, racing to keep ahead on job progress, and solving a myriad of emergency problems, some of which reflect earlier failures to plan in detail. If a field engineering staff exists, it is usually already occupied in doing other tasks so that it cannot undertake an added chore. Thus to have preplanning capability does often require additional staff. The difficulty is that this additional staff costs money, and in the usual project cost-reporting system, these added costs will appear in an overhead account that is highly visible and will be the first target for cost reduction if the job seems to be in trouble. On the other hand, the cost of false starts, delays for materials or for other crews to clear working space, and inefficient operations are not measured separately in the cost system and will receive attention only when the average re- ported costs for that particular work classification are substantially above the estimate or budget. Preplanning can then be sacrificed because management unwisely looks only at reported costs while disregarding unquantified savings.

One objection voiced against formal preplanning done some time before the operation is to be carried out is its sense of unreality. To have the real world of actual tasks in one's mind and then reduce those thoughts to a written plan that actually portrays the operation is difficult. This may be particularly hard for engineers who have been educated in design but lack the feel developed by action-oriented construction people who are accustomed to dealing with real materials, equipment, and people reacting to crises. To develop an entirely new way of reasoning is not easy.

There also are field managers at all levels who reject the concept of preplanning because they think of the preplanner as a management spy or as an outsider who will preempt their right to make decisions. Actually, the preplanner's role as a part of the management team is to solicit management's ideas and then to plan in detail. Afterward, the scheme is reviewed by field management and finally reduced to detailed instructions. With these details taken care of, field managers can devote their attention to executing the work. Among preplanning's advantages, then, is that line supervisors should not have to improvise or search out and find necessary elements such as equip- ment, tools, materials, instructions, or working space. They should be at hand so that the field manager is free to fine-tune the operation rather than contin- ually having to pick up loose ends.

From a human-behavior point of view, preplanning must not take authority away from field management. If done properly, it instead results in a team ef- fort—a pooling of ideas and a chance for all involved to to be more effective and creative. Given successful preplanning, the final execution is less likely to suffer from the effects of overlooked details of the sort which lead to disrup- tions of the work, delays, or postponed tasks, all of which can be very costly.

DOES PREPLANNING PAY?

It was pointed out above and earlier in Chap. 3 that the usual cost-reporting sys- tems employed by contractors do not and cannot measure the savings that result from preplanning. Moreover, it would be difficult to devise specific dollar mea- sures of preplanning's benefits, because it affects so many parts of the overall construction process in so many ways. In any event, most contractors are reluc- tant to accept reports such as, "The savings from preplanning will be 4 to 8 times the cost or, even 15 times the cost," or in another instance, that "preplanning turned a project around. The expected loss was a million dollars, but because management introduced preplanning, the job in the end made money."

Certain owners who believe in preplanning insist that their contractors em- ploy it on their cost-plus work. For example, Proctor & Gamble reimburses contractors for the expenses involved in preplanning and requires them to report all job expenditures and unit costs through an owner-supplied accounting and cost system. This approach permits a small owner's staff to monitor a number of projects in detail. It also makes possible cost comparisons among the contractors doing different projects and thus provides input for selecting contractors for future work. The authors have been told by owners who follow this procedure that it alone saves $4 to $8 for every $1 spent on preplanning.

Today, it is the unusual contractor who does formal preplanning. But evidence such as that quoted above indicates that it should be tried. Furthermore, because of the findings of the Business Roundtable Cost Effectiveness Study, more and more owners will be insisting that the contractors they hire preplan, at least on cost-reimbursable projects.

STAFFING THE PREPLANNING EFFORT

If preplanning offers an opportunity to make projects more efficient and profitable, the next question is, "What sort of person will probably do preplanning best?" In an ideal situation, it might be someone who knows construction well and who challenges present practices and thinks creatively (see Chap. 10 for further discussion). In the real world, however, this is often not possible and the best sources of ideas for finding such people are the contractors and owners who do preplanning successfully. Among these firms, the assignment is usually given to young engineering or construction graduates who are being groomed for positions in higher management. They work under or in close liaison with superintendents or other field managers. Two reasons are given for using such inexperienced people rather than relying entirely on field supervision. First, they usually have open, inquiring minds and are willing to consider all ideas, even those that at first seem unworkable. They do not feel that they know all the answers or that they have a proven method, one that is the only way to do a particular task. In the words of our computer-oriented society, they are not already programmed to use a given solution. Second, the task of planning, which involves searching for effective combinations of materials, equipment, labor, and methods is an excellent way to give such people a broad background and feel for construction. Furthermore, they become intimately involved and learn to communicate with managers and craftsmen alike, which is an essential skill for good management.

The number of preplanners that can be gainfully employed on a project depends in part on the degree of repetitiveness in the work, which varies widely. One company, which does highly technical projects involving the mechanical trades, provides one planner for every 20 field workers and adds an experienced coordinator for every 5 planners. This figure is probably far too high for less-complicated projects.

There is no magic rule for the percentage of the work that should be preplanned. Weldon McGlaun, who directed Proctor & Gamble's operations for many years, estimated that 95 percent of the firm's construction activities could be preplanned to good advantage. Much of this work was in industrial plants, which are quite complex. Certainly the percentage would be lower when the work being done is duplicated again and again. But even a substantially lower percentage does not represent industry practice, since a majority of contractors do no formal preplanning at all. Staffing for preplanning has another and subtle dimension, that of gaining management's participation in and acceptance of the plan. Numerous studies of management behavior have made clear that neither those at the top or in the line below them can or will devote large blocks of time to detailed planning, which is one reason why planning has developed as a staff function. Therefore, the effective planner must somehow gain sufficient management attention at the appropriate levels to (1) gain access to the data which the line has and controls, (2) secure its inputs as the plan develops, and (3) get management to accept and implement the plan as if management had developed it in the first place. This is no easy task.

TOOLS EMPLOYED IN PREPLANNING

As indicated earlier, the philosophy underlying formal preplanning is that all but the simplest operations on a construction project should be carefully thought through and discussed. Then the conclusions should be written down and made available in usable form to those who are to carry through with the operation. The examples of preplanning given later in this chapter will show the wide range that these activities can take. But before giving the examples, the authors feel that a discussion of some of the tools that can make the various stages of preplanning more effective is necessary. The tools include estimates, schedules, block or task diagrams, physical models, and computer-based physical models.

Estimating as a Preplanning Tool

Some form of cost estimate is the tool employed successively by owners, designers, financial interests, and constructors to forecast the expected cost of a project. Cost, along with time, is usually the most important factor an owner must consider in deciding whether a project as conceived should go ahead. Cost is determined by a preliminary and usually approximate estimate. After design is well along or completed, another and more refined estimate is made for the owner by the designer, contractor, or construction manager if one has been engaged. Contractors make a more detailed estimate at the time of negotiation or bidding for a project. If successful in obtaining the work, they prepare an even more refined estimate, sometimes called a budget. As a project moves forward, if the scope of the work or other features of the project change, the cost of modifications is estimated as a basis for adjusting the contract amount. In sum, some form of cost estimate is an essential element in every step required to bring an owner's concept of a construction project to completion. Estimating is also an essential tool in preplanning, since the costs of doing operations in various ways is a very important and often the sole determinant in choosing among them.

Preplanning usually calls for yet another estimate. It includes listing and possibly pricing the labor and other items required under each proposed scheme for carrying out an operation. As with the original one, its reliability varies according to how it is made. In all cases, the process of estimating can be any one among those listed below and is, of course, subject to the weaknesses of each.

Often preplanners are engineers with little field construction background or experience to guide them. Sometimes the methods they propose for doing a task may be different enough from the way a similar operation was done to make the available historical job records an unsatisfactory guide. In such cases, it is often helpful to have experienced field personnel review the preplanner's estimate.

There are four common methods for determining or setting cost and time allotments and other variables to be used in estimates. These are (1) "horseback" estimates, (2) judgment of a qualified expert, (3) the statistical or historical method, and (4) time studies based on computed or engineered standards such as methods time measurement (MTM). Each has its place, but if construction follows the example of other industries in this as it has in other management areas, it will gravitate toward the more advanced approaches represented by 3 and 4.

Method 1. Horseback, or "Ballpark," Estimates   At the time when a project is in the conceptual stage, an owner or others with financial or similar interests will often need to know its approximate cost and completion time. In such cases they will draw on knowledge of similar projects done in the past and apply adjustments to recognize factors such as inflation and relative costs between geographical areas to get very rough forecasts of cost. Results of this procedure are stated in terms such as: for an office or commercial building or dwelling, X dollars per square foot of floor area; for an industrial plant, Y dollars per unit of output; or for a road or street, Z dollars per lineal foot or square foot of area.

Data for making these approximate estimates can come from a variety of sources. Among these are contractors who have built similar projects, financial institutions such as banks or bonding companies, or public records such as building permits or bid tabulations. Trade magazines, for example, ENR (*Engineering News Record),* have cost reports or news articles that are helpful. There are also several commercially published cost guides.

Owners will frequently decide whether or not to go forward with a project basing their decisions on such horseback estimates or, alternatively, deciding whether to make a more detailed estimate by one of the methods discussed below.

Method 2. Estimates By Qualified Experts   Method 2, which relies on the personal knowledge and skill of an expert, usually called an estimator, is the one that many constructors use to predict costs for bidding purposes and for pricing changes. From past knowledge, these experts first build the project step by step in their heads. Using these images, they then estimate the amounts and costs of materials, equipment, and labor hours needed to complete each element of the project. In a similar manner, they also set completion times. Examples of this approach for the labor estimate are:

Case 1: Three carpenters and a helper will require one day to build and erect forms for a concrete wall 6 feet high and 20 feet long.

Case 2: A mechanic can overhaul a small transmission in 4 hours.

Case 3: A driller and a half-time helper will put down 40 feet of hole per hour.

This approach relies on the experience of the persons making the estimates and their views of the conditions under which the task will be done. These opinions will obviously vary among experts because of their differing assessments of the performance of the individuals or crews as compared with others with which they have been familiar.

A common practice is to involve field personnel, sometimes down to the foreman level, in making estimates by expertise. These people have a depth of knowledge in particular areas such as those just named that a generalist estimator cannot possibly have. For pricing operations involving large quantities of work, the estimator will often bring together a group of field people with experience in a given area and ask their opinions or even brainstorm (see Chap. 3) about the operation before deciding on the performance figure to use in the estimate.

Method 3. Historical or Statistical Estimates   This form of estimate relies on records from earlier projects to determine the cost and time required for the appropriate individuals, crews, and machines to do similar work under like conditions. Estimated accomplishment is usually adapted from historical production rates or unadjusted work-sampling data if they are available. In the usual construction company, production or cost records from similar projects provide the data base. This accumulated information is adjusted by the estimator to allow for local conditions. Examples of information from such a data base, when available from a recent comparable project, might include:

- Rock excavation: $5.20 per cubic yard
- Installing, aligning, stripping, and cleaning straight wall forms: $4.20 per square foot of contact surface
- Carrying, placing, and tying straight no. 8 reinforcing bars: 250 pounds per hour per ironworker[2]

For the examples given under method 2, the method 3 breakdown for them probably would be as follows:

---

[2] For those who do not have in-house information or wish to check their data, a wide variety of books and other publications giving unit performance rates for the more common construction operations is available.

Case 1: Formwork, flat, and ordinary: output, 8 square feet per carpenter-hour

Case 2: Repair transmissions: either 0.01 mechanic-hours per hour that the transmission operates or, for small transmissions, 3 to 6 hours per overhaul

Case 3: Drill production, in granite, with a 3-inch drifter: 37.6 feet per hour

Almost all the estimating done by contractors and construction managers is based on a combination of methods 2 and 3, that is, expert opinion and historical records.

Estimating by Computed, or Engineered, Standards   Computed standards utilize measured values of the times required for trained individuals to carry out basic movements of fingers, hands, limbs, and body. In turn, the data for each movement are combined to give the time required to carry out a given maneuver which can then be extended to give the time required to complete a given task. The times are subsequently adjusted to reflect productivity in a given geographical area as compared with the standard. To these are added allowances to cover travel to or on the work site, materials handling, job preparation, and personal time. From these times and the unit costs of labor, materials, and equipment, a cost estimate is then created.

Tabulations giving the times required for the individual movements and for many of the customary combinations of movements called for by some of the common tasks are available in manuals and books. They represent a trained technician's reconstructions of the times required to carry out all the body movements for a single act or a combination of acts. A few examples of such data are given in Fig. 5-1 and App. A.

In theory, MTM or procedures based on this approach can have a particular value to preplanners who wish to compare the times required to do a new and repetitive task in each of several ways before it is actually undertaken for the first time. Rather than starting with a method arrived at by a field crew and then analyzing it by trial and error, planners can develop an efficient method in advance, subject to further refinement in the field. The disadvantage is that the procedures are very time consuming for the planner.

For the tasks such as those listed under Cases 1, 2, and 3 above, estimating would involve referring to tabulations showing the normal times required to do each of the several steps in them and summing up the listed times. In some of the MTM manuals, this summation has already been carried out for certain common construction tasks. The summations however obtained are then corrected to reflect local productivity compared with established standards. Allowances would be added to cover travel to the work site, material handling, job preparation, craft adjustment for extra travel and preparation, and personal time. The tabulations of required times for the operations themselves are from a manual or set of manuals which cover a wide variety of operations, the allowances can be assembled and modified as appropriate. For Case 1 given above for carpentry work, the estimate would be computed as follows. Forms: fabricate, install, and remove, any wall 4 to 6 feet high (per linear foot), 0.50

METHODS-TIME MEASUREMENT APPLICATION DATA

SIMPLIFIED DATA

(All times on this Simplified Data Table include 15% allowance)

HAND AND ARM MOTIONS	BODY, LEG, AND EYE MOTIONS
REACH or MOVE TMU 1″ .... 2 2″ .... 4 3″ to 12″ 4 + length of motion over 12″ 3 + length of motion (For TYPE 2 REACHES AND MOVES use length of motion only)	TMU Simple foot motion....... 10 Foot motion with pressure 20 Leg motion .... 10 Side step case 1........ 20 Side step case 2........ 40 Turn body case 1....... 20 Turn body case 2....... 45
POSITION Fit Symmetrical Other Loose 10 15 Close 20 25 Exact 50 55	Eye time.............. 10 Bend, stoop or kneel on one knee............. 35 Arise.................. 35
TURN—APPLY PRESSURE TURN.............. 6 APPLY PRESSURE.. 20	Kneel on both knees..... 80 Arise.................. 90
GRASP Simple.............. 2 Regrasp or Transfer... 6 Complex............ 10	Sit.................... 40 Stand................. 50 Walk per pace.......... 17
DISENGAGE Loose.............. 5 Close.............. 10 Exact.............. 30	1 TMU = .00001 hour = .0006 minute = .036 second

FIGURE 5-1
Methods-time-measurement (MTM) simplified data
(*From Harold B. Maynard, Ed.*, Industrial Engineering Handbook, *2d ed. McGraw-Hill, 1963.*)

worker-hour per foot; local productivity factor, 0.95; travel to a specified point, 0.35 hour; material-handling allowance, 0.11 hour per 100 pounds; job preparation for carpentry, 0.30 hour; a craft adjustment of 34 percent; and personal time 10 percent of total craft time. Tabulated, the estimate is:

Craft time: 2 sides × 20 feet × 0.50 worker-hour per foot × 0.95 allowance	19.00 hours
Travel: 4 men × 0.35 hour	1.40 hours
Material handling: 240 square feet × 10 pounds per foot × 0.11/100	2.64 hours
Job preparation: 4 men × 0.3	1.20 hours
	24.24 hours

Craft allowance: 34 percent; 1.34 × 24.24 = 32.4 hours

The time and cost required to develop the data and arrive at estimates made by the three methods varies from a few minutes for the technical estimate to weeks and months for developing data for engineered standards. Many construction companies maintain records that make statistical and historical data readily available although its reliability may be questioned. However, others have no records that are useful. Many manufacturers and service organizations, as well as authors of texts and reference volumes, publish data that are largely statistical. Private companies tend to guard their data jealously, not wishing to give their competitors the benefit of their years of work and dollars of expense.

Although there are several different sources of the engineered standards employed by industrial manufacturers, the one that appears to be in broadest use is known as methods time measurement, or MTM. The basic text was published in 1948 as a culmination of many years of research and application by its authors.[3] Very shortly thereafter a national Methods-Time-Measurement Association was formed, and the basic data have been made freely available to all interested persons. This free dissemination of data, together with the excellent basic research that preceded its published findings, has done much to make this a widely accepted and used standard.[4]

The MTM system is based on the concept that methods determination must precede time determination. Once a method is determined, a time can be very accurately compiled by adding up the times needed for each of the elemental steps. Basic elements making up body movement are reach, move, turn, apply pressure, grasp, position, release, disengage, eye travel and focus, and/or body, leg, or foot motions. Each of these in turn is subdivided into as many as 18 different actions. For each of these actions there is a standard time, commonly stated in TMU (time-measurement units), where one TMU equals 0.00001 hour, 0.0006 minutes, or 0.036 seconds. Figure 5-1 gives some very simplified data on the MTM system. The reader is encouraged to refer to the excellent texts available.[5]

The service organizations of the U.S. military have been using MTM time standards for facility maintenance management since the 1950s. The data manuals for ''Engineered Performance Standards for Real Property Maintenance Activities'' for the Army, Navy, and Air Force are contained in a library of 20 manuals, including 3 manuals for teaching the system and 17 covering the gamut of trades from carpentry through wharf building plus subjects such as emergency and janitorial service, pest control, and preventive maintenance.[6]

[3] Harold B. Maynard, G. J. Stegemerten, and John L. Schwab, *Methods Time Measurement,* McGraw-Hill, New York, 1948.

[4] According to Delmar Karger and Franklin Bayha (*Engineered Work Measurement,* Industrial Press, New York, 1957, p. 60), more than 3,500 companies use this method.

[5] Maynard, Stegemerten, and Schwab, *op. cit.;* Karger and Bayha, *op. cit.;* and John L. Schwab, ''Methods Time Measurement,'' in Harold B. Maynard (ed.), *Industrial Engineering Handbook,* 2d ed., McGraw-Hill, New York, 1963, pp. 5-13–5-38.

[6] These manuals are as follows (Navy, Army, and Air Force have different publication numbers; those given here are from the Navy): P-700 *Engineer's Manual;* P-700.1 *Instructor's Manual;* P-700.2 *Student's Workbook;* P-701 *General;* P-702 *Carpentry;* P-703 *Electrical & Electronic;* P-704 *HVAC & Refrigeration;* P-705 *Emergency & Service;* P-706 *Janitorial & Custodial Services;* P-707 *Machine Shop & Machine Repairs;* P-708 *Masonry;* P-709 *Moving & Rigging;* P-710 *Paint;* P-711 *Pipefitting & Plumbing;* P-712 *Roads, Grounds, Pest Control, & Refuse Collection;* P-713 *Sheet Metal, Structural Iron & Welding;* P-714 *Trackage;* P-715 *Wharfbuilding;* P-716 *Unit Price*

The performance standards just described give the average times necessary for qualified workers to carry out defined tasks of specified quality while following acceptable trade practices. It is assumed that they proceed at a normal pace under capable supervision and experience normal delays at the work face. In addition to the time estimates for the actual tasks, the manuals give adjustments for various quantities of work. Other allowances are given for job preparation; travel; material handling; difficult conditions; and delays for planning, personal time, unavoidable incidents, and balancing (i.e., the number of workers and partial-day influences). Still other adjustments can be made for special safety requirements, recurring or standby work, and additional per-day preparation time.

Table 5-1 is a simple and incomplete illustration of the estimating method based on the MTM approach, as employed by the U. S. Navy. In this instance, additional task times would be added for job preparation, travel, delays including balancing for multiperson crews, and partial-day influences, as well as additional-day-preparation time.

It is quite apparent from the work involved in establishing a standard time that highly repetitive jobs are the only ones for which such standards could be computed. But it is important to remember that while the end product of a given construction task may be unique, many of the elemental tasks that make it up are repeated many times and these repeated tasks can be analyzed for method and a time computed. The data given in Fig. 5-2 make this point in a striking manner. It shows that it takes a carpenter 9.3 seconds to make a simple measurement at waist level and 17.6 seconds to stoop, make and mark a measurement at floor level, and stand up again. At an hourly wage of $15, the costs on the average of these simple tasks are, respectively, 3.9 and 7.4 cents. And operations such as these are done repeatedly on most construction projects. With knowledge such as this, there should be strong motivation to find a substitute for measuring or marking out distances with a tape. With a little advance planning, it might be possible to eliminate it by precutting or other strategies.

No contractor or potential user has to start from the beginning in using MTM, since so much applicable material is already available. Any organization can develop a system of standard times and constantly add to them.

MTM can be a very useful tool in estimating the times required to do highly repetitive tasks in several ways before they are undertaken at the work face. A promising application is highly repetitive benchwork, which is usually given little management attention. Such operations would be logical starting places for MTM planning; and as capability and confidence are established, additional tasks could be added.

One word of caution concerning MTM: there are instances where assumptions must be made which can drastically affect the results of an MTM analy-

---

Standards; P-717 *Preventative/Recurring Maintenance.* In addition to these manuals, the Navy is also using a computerized version called FEJE (Facility Engineering Job Estimating System) and preparing one for a microcomputer. Appendix A gives an example of the synthesis of MTM Engineered Performance Standards.

TABLE 5-1

EXAMPLE OF THE NAVFAC METHOD FOR ESTIMATING THE LABOR-HOURS REQUIRED TO CONSTRUCT A STUD WALL COVERED WITH GYPSUM BOARD*

# Reference	Work Unit Description†	Hours	Units
	CT-207		
1 PWMU-1-8136	Layout and install bottom plate	0.08140	job
		0.04335	l.f.
2 PWMU-1-8136	Layout & install bottom plate, first stud	0.05780	job
3 PWMU-1-8137	Layout and install top plate	0.16180	job
		0.04185	l.f.
4 PWMU-1-8137	Layout and install top plate, first stud	0.05580	job
5 PWMU-1-8138	Measure, cut, & install studs to plate	0.07868	l.f.
6 PWMU-1-8138	Measure, cut & install studs to plate, first stud only	0.10490	job
7 PWMU-1-8140	Layout, cut & install 1 row of fireblocking	0.00100	job
		0.04808	l.f.
8 PWMU-1-8139	Layout, cut & install 1 row of fireblocking first stud only	0.06410	job
9 PWMU-5-II	Material Handling	0.01761	l.f.
	0.52680 hrs per job + 0.22957 hrs per ft		
	CT-181(2)		
1 PWC-12-V	Install gypsum wallboard	0.05336	l.f.
2 PWC-14-II	Install ceiling molding	0.00948	l.f.
3 PWC-12-IV	Install baseboard	0.01407	l.f.
4 PWC-14-I	Install shoe molding	0.00784	l.f.
5 PWMU-1-8355	Move, climb up and down scaffold	0.00520	l.f.
6 PWA-5-II	Material handling	0.00932	l.f.
	(2) of 0.09926 hrs per ft		
Total: CT-209 =	0.52680 hrs per job + 0.40953 hrs per ft		

Source: "EPS: Carpentry Handbook; NAVFAC P–702.0 (also ARMY TB 420–4, Air Force AFM 85–42) dtd March 1982."

† Full Wall, studs @ 16" O.C., 8' high, one row of fire blocking, fabricate and erect, install gypsum wallboard, baseboard, and molding 2 sides, per lin ft. (Task CT–209=CT–207 and 2 of CT–181)

sis. What, for example, is the number of hammer blows necessary to drive a nail? Given a carpenter's hammer and an 8-penny nail, six blows is a common assumption. Yet carpenters nailing roof sheeting on relatively flat slopes often drive a nail in one blow from a heavy weighted hammer. They tip the nail to set it on the upswing and drive it with a single heavy downward stroke. This example and others like it should warn the user of MTM to look carefully at the assumptions underlying the analysis.

Regardless of the procedure employed, estimators must be cognizant of the effects on productivity of learning by repetition. This is to be expected for tasks where the procedure is not highly developed or is not controlled by factors such as machine pacing. Many and perhaps most construction activities certainly fall into this category and therefore may be susceptible to marked improvement if studied carefully after being carried out in the field for the first time. Generally,

			TMU	Seconds
A. A carpenter makes a measurement; measuring tape in carpenter's pocket, measurement made at waist height				
Reach into pocket R-12	3 + 12		15	
Grasp tape			2	
Move tape to measuring position	3 + 24		27	
Grasp end			6	
Pull tape out			24	
Position zero end, exact			55	
Position other end, Read tape: eye time, close fit	10 + 20		30	
Retract tape			30	
Return tape to pocket, hand back to work			70	
Total			259	9.3
B. To mark a measurement; in addition to above				
Reach for pencil—pocket 12" reach			15	
Grasp pencil			2	
Move pencil to work 18"			21	
Mark measurement, eye time + exact position			70	
Return pencil and hand to pocket			21	
Return pencil to pocket, position pencil			10	
Return hand to work			21	
Total			160	5.8
C. To do measurements at floor level				
Kneel on one knee			35	
Arise			35	
Total			70	2.5
Total: Measure			259	9.3
Mark			160	5.8
Kneel			70	2.5
Grand total			489	17.6*

*At an employee cost of $16 per hour for wages, fringe benefits, insurance, taxes, but not costs of supervision or overhead, each second costs 0.44 cent. A measurement at floor level then costs 17:6X0.44=7.7 cents

FIGURE 5-2
An example of a time and cost analysis based on the MTM approach, utilizing the simplified data.

the more difficult the work and the less experienced the supervision and crew, the steeper this learning curve and therefore the greater the potential for improvement. This topic is discussed in greater depth in Chap. 6.

Scheduling as a Tool for Preplanning

Estimating as just described is done at the time bids are prepared and again by the contractor selected to do the work as a more detailed project budget is formulated. These estimates are primarily in terms of the costs entailed in completing the individual operations. Although estimators often examine the times required for and sequencing of the major elements of the project, they are usually not concerned with establishing the times when individual operations will be carried out or the time sequencing among them. Except for very complex operations, these activities, or at least the details of them, are carried out by management in the field and are often adjusted to fit changes that develop as the work progresses. This preplanning function goes under the heading of scheduling, an essential element in project control. Preplanning without scheduling has little merit. Only with a suitable schedule in appropriate detail and worked out sufficiently in advance can job management function effectively, because without a suitable schedule, many questions about what, why, when, where, how, and who will go unanswered.

Traditionally in construction, the bar or Gantt chart (see Fig. 5-3 for an example) has been the tool for scheduling the timing and overall duration of individual operations. Its merit is that it is easily understood. But it is of little use as a tool for any but the most rudimentary planning. Among its many weaknesses are that it cannot portray the interrelationships among the various activities or the resources demanded by each. Neither does it distinguish among those activities that are critical or noncritical in determining overall progress toward completion or alert management to potential conflicts among activities.

Beginning in the 1960s, networking techniques described in terms such as critical path method (CPM), precedence diagramming, and PERT came into fairly widespread use, particularly on larger projects. Today, with microcomputers and user-friendly software available for almost any job site, these networking techniques have become more commonplace (see Fig. 5-4 for an example of a network). There are, of course, protesters who claim that the networking forms of scheduling are too complex, too expensive, too time-

FIGURE 5-3
Bar-chart schedule for a concrete gravity arch dam. (*From D. S. Barrie and B. C. Paulson, Jr. Professional Construction Management, McGraw-Hill Series in Construction Engineering and Management, 1984.*)

Item No.	Description	First Year												Second Year											
		J	F	M	A	M	J	J	A	S	O	N	D	J	F	M	A	M	J	J	A	S	O	N	D
M-10	Mobilization	▓	▓																						
E-10	Foundation excavation			▓	▓	▓	▓	▓	▓	▓	▓	▓		▓	▓										
D-10	Diversion stage — 1						▓	▓	▓																
D-20	Diversion stage — 2															▓	▓	▓	▓	▓	▓	▓			
G-40	Foundation grouting							▓	▓	▓	▓	▓													
C-10	Dam concrete								▓	▓	▓	▓	▓	▓	▓	▓	▓	▓	▓						
I-20	Install outlet gates													▓		▓									
I-30	Install trash racks																		▓	▓					
P-10	Prestress																			▓	▓	▓	▓	▓	
R-80	Radial gates																					▓	▓	▓	
S-50	Spillway bridge																	▓	▓						
G-60	Curtain grout																	▓	▓	▓	▓	▓			
L-90	Dismantle plant, clean up																							▓	▓

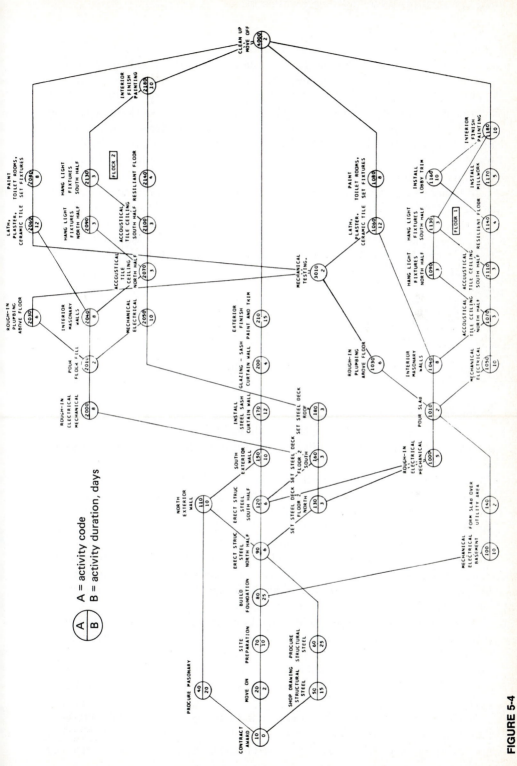

FIGURE 5-4
Sample network schedule for a small office building (precedence diagram with activities on nodes).

A = activity code
B = activity duration, days

101

consuming, or not worth the effort. But today these protests must be viewed as excuses, not reasons. Given the basic data about the individual operations and their durations and sequencing, which management should develop in any event, a planner can quickly call on the computer to carry out the required computations, plot the network, display the input data, and work out and show the time relationships for any proposed set of job conditions. Also, with the original network developed, job management can easily explore the effects of proposed changes in the sequencing or durations of operations and have an up-to-date schedule readily available to all on-site managers.

In today's construction world, on-site managers on all projects, regardless of size, should be expected to develop, maintain, and follow network-based schedules. First of all, they are a valuable and time-saving tool to the manager for running the work. In addition, representatives of owners, subcontractors, bankers, and bonding companies are becoming more knowledgeable and will therefore demand detailed schedules from the contractors with whom they deal.

This book will not go into the details of scheduling. It is a topic treated exhaustively in many books and articles and one that is taught as a separate subject in many construction curricula. But its importance cannot be overemphasized. For, without a clearly spelled out schedule, a manager or preplanner cannot outline and analyze the alternative courses of action that are available.

Block and Task Drawings

A very common form of preplan for an individual operation is the block or task drawing. It is typified by the individual lift drawings that many contractors employ in preparing for concrete placements in dam and powerhouse construction (see Fig. 5-5) or for industrial installations or high-rise buildings. This concept has a wide application. On such lift drawings, the preplanner will combine on a single sheet all the relevant information from the structural, electrical, piping, machinery installation, and other detail drawings, along with references to the pertinent specifications and addenda. Also given on this sheet are bills of material that must be supplied, notations about which items may require special attention (e.g., x-ray inspection of certain welds, pressure testing of hydraulic lines), and summaries of all quantities (for ordering or requisitioning tools or materials and for cost accounting).

Lift drawings save time for field supervisors and crews, who do not have to search out the information from many sources. In addition, they provide an important check against omissions, errors, and discrepancies between the separate drawings so that any deficiencies can be rectified, or the operation can be rescheduled if necessary, thereby avoiding delays to or idleness of crews in the field.

Since the lift drawing is made to be an aid both to supervision and crews in the field, several modifications to usual detailing procedures may be in order. For example, the plan view may be placed in the center of the sheet with a face elevation for each side. The elevations are oriented so that up is toward the plan view; therefore, the drawing must be rotated through a full circle to read all of the elevations. However, since this is the normal way to view an elevation, interpretation by field personnel is much simpler. Only information relevant to the task at hand is included, and hidden lines are shown only if they are required to ensure clarity. Only essential dimensions are given, and they are referenced to the base line or control that is being used by the field forces. With this procedure, all calculations are made and checked in the office, which reduces the chance for field errors in adding or subtracting or in laying out a series of consecutive dimensions.

Block drawings, often prepared as a part of the preplanning effort, provide on a single sheet all the information needed by a field crew when it is undertaking a task. Among other features, block drawings will show all the special items, blockouts, or other details called for on the separate drawings of the various crafts. Note that side views are oriented in a natural relationship to the plan view. Block drawings are illustrated in Fig. 5-6. This example details all the preparations necessary before concrete is placed in the form for a precast element for a building.

Physical Models[7]

Physical scale models have been widely used for some 40 years as a planning and design tool for industrial projects. These models, which are replicas of the proposed project or individual elements in it, are built to scale in three dimensions and thus correctly show the spatial relationships of the parts of the project. Such models may be of substantial size (500 square feet) and are often the basic layout document for the project or its major elements (see Fig. 5-7). Detailed drawings are made only after the physical relationship of the parts have been verified on the model. A photograph of a more detailed model of an element of an industrial plant is shown in Fig. 5-8.

Individual pieces for models which represent common elements such as pipe, valves, fittings, and structural members to several scales can be purchased from firms that manufacture and sell them. Reproductions of special large elements such as vessels, tanks, and other very large pieces are often made to scale in a model shop. Often the designer will first develop site or individual layout schemes by putting a model together at a large work table. After a final arrangement is determined, the model will be glued together for use in detailed design and for transporting to the construction site.

In addition to being a design control, physical models are also used to plan erection and construction sequences, to satisfy the owner's maintenance group

---

[7] This section of the book draws heavily on an excellent study titled, "The Use of Scale Models in Construction Management" by Edward C. Henderson, Jr., published as *Technical Report 213,* Department of Civil Engineering, Stanford University,Stanford, Calif., November 1976. This report describes in detail the uses and advantages of physical models in design and construction.

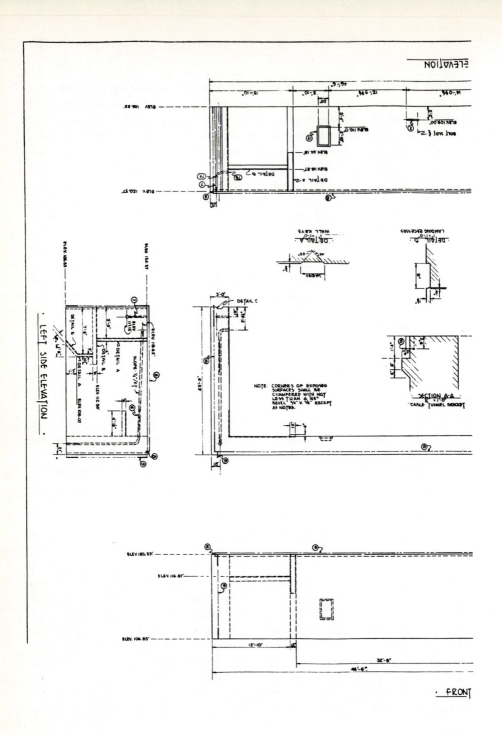

FIGURE 5-5
Lift drawing for a dam.

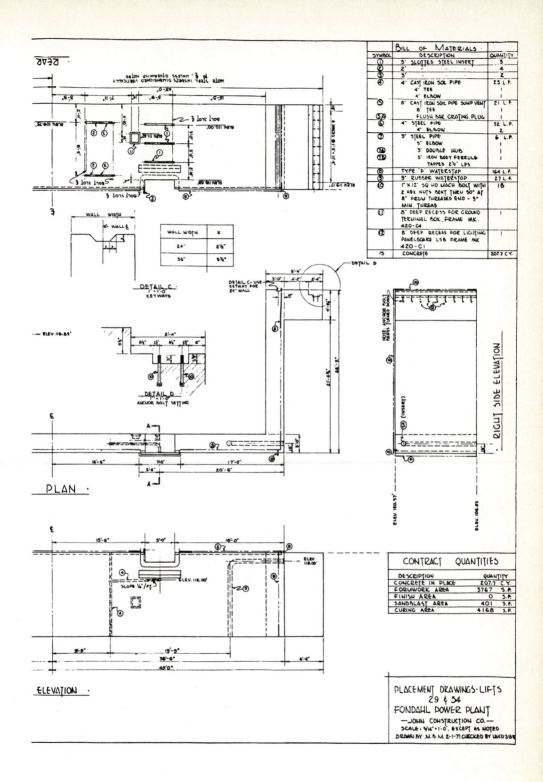

BILL OF MATERIALS

SYMBOL	DESCRIPTION	QUANTITY
①	3" SLOTTED STEEL INSERT	3
②	2"	4
③	3"	2
④	4" CAST IRON SOIL PIPE	23 L.F.
	4" TEE	1
	4" ELBOW	1
⑤	8" CAST IRON SOIL PIPE SUMP VENT	21 L.F.
	8" TEE	1
⑤A	FLUSH BAR GRATING PLUG	1
⑥	4" STEEL PIPE	32 L.F.
	4" ELBOW	2
⑦	3" STEEL PIPE	6 L.F.
	3" ELBOW	1
⑦A	3" DOUBLE HUB	1
⑦B	3" IRON BODY FERRULE TAPPED 2½" LPS	1
⑧	TYPE "F" WATERSTOP	164 L.F.
⑨	9" RUBBER WATERSTOP	27 L.F.
⑩	1"×12" SQ HD MACH BOLT WITH 2 HEX NUTS BENT THRU 30° AT 8" FROM THREADED END - 3" MIN. THREAD	18
⑪	8" DEEP RECESS FOR GROUND TERMINAL BOX, FRAME MK. 420-C4	1
⑫	8" DEEP RECESS FOR LIGHTING PANELBOARD LSB FRAME MK 420-G1	1
13	CONCRETE	207.7 C.Y.

REAR

NOTE: STEEL INSERTS DIMENSIONED VERTICALLY TO ₵ UNLESS OTHERWISE NOTED

WALL WIDTH

WALL WIDTH	X
24"	2½"
36"	3½"

DETAIL C
KEYWAYS

DETAIL C - USE KEYWAY FOR 24" WALL

DETAIL D

DETAIL D
ANCHOR BOLT SETTING

PLAN

ELEV 118.85'

RIGHT SIDE ELEVATION

ELEVATION

SLOPE ¼"/FT.

CONTRACT QUANTITIES

DESCRIPTION	QUANTITY	
CONCRETE IN PLACE	207.7	C.Y.
FORMWORK AREA	3767	S.F.
FINISH AREA	0	S.F.
SANDBLAST AREA	401	S.F.
CURING AREA	4168	S.F.

PLACEMENT DRAWINGS·LIFTS
29 & 34
FONDAHL POWER PLANT
—JOHN CONSTRUCTION CO.—
SCALE: ¼"=1'-0", EXCEPT AS NOTED
DRAWN BY M.B.M 2-1-71 CHECKED BY

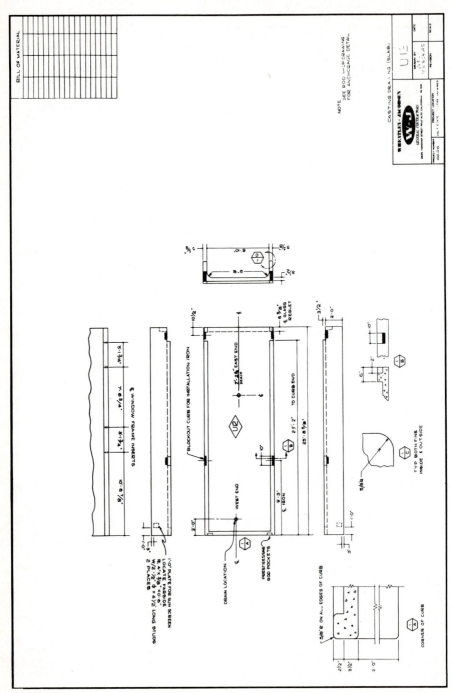

FIGURE 5-6
Block or task drawing for a precast concrete floor panel. (*Courtesy, Wheatley, Jacobsen, Inc.*)

FIGURE 5-7
Photograph of a model of a wastewater treatment plant. (*Courtesy, Bechtel Corporation.*)

that all elements are accessible for later repair or changes and to use for training operating personnel. The design model is customarily shipped in its entirety to the construction site for use by the constructor and later by the operating staff.

Models constructed for the sole use of construction are less common. When employed, they have invariably proven to be effective in saving time and money. Figure 5-9a is the photograph of a preconstruction model made by the project manager constructing a precast building for the Wheatley, Jacobson Co. Figure 5-9b is a photograph of the completed building frame. The model, constructed of balsa wood, had a separate piece for each precast element. Working with this together with a model of the crane, the manager and his staff were able to simulate several schemes for erecting the building and determining the least costly procedure. In addition, by dissembling the model piece by piece they planned and laid out the fabrication sequence and the casting yard to fit the erection plan. In this instance, an estimated $30 of balsa wood and 50 hours of time for construction and manipulation of the model were involved. An unexpected bonus was that after studying the plan, the erection subcontractor lowered his bid by almost 50 percent. Furthermore, there were unmeasured savings in more efficient and orderly job execution and shortened job duration.

In a number of instances, constructors of complicated structures, dams, and power structures have used models to good advantage. Figure 5-10a and b are

FIGURE 5-8
Photograph of a model of a small element of an industrial plant (*Courtesy, Engineering Models Associates, Inc.*)

photographs of models of the Wells Hydrocombine project on the Columbia River. These were made in individual small pieces, each of which represented a single concrete placement. The model was used not only for overall planning, but also for detailed staff discussions of the sequencing and working arrangements for placing the individual blocks of concrete.

In instances such as these, it is easy to determine the cash outlay to construct a model. Unfortunately, savings resulting from its use occur in many ways that are not easily measured or recorded. In each instance cited here, management at all levels agreed on the great value of the physical model as a tool both for planning and communications, and the consensus was that the model paid for itself several times over. Even so, because records of savings in hard dollars are almost nonexistent, many construction managers who have not used models are unwilling to try to find out about their value.

Henderson (op cit.) cites one owner who estimated the cost and gains from using physical models in a somewhat different way. He states: "If you elect not to model, add from one to two percent to the total field cost and ten percent to the piping cost alone." Another study reported by Henderson shows that 2.2 percent of the cost of construction was saved by use of the model on the job site. Some of the nonquantifiable benefits reported are that workers can more easily understand and read the three-dimensional model than the standard sheafs of hundreds of drawings, that superintendents and foremen

FIGURE 5-9a
Cafeteria building constructed of precast concrete elements, preconstruction model.

FIGURE 5-9b
Cafeteria building constructed of precast concrete elements, photograph of building after concrete elements were erected.

can plan their work more effectively and more quickly around the model, and that erection sequences are easier to plan.

Physical Modeling by Computer (CAD)[8]

A logical evolution of physical models ties the gains from using physical models and computers together in three-dimensional computer-aided drafting and design (CAD) models. After the physical details of the structure or plant have been entered into the computer, one may call up any orthographic view and rotate it in any way. In addition, construction equipment can be put on the screen to demonstrate erection sequences or how blocks or cells may be ma-

---

[8] See Chap. 13 for a discussion of several other computer applications.

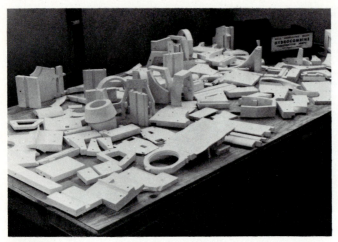

FIGURE 5-10a
Model of a portion of the Wells Hydroelectric Project on the Columbia
River, disassembled pieces. (*Courtesy Bechtel Corporation.*)

neuvered to simulate erection. Although having these computer applications available for superintendents or foremen at the job site may not be feasible, they can play a valuable role at the constructibility review stage, where construction people evaluate preliminary designs, as discussed below.

Another gain offered by computer-aided design models is that takeoffs of materials by types and quantities can be made directly from the model and tabulated and summed in the computer. These are very valuable for both estimation and procurement. Furthermore, if the computer model is accessible, field people can develop a material list from the screen and, if purchasing and inventory records are adequate, find the status and location of the items necessary to do a given block of work.

In summing up this discussion, because computer-based modeling has the capacity to make planning easier and more effective, it is another construction-management tool that is here to stay. But its effectiveness as a complete replacement for physical models is still debatable, since visualization or the actual structure may be more difficult on the computer screen than on a model, particularly for field personnel.

EXAMPLES OF PREPLANNING

As noted earlier, construction in all its aspects involves many sorts of operations and many skills. It is, therefore, impossible to give more than samples of successful applications of preplanning. Consequently, in practice, details of preplans of importance to individual contractors must be developed by them to fit the sorts of operations they carry out.

FIGURE 5-10b
Model of a portion of the Wells
Hydroelectric Project on the Columbia
River, fully assembled model.
(*Courtesy Bechtal Corporation.*)

A few examples of preplanning, some involving off-site and some on-site activities, are given in the following paragraphs.

Preplanning to Improve Constructiblility

Constructibility is defined by the task force on constructibility of the Construction Industry Institute (CII) as follows:

> The optimum integration of construction knowledge and experience in planning, engineering, procurement, and field operations to achieve overall project objectives.

Efforts toward constructibility come about because of a management awareness that planning, design, and construction are in many ways separate disciplines. Even the best people in one field cannot know the other disciplines in great detail. This concern has been given increasing attention following publication in 1982 of *Business Roundtable Cost Effectiveness Study Report B-2* titled "Integrating Construction Resources and Technology into Engineering."

Research on constructibility has been an important early activity of the Construction Industry Institute and its task force on the topic. Its July 1986 report, *CII Publication 3-1,* titled "Constructibility, a Primer," outlines its aims and gives several examples of past efforts in constructibility. More recently, CII publications 3-2 and 3-3 titled respectively "Guidelines for Implementing a Constructibility Program," July 1987, and "Constructibility Concepts File,'' August 1987, have been issued. (These and other CII publications are available from CII, 3208 Red River Road, Austin, TX, 78705-2650).

One of the research efforts sponsored by the constructibility task force was titled ''Constructability Improvement During Engineering and Procurement.''[9] It analyzed 335 ideas collected on a large refinery expansion project. The strategies developed to improve constructibility included the following:

- Make use of improved construction systems
- Simplify the design and/or combine elements
- Standardize design and/or repeat elements
- Improve information availability
- Improve information understandability
- Improve the construction sequence
- Improve the use of equipment or tools
- Improve constructor-designer communication

Methods employed to gain these improvements included an increase in the following:

- Required field supervision and planning
- Required shopwork (as contrasted with work-face activities)
- Equipment usage
- Tool usage
- Required design effort
- Required communication between engineer and constructor

Cost saving impacts included a decrease in the following:

- The likelihood of delays
- The amount of required direct construction manpower
- The duration of construction activity
- The amount of work at high elevations
- The amount of material required
- The likelihood of labor problems

The gains were not without costs. Among them were more manpower for design and procurement and more communication within engineering and between engineer and constructor, engineer and vendor, and owner and constructor.

The task force on constructibility has also sponsored a research project titled ''Constructibility Improvement During Conceptual Planning.''[10] This study reported the results of in-depth interviews with 68 construction and en-

---

[9] The underlying Source Document is CII publication SD 5. See also J. T. O'Connor, *Journal of Construction Engineering and Management*, ASCE, vol. 111, no. 4, December 1985, pp. 404–410, and, with Richard Tucker, *ibid*, vol. 122, no. 1, March 1986, pp. 69–82. For a discussion of the data-collection procedures, see these authors and M. A. Larimore, *ibid*, vol. 112, no. 4, December 1986, pp. 463–475.

[10] See CII Source Document SD. 4, March 1986, by C. B Tatum, J. A. Vanegas, and J. M. Williams. A condensation by Tatum has been published in *Journal of Construction Engineering and Management*, ASCE, vol. 113, no. 2, June 1987, pp. 191–207.

gineering executives in 32 participating firms involved in building, industrial, commercial, and public-works construction. In addition, penetrating studies were made of the conceptual planning activities on 15 projects and of the formal conceptual-planning programs of 8 firms.

This conceptual-planning study reported two types of benefits from construction involvement in early planning: (1) a better plan resulted because of increased support from those involved in design and procurement, and (2) design changes at the conceptual stage circumvented potential constructibility problems. The two areas in which these were most rewarding were site layout and selection of major construction methods. The study also outlined the following five types of actions that it found managers must take to implement successful constructibility programs:

1 Managers must develop, maintain, and inculcate a strong commitment to increased cost effectiveness. In other words, more construction for the money.

2 Managers must use constructibility to meet project objectives, that is, focus on ways to reduce costs and improve schedule and quality and keep these in mind when reviewing the design for function, reliability, and aesthetics.

3 Managers must bring construction aboard early, and this must be a major consideration with respect to overall project organization and the terms and conditions of all contracts. Furthermore, the qualifications and attitudes of the individuals involved are important; experienced personnel with a team approach are essential.

4 Planning must be done with passion. Only in this way can construction people make inputs to design and argue effectively for the construction viewpoint.

5 The designer must be open-minded and receptive, seeking construction inputs, putting project rather than design-firm priorities first, and playing team ball.

Research on another important aspect of constructibility sponsored by CII addressed the question of employing special construction methods off site or at least away from the work face. This involves the following three main approaches:[11]

1 Prefabrication. This is a manufacturing process, generally carried out in a specialized facility, in which materials are joined to form a component of a final installation.

2 Preassembly. This is a process by which various materials, prefabricated components, and/or equipment are joined together in a remote location to form a subunit of a complete installation.

3 Modularization. A module is created from a series of assembly operations conducted away from the job site or work face. It is usually the largest self-

---

[11] See C. B. Tatum, J. A. Vanegas, and J. M. Williams, *CII Source Document No. 25,* February 1987. Practices on some 10 industrial and 20 building projects were examined and analyzed in depth. From this came recommendations and guidelines concerning their use.

contained unit that can be transported and installed and is customarily fitted with all the structural elements and process components that will occupy its space in the completed facility. Examples range from a module that would be set into a bay in a building frame up to a complete oil collection and processing facility assembled on a barge and floated to the site at Prudoe Bay, Alaska, on the Arctic ocean, during the short time window when the North Shore is accessible by water.

Any of these three arrangements can lead to cost or time advantages. If they are to do so, they must accomplish one or more of the following:

• Mitigate the effects of such factors as adverse site locations or conditions or local area problems related to labor or competition
• Provide better use of specialized design, building, or process technology
• Offer indoor and possibly assembly-line conditions like those common in manufacturing
• Permit an accelerated schedule by compressing front-end activities and decoupling sequential operations into parallel ones by creating multiple work faces
• Reduce either the number of workers or the skills needed on site

Space in this book does not permit presentation of data about or descriptions of the many applications of these special construction methods. Employed properly, they can save both money and time and sometimes make projects possible that could not otherwise be carried out. But, as the research indicates, careful preplanning for them is essential because they do not follow many of the patterns established for on-site construction. Among the possible difficulties from a management point of view can be that (1) they place new and different demands on managers; (2) they may alter the project organization; (3) they may change planning and monitoring procedures; (4) they add new operations and change conventional ones; and (5) they require greater coordination. With respect to "people," they may create conflicts on site and among on-site and off-site personnel. In addition, if the procedures result in assembly lines or other dehumanizing arrangements, morale and possibly productivity may suffer.

Experience has shown that at times, some of the work that might be done as a part of the prefabrication, preassembly, or modularization operations can better be left for completion on site. One example might be applying and finishing wallboard. If done earlier, last minute alterations in plans for electric devices or plumbing connections may involve cutting new openings and repairing work put in earlier but damaged in storage or transit.

Without question, introducing construction thinking into design and early planning, including decisions to do some of the work off-site or away from the work face, is a valuable concept. However, to introduce and implement it in the traditional design-construct progression is neither simple nor easy. In addition to getting the commitments of time and money needed to carry through such plans, many present-day business and contractual practices, along with the prejudices and human-relations problems that accompany them, will have to be overcome. Some of these problems are addressed in Chaps. 1, 3, 10, and 11 of this book. CII Publication 3-2, July 1987, titled "Guidelines for Implementing a Constructibility Program," offers many helpful suggestions.

Preplanning the Work Site

Provisions for temporary facilities on small- to moderate-sized construction work sites just grow, like Topsy, with little planning or coordination among the parties who must have access at different times to various areas or places to store materials or machinery temporarily. Only where severe restrictions on access or space are foreseen will special studies be made. Too often, a quick sketch locates the first several buildings, with little thought given to the needs of the various parties for future access, storage, parking, or fabrication yards. Also, early installation of utilities, temporary surfacing of traveled areas, or installation of permanent pavement are not considered to be factors which will make the work easier for all concerned. On the other hand, site planning for very large projects with a multimillion-dollar construction plant is usually carefully thought out, with close integration of major production elements such as batch plants and associated aggregate- and concrete-delivery systems.

Preplanning the work site is usually straightforward, once the general work program and schedule are developed. Characteristics of the job itself will pinpoint the locations of major temporary facilities. For example, if a concrete batch plant, with provision for waste- and washwater disposal, is central to the operation, its location and space requirements will be given high priority. In other situations, primary attention will be given to space for storing raw materials or finished products and for housing equipment repair shops with adequate parking for equipment, fuels, lubricants, wire rope, tires, and unused equipment or a carpenter shop with form lofts and storage for lumber and finished products. With an adequate knowledge about these and other space-consuming activities, preliminary layout drawings for the various phases of construction can be made, followed by adjustments to them to give a workable overall plan.

For jobs that are widespread, the first consideration is to find the logical center of activity. Support facilities should be located so that total job travel is minimized. Although there may be several subcenters of activity, usually there is one central location that can serve continually as the site for job office and yard. Usually it makes little sense to locate these support activities for extended projects such as canals or highways at one end unless there is a heavy concentration of work that brings the center of activity nearby. Major decisions such as these can well be supported by comprehensive studies of the costs and operating constraints associated with the movement of personnel and materials during the life of the project.

Once the location of a centralized site is set and access points are determined, other controls become important, such as space and relative locations among the various activities, prevailing winds, and drainage patterns. For example, noisy and dusty operations should be downwind from office, repair shops, warehouse, and quiet activities. Wet operations such as concrete batch and aggregate plants and vehicle washing facilities should be located where their drainage will have little or no adverse effect on other activities or on neighboring property owners. A favorable location with easy access and dust control for employee parking lots is a must. A detail easily overlooked is that this parking must not be downhill from wet activities or downwind from dusty ones.

Temporary buildings and the adjacent areas needed for access and storage should be laid out to scale on topographic maps. Equipment, material, and personnel flow should be set out on them to ascertain possible congested and conflict areas during all phases of the project. Ample space should be provided for short-term parking and storage, together with a pattern of flow that will ensure easy access at all times to all storage. Individual storage access should be carefully planned to ensure the capability of first in–first out or free access and egress at all times.

A completely different set of problems in site planning for temporary access and storage exists, for example, for a high-rise building on a constricted downtown site. In this instance, the help of local authorities who control street space and traffic is essential. Then the needs of the succession of specialty trades for on-site work and storage space, and for time on elevators, hoists, or cranes must be developed and adjudicated. This form of preplanning is both important and time-consuming.

This brief discussion is intended to indicate that preplanning site layouts for temporary facilities is important and to cite a few examples of the many specific complexities that should be considered. As with so many other preplanning efforts, it is difficult to quantify their value in dollars saved. But once contractors have seen the advantages of attending to such details early, they almost always continue doing so.

Materials Management as an Element in Complete Preplans

The goal of a complete materials preplan should be to integrate, coordinate, and monitor all materials takeoff, purchasing, expediting, receiving, warehousing, distributing, testing, and even disposing of residual items. In the context used here, materials includes all the items that are needed by a project, whether part of the permanent installation or the consumables or reusables required to carry out the construction process.

Deficiencies in many of the present-day procedures for managing materials and a discussion of an overall computer-based materials-management scheme which addresses all the activities just listed and its accompanying benefits is offered in Chap. 13. With such a scheme in place and working well, many of the difficulties related to the materials aspects of on-site preplanning can be circumvented. Unfortunately, such a situation seldom if ever exists on present-day projects. With this thought in mind, the following paragraphs are offered to illustrate a few of the problems and possible solutions in the area of materials that frequently occur and that joint preplanning by the affected parties can help to alleviate. The discussion focuses on problems that can arise in purchasing, although preplanning can help in almost all other areas where materials are involved.

Case 1: Unless instructed to the contrary, purchasing agents rightly try to buy at what to their knowledge is the best price. In one instance, 2-inch planking in random lengths was procured for a roof with supports laid out at 4-foot centers. The price was $15 per thousand board feet less than that for specified lengths. However, considering handling, cutting, and waste, field costs in place came out to $25 per thousand board feet higher. Often, of course, buying cheaper grades of lumber in random lengths makes good sense to a purchasing agent, and may sometimes be logical. But the choice is wise only if the final cost per board foot in place is lower, which depends on whether the crews can slap the lumber into place or must sort, rehandle, cut, and even discard unusable pieces.

Case 2: Attention to shipment weight, bulk, and means of handling can be important and requires advance planning. As an example, many field sites do not have the kinds of unloading docks and materials-handling equipment that are commonplace at suppliers or at high-capacity freight depots. Consequently, a heavy pallet of machinery or parts that slipped so easily into a closed van must on site be jacked up and rolled to where field hoisting equipment can handle it. Related to this handling problem is the necessity to provide for lifting all heavy machines or vessels. If the designer specifies lugs or eyes for lifting, it almost always saves field time and cost. Note that in this instance preplanning should be the designer's responsibility.

Case 3: Planning to ensure that major purchases arrive at the job site when needed rather than leaving delivery at the vendor's convenience is often a good practice, so that costly storage and rehandling costs can be avoided. For example, pipe delivery might be specified to be "as directed by the contractor." Then it can be strung along the trench directly from the delivery truck. Again, large vessels or machines delivered when directed by the contractor can often be put into their final position without rehandling. One difficulty with a deliver-when-instructed procedure is that field managers often resist it; they worry when critical items are not on hand.

Case 4: Stipulating the form of packaging for major items can be cost effective. The costs of unwrapping those that are too well packed and disposing of the wrappings can result in significant expense that often can be avoided.

To repeat what may seem obvious but is too often neglected, materials and their handling are important controls over both cost and time in the field. But those who supply materials often lack detailed knowledge of the effects of their decisions on project costs. Extending preplanning to cover materials supply and handling can do much to eliminate such difficulties.

An Example of Step-by-Step Preplanning—The Acquisition and Installation of Equipment

To make detailed preplans for the acquisition and installation of a major item of equipment is time consuming and often plagued by uncertainties. Consequently, making such plans is unusual, even though there is strong evidence that if they are properly devised, cost and time on the project can be reduced and waste in resources minimized.

The following outline of a preplan (standard method) for the acquisition and installation of an item of equipment is given to illustrate the detail that must be included. It starts with a statement of purpose and scope and then proceeds through requisition, purchasing, receiving, storage, site preparation, preassembly, installation, and testing.

Outline of a Preplan for the Acquisition and Installation of Equipment

1 Purpose and scope of the effort to be undertaken
   a Describe the equipment itself and the major controls affecting its acquisition and installation
   b List any special considerations, such as lead times for particular or unusual equipment or tools that will be needed
2 Requisition and purchase
   a Specify grade, type, special requirements; provide drawings
   b If required, specify testing needed prior to shipment
   c Specify packaging instructions such as minimum or maximum package size or weight, pallets, lifting lugs, labeling, etc.
   d Establish firm delivery dates, including hours of day when delivery can be handled, or stipulate storage by vendor and delivery when directed
   e Specify delivery method—enclosed van, flatcar, trailers, etc.
3 Receiving and storage
   a Prescribe that delivery to job site is to be coordinated with need in order to preclude costs of rehandling and storage
   b Specify types of handling equipment and rigging needed, kinds of support, protective covers, etc.
   c If storage is required, specify best way and conditions
   d Give data covering special instructions or inspections
4 Preparation on site
   a State any special access conditions, roads, clearances, weight limits, etc., for hauling or handling equipment
   b Specify required dunnage for handling and placing equipment
   c Stipulate any preestablished conditions to be met prior to receiving or in-

stalling equipment (trench, paint, minimum strength of support, bracing removed, cleanliness, etc.)

d Clarify whether a special work area plan or arrangement is needed

5 Preassembly

a List items that should or might be prefabricated or preassembled and determine whether or not special work area is needed

b Detail items and supplies needed but not normally available for preassembly, including tools, jigs, etc.

6 Installation

a Provide step-by-step requirements for installation, including:

(1) Supplying diagrams, photographs, etc., that are relevant

(2) Establishing crew size, with specific individual assignments, in sufficient detail to preclude misunderstanding

(3) Supplying data on flow patterns and progressive assembly

b Give equipment and tool requirements, including reusable items

c Integrate safety instructions and cautions into the work plan

d Specify steps of the installation process in detail

e Give testing sequences and checks that are an integral part of installation, together with instructions for correcting elements that fail any particular test

7 Acceptance testing

a Give inspection test procedures to verify successful completion of elements or whole

b Provide forms to record and report inspections and tests

Preplanning On-Site Operations

Not all the items in the outline just given would apply when preplanning on-site activities. Rather the plan would follow an outline similar to the following:

Outline of a Preplan for an On-Site Operation

1 Description and scope of the task

a Describe the location, purpose, and parameters of the task

b Provide simplified preplan drawings

c List the equipment and special tools that will be used and describe their application and peculiarities

d Describe any and all safety procedures beyond normal daily practices that are required by virtue of the method, equipment, materials, or tools

2 Work plan

a Prepare sketches if needed to supplement preplan drawings

b List and give sources or locations of materials to be received or procured

c Describe in detail the step-by-step process for obtaining staff (crew size and skills), materials, tools, and equipment to accomplish task; use pictures or free-hand sketches to convey ideas or to show details of job-built jigs and devices; describe separately the processes that will be accomplished sequentially (for example, for a manhole: digging hole, placing concrete for base, installing prefabricated manhole elements, mounting cover, backfilling)

3 Inspection, testing, quality control

   a List sources of data, tables, and references or further detailed information on testing equipment, techniques, and methods

   b List applicable steps of inspection and testing

4 Provide get-ready instructions for the next task

Appendix B gives 10 examples of well-conceived preplans for on-site operations.

Assigning Management Responsibilities

If a project is to run well, responsibility for every on-site activity must be assigned to an appropriate and competent individual, one who can be depended on to carry out and report back the details of how and when an activity was accomplished and at what cost for labor and materials. On small simple jobs, responsibilities for a given activity or group of activities might be assigned to the same individual for the entire duration of the project. However, on larger, longer, and more complex projects, assignments should be reviewed at appropriate intervals.

Research by Victor E. Sanvido[12] has led to a workable and proved method for designing and analyzing assignment systems to determine their effectiveness. The basic premise underlying this method is that there is a proper level in the job hierarchy at which responsibility for each act or activity should be placed. He demonstrated that jobs on which most of these assignments had been made properly were profitable; those where most of the assignments were at the wrong levels lost money. The rules for selecting the level at which responsibilities should be assigned are as follows:

• If a single individual (craftsman) is involved in carrying out a particular assignment, responsibility for it should be given to that individual.

• If more than one individual is to be involved and affected by decisions, the supervisor (foreman) for that level should have the responsibility.

• If more than one crew is affected, responsibility is then assigned one level in the hierarchy above those directing the activity. For example, the responsibility for providing scaffolding for several crews would not be given to anyone among the foremen but placed with the superintendent, who is one level

[12] See "A Framework for Designing and Analyzing Management and Control Systems to Improve the Productivity of Construction Projects," *Technical Report 282*, Department of Civil Engineering, Stanford University, Stanford, Calif., June 1984.

above any of them. In instances where crews are directed by several superintendents, responsibility would then be above them. Only in that way can the person in control be impartial and deal with all users evenhandedly.

Preplanning the assignment of responsibilities for the various management functions first requires determining which functions to consider. Those chosen by Sanvido in his study totaled 19 in all. These are shown along the bottoms of Figs. 5-11 to 5-14. As shown on the figures, the 19 activities can be further classified into broader categories, such as tools, which in turn comes under the more general heading of resources. As a whole, the chosen items encompass most of the activities that will be found on a typical construction project. They clearly fit the four examples depicted by the figures: two high-rise buildings, piping for a refinery, and pipe hangers in a nuclear plant.

To follow Sanvido's model, one must think through clearly each function for which responsibility is to be assigned and determine the user in each instance. As already indicated, with only one user, that individual is also the controller. Where there is more than one user, the controller should be in the first management level above any of the users. These choices have been marked in the figures with heavy squares. For example, in Fig. 5-11, which was developed for a high-rise building, under the classification tools, the users of individual tools are the craftsmen, who are assigned responsibility for them, as the heavy square indicates. Responsibility for tools shared by a crew, although used by individual craftsmen, is given to foremen, since they are one level higher. In similar manner in the model, special tools used by several crews are the responsibility of the area superintendent.

In some instances, the model shows with the heavy squares responsibility for activities shared among two levels or classifications of management. For example, both the project engineer and area superintendent are involved in permanent engineered materials and area management. One would be primarily concerned with planning and the other with supervison.

Sanvido's approach can serve both planning and monitoring roles by permitting comparisons between the ways responsibilities should be and actually have been assigned on a project. The actual assignments found on each of the four projects are shown in the diagrams by an X. If the X markings fall directly on top of the squares, the assignment of responsibility is correct; but if they fall above or below, they are wrong. To illustrate, in Fig. 5-11, special tools used by several crews are shown to be controlled by individual foremen, which is inefficient and can lead to delays and may generate conflict and bad feelings.

Figure 5-12, developed by Sanvido from observing an unprofitable high-rise building project, shows even more clearly lapses in management planning. Not only is responsibility for some activities misassigned (for example, special tools), but responsibilities for several others are not assigned at all, as illustrated by "area look ahead" and "feedback to crews" and "feedback to individuals."

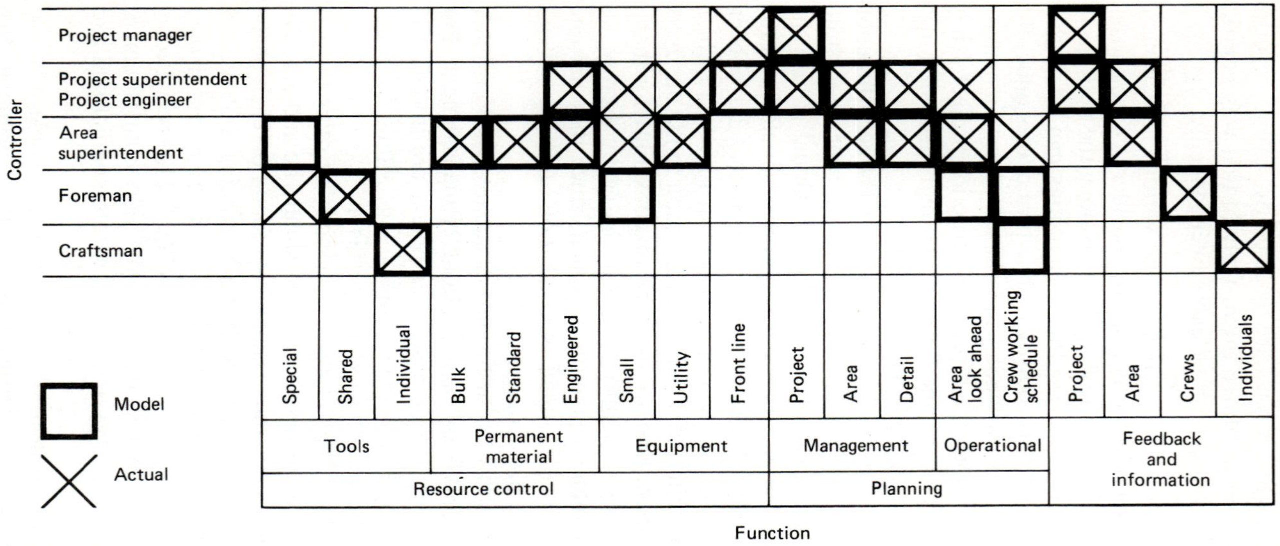
Controller
Project manager
Project superintendent
Project engineer
Area
superintendent
Foreman
Craftsman
Model
Actual
Special
Shared
Individual
Bulk
Standard
Engineered
Small
Utility
Front line
Project
Area
Detail
Area
look ahead
Crew working
schedule
Project
Area
Crews
Individuals
Tools
Permanent
material
Equipment
Management
Operational
Feedback
and
information
Resource control
Planning
Function

FIGURE 5-11
Controller-function chart for a well-managed high-rise building. (*After V. E. Sanvido.*)

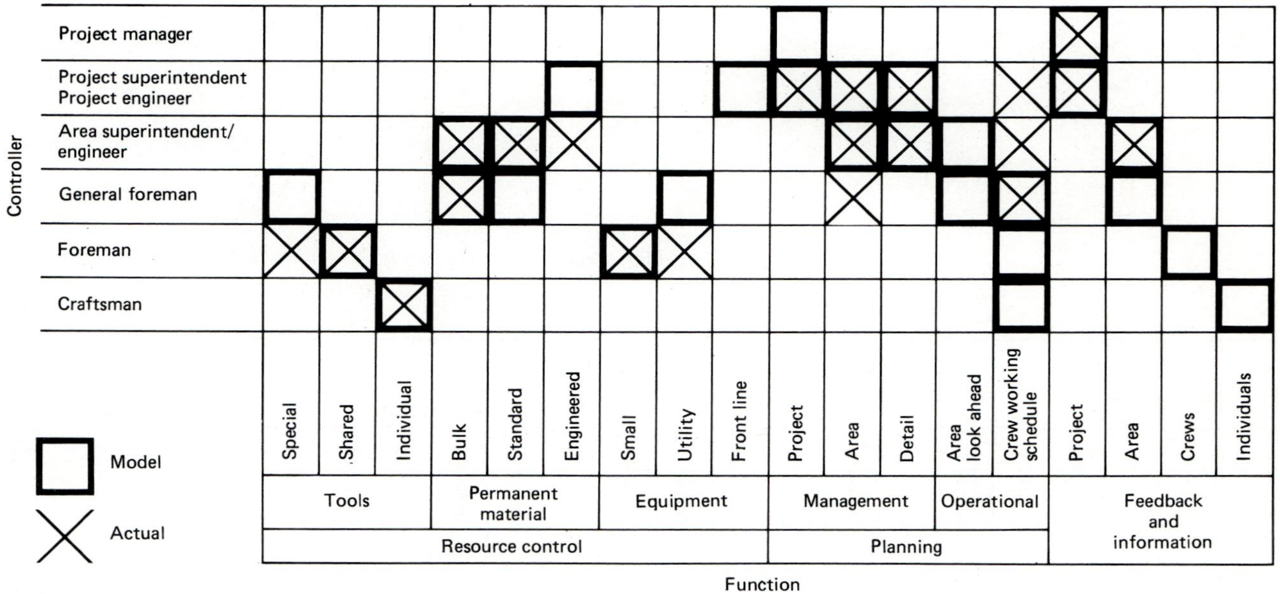

FIGURE 5-12
Controller-function chart for a badly managed high-rise building. (*After V. E. Sanvido.*)

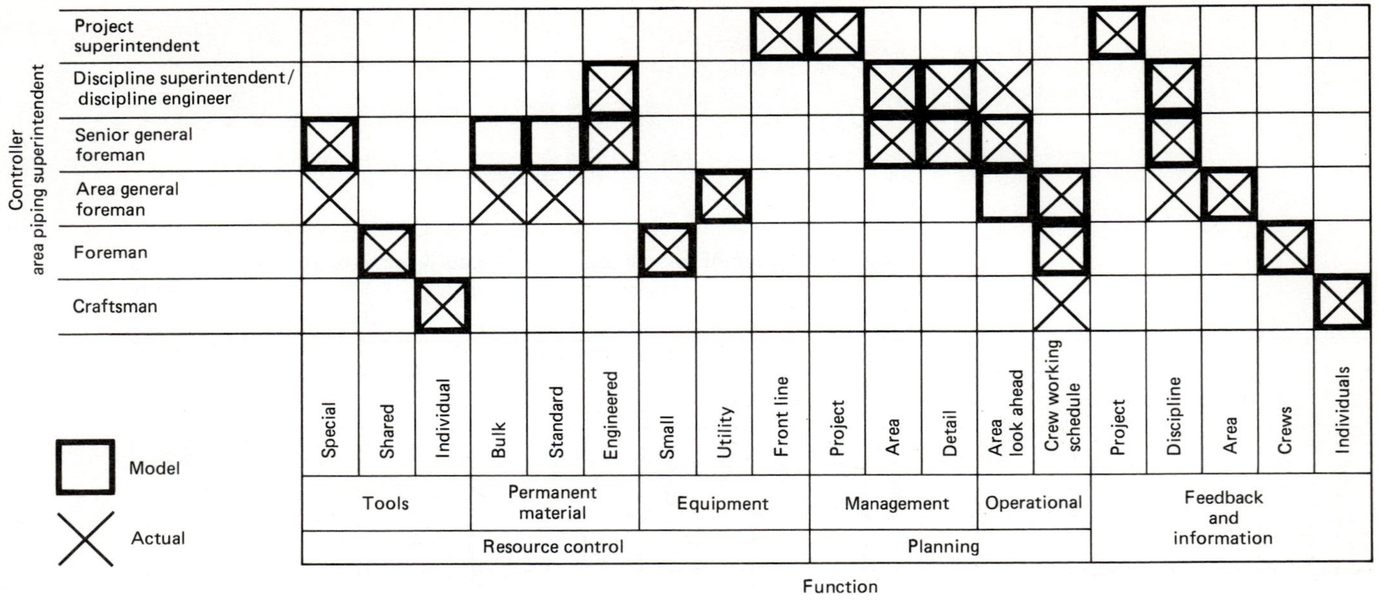

FIGURE 5-13
Controller-function chart for piping operations on a well-managed project. (*After V. E. Sanvido.*)

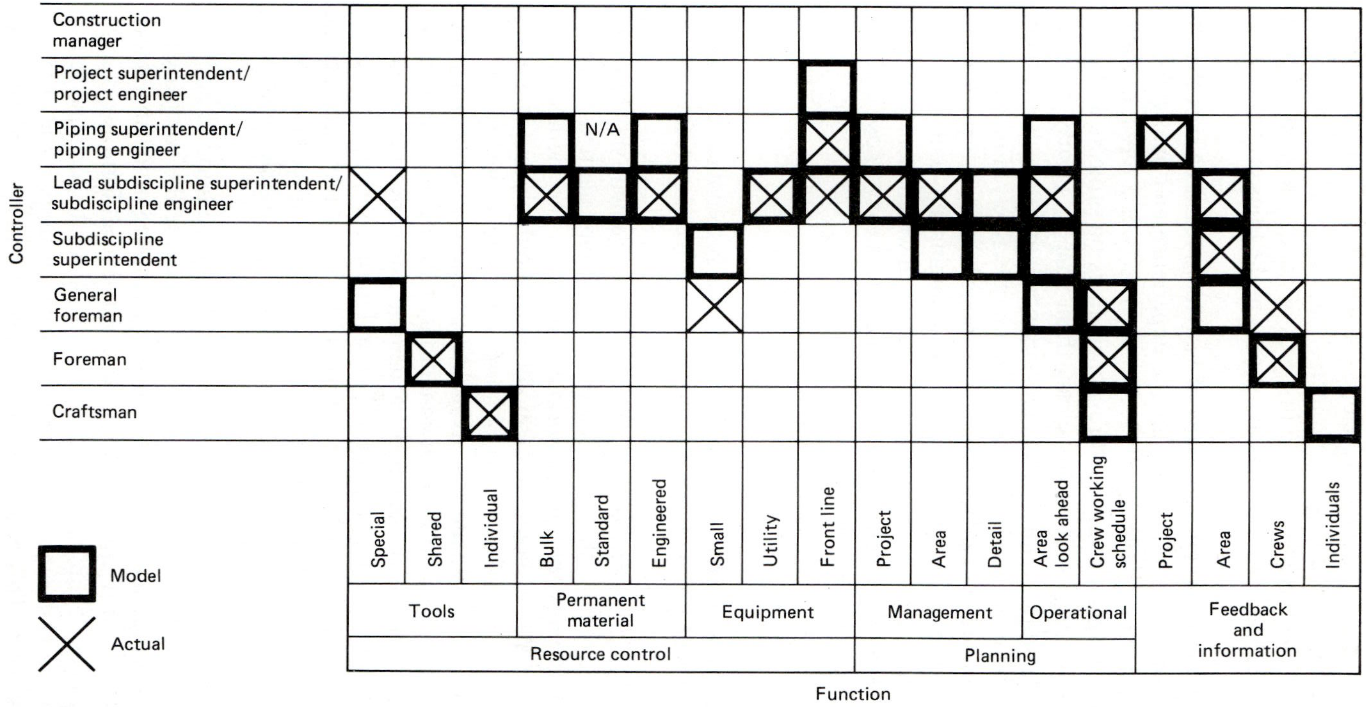
Construction manager
Project superintendent/ project engineer
Piping superintendent/ piping engineer
Lead subdiscipline superintendent/ subdiscipline engineer
Subdiscipline superintendent
General foreman
Foreman
Craftsman
Controller
N/A
Model
Actual
Special
Shared
Individual
Bulk
Standard
Engineered
Small
Utility
Front line
Project
Area
Detail
Area look ahead
Crew working schedule
Project
Area
Crews
Individuals
Tools
Permanent material
Equipment
Management
Operational
Feedback and information
Resource control
Planning
Function

FIGURE 5-14
Controller-function chart for installation of hangers for a large-diameter pipe on a marginally managed atomic power plant. (*After V. E. Sanvido.*)

As a part of his study, Sanvido developed a numerical scale for rating responsibility-assignment practices for projects. The details of it will not be given here. He also interpreted that scale in terms of good, marginal, or poor management practices, and these ratings are shown in the titles of the four figures.

One of the most important contributions of Sanvido's research was that of testing his model using his professional on-site experience and data gathered from site visits. He demonstrated conclusively that there was a direct correlation between how management assigned responsibilities on projects and their success, measured by profits. It seems clear, then, that by clearly defining management responsibilities for a project at its beginning and reassessing them from time to time, the work can be carried out more effectively. Furthermore, by evaluating how on-site management assigns responsibilities, it is possible to get an overall measure of its effectiveness. In summing up Sanvido's approach, it can be said that here is a preplanning approach which clearly defines how management responsibilities should be assigned. It is easy to use and one that management ignores at its peril.

Standard Methods

A natural follow-up to a company's preplanning efforts is to collect and make available for subsequent use methods that have been effective. A few examples from a large battery of such standard methods developed by one company is given in App. B. In providing such a collection, management should make clear that it does not require that the recorded or standard method be used. But it can suggest strongly that under normal conditions, the method which has been proved in practice should be carefully considered, since it is already known to be efficient and safe and to produce satisfactory quality. A preplanner should therefore evaluate any available standard method under the conditions at hand, modify it to fit local conditions, and submit it for evaluation along with any other likely schemes. A general company rule would be that the standard method would be used unless an improved method were developed. In this case, the improved method would be written up and become an addendum to or supercede the earlier standard method. Procedures for introducing these proposals could be one of the elements in a broader suggestion system, as described in Chap. 11.

Some of the standard methods given in App. B do not conform to conventional construction practices. They were substituted because the company felt that the conventional practices were more costly or resulted in unsatisfactory quality. To illustrate, in Exhibit B-6 for integrated wall-form work, the plywood-wale-strongback combination has no studs. Also, the long and strong direction of the plywood is running with the wales which support the plywood rather than across them. This conformation is cheaper and is permissible because in this particular application, plywood in its weak direction has sufficient strength and the deflection of the plywood between wales is not sufficient to adversely affect the final appearance of the wall.

Although very few construction organizations have developed a cross-reference file of standard methods, some are recognizing the tremendous waste inherent in starting every operation on each new project from scratch (reinventing the wheel) or relying on the chance that new innovative ideas will be transferred from project to project by management personnel. Standard methods are a partial solution to this problem. Another advantage of employing them is that schemes such as the design of formwork and the accompanying hardware can be standardized for company-wide use and possibly made available through the home office. Furthermore, the old argument that the development and publication of standard methods is too costly may no longer be valid. Computers are becoming prevalent on every large construction site, and using them, employees will have both standard designs and drawings available for the asking without burdensome paperwork.

To get good results from either standard methods or standardized designs requires (1) a commitment from top management, (2) sufficient money, put up early, (3) appreciable lead time, and (4) rethinking of office and field functions. The task will not be easy.

Job-Assignment Sheets

One of the simplest forms of preplanning is a job assignment sheet. It gives written rather than oral instructions to a worker and is intended to answer most questions before they are asked. At the same time, it provides an opportunity to clarify any uncertainties. After the task is completed, the sheet serves as a record for the foreman or superintendent. A sample of a preprinted job assignment sheet, prepared for a particular project and craft is given in Fig. 5-15.

Following is a partial list of the craftsman's or crew's questions that should be answered by the job assignment sheet.

1 Where do you want me to go?
2 What do you want me to do?
3 Whom do I take with me?
4 What are the dimensions, details, and positioning of the work?
5 What procedure should I use in accomplishing the job?
6 To whom and where do I go if I need answers to questions?
7 How about drawings? Are they needed? If so, are they available and are they accurate? On which sheet or sheets do I find the details I need to know?
8 What materials should I use and where are they?
9 What tools should I take with me?
10 When do you want me to go there?
11 Where will you be if I need you?

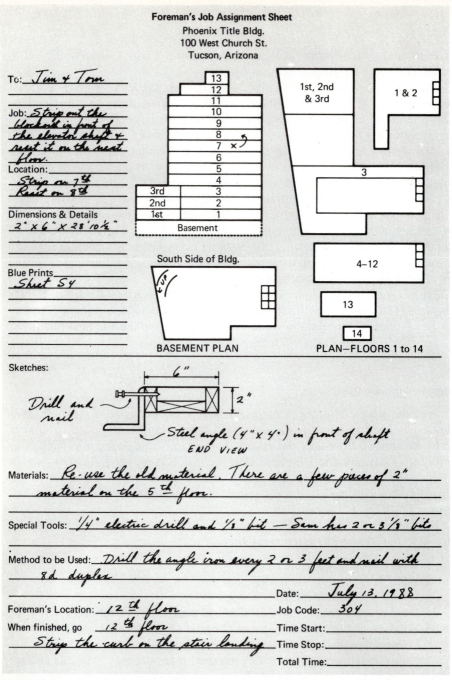

Foreman's Job Assignment Sheet
Phoenix Title Bldg.
100 West Church St.
Tucson, Arizona

To: _Jim & Tom_

Job: _Strip out the blockoub in front of the elevator shaft & reset it on the next floor._

Location: _Strip on 7th Reset on 8th_

Dimensions & Details
_2" x 6" x 28' 10½"_

Blue Prints
Sheet 54

Sketches:

Drill and nail

Steel angle (4" x 4") in front of shaft
END VIEW

Materials: _Re-use the old material. There are a few pieces of 2" material on the 5th floor._

Special Tools: _¼" electric drill and ⅛" bit — Sam has 2 or 3 ⅛" bits_

Method to be Used: _Drill the angle iron every 2 or 3 feet and nail with 8d duplex_

Date: _July 13, 1988_

Foreman's Location: _12th floor_

Job Code: _304_

When finished, go _12th floor_

Strip the curb on the stair landing

Time Start: _____

Time Stop: _____

Total Time: _____

Basement plan labels: South Side of Bldg. / BASEMENT PLAN / PLAN—FLOORS 1 to 14

FIGURE 5-15
Foreman's job assignment sheet for a multistory building. (*Courtesy, James Barber.*)

12 Where do you want me to go when I have finished this task?

13 What is the time allowed to do the task?

After the task has been completed, the supervisor will need to know the following:

1 What day did we do this job?

2 How long did it take?

3 What is the cost-reporting code and what do I charge to it?

Such an assignment sheet not only gives instructions but also serves as a record. Far more important, for a foreman to prepare it requires that the task be thought through in advance so that it can be done in the most logical, economical, and/or easiest way. Such smooth and economical operations are seldom possible when job assignments are little more than spur-of-the-moment verbal directions.

SUMMARY

This chapter has addressed several problems among the many that a preplanner encounters in prethinking a few of the many procedures and operations that make up on-site construction. It also briefly describes some of the tools that can be useful. It has tried to emphasize that to be successful preplanning must be a cooperative venture in which the knowledge and skills of field managers and the planner are pooled to make field operations run more smoothly, safely, and profitably.

CHAPTER 6

CURRENT PRACTICES FOR HANDLING AND EVALUATING WORK-FACE ACTIVITIES AND CRISES

INTRODUCTION

The earlier chapters on construction as a system and preplanning have dealt primarily with activities away from the work face that must be handled well if individual craftsmen or foremen and crews at the work face are to be productive. For example, Chap. 4, and particularly Figs. 4-1 to 4-4 show the many activities and indicate some of the interplays among them. Examples of preplanning in Chap. 5 and App. B demonstrate some of the preparatory work that goes on in getting ready for the actual activities at the work face. This section of the book (Chaps. 6 to 8) deals with the actual activities at the work face. Chapter 6 discusses how they are done now, Chap. 7, how data can be gathered on which to base possible improvements, and Chap. 8 how that information can be analyzed and presented in a manner that will make it understood, salable, and usable to managers and craftsmen alike.

   Those who have not been involved with the construction industry sometimes criticize its approach to work-face operations as being unplanned, disorganized, and based almost entirely on doing it the same way as last time, rather than by developing new approaches. They fail to recognize that, as pointed out in Chap. 2, construction, in contrast to manufacturing, involves several parties and performs custom work to create unique, one-of-a-kind designs. Moreover, its management must assemble an organization and work force of specialists which will disband when the job or a particular segment of it is completed. In addition, many of the multitude of individual tasks are done only once or a few times and are carried on out-of-doors where climatic influences can lead to uncertainty and difficulty. Under such circumstances, as well as the other pressures that construction managers are under, their rule of thumb approach and lack of attention to what are sometimes called niceties are understandable. Therefore, it seems wise to begin our discussion with a description of current work-face practices.

CURRENT PRACTICES FOR DEVELOPING WORK METHODS

Figure 4-2 summarizes the elements that management must bring together at the work face at a designated time if a construction operation is to take place. As shown on the left, management supplies specific resources. Then, as shown in the center, it arranges to have workers and supervisors in sufficient number and with the necessary skills on hand. If the appropriate resources are available, they accomplish the task. The method chosen depends on how, in the opinion of those responsible, these two can be fitted together most successfully.

To describe in detail the work methods or set the times required to accomplish work-face tasks for the many different kinds of operations that make up even a single construction project would be impossible. A quick rereading of the section of Chap. 2 titled "The Knowledge and Skills Demanded by Construction" can help to make this point clear. But looking at some of the factors that individuals who select methods must consider may be helpful. These are discussed next under the headings of characteristics of the task, resources available, site conditions, and relationships between the task and the labor force. In almost every instance, the most important control over method selection is the knowledge of experienced construction people coupled with cost estimates or engineering calculations when appropriate.

Characteristics of the Task

Hand in hand and intertwined with scheduling a designated task and lining up the necessary material, instructions, and human resources goes the selection of a work method. In a very simple case where the task is to dig a ditch of a particular width and depth between two stakes previously set by an engineer, the worker or foreman will probably choose the method and tools. In contrast, when a very large vessel in an industrial plant is to be set and aligned, the method and work assignments, which will involve several pieces of equipment and more than one crew, must be thought through in considerable detail. Either a manager will issue verbal instructions or a conference of supervisors will approve a detailed written preplan and augment it by adding the other specifics needed to flesh out the procedure. Even these actions usually do not cover all the steps, and field supervision must follow through by deciding the details such as the sequencing of operations or assigning individual tasks and tools or equipment to specific craftsmen or crews.

On any well-run project, where management and workers have common objectives, many details of a work method that is to be repeated several times will first be developed from the pooled experience of management, foreman, and crew. Suggestions for improvement will then be solicited, considered, and often adopted in subsequent iterations. This common practice is sometimes formalized under the learning-curve procedure discussed later in this chapter.

Resource Availability

Resource availability and characteristics will, in many cases, have a strong or controlling influence on the selection of the work method. For example, if an air-actuated spade and a compressor are available to ditch diggers, they will be able to loosen the soil more easily and quickly than if picks are the only applicable tool. Again, at the other extreme, procedures for setting a large vessel will be primarily dictated by the lifting ability of the available cranes or derricks that can enter the working space.

As mentioned earlier, good construction people are very resourceful and will often improvise to make up for constraints such as material shortages or the lack of the most satisfactory equipment. This can-do attitude typified by the Sea Bees in World War II is valuable and should be employed where it seems to be the only way out. But as emphasized throughout this book, such improvising carries with it penalties in increased cost, lost time, and sometimes dubious quality.

Environmental and Site Conditions

The choice of work method and sometimes the ability to work at all is heavily influenced by such weather-related factors as rainfall, temperature, wind, snow, and relative humidity. At times, jobs may be shut down because of rain or snow or temperatures so high or low that productivity or quality will be adversely affected. These conditions, in turn, may have a strong effect on labor's willingness or ability to work efficiently or even to work at all (see Chap. 9 for a detailed discussion). Often the terrain or access to a project site will govern the equipment, materials, or supplies that can be brought in.

An observer visiting construction sites is often appalled to see situations that seriously affect project efficiency and progress but could have been avoided. In most of these instances, management had not thought through the implications of site layout or weather in advance or had not taken sensible precautions. One of the most common difficulties is that of finding in the rainy season work shut down or vehicles plowing through or mired in the mud because drainage or hard surfacing was not provided. In other words, in selecting work methods, busy management did not devote sufficient attention to the far-reaching effects of environmental factors on work methods.

Some construction managers do make strong efforts to take site and environmental conditions into their planning of work sequencing and work methods. For example, as discussed in Chap. 9, winterizing buildings under construction in cold climates is becoming a common practice.

Relationships between Tasks and the Labor Force

Each individual worker or the foreman and members of a crew that has been assembled to work on a task brings to it certain skills. Matching these skills to the task to be done is a very important detail of any work plan. Where the work is done by union members or by classes of craftsmen on nonunion jobs, some of this matching may be controlled by established work rules, as between bricklayers and hod carriers or carpenters and laborers. But in specific situations, management (usually the superintendent or foreman) may give assignments to certain individuals or crews because they are more highly skilled or experienced or have some other valuable traits such as a desire or willingness to lead or work hard. Even the detailed work plan may at times have to be fitted to skill levels. For example, on large projects, certain carpenter crews having greater abilities may be assigned to doing finish work or assembling forms for exposed concrete surfaces, leaving the other tasks to those with lesser abilities.

Not only skill levels but individual and group goals and motivations may affect the work plan and the level of productivity associated with it. Personal goals as diverse as a determination to keep warm or cool, to avoid heavy physical labor, to escape exhaustion, or to keep from getting hurt or being a rate buster must be considered in making a work plan.

On job sites where management leaves the selection of the details of the task to the working crew (not an unusual pattern on busy jobs), the personal goals of the workers combine with other perceptions about the work situation to produce results different from what management may have anticipated. For example, when the work method is not precisely defined by management, the foreman or crew will usually select it based on the assumptions (right or wrong) that either makes about the situation. Either one may decide, based on past experience with management's unkept promises, that the probability is low of getting the necessary materials, tools, equipment, and support. In such a situation, management would be foolish to expect the crew to develop better methods or to have any sense of urgency. Furthermore, the crew has probably also learned that management will accept less than the anticipated rate of production, which has no doubt been the usual case. In contrast, if the conditions and arrangements make it possible for the crew to be productive and management's expectations are clearly stated, the probability is high that the crew's goals will be met. These physical and psychological factors are discussed in more detail in Chaps. 9 and 10.

It may be useful to summarize here some rules cited in Chap. 4, since they offer a means for judging whether a work plan will be successful or not from a human relations point of view. A work plan will be successful only if it meets all the following conditions:

1 It is clearly understood.
2 It can be accomplished both as a human task and in terms of physical arrangements.

3 All, supervisors and workers alike, believe that it can be done.

4 It is to the advantage of the group to which each of the parties owe their loyalties.

5 It is in each affected individual's personal interest.

To some managers, involvement with the workers as people rather than as an element in production may seem unnecessary or unwarranted, inasmuch as workers are there to do what they are told to do, from the managers' point of view. Today, this approach is considered to be naive. There is now a realization that lower-level supervision and craftsmen can make valuable contributions to efforts to find methods that are efficient, timely, and safe and will result in suitable quality. This can be achieved first of all by creating a climate in which workers feel free to make proposals for methods development either in an informal way or by means of some kind of organized program (see Chap. 11) and by providing meaningful rewards which can range from a pat on the back to more formal recognition.

CURRENT METHODS FOR ASSESSING WORK-FACE PRODUCTIVITY

As a rule, on-site construction managers look to the future—what is ahead tomorrow, next week, next month, or next year, if the project is that long—which means that once the task is completed, work-face operations and the methods employed to get them done are gone and forgotten. Neither do they make a practice of recording the methods employed in usable and detailed written form or on film or videotape so that the best ones can be used by their own or other crews in the future. In most cases, energies are focused on new operations. The tasks already underway, ones that are continuing, or those that are to be repeated again on the site will also get attention if, through the available measures, they are seen to be losing money or causing delays that will seriously affect the schedule. Generally, however, unless cues raise danger signals on site or in the home office, managers turn their attention to more pressing matters—of which there are many on all projects. Work-face activities often get little attention.

The cues or assessment methods that permit management to judge work-face activities and to find inefficiencies fall into three categories, covered in the paragraphs which follow under the labels of informal methods, formal methods, and learning curves.

Informal Assessment Methods

All construction managers from top to bottom of the field organization will tell you that they can judge how well a work-face task is being carried out merely by watching it for a short time, which could no doubt be true if they devoted their entire attention to this watching for long enough intervals to actually know in detail what was going on. But these same managers will tell you that they do not have that amount of time available. In reality, and despite management claims to the contrary, once an operation is underway, it will seldom get detailed scrutiny with respect to such questions as, "Are materials and tools available and suitable?" "Is the work procedure and its sequencing the most efficient?", "Is the crew of suitable size and makeup?" and "Have tasks been assigned among members of the crew in a way that best uses the available skills and keep all hands busy?" Answers to these kinds of questions can be obtained only by someone assigned full time to make such observations.

The charge that managers do not do a good job in appraising the efficiency of work-face operations is not unfounded. Many reports from work-study specialists support this view by demonstrating repeatedly that operations which field managers said were being done well and were on time and below budget were instead ripe for improvement. Usually these particular tasks were chosen for study by using one or more of the methods described in Chaps. 7 and 8 because management was not on the defensive about them as it might have been about operations in which there was anxiety about costs or schedule. When not feeling threatened, managers and crews were usually cooperative and interested. Given the study results which they had helped to generate, managers were usually willing to implement them, but customarily only after saving face by explaining that the operation as studied was not typical of the way the job generally operated. In other words, there was cooperation because no one was put in the position of being considered a scapegoat.

Even if managers, and particularly higher-level ones, observe operations carefully, there is the added problem that workers will do things simply to look busy that are not really productive. They and their foremen have learned that to appear to be idle angers management, so that it is far better to make work or hide. A cursory look will usually not reveal these behaviors, so that the observer assumes that all is going well. One of the pitfalls that those doing work sampling (see Chap. 7 for details) must guard against is that workers learn the rules by which they are judged to be working or idle and pattern their behavior to beat these rules.

For the reasons just cited, the informal methods that many managers employ to assess work-face methods and efficiency are weak at best and may at times be misleading.

Formal Assessment Methods[1]

On the usual projects, only two formal assessment methods—slippages in schedule and cost overruns—are used to alert managers to the fact that some thing is awry with work-face operations. If an important operation takes considerably longer than the time given for it in the schedule or if the reported unit costs overrun the budgeted amounts substantially, management is then

---

[1] For a discussion of the common methods for measuring construction productivity as they are applied to industrial piping, see H. R. Thomas, Jr. and C. T. Mathews, *CII Source Document 13,* May, 1986.

alerted. This assumes, of course, that data from the scheduling and cost systems are available, workable and working, and timely and come to the attention of management at the proper levels and at the proper times.

Numerous reasons can be cited to explain the weakness and sometimes the danger in relying almost exclusively on schedule and cost reports, among them are the following:

1 They are based on often misleading after-the-fact information, which means that for any but continuing or repeated operations, it often may be too late to put the findings to good use on the current project. Also, because subsequent jobs will be different and often staffed by different people, the reports may be of little use in planning work-face activities on later jobs.

2 They may be inaccurate. Costs are generally reported and charged against the various cost codes by the foremen at the end of each shift. These individuals may be tired and careless or may not care very much, so that the reporting and coding will often be wrong. Some companies assign this reporting task to a cost clerk or engineer to avoid such errors.

3 They may not provide the information needed to evaluate work-face activities. Among the reasons for this are (*a*) the way the data are broken down may be too coarse-grained to permit an accurate diagnosis of what the productivity problem is or how practices might be altered or (*b*) the reporting scheme may be ambiguous or not carefully monitored so that the recorded costs may understate or overstate the true situation for the item being examined. There is the common joke, but one that is often all too true, that if an item is running above the budget, charge some of its costs to OVERHEAD. It is for this reason that some cost codes do not carry an overhead item.

4 They can be unreliable, because of errors such as those mentioned above or data that have been deliberately falsified. As pointed out earlier in Chap. 2, falsification can be expected if management uses the cost and schedule reports to assign blame or establish rewards. Some contractors with computer-based reporting systems have built in special checks to detect fudging of cost records, such as special data-analysis programs to analyze the periodic (weekly or monthly) reports for wide variations in the reported unit costs or to plot trend lines for continuing operations. Readers should be reminded that reports such as these can be used for one of two purposes—either to get useful information or to reward or punish. To have both is impossible. Smart people will quickly detect management's intentions and respond accordingly.

The Learning Curve (Productivity Improvement through Repetition)[2]

Learning (experience) curves have been used in industry for many years to give a mathematical means for predicting or measuring improvements in pro-

[2] General references on learning curves include D. Karger and W. Hancock, *Advanced Work Measurement,* Industrial Press, New York, 1982 and A. Belkaoui, *The Learning Curve,* Quorum, Westport, Conn., 1986.

ductivity when a process or task is done over and over again. They offer graphical portrayals by means of which management's and labor's effectiveness on different but similar repetitive tasks can be predicted or judged. In general, complex and labor-intensive tasks free of external constraints such as shortages of space, workers or equipment are the best subjects for learning-curve applications.

Any person who has done the same task several times in succession under the same general environmental conditions recognizes that barring significant changes, a shorter time than when first done will be required to perform the task the second time. This continues at a decreasing rate with additional repetitions until there is finally a leveling off. The improvements are a result of several factors, including greater familiarity with the task, better coordination, more effective use of tools and methods, and, most important, more attention from management and supervision.

Some writers on learning curves try to make a distinction between experience and learning, but most authorities recognize no such difference. There are, however, two or three distinct phases that are recognized by most researchers. The first is that which takes place through an operational learning phase, when workers or crews are developing familiarity with the process. The second phase involves learning the routine, which leads to better coordination and minor site and environmental improvements. A third phase can result along with or after the first two if management and craftsmen make a concerted and continuing effort to improve by applying a combination of techniques, including those presented on this book.

Failure to improve output with successive repetitions, if conditions remain fairly constant and there are no external limiting factors, may indicate that there is a lack of continued attention or effort. In some cases, particularly near the end of the string of repetitions, productivity may actually fall off because of complacency or boredom or, in some instances, a desire on the part of those involved to make the task last longer.

Anticipated or actual improvements in output or decreases in cost can be described mathematically or graphically as experience, or learning curves. Mathematically, learning curves can be expressed by means of several different equations.[3] The one portrayed graphically in Figs. 6-1*a* and 6-1*b* and discussed here is based on the assumption that the percentage ratio between succeeding and preceding costs forms a geometric series with a common ratio of 2. Thus the percentage change in costs resulting between the fourth unit and the second, the eighth and the fourth, the fiftieth and the twenty-fifth, and so on, would express the reduction. Note that although the second and first repetitions have a ratio of 2, calculations are seldom based on the first unit because it rarely falls on the same curve as do succeeding units. In situations that

---

[3] H. R. Thomas, C. T. Mathews, and J. G. Ward, *Journal of Construction Engineering and Management*, ASCE, vol. 112, no. 2, June 1986, pp. 245–258, describe five of them and discuss their relative merits.

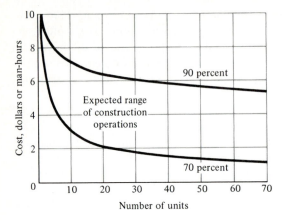

FIGURE 6-1a
Typical learning curves when plotted arithmetically.

are machine-paced so that there can be no increase in output, the learning-curve ratio would be 100 percent, i.e., no increase in productivity with successive repetitions. For complicated tasks without outside constraints, the improvement ratio might be as much as 70 percent. In this instance, the fourth unit would take only 70 percent as long as the second, the fortieth only 70 percent as long as the twentieth, and so on. This form of learning curve can be expressed mathematically as

$$r/s = (j/i)^n$$

where i = unit sequence number of earlier unit
$\quad\; j$ = unit sequence number of later unit
$\quad\; r$ = variable for unit i (measured in dollars or manhours)
$\quad\; s$ = variable for unit j
$\quad\; n$ = exponent that describes the variation

When the expression above is plotted arithmetically, the curve is hyperbolic; when plotted on graph paper with logarithmic ordinates and abscissa (double log or log-log graph paper), the curve becomes a straight line. Normally the log-log plot is used because straight-line comparisons are easier to visualize. The slope of the line of a log-log plot is expressed by

$$n = \frac{\log r - \log s}{\log j - \log i}$$

The relationship of the slope n and the percent cost curves as expressed by the above formula is calculated by the expression

$$\text{Percent of the curve} = 2^{-n}$$

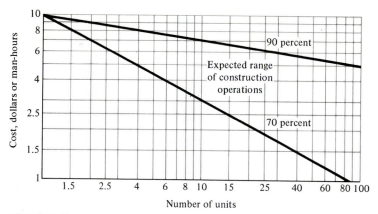

FIGURE 6-1*b*
Typical learning curves when plotted on logarithmic coordinates.

Relations between percent and slope are as follows:

Percent	100	95	90	85	80	75	70	65	60
Slope	0.0	0.074	0.152	0.234	0.322	0.415	0.515	0.621	0.737

Mathematical relationships such as these are only an approximation of the real-life situation. To place too much emphasis on formulas may obscure the true situation. But if an analysis of past activities of similar nature has established a pattern, it can be useful both in estimating and in analyzing performance.

Typical learning curves for construction fall in the 70 to 90 percent range. Figure 6-2*a* shows an arithmetic plot of the recorded manhours (a measure of costs) versus the number of generator units installed in a hydroelectric power plant. Such a curve is difficult to evaluate because of its changing slope. The same information plotted on log-log coordinate paper in Fig. 6-2*b*, shows the straight-line characteristic of typical learning curves. In this case, the slope of the curve is well described by a line joining units 2 and 8, and the slope for the best fit between these two units would be 67 percent, showing an unusually high improvement rate.

In Figure 6-2*b*, the plotted value of the hours required for the first unit falls below the best-fit line and has been ignored in drawing that line. This practice is common, recognizing as it does what is sometimes referred to as the B factor, which acknowledges that such an early change in slope is not unusual. In fact, three of the more commonly used learning curves exhibit such a hump.

The decreasing hours shown by a learning curve are predictable only when outside factors do not interfere with the continued production of units. When such changes are made, the predictability of continued gains is subject to serious error. In the real-life example plotted in Fig. 6-2*b*, the crew, after com-

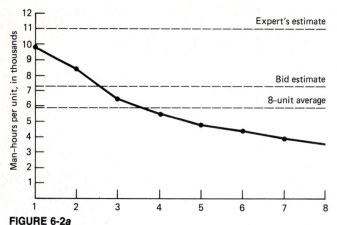

FIGURE 6-2a
An arithmetic plot of worker-hours expended for installing successive generator units.

pleting the eighth unit, was unable to maintain its momentum because of failures in the delivery of materials, lack of a prepared work site, and a partial change in personnel. This difference between the projected experience curve and the ascending points indicates the penalty suffered by losing continuity.

A second and recent example of changes in productivity with repetition occurred at the Baker Ridge Highway Tunnel built by the Guy F. Atkinson Construction Company for the State of Washington Department of Transportation. This case clearly illustrates learning-curve influences on productivity over time—in this case as a result of the combined efforts of management and workers.

This project involved a unique 63-foot-diameter soft-ground tunnel with a lining consisting of 24 tangentially joined small-diameter concrete elements (see the inset in Fig. 6-3 for a cross section). Each of these small tunnels was drilled separately with a mole and filled with concrete before the adjacent tun-

FIGURE 6-2b
A logarithmic plot of worker-hours for installing successive generator units.

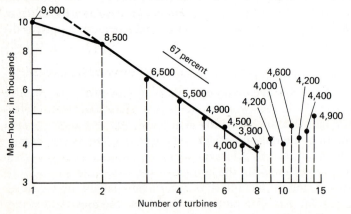

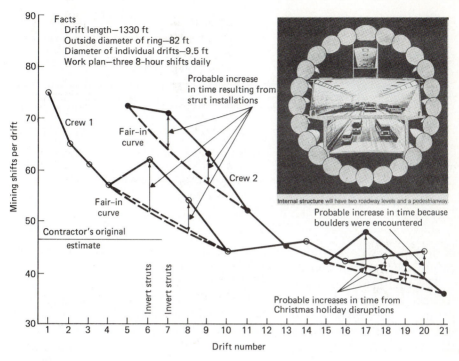

FIGURE 6-3
Excavation rates at Baker Ridge tunnel. (*Data courtesy Washington State Department of Transportation; inset courtesy ENR.*)

nel was excavated. After all the 24 tunnels had been driven and filled with concrete, the enclosed cylinder of earth was excavated to give a 63-foot-diameter tunnel which was fitted with a three-level automobile and pedestrian way. Each tunnel drift was 1330 feet long and was excavated in three 8-hour shifts per day, 5 days per week. The invert drift was completed first as no. 1, with odd-numbered drifts rising on one side and even-numbered drifts on the other. No. 24 was the "key" at the top.

Figure 6-3 shows for the first 21 drifts an arithmetic plot of the number of shifts required to excavate each drift vs. its drift number. The first crew excavated drifts 1 through 3; the second started drift 5 at the same time that crew 1 drove drift 4. The plot shows that progress was not always smooth. For example, the higher plotted points shown for drifts 6 and 7 reflect the interferences and additional times required to place invert struts that were required by the design. Other performance hitches were in drift 17, caused by a prolonged shutdown and loss of momentum at Christmas, and in drift 20 where boulders were encountered. Progress was slower in drifts 22 and 23 because of the fragility of the wall of earth between them, which was to be drift 24. Completion

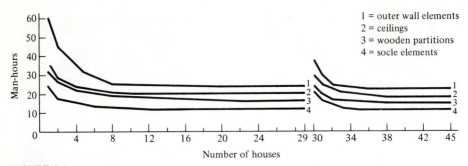

FIGURE 6-4
Effects of breaks in time on labor productivity in the erection of 45 one-family houses.

of drift 24 was slower still (90 shifts) because the varying space between the two adjacent already concreted drifts required hand mining.

The contractor's original estimate for average progress in mining was 85 feet per day; or 47 shifts per drift. Maximum progress achieved was 164 feet per day, which would equate to 24 shifts per drift. This demonstrates that, as with many tasks in construction, the rate of construction can be affected by the supply and/or functioning of any of the scores of elements that go together to make up a task. The fact that the curve seemed to level off at 42 to 44 shifts per drift and then plunge to 36 shifts might indicate only that at the earlier time there were glitches in the delivery of elements and that the crews were continuing to improve their systems and methods. Also, as shown by the plotted curves, crew 2 apparently learned from crew 1. Stated in percentages, crew 1 showed a 76 percent rate up to drift 10, leaving out the delays caused by the struts. Crew 2 had a rate of 67 percent for drifts 9 through 15 and 76 overall, when the effects of delays for the struts and holidays were excluded.

An early United Nations study[4] on the effect of repetition in construction summarizes a number of reports from various European countries and discusses the cost effects of repetition and the decrement in costs caused by interruptions to it. This report clearly shows that the more complicated tasks such as setting forms are more susceptible to improvement than are simpler tasks such as concreting and form stripping (75 percent for setting forms, 90 percent for concreting and form stripping).

The man-hour cost of interrupting an operation, based on a Finnish study as reported in the previously cited U.N. report, p. 105, is shown in Fig. 6-4. The report indicates that the duration of the interruption, the length of time that the operation had been going before interruption, the complexity of the operation, and the kind of work the crews do during the interruption will all have an effect

[4] See "Effect of Repetition in Building and Planning," *Doc. ST/ECE/HOU/14,* United Nations, Economic Commission for Europe, Committee on Housing, Building and Planning, 1965, p. 45.

on the decrement in cost and productivity. A Swedish paper in the U.N. report, p. 101, gives a theoretical analysis of the cost of breaks in the sequence of repetitive work. Figure 6-5 shows the conclusions.

Costs will not automatically decrease as repetitive operations are repeated, as might be suggested by this discussion of learning curves. The curves merely portray experiences resulting from the continued and dedicated efforts of management and workers to decrease costs. And this comes about only when the human climate of the project is conducive.

REACTING TO CRISES

Under the usual circumstances, if the performance measures just discussed show that a project is behind in schedule or over in cost but the situation is not extreme, the chances are that management will keep the job plugging ahead and accept and deal with the consequences that come with the delay or cost overrun. But suppose that the owner or the contractor's management, either

FIGURE 6-5
Theoretical analysis of the effects of breaks after the ninth and nineteenth repetitions of identical operations.

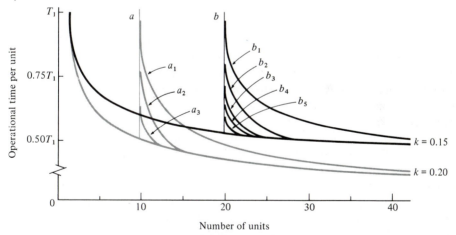

a = break between 9th and 10th units
b = break between 19th and 20th units
a_1 and b_3 = a 12-week interruption
a_2 and b_4 = an 8-week interruption
a_3 and b_5 = a 4-week interruption
b_1 = a 20-week interruption
b_2 = a 16-week interruption

k = slope of curve given as the
improvement from T_1 instead
of the percentage of T_1

on site or in the home office, decide that drastic action must be taken. What may then be the consequences?

Under these circumstances, the first to be blamed are usually the work-face people, foremen and crews alike. Certainly the records on schedule, cost, and learning will point at them, because the measures that management may use directly reflect their performance. Seldom are there firm data at hand to pinpoint the many other ways that things could have gone wrong, sometimes weeks or months earlier. Among the other possible but undisclosed villains can be changes in the national or local economy or other conditions, owners who have unrealistic expectations or who make frequent changes, designers who create almost unbuildable schemes or details, buyers who procure the wrong or inappropriate materials, home-office planners who specify permanent or construction equipment that may not meet requirements and that will be scheduled for delivery at the wrong time or arrive in damaged condition, the personnel department that supplies insufficient unskilled or improperly balanced numbers of workers, or job management that has set up a poorly scheduled or sequenced game plan. Is there any wonder under these circumstances, given the measures used, that critics blame poor or falling productivity on the work force? Probably the next most common scapegoat when things go wrong is job management, since it directed the work force; so it as well as the work force often carries the blame for factors far beyond its control.

Possibly the most common corrective action taken by higher management when projects are in trouble is that of firing the job managers, since the work force can seldom be replaced. The new management comes in and attempts to turn things around, usually without getting better results since they are faced with a myriad of problems but know little about their origins or what attempts have been made to solve them. There are, of course, exceptions; cases can be cited where the on-site representatives of owner, constructor, and sometimes labor were at odds, working at cross purposes and blaming the other parties. Their replacements, with the common objective of turning the job around, could possibly work together on problems of all sorts, including details of work methods, and secure good results.

In attempts to get a job back onto schedule, an all-too-common approach, often under pressure from owners, is to increase the size of the work force or the number of crews, by going to multiple shifts, or by increasing the duration of the work week with scheduled overtime. Usually these approaches have had disastrous results (see Chap. 9 for added discussion). One of the reasons is that the added responsibilities are piled onto an already overloaded supervision and it cannot use the added people to good advantage, so that costs pile up but progress is unchanged.

The lesson underlying this discouraging discussion is that to correct past mistakes in planning and selecting work-face methods is difficult at best. First of all, the present warning systems described in this chapter tell too little too late. Then, correcting past blunders is not easy. This effort is like closing the barn door after the horse is out.

SUMMARY

The first part of this chapter discussed briefly how constructors develop the procedures they employ in determining work-face methods and schedules. These procedures are primarily based on the knowledge and past experience of those who will direct the operation or actually carry it out. Continuing or repeated operations will at times be refined, given a suitable job climate and the ability and willingness of management to generate and implement changes. But once an operation is over, it is largely forgotten.

The second part of this chapter examines the informal, formal, and learning-curve methods employed today by some constructors to evaluate the productivity and timeliness of work-face activities. Unless used sensibly, these all have weaknesses from both a technical and a human-relations points of view.

The last section of the chapter reports on how construction management commonly reacts to crises, and the weaknesses of its approaches for solving them.

The authors of this book argue that there are ways to improve productivity that go beyond those in common use today. The chapters which follow outline and demonstrate some alternatives. At present, many of these approaches are little known to the industry. But those who make use of these helpful methods will testify enthusiastically that they will lead to improvement in productivity.

CHAPTER 7

DATA GATHERING FOR ON-SITE PRODUCTIVITY-IMPROVEMENT STUDIES

INTRODUCTION

A primary aim of this book is to show how on-site construction operations can be planned and done right the first time. This is the reason why construction as a system (Chap. 4), formal preplanning for on-site construction operations (Chap. 5), and current practices for assessing and handling work-face activities and crises (Chap. 6) have been discussed first. This chapter deals with data gathering for productivity improvement-studies, or, stated differently, how to gather data for improving on-site operations that are already underway. Attention to this added dimension of construction productivity improvement is absolutely necessary because construction involves many very complex operations and relationships. To quote the ancient proverb sometimes attributed to Homer, ''There is many a slip 'twixt the cup and the lip.'' This chapter and the one which follows deal with techniques for finding those slips and for correcting their damage, or at least mitigating their effects.

There are many ways to get information that can be helpful in productivity improvement. Two of the better ones which will be discussed in this chapter are:

1 Ask those who are involved.
2 Observe the process to develop factual records of how it is now being done.

There are no fixed rules about which of these two approaches with their many variations is better. Each has its advantages. But, even when the second, or data-gathering, approach is chosen, the first should not be ignored. Asking can often help in the detection of weaknesses or lead to suggestions for improvements in the support system and in work-face plans, as well as disclose the frustrations of those who are to carry out tasks but lack the essential elements to do so properly and effectively. Again, those asked may have personal concerns, and their having an opportunity to voice them may lead to valuable consequences not only for the individuals but for the project as well.

Observation is a valuable tool for gathering information on a work process or method, because the actions of people or the movement of materials or machines can be recorded for detailed study. It can answer questions about the what, when, where, how, and who of present practices. Only asking will explain the very important "why."

When either asking or observing to get data that can lead to productivity improvement, one must enlist the attention and concern not only of managers but also of every worker. The workers have the knowledge, the eyes, and the other senses that can appraise the operation and make valuable suggestions for improving it. Worker involvement can be accomplished through strategies such as informal discussions, questionnaires, and interviews, as discussed in this chapter, and by more formal organizations such as work teams (quality circles), as already discussed in Chap. 3, or suggestion plans, as described in more detail in Chap. 11. Unfortunately, in many construction organizations, enlisting ideas from craftsmen is frowned upon either because it takes time away from tasks that are to be done or because it is considered as an infringement on management's right to control the work. Whatever the reason given for neglecting this resource, the result is an outmoded approach that can cost dearly.

This chapter addresses the topic of data gathering under the following headings:

1 Statistical aspects of data gathering
2 Shortcut ways to improve productivity
3 Questionnaires
4 Interviews
5 Activity sampling

- Field ratings
- Productivity ratings
- Five-minute ratings

6 Recording present work-face practices

- Acquiring data from records
- Still photographs
- Time studies
- Recording with photographic or video methods

The rest of this chapter and Chap. 8 will discuss in greater detail the techniques used in organizing data gathering and analysis, identifying problems, and searching out and implementing improved methods.

In this chapter, the place of statistics is discussed first, since its principles, whether employed directly or indirectly, provide a basis for correct thinking and are of utmost importance.

STATISTICAL ASPECTS OF DATA GATHERING— WORK-SAMPLING PRINCIPLES

It is physically impossible to observe and record all the minute details of every repetition of any construction operation. When gathering data for productivity improvement studies, the aim is rather to have the observations approximate the reality that has been observed—but within acceptable limits. This is done by sampling. If it is to be relied upon, sampling must adhere to certain statistical principles and rules.

Sampling, as used for productivity-improvement applications, involves observing and classifying a small percentage of some whole to get a representation of that whole. The results of sampling can thereby form a basis for judgments about productivity problems, whether the sampling is done by questionnaire, interview, or observation. With a representative sample large enough to be statistically valid, a given characteristic involving an entire project or a single element of that project can be predicted. This prediction is not exact, but if the sample is representative and large enough, the results are close enough to the real situation to serve as a basis for analysis and possible action.

Fundamental to any sampling effort is the fact that as the number of observations increases, the accuracy of the prediction improves. But sampling takes time and costs money. Therefore, the desire for accuracy must be balanced against the time and cost of more complete sampling.

A simple illustration will be used here to show the reasoning behind the statistical theory of activity and work sampling, and also will acquaint the reader with some of the terms that will be used in the applications which follow. If, for example, a shipment of 4000 concrete blocks were to be inspected for cracks by a spot-check system, assuming that the whole lot of blocks was unacceptable if the first one examined proved to be cracked would be unwise. It would be equally unwise to assume that all blocks were perfect if the first block was perfect. Thus, it is evident that the number of acceptable blocks in this lot cannot be determined by looking at only one. In fact, the total number of good and bad blocks would not be known for certain until each one had been inspected. However, as more and more of the blocks are inspected, there will be increasing certainty that the relative number of good and bad blocks in the shipment follow the prediction offered by the sample.

The statement just made is, of course, based on the requirement that the samples be from a supply that has consistent characteristics, that is, (1) the condition of each unit inspected must be independent of that of any other unit, (2) each unit must have an equal opportunity or chance of being selected for inspection, and (3) the basic characteristics of the lot being sampled must remain constant.

Many examples can be given of sampling procedures that violate the rules just cited. For instance, if inspection were limited to blocks that had been dropped, there would probably be a larger proportion of broken ones than there would be in the complete shipment. Again, if only blocks from the top, bottom, or corners of the pallets were selected, one might find that the sample was not comparable to the entire lot, since these may have had a greater exposure to possible damage. And if the shipment were dumped from the truck after inspection, the characteristics would be changed. In general terms, for a sample to be representative, it must not reflect some special condition or situation that could have an effect on the characteristic being checked.

As pointed out, the larger the percentage of the total lot of blocks examined, the more closely the results from sampling will correspond to the actual situation. To express the degree of certainty between the results of sampling and the findings if the whole lot were examined, three statistical terms are used: confidence limit, limit of error, and proportion of the sample having (or failing to have) the characteristic that is being observed. The first two terms are purely statistical: the last is a physical condition and is noteworthy only because it affects the relationship between sample size and the values of the other two factors.

Confidence limit has to do with the dependability of the result. Thus, to say that the confidence limit is 95 percent is to say that purely as a matter of chance, the answer can be relied on 95 percent of the time or, conversely, that the answer may be wrong 5 percent of the time (one time in 20 determinations). The confidence limit is very often chosen according to the purpose of the sample: When human life is involved, the tolerable limit for acceptance may be 1 in 1 million, i.e., a confidence limit of 99.9999 percent. When sampling concrete blocks, on the other hand, to be wrong 1 time in 20—i.e., to have a confidence limit of 95 percent—may be acceptable. The higher the confidence limit (the less chance of being wrong), the larger the number of observations or samples that must be made. If certainty (i.e., 100 percent confidence limit) is required, every item must be inspected. Thus, in the concrete block example, the only way that one could be certain (have 100 percent confidence) about the quality of every one of the blocks is to check each of the 4000 one by one.

Limit of error expresses the accuracy of the estimated result. It is the percentage variation on either side of the value obtained by sampling within which the results of the true value can be expected to fall, given a prescribed confidence limit. Thus, it could be stipulated that the estimate of damaged blocks based on sampling will fall within plus or minus 10 percent of the total number of blocks that were actually damaged and that this result could be depended on 95 times out of 100.

Category proportion, the proportion of the sample that is expected to have a given characteristic, is a physical matter. It is the portion of the sample that is expected to have the characteristic being measured; it affects (mathematically) the size of the sample required to meet the preestablished confidence limit and limit of error. For the block inspection example, perhaps 1 out of 10 cracked blocks in each lot would be acceptable, so category proportions of cracked versus uncracked blocks are set at 10 and 90 percent, respectively. Again, if power shovels are doing productive excavating during 37 percent of

TABLE 7-1
SAMPLE SIZES (NUMBER OF OBSERVATIONS) FOR SELECTED CONFIDENCE LIMITS
AND CATEGORY PROPORTIONS*

Sample sizes required for 95 percent confidence limits						Sample sizes required for 90 percent confidence limits					
Category proportion percent	Limits of error, percent					Category proportion percent	Limits of error, percent				
	1	3	5	7	10		1	3	5	7	10
50	9,600	1,067	384	196	96	50	6,763	751	270	138	68
40,60	9,216	1,024	369	188	92	40,60	6,492	721	260	132	65
30,70	8,064	896	323	165	81	30,70	5,681	631	227	116	57
20,80	6,144	683	246	125	61	20,80	4,328	481	173	88	43
10,90	3,456	384	138	71	35 †	10,90	2,435	271	97	50	24 †
1,99	380 †	42 †	15 †	8 †	4 †	1,99	268 †	30 †	11 †	5 †	3 †

*As a practical rule of thumb, the product of sample size and category proportion should be 5 or more. The sample sizes marked with daggers (†) do not meet this criterion. The minimum number of observations for category proportion 10,90 percent should be 50 (50 X 10% = 5) and for 1,99 percent should be 500 (500 X 1% = 5).

their work period, the category proportions of productive excavating versus lost time would be 0.37 versus 0.63, respectively. The range of percentages of productive work usually found in construction-labor sampling is between 40 percent productive (60 percent unproductive) and 60 percent productive (40 percent unproductive). A category proportion of 50 percent requires the maximum sample size to attain a stated confidence limit and limit of error. Category proportions of either 40 or 60 percent result in sample sizes only 4 percent smaller. Thus category proportion is usually a relatively insignificant variable for most construction sampling, and it is easier (and slightly conservative) to consider the category proportion as a constant 50 percent.

As noted, within given values of confidence limit and limit of error, the category proportion determines the size of the sample to be examined. Consequently, a smaller sample is required as the category proportion falls away from the 50 percent–50 percent split in either direction. Figure 7-1 shows that for proportional categories of 40 to 60, the sample size is approximately 4 percent smaller; and that for 30 to 70, 20 to 80, and 10 to 90, the sample sizes are approximately 16, 36, and 64 percent smaller, respectively. A category proportion of 15 or 85 percent (i.e., 35 percent from the 50–50 split) requires only half as many items in a sample as with a 50–50 split for the same confidence level and limit of error.

Table 7-1 or the nomograph of Fig. 7-1 permits easy determination of the required sample size if the category proportion can be estimated in advance, or of the limit of error if the category proportion is different from that estimated. Normally, in sampling worker activity, the sample size is chosen using the category proportion of 50 percent; and no additional considerations are given to the proportion.

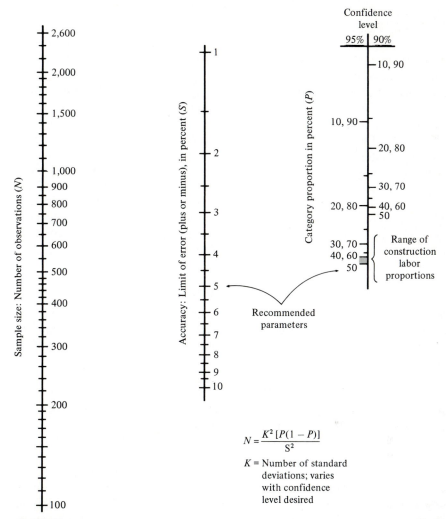

FIGURE 7-1
Nomograph for relating sample size, category proportion, and limit of error for confidence levels of 90 and 95 percent.

To illustrate the concepts just explained, assume that up to 10 percent of damaged blocks will be tolerated and that a confidence limit of 95 percent together with an answer (limit of error) accurate to plus or minus 5 percent will be acceptable. In this instance, the sample size must be at least 138 blocks. If it is desired to use confidence limits of 95 percent and a very restrictive accuracy (limit of error) of 1 percent, it then becomes necessary to check 3456 blocks. If the lot were smaller than this, 100 percent inspection would be required.

For sampling construction operations, there is a general industry consensus that a confidence level of 95 percent and a limit of error of plus or minus 5 percent give a good indication of the overall effectiveness of an organization or of an operation. As noted, the "working" proportion of activities usually falls within the range of 40 to 60 percent of the whole for most activities. Given these limits, and with the possibility that the split may be 50–50, the minimum sample size is 384. Since this is the maximum sample size required for any combination of working and not working with the stated confidence limit and limit of error, it is often cited as acceptable practice for observations made on construction sites.

The concepts given in this brief discussion of sampling apply to any situation where data are being taken or findings analyzed. In the context of this book, the emphasis is on data gathered from interviews, questionnaires, or field observations. Studied carefully, the information given will make clear that those who jump to conclusions based on sample sizes that are too small can be seriously in error.

SHORTCUT WAYS FOR FINDING IMPROVEMENTS

The formal steps leading to changes in construction operations aimed at improving productivity are (1) gathering sufficient data, (2) analyzing that data, (3) identifying the problem, (4) developing a new or revised method, and (5) implementing the new method. While this is the usual approach, experience has shown that much of the time this process is short-circuited and never completed because the problem is solved in the data-gathering, data-analyzing, or problem-identification stages. The problem has not disappeared or been resolved, but during the process of gathering and analyzing the information and discussing the situation, an understanding among those who are involved has led to a solution.

Intuitive and creative solutions during the data-collecting and analysis-and-discussion stages usually result from the communication they generate. In the hustle and bustle of most jobs, there is unfortunately too little time to communicate about job problems, to exchange information, and to develop cooperation among the people that should be involved. On most sites, unless there is a crisis, communication is limited to the minimum necessary to get one task underway and done and the next started. There never seems to be enough time to brainstorm about the current task or to discuss the fine points, even though these discussions can often make the difference between getting a job done somehow and getting it done a better way.

A classic example of the consequences of poor communication is illustrated by a situation a work-study specialist found on a concrete dam. A special wrench was required to attach and disconnect the large gang forms. There was only one such tool on the job site, and it was shared among three crews. Costs were very high and the schedule was slipping. Higher management had sent the specialist to the site to film the operation. Analysis clearly indicated what the problem was, and more tools were ordered. But when the foremen and crews were asked about why they had let the situation persist, they responded, ''No one asked us what was the matter.''

Lack of communication, which inhibits problem finding and solving, is not limited to that between individuals or crews but also that between specialties or divisions in a company, owners and contractors, contractors and engineers, architects and owners, engineers and accountants, unions and contractors, unions and unions, and suppliers and owners or contractors. And often the key to solving a work method or delay problem comes from an unlikely party, one who was considered to be outside the immediate situation or who was judged to be uninterested or uninformed. As an example, tasks sometimes have almost come to a halt because buyers or warehouse managers believed that management wanted them to be frugal. For this reason, a shortage would develop in item x because 100 had been purchased and issued while the budget called for only 75. This can happen regardless of what a given item costs. But the purchasing agent or warehouse manager may not know or possibly does not care that the failure to have enough of x when needed can cost the project thousands of dollars a day in direct, indirect, and ripple costs. Clearly, in this case, management communicated the wrong signals by making an issue of materials costs and by not making clear the costs of a scarcity. Most readers can substitute their favorite item for x; and no matter whether it is chuck keys, scaffolding, grinders, extension cords, concrete vibrators, hacksaw blades, welding machines, or water buckets, lack of such items can effectively shut down a job. In gathering data and discussing problems such as these, the solution will often become obvious and the problem will be corrected as soon as those who have the authority to solve it understand the situation.

There are other instances when the solution to a job problem is equally obvious and yet no change is made to effect an improvement. In some of these cases, there is a situation in which the authority of a power figure is threatened and the suggested change is blocked. If such a stone wall is encountered, using an indirect and judicious approach rather than confronting the individual directly may accomplish the same end.

Whether a problem is solved by going through the usual steps or by short-circuiting it, an essential element is mutual trust and respect by all parties. Also, for most people, change involves risk, which, in turn, can lead to resistance that must be overcome. (See Chap. 10 for added discussion.) Fortunately, in this regard, construction field people are accustomed to uncertainty, since this element is almost always present, and they probably will be more amenable to change than those in other industries.

QUESTIONNAIRES AND INTERVIEWS

One of the best ways to find the problems in an on-site construction organization or with individual operations is to ask those who are involved day after day. They are often knowledgeable, know what is going on, and have excellent ideas which they are anxious to share with management. The difficulty is that of devising ways to get the information from them. Questionnaires and interviews are two among several useful and relatively formal and organized strategies for doing so. As with any other approaches, these must be carried out skillfully and in a manner that is not perceived by those involved in administering or responding as threatening to their well-being. Also, the answers must be weighed to recognize that people respond on the basis of their perceptions of their situations and feelings. Some are satisfied and pleased with the conditions being appraised; others can be frustrated or unhappy with them. Some are old-timers; others are new hires. Some want to cooperate with management; others want or feel pressured to resist it, based on their own views of the work and the work environment.

Basically, questionnaires or interviews are employed because experience has shown that workers or foremen often have a better perception and greater knowledge of situations on site than does higher-level management, particularly if management's principal source of information is through a formal reporting system. The aim, then, is to tap that very important resource.

The problem with the usual formal and informal reporting systems instituted by management is that the traditional channels give management highly selective and often censored and filtered information that is biased by influences acting on both senders and receivers. (see Chaps. 2 and 10 for more detail). Usually, managers hear the good news—in full detail—from those reporting to them. The bad news, if reported at all, consists of inaccurate, fragmented accounts of failures that cannot be entirely hidden. Questionnaires and interviews, done and interpreted skillfully, can offer a useful means for gaining helpful and unbiased information from sources that fall outside these usual reporting channels. They can often help identify good or bad managers, management policies, job situations that support or constrain efforts to improve productivity, or conditions that are motivators or demotivators. Information such as this is basic if management wishes to provide a workplace where committed workers can be productive.

As explained in Chap. 2, written and oral on-site management information systems are structured to provide data that managers believe are important for staying on top of a project. The information is collected in the field, passed on to the office, summarized, and provided to management for distribution to others that management feels should have it. They, in turn, identify problems and when they find it appropriate, issue orders to lower-level managers to continue as before or to make specified changes in procedures or operations. Through these and other actions they communicate to the work force what management wants done, what it thinks is important, what it expects in the way of performance, and how that performance is recognized through formal or informal rewards or penalties. This, then, is the traditional management system, by means of which information that management says it wants, accurate or otherwise, flows upward; directives based on that information flow downward. The weakness in this system is that there may be no formal or informal way that workers or lower-level supervisors can send messages to management. On small jobs, no formal system is needed if management is perceptive; bosses are not far removed from the field, and workers can and do talk with them every day. However, on larger jobs this informal communication is impossible, and any messages from workers intended for the controlling level of management pass through many individuals. Those at each level screen them and modify the messages in light of their own interests. Thus, if the information reaches higher management at all, it has been filtered or censored to "cover the senders' fanny" or "please the boss" at the next level up.

A typical real-life situation wherein the formal information system failed involved a combination of tool shortages coupled with long waits to check out and return those that were available. This situation was common knowledge among the craftsmen and was unearthed by a consultant who had been chatting with them. In the follow-up, the consultant first queried the various levels of management. The project manager was unaware that any problem with tools existed; reports from the toolroom supervisor indicated that it was operating smoothly and under budget. The project manager guessed that the ratio between tools not available (turndowns) and tools obtained was about 1 in 40. Going down the management hierarchy, the job superintendent thought the ratio was about 1 in 20; the craft superintendent about 1 in 10; the foreman 1 in 5. A questionnaire filled out by the craftsmen said that about half the time they could not get the tools they requested. To determine the actual situation, the consultant made a detailed count at the toolroom over a period of several days. The actual tally showed 140 turndowns in 284 requests, a ratio very close to what the workers had reported and vastly different from the perceptions of the various levels of management. In sum, this was a situation in which management relied on a faulty information system which showed that all was well in the toolroom; it was operating under budget in terms of both personnel and tool cost, but nothing in the reporting system measured the actual costs of the tool shortage to the project with respect to delays and inefficiencies. Only through questionnaires did management find the true situation in time to take corrective action.

On projects such as these, the workers are often cynical about management's intentions and abilities. They resent being asked to work efficiently when, for example, (1) they do not have clear direction or adequate tools and materials, (2) they are sent to work stations that are not ready for them or are already occupied, (3) they often must redo work because of changes, (4) they are required to work in unreasonable sequences or using inefficient methods, and (5) to add insult to injury, they are called inefficient and lazy by management. Worst of all, they feel helpless, since there are no open channels of communication.

A system where questionnaires or interviews are required periodically can offer one way to ensure badly needed upward communication about job conditions. And if higher management insists that these surveys be carried out and the results reported, it not only will find inefficiencies and ways to correct them, but also will send a loud and clear message that listening to the work force is expected and important.

Questionnaires for Craftsmen or Foremen

If questionnaires are to bring solutions to problems such as those just discussed, they must be carefully designed to fit each situation. Among the criteria to be considered are their general purposes, the time that can be devoted to filling them out, and the attitudes and verbal abilities of the respondents. A variety of ways for getting information might be considered, from questions that can be answered by a check in a box to multiple choice or written statements. Regardless of the form employed, what is sought are the perceptions that the respondents have of their jobs and the conditions that surround them. Figure 7-2 is an example of a craftsman's questionnaire that is suitable for many construction projects. It should not be used without careful review to fit it to job conditions.

Questionnaires can be a valuable source of specific facts; and, if properly designed, they can also:

1 Raise questions that management should be asking itself
2 Identify the sources of inefficiencies and delays and possibly suggest ways to mitigate them
3 Provide job-to-job comparisons of methods, materials, management, and working conditions
4 Identify demotivating policies and their sources
5 Verify the accuracy of or raise questions about other management reporting procedures

Because the information obtained from questionnaires has not been filtered by those higher up in the job hierarchy, it can often be more trustworthy and revealing than that which comes up through the management chain.

Using questionnaires requires thought, planning, and careful introduction. Many lower-level managers will view them as prying and will resist their introduction and challenge their findings. For this reason, it must be made clear that the questionnaires will be used not to punish or discipline but to help make the job go better. Workers may also feel threatened, so confidentiality must be guaranteed, because if they think their statements will be disclosed and used against them, their answers may not be truthful. Also, feedback of some kind must be provided to both managers and workers. Action on problems disclosed by the questionnaires is an ideal way to demonstrate that the answers have value. On the other hand, if nothing happens, future questionnaires will probably not produce helpful responses.

There may be instances where replies to questionnaires show a wide diversity of opinions, but usually there is an amazing similarity among them. For example, there are often cases where almost all the workers of a particular craft working in the same area of the job-site report the same problems or

CRAFTSMEN'S QUESTIONNAIRE

This voluntary, anonymous questionnaire is a follow-up to a job-wide study being conducted on this project. The purpose of this study is to find areas where you think the job can be improved. The questions are designed to help us understand the nature of and problems in your daily work so that you can be better satisfied with it. In other words, we want to identify what you feel is good about your job and what you feel needs improvement.

We need your input to help us be better managers and only you can supply the needed information. Similar studies have been conducted on many other major projects and the information gathered has helped to identify problems on those jobs and we expect that it will help here too.

The questionnaire asks you to estimate the hours you lose in a week because of the delays you encounter in getting various resources. Please answer the questions from your experience on this project over the last two weeks.

Please do not leave any blanks in the numerical questions. For example, if you answer "Yes" to a question about a delay source, then place your best estimate of the hours spent waiting in the blank for the related number of hours lost. If you answer "No", place a zero (0) in the space for number of hours lost.

If you feel some important concerns of yours has been left out of this questionnaire, please add your comments. Use the back of the questionaire if you need more space.

We will be very careful not to disclose your individual identity, but general information about your job is necessary to organize the data logically and make it more useful to us. But we assure you that no one but we as consultants will see the questionaires. Please do not pull any punches! Thank you in advance for your help.

CRAFTSMEN'S QUESTIONNAIRE

1. Under what craft designation or jurisdiction do you work?

   _____ Carpenter   ____ Laborer   _____ Oper. Engr.

   _____ Instr. Fit'r   ____ Iron Worker   _____ Electrician

   _____ Pipe Ftr   ____ Other: specify_____

2. In what area do you work?

   Area A _____   Area B _____   Offsite _____

   Fab Shop _____   Other _____

FIGURE 7-2
Example of a craftsman's questionnaire.

3.   What kind of work does your crew do?

4.   What do you like best about your job?

5.   What do you like least about your job?

MATERIALS

6.   Do you often have to stop work and wait or move to another task because you don't have the materials needed?

_____ Yes   _____ No

7.   How many hours per week would you estimate you spend waiting for materials, getting materials from somewhere else, or moving to a different area because materials are not available at all?

_____ Hours per week

8.   In your opinion, what materials are in short supply at present?

TOOLS

9.   Are the tools supplied to you the best ones for the job?   Yes__   No__
     If the answer is No, What tools would you recommend?

10. Do you often have to stop work, wait, or move to another spot because you do not have the tools you need?

_____ Yes   _____ No

11. How many hours per week would you estimate you spend waiting for suitable tools, getting tools, or moving to a different area because no tools are available?

_____ hours per week

12. What specific tools are in short supply?

FIGURE 7-2 (*Continued*)

13.   What consumable items are in short supply?

       EQUIPMENT

14.   Do you often have to stop work and wait or move to another spot because you do not have the equipment you need?

       _____ Yes   _____ No

15.   How many hours per week would you estimate you spend waiting for equipment, getting equipment, or moving to a different area because equipment is lacking?

       _____ hours per week

16.   Which equipment that is not available would help you work more efficiently?

       SCAFFOLDS

17.   Do you often have to stop work and wait or move to another spot because you do not have scaffolds?

       _____ Yes   _____ No

18.   How many hours per week would you estimate you spend waiting for scaffolds, getting scaffolds, jerry-rigging scaffolds, or moving to another spot because you do not have scaffolds?

       _____ hours per week

19.   How could the material, tool, equipment, and scaffold situation be improved?

       WORK REDONE

20.   Do you often spend time doing completed work over again?

       _____ Yes   _____ No

21.   How many hours per week would you estimate that you spend doing work over again?

       _____ hours per week

FIGURE 7-2 (*Continued*)

22. Rank from 1 to 3 the items listed below that most frequently cause rework?

_____ Changed drawings or specifications.

_____ It was not done right the first time

_____ It was caused by work to be done or damage caused by other crews.

23. Please list other causes of rework:

CREW INTERFERENCE

24. Do you often have to stop work and wait or move to another spot because another crew has not finished or was moved in ahead of your crew?

_____ Yes _____ No

25. How many hours per week would you estimate that you lose because you are waiting or moving from one spot to another because of another crew?

_____ hours per week

OVERCROWDED WORK AREAS

26. Do you often have to work in an area that is so overcrowded with people that it slows you down from doing work as efficiently as you could have done under normal conditions?

_____ Yes _____ No

27.. How many hours per week would you estimate that you lose because an area is overcrowded with people?

_____ hours per week

INFORMATION

28. Do you often spend time waiting for someone to give you the information that you need to do your job?

_____ Yes _____ No

29. How many hours per week would you estimate that you spend waiting to get information you need to do your job?

FIGURE 7-2 (Continued)

_____ hours per week

30. What can be done to reduce delays because of crew interference, overcrowding, and lack of information?

INSPECTION

31. Do you often have to stop work and wait or not begin work because of inspection?

_____ Yes _____ No

32. How many hours per week would you estimate you lose because of waiting for or moving to another area because of inspection?

_____ hours per week

33. In your opinion, are delays because of the listed factors decreasing, staying the same, or increasing? (Check one for each factor).

	Decreasing	The Same	Increasing
A. Materials	_____	_____	_____
B. Tools	_____	_____	_____
C. Equipment	_____	_____	_____
D. Scaffolds	_____	_____	_____
E. Rework	_____	_____	_____
F. Interference	_____	_____	_____
G. Overcrowding	_____	_____	_____
H. Information	_____	_____	_____
I. Inspection	_____	_____	_____

WORKING CONDITIONS

34. How would you describe the morale and attitudes of the workers on this project?

Good _____ Fair _____ Poor _____

35. How do you rate the safety program and record on this job, as compared with other jobs

FIGURE 7-2 (*Continued*)

you have been on?

Good _____   Fair _____   Poor _____

36.   Can you suggest safety improvements that could be made on this project?

37.   Please list, step by step, what you would do in case a serious accident happened to a fellow worker?

38.   Where is the first aid station or other source of medical atttention?

39.   What improvements could be made in working conditions? Please be specific about such items as parking, toilets, drinking water, etc.

40.   What do you see as the most common reason that planned acitivities are interrupted?

41.   Do you receive all the information you need to do your job in the most efficient way?

42.   Are you given clear priorities and completion dates?

43.   When you ask, do you get what you need to do the work necessary to meet the given target costs and dates?

44.   What courses or training would help you do a better job?

45.   What courses or training would help your foreman?

46.   How well do you like working on this job? Circle one.

Very	I like	It's a	Not	Not
much	it	job	much	at all

EMPLOYEE RELATIONS PROGRAMS

47.   Do you read the __job__ newsletter? Why or why not?

48.   What do you like about the newsletter, if there is one?

FIGURE 7-2 (Continued)

49. What do you dislike about the newsletter?

50. What would you like to see more of in the newsletter?

51. Have you ever used the bus service that is available to the job? Why or why not?

52. Are you aware of the orientation program for new hires? If so, do you think it is effective?

53. Are you familiar with the Suggestion Box Program? Have you ever submitted a suggestion for job improvement?

54. What are your feelings regarding the Job Improvement Program on this project? Do you feel that it has been effective? Why or why not?

55. Do you have any other comments or suggestions? If so, please write them here.

THANK YOU FOR YOUR TIME AND EFFORT

FIGURE 7-2 (*Continued*)

shortages, which means that there is a very high probability that such a condition exists.

In construction, worker and foreman questionnaires have been used since the mid-1960s and have been found to be an effective means for getting information on particular subjects. They can be simple, requiring only 15 minutes of an employee's time, or longer and in-depth, which takes 45 minutes or more. On lengthy projects, questionnaires given to craftsmen are usually administered to large groups (say 10 to 50 people) at the same time, thereby requiring that an employee be absent from work for an hour or more. In every instance, an explanation of the purpose of the questionnaire is given, stressing that management is interested in what each person has to say; and that to protect the employees, their individual responses will remain confidential and only summaries of the replies released. A question period allows the workers to satisfy their curiosity and clears up any ambiguities. If there are people who cannot complete the questionnaire because they are not sufficiently fluent in the language, they are offered the assistance of a clerk or are excused. Any person who may feel uncomfortable should also be given the opportunity to leave. (According to those who have administered many questionnaires, almost no one leaves for this reason.) On large projects, questionnaires may be administered when there are special circumstances or, in any case, once or twice a year. On short-duration jobs, less formal ways of administrating questionnaires may be suitable and may provide valuable information.

The questions that are usually asked in craft questionnaires are specific for each job site and each craft. They generally consist of 50 or more short questions with blanks for the answers. Among the areas that might be covered are:

- Craft, area, type of work
- General level of satisfaction with work situation and supervision
- Provisions for personal needs
- Materials: availability, suitability, and condition
- Contractor-supplied tools: availability and condition
- Equipment: availability and condition
- Scaffolds: availability when needed, suitability, and safety
- Rework: amount, why necessary, and who directed that it be done
- Crew interferences: with what craft and reason
- Overcrowded work areas: cause and conflict with what craft or crafts
- Availability of instructions and other information: what is missing and why
- Inspection: reasonableness and whether or not done at proper time
- Hours per week lost from each of the above causes, by category.

Reports from those who administer questionnaires of this sort often find that workers are amazed to learn that management is really interested in their opinions and perceptions. In filling out the questionnaires, they often go to great lengths to explain their answers and to suggest helpful solutions to problems.

After questionnaires are collected, sufficient staff time should be made available to read and assemble all the information within a day or two. To avoid embarrassment to and possibly alienation of supervision, arrangements should then be made to review the results carefully with them and to develop a course of action based on the review. Soon thereafter, pertinent findings should be shared with the craftsmen or their representatives.

Another approach to worker questionnaires has been called craftsmen questionnaire sampling, whereby the foreman pulls some of the members of the crew together and interviewers question them about an activity in which they were engaged recently. These questions in short form will cover the same topics as those of the craftsman's questionnaire just described. The approach has the advantage of getting reactions while the activities are still fresh in the craftsman's mind. They are also far less time-consuming than are regular questionnaires. However, unless a highly favorable job climate exists, attempting such an approach may create serious frictions within the crew and/or between the crew and the foremen.[1]

Questionnaires for foremen generally cover topics such as those given in the craftsmen's surveys discussed above. Another possibility is to have foremen respond to the questionnaire provided to the craftsmen, since it forces foremen to think through questions that are important to members of their crews.

Foreman Delay Surveys

Foreman delay surveys serve a different and more specific purpose than that of questionnaires. They involve periodic reporting by foremen about factors that affect the productivity of their crews. A form for introducing foreman delay studies is shown as Fig. 7-3. Figure 7-4 shows a typical foreman's delay survey form for a single crew.

Depending on circumstances, foremen delay surveys can be conducted in several ways, such as: daily, with weekly reporting; daily for several weeks; or as-needed (for example, to measure the change in productivity before and after holidays). As with any other questionnaire, there must be an introductory session preceding its first use. Figure 7-3 is a copy of a handout distributed at the introductory sessions of a highly successful program.[2] General foremen and superintendents should attend this orientation session to preclude misunderstandings on their parts and to make clear that they support the procedure.

---

[1] For further discussion and comparisons of this procedure as well as craftsman's questionnaires and work sampling, see L. Chang and J. D. Borcherding, *Journal of Construction Engineering and Management,* ASCE. vol. 111, no. 4, December 1985, pp. 426–437, and vol. 112, no. 4, December 1986, pp. 543–556.

[2] See R. L. Tucker et al., *Journal of the Construction Division,* ASCE, vol. 108, no. 4, December 1982, pp. 577–591, and D. F. Rogge and R. L. Tucker, ibid, pp. 593–604 for a detailed description and report on the results of an extensive study done on two large industrial sites. These authors see foreman-delay studies as particularly applicable to large projects. They find them advantageous in the following ways: (1) fast and relatively easy studies to carry out, (2) one way to reduce delays, (3) a trouble-shooting device for projects that have serious productivity problems, (4) a way to identify and quantify obvious problems, and (5) an excellent communications link between foremen and higher levels of management.

FOREMAN DELAY REPORTS

I Introduction
  A *What is a foreman delay report?* The foreman delay report is a tool for management to help the foreman do his job. It is a simple daily account of problems that create delays in the work, and which the foreman may not be able to directly control.
  B *What does the foreman delay report do?* It makes sure the project manager has a chance to see what problems are causing delays for the crews. It puts a dollar figure on delays so that effort and money may be spent to fix the problems causing the delays.
II Using the foreman delay reports
  A You will be required to report only about five days per month, as directed by your superintendent.
  B *The delay report form is attached.* It provides a check-list to record manhours lost to common problems. Space is provided to write in other problems. When a report is required, you should put down the total manhours lost each day for each problem in the last column of blanks. The other blanks are provided as a worksheet if you want to use them.
  C *During the day,* try to notice every time anyone on your crew has a serious delay (more than 15 minutes). At the end of the day, record the total lost time for each problem on the report form.
  D *Foreman delay reports will work* only if you take a few minutes at the end of the day to think about your delays and to record them. Otherwise, they will be just another piece of paperwork.
  E The summary of your reports will show which problems are biggest across the site, and will put a dollar figure on the cost of those problems. Management can then justify spending effort and money to do something about them.
III Examples of problems to report
  A Your crew needs a chainfall. Because of delays in getting it, five men in your crew wait ½ hour each. Record 2 ½ hours lost for "Waiting for construction equipment."
  B Your entire eight man crew spends the entire 8-hour day reworking pipe spools because of off-site fab errors. You record 64 manhours lost for "Rework-engineering." While doing this rework, six men are delayed ½ hour each because the chainfall breaks. You record three hours lost to "Construction equipment breakdown."

IV Goals of foreman delay reports
  A *Short term:* Fixing of day-to-day problems by project management, when the cost of the fix is less than the cost of delays.
  B *Long term:* (1) Improvement in procedures based on foremen's comments. (2) Improved engineering packages as engineering personnel become more aware of their impact on construction costs.
V Why should I fill it out?
  A Now is your chance to talk to top management, directly.
  B Problems that are really bad can be fixed right away if possible.
  C Top management cannot do anything without the support of the foremen, so communicate your thoughts, ideas, even wishes back to them!
  D *You have got nothing to lose, but a few minutes of time.*

(*Source*: R. L. Tucker et al., *Journal of the Construction Division,* ASCE, December 1982, pp. 577–591)

FIGURE 7-3
Explanation of a foreman's delay study.

Date: _____   Name: _____

Number in crew: _____   General Foreman: _____

                            Foreman's name: _____

Problems causing delay	Manhours lost		
	Number of hours X	Number of workers =	Labor hours
Changes/redo (design error or change)	_____ X	_____ =	_____
Changes/redo (prefabrication error)	_____ X	_____ =	_____
Changes/redo (field error or damage)	_____ X	_____ =	_____
Waiting for materials (warehouse)	_____ X	_____ =	_____
Waiting for materials (vendor delay)	_____ X	_____ =	_____
Waiting for tools	_____ X	_____ =	_____
Waiting for construction equipment	_____ X	_____ =	_____
Construction equipment breakdown	_____ X	_____ =	_____
Waiting for information	_____ X	_____ =	_____
Waiting for other crews	_____ X	_____ =	_____
Waiting for fellow crew members	_____ X	_____ =	_____
Unexplained or unnecessary move	_____ X	_____ =	_____
Other: _____	_____ X	_____ =	_____
_____	_____ X	_____ =	_____

Comments: _____

_____

_____

FIGURE 7-4
Foreman's delay survey questionnaire.

These first sessions can last from 15 minutes to an hour and should be planned to allow questions and discussion. No more than 25 people should be in any one meeting, since larger groups are not effective when ideas are to be exchanged. The first survey should be completed the same day as the orientation session, to fix the procedure in all minds.

The primary purpose of foreman delay surveys is to highlight problems that are outside the responsibility and control of individual foremen. Findings derived from these questionnaires will often put management at higher levels in an unfavorable light, so foremen must be assured that there will be no penalty for honest reporting. Moreover, superintendents, general foremen, and, if possible, the project manager should be at the orientation meetings, so that they can put the foremen's minds at ease.

Where foremen delay questionnaires are employed routinely or for a period of time, those involved should be told when and how they will get the forms, when the forms should be completed, and how they will be collected. It is generally more successful to have them handled just as other foreman reports are, that is, at the same frequency and through the same channels as time cards, production and materials reports, and the like. The summary of responses should be shared with the foremen as soon as it is completed and discussed at a group meeting as soon thereafter as possible.

For foreman delay questionnaires to be most effective, each must be discussed line by line at informational staff meetings. Top managers should attend these meetings so that foremen can get first-hand responses to and action on their concerns and questions. This higher-level attendance is also important because it shows that management values the foremen's inputs and that something will be done with the information to help correct the reported problems. Insofar as possible, to show good faith, higher management should make a real effort to correct or begin to correct some of the items on the list.

If foremen delay questionnaires are filled out at the end of each day, when accomplishments and frustrations are still fresh in the foremen's minds, foremen can usually estimate with reasonable accuracy the time lost for various reasons. Thus, the questionnaires can provide a means for measuring and possibly improving performance and for quantifying time losses that occur in the daily routine.

A form summarizing the results of a foreman delay survey is given in Fig. 7-5. It is helpful if forms of this kind, fitted to a given project, list 6 to 12 of the most common reasons for delay in the kind of work being done. The foreman merely jots down the duration of the delay attributable to each cause and the number of men involved. The findings can be summed over the reporting period to give a very useful record and basis for corrective action. The difficulty will be to get the foremen to keep such records, because (1) they already feel overloaded with paperwork and (2) their fears, often justified, that if they report items that reflect adversely on higher level management, there will be unfavorable consequences for them. As with so many other approaches to productivity improvement discussed in this book, this one will be successful only when there is a high level of cooperation and trust at all levels of management.

Other Construction-Related Questionnaires

In addition to questionnaires developed to deal with project-related situations, others have had wide use as data-gathering devices for those who wish to use the results of research on construction-related problems. A recent study, de-

Project: _____   Date: _____

Superintendent/area: _____   Craft: _____

	Electrical	Equipment	Hangers	Instruments	Insulation	Ironwork	Mechanics	Paint	Total	Percent
Changes/redo (design error or change)										
Changes/redo (prefabrication error)										
Changes/redo (field error or damage)										
Waiting for materials (warehouse)										
Waiting for materials (vendor delay)										
Waiting for tools										
Waiting for construction equipment										
Construction equipment breakdown										
Waiting for information										
Waiting for other crews										
Waiting for fellow crew members										
Unexplained or unnecessary move										
Other: _____										
Other: _____										
Total:										
Percent:										

FIGURE 7-5
Summary of foreman's delay questionnaire.

scribed briefly here, illustrates such efforts. This study[3] develops a methodology for differentiating between more-productive and less-productive foremen working on industrial construction sites, based on their answers to a series of questions. The answers show that better foremen:

• Planned their work further ahead, often in their heads rather than by formal methods
• Ordered items such as materials, tools, equipment, and scaffolding earlier

[3] See G. J. Lemna, J. D. Borcherding, and R. L. Tucker, *Journal of Construction Engineering and Management,* ASCE, vol. 112, no. 2, June 1986, pp. 192–210.

• Were more honest in giving their crews information on where they stood with regard to the schedule

Another interesting finding of this study was a rank ordering by foremen and craftsmen of the important qualities of good foremen. The foremen's responses are listed here in their order of importance; craftsmen's ratings are in parentheses:

1 Has work well planned (2)
2 Gives clear directions (1)
3 Stresses safety (5)
4 Is good at motivating the crew (3)
5 Is willing to accept suggestions (4)
6 Is proud of the work and crew (6)
7 Accepts responsibility for work (9)
8 Plays no favorites among workers (8)
9 Has a lot of self-confidence (12)
10 Is creative when it comes to problem solving (7)
11 Makes decisions quickly (11)
12 Takes into account preferences of workers (13)
13 Lays out work quickly (10)
14 Is even-tempered (14)

From these ratings it can be seen that foremen and workers find common ground in the way that good foremen should perform, which is basically in an organized, humane, and fair manner.

Arranged Interviews

Arranged interviews, as contrasted with the informal discussions that are a part of every work situation, can be a very useful way to gather information or to gain more detail about topics such as personal concerns, productivity, quality, or safety. They can be one on one, involve several workers informally gathered at the work face, or include a group assembled for the purpose. In most cases, the interviewer has a specific topic or set of topics to explore and comes with a plan and a series of questions to get the desired information or to start a discussion.

Conducting successful interviews requires skill in communication and sensitivity to the concerns of the interviewees. Often at the beginning of the interview, it is necessary to let the discussion stray a bit away from the intended direction to dispose of the hidden agendas that so often exist. Such departures from the interview plan can be very useful. For example, a discussion set up to explore material or tool shortages might digress to disclose information about an organized theft ring. Again, a discussion on scaffolding might turn to one concerning safety.

Interviews can also provide a safety valve for individuals frustrated by job situations. For example, on jobs that are having troubles with productivity, individuals or small groups of craftsmen or foremen sometimes describe their difficulties with tears in their eyes. In such cases, the interview has given them an outlet to express the feelings that result when management actions or inactions prevent them from doing a good, personally rewarding day's work.

Interviewers should carefully think through the intended purpose or purposes of the interview and prepare a list of topics that they wish to cover. These may be much the same as those on a questionnaire dealing with the same subject (see Fig. 7-2).

Interviews take time; and since they are almost always conducted during working hours, their costs are highly visible, which often causes employers to resist them. And as is so often the case in productivity studies, the benefits are less easily quantified. For this reason, they must be used sparingly, at least until their value is demonstrated. They are also time-consuming for the interviewer, since only three to five can be done in a day, For this reason, they are more commonly used as a supplement to questionnaires to expand the data base on items that were reported or suggested by it.

The discussion here has dealt with interviews aimed at on-site or workface problems. It has also provided highly successful approaches to data gathering by researchers concerned with the human-behavior and safety aspects of construction. These are discussed and references are given in Chaps. 9, 10, and 12.

ACTIVITY SAMPLING[4]

One responsibility of managers and foremen is to see that workers are doing tasks efficiently and are not merely acting busy doing unproductive things or doing nothing at all. This is particularly true in areas where there is a relatively high cost of labor. However, examples abound in these high-cost areas as well as other locations where such practices waste energy, time, and money. For example, on many construction sites, form-yard carpenter crews spend 75 percent of their time carrying or rehandling materials. Much of this activity could be eliminated if careful attention were given to yard layout, material delivery, work flow, and the use of roller conveyors. Again, laborers often hand-carry

---

[4]There is no "right" way to do activity sampling. The authors have selected among the several approaches and the various ways of classifying activities. Those who wish to explore the topic further will find the following general and construction-specific references helpful. General references: E. J. Polk, *Methods Analysis and Work Measurement*, McGraw-Hill, New York, 1984. Construction references: From the *Journal of the Construction Division*, ASCE, as follows: H. R. Thomas, Jr. and M. P. Holland, vol. 106, no. CO 4, December 1980, pp. 519–534; H. R. Thomas, Jr., vol. 107, no. CO 2, June 1981, pp. 263–278; H. R. Thomas, Jr., M. P. Holland, and C. T. Gustenhoven, vol. 108, no. CO 1, March 1982, pp. 13–22; R. C. Olson, vol. 108, no. CO 1, March 1982, pp. 121–128. D. F. Rogge and R. L. Tucker, vol. 108, no. CO 4, December 1982, pp. 592–604. From the *Journal of Construction Engineering and Management*, ASCE, as follows: H. R. Thomas, Jr. and J. Daily, vol. 109, no. 3, September 1983, pp. 309–320; H. R. Thomas, Jr., J. M. Guevara, and G. T. Gustenhoven, vol. 110, no. 2, June 1984, pp. 178–188; F-S Liou and J. D. Borcherding, vol. 112, no. 1, March 1986, pp. 90–103.

or use wheelbarrows to move materials when a little advance planning would have permitted truck delivery to the immediate work site. Workers who use a wide variety of tools and install special materials often spend too much time chasing down both tools and material. For example, on a large project in which thousands of dollars were being spent to make improvements in productivity, a particular assembly operation required that a highly paid craftsman walk 200 feet every few minutes to get and return a small attachment that was used by him in every cycle but shared with others. In this instance, $20 for another attachment could have increased productivity by 25 percent on a task that was repeated over and over for several months.

Inefficiencies of this sort go on because job management is too busy to see them or is too involved in minute details of the work. At times, inefficiencies might be picked up by other managers or higher-level supervisors who are more detached from a particular operation, but such an observation would be at best hit or miss. What is needed, then, is a formal way that detection can be made on a planned and consistent basis. Activity sampling offers several techniques that alert constructors have found to be very effective in finding inefficiencies. It involves making and analyzing the results of field observations to determine what individual workers are doing at specific instants in time. Findings from these individual observations are compiled to give a picture of the activity level of a specific construction craft or operation or for an entire project. Because most of these approaches involve sampling, the results must meet certain criteria for statistical reliability, as discussed earlier in this chapter.

The three approaches to activity sampling discussed here are (1) field ratings, where the observations are simply working or not working; (2) productivity ratings, where activities are recorded in more detail and then reported as effective, contributory, and not-useful work; and (3) 5-minute ratings, where the activities of a crew are recorded for short intervals.

There is a general consensus among specialists that for activity-sampling results to be considered reliable in construction situations, a combination of confidence level of 95 percent and limit of error of 5 percent is satisfactory. At these levels the results give a good indication of the overall effectiveness of an organization or the individual crews that make it up. Regardless of the rating scheme, the working portion of activities usually falls into the 40 to 60 percent category. With these limits, and with the possibility that the split might be nearer 50–50, it is reasoned that the sample size should be about 384.

In sampling for construction activities, there is not a specific discrete item such as a concrete block to evaluate. Instead, for each person being observed, whether working alone or in a group, there are recurring blocks of time that must be evaluated. If a single observation of a worker involves a 3- to 6-second snapshot or sample of an individual's activity, then there can be a very large number of samples for every worker. For a project, this number is the product of the number of workers times 10 to 20 possible samples per minute times 60 minutes per hour times the length of the working day. Leaving out start-up and clean up time as irregular activities, the activities of each worker could be sampled a possible 4000 to 8000 times during a working day. For a 100-person job, the total would be 100 times as many or 400 to 800 thousand. Since it would not be either feasible or economical to have a 100 percent sample of all activities, the statistical estimate from a sampling study becomes the working tool by which management can monitor job performance. Properly used, it can be a powerful one.

Sampling can be applied to crews or projects of any size, since sample size is not related to the number of individuals observed. However, it would be necessary to sample a crew of 100 men 4 times or a crew of 10 men 39 times to meet the minimum of 384 observations, which, as stated above, must be taken if the results are not to fall outside the prescribed criteria (confidence limit of 95 percent, limit of error plus or minus 5 percent) purely by chance.

To ensure that an independent relationship is maintained and that the sample has the same characteristics as the universe from which it is drawn, the observations or other data should be collected at random times and in different sequences. For example, to set the starting time and location for activity sampling, it is common practice to use a random-number table, a deck of cards, or dice to ensure that start time and location do not introduce a bias.

Certain general rules for activity sampling must be observed in sampling construction operations, such as:

1 The observer must be able quickly to identify the individuals to be included in and excluded from the sample. If sampling is to be by crafts, hard hats or shirts coded by color to distinguish among them are very helpful.

2 A sample shall contain no less than 384 observations.

3 There must be an equal likelihood of observing every worker.

4 Observations must have no sequential relationship.

5 To preclude any bias, the rating must be made at the instant each person is first seen; the observer must not rationalize on what tasks the workers have just finished or what they are about to do.

6 The basic characteristics of the work situation must remain the same while the observations are being made. This means that comparisons among sets of observations are valid only if the work situation is substantially the same.

Conforming to some of these criteria may at first seem difficult because of the fear that workers will modify their behavior or react adversely to the presence of the observer. However, there is strong evidence that this is not a serious problem. Trained observers have stated that without trying to catch the workers off guard, they make 85 to 90 percent of their evaluations before those observed are aware of the observer's presence. In instances where this problem seems to exist, it can be minimized if observers have other duties that often require their presence throughout the project at unpredictable times.

To illustrate the work-sampling approach, assume that a group of 384 workers on a project were observed one at a time. These workers were classified as working or not working by the observer, with working defined only as not being idle. Assuming that the results of the observations showed that 278 of the 384 workers observed (72.5 percent) were classed as working by the observer, one could predict from Fig. 7-1 the expected range of percent working for a chosen confidence level. Using the category proportion of 72.4 percent and the sample total of 384, one can determine that the limit of error for 95 percent confidence level would be plus or minus 4.6 percent. In other words, the percent working would be 72.4 plus or minus 4.6 or in a range of between 67.8 to 77.0 percent during the time of sampling. Stated differently, there would be only 1 chance in 20 that the activity level of the entire crew was less than 67.8 percent or more than 77.0 percent.

If another check were made of the crew so that there was a total of 768 observations and if the cumulative work count were 664, then one would know that in 19 cases out of 20, between 84.0 and 89.0 percent (i.e., 86.5 plus or minus 2.5 percent) would be working, with a confidence limit of 95 percent. By accepting a confidence limit lower than 95 percent (i.e., agreeing that being wrong a larger percentage of the time is acceptable), the analyst would find that the samples indicate a slightly smaller limit of error. For example, for the 384-worker sample with 72.4 percent working, Fig. 7-1 shows that between 68.5 and 76.3 percent (i.e., 72.4 plus or minus 3.9 percent) would be working, with 90 percent confidence limit.

It is beyond the scope of this book to present more than the bare bones of the theory of probability that underlies work sampling. Even this brief presentation should warn possible users that conclusions drawn from samples that are too small may well be wrong. Those interested in going more deeply into probability theory can find many books on the subject.

Field Ratings

As noted, several statistical approaches to activity sampling are in use, of which two, field and productivity ratings, which vary mainly in their degree of sophistication, are discussed here. The third, 5-minute ratings, which is not statistical, is discussed later.

Field ratings require that the activity of workers be grouped at the moment of observation into two classifications, namely, working and not working, or, more specifically, engaged and not engaged in a useful activity. The basic operating rules for a simple field rating, in addition to those given above, are roughly as follows:

1 Mechanical counters mounted on a clipboard on which specific observations can be recorded are very advantageous. One counter records the active personnel; a second records the total number that have been observed.

2 To the greatest extent possible, all those to be covered by the survey should be observed. At least 75 percent of the personnel must be in the sample to get dependable results. When greater detail is desired, counts should be made and reported by crafts, areas, or crews.

3 The individual making the count should devote full time to the count while it is being made and avoid distractions of any kind.

4 The rating should be taken at the first instant of observation. The observer should not bias the result by speculating about whether or not the subject was or will be active a moment before or after observation.

5 The person doing the counting must understand the reasons for making the count and should be drilled in correct procedures.

6 To record normal activity for a project or crew, counts should not begin until at least ½ hour after the workers start (or return to) work or closer than ½ hour until quitting time (lunch or end of day). This rule does not preclude taking special-purpose counts at the beginning and end of shifts to determine whether or not activities get under way quickly or if activity tends to slack off just before quitting time.

7 No counts should be discarded. Two counts are more representative than one.

To qualify as working, personnel should be engaged in such activities as:

1 Carrying material or holding or supporting material
2 Participating in active physical work, including:
    a Measuring; laying out; reading blueprints; filling in time cards; writing orders; giving instructions
    b Holding a tag line or supporting a ladder
    c Operating a machine or piece of equipment, but only while actively engaged
3 Discussing the work, provided it can be positively determined that such is the case

Activities such as the following would be listed as not working:

1 Waiting for another to finish work, such as laborers waiting for their wheelbarrows to be loaded or waiting for a hoist
2 Talking while not actively working
3 Attending self-operating machines, unless engaged in a useful task
4 Walking about empty-handed
5 Riding

It should be clearly understood that rules such as these are not absolutes and that management should adjust them to provide the desired information. For example, if union working rules require that a full-time attendant stand by a compressor, little is gained by including him in the count at all, since his being active or idle has no effect on cost and supervision has no control over his productivity.

The result of the count will then be the total number of employees observed and the total number classified as working. The percentage working is the number working divided by the total observed. To cover foreman and personal time, 10 points are sometimes added to the percentage as computed to find an adjusted field-rating index. If this overall index is less than 60 percent, job activity is often considered unsatisfactory. For specialized crews, satisfactory performance might be considerably higher than the 60 percent figure. Painters, for example, probably would develop a higher percentage while casual labor doing cleanup might show considerably lower ratings.

A sample report on a field rating might be as follows:

Number of workers on job site	132
Number of workers observed	122
Number of workers classified as working	59
Percentage working (59/122)	48 percent
Add for foreman and personal time (optional)	10 percent
Adjusted field-rating index	58 percent

A single field rating of the size shown here is merely an indication of probable conditions; additional ratings are necessary to make the results reliable. As a rule of thumb, it may be stated that 100 observations may identify situations that are seriously awry; but at least 400 observations are required to give reasonable certainty.

Field ratings may be made with work forces of any size. Where a large project is being observed, only a few trips around the project will be required to secure the necessary number of observations. If a single crew is being rated, its members must be observed repeatedly. Of critical importance is the accumulation of enough individual observations for the results to be statistically reliable.

Productivity Ratings

Productivity ratings carefully define the individual activities that workers do and, in turn, classify them into the three categories of effective, contributory, and not-useful work, or idle. There is no right way to categorize the multitude of activities for productivity-rating purposes. Rather it is necessary only to make clear the activities or conditions that are to be measured and how they are to be classified. For example, for a crew of mechanics, effective work could include cleaning, disassembling, or reassembling equipment; contributory work could include rebuilding parts, doing paperwork such as ordering parts, waiting for another worker to finish a task, talking to the foreman, and taking personal time; idle might include standing or sitting around, walking empty-handed, or talking to anyone but the foreman. One general breakdown that can be used in productivity ratings for almost all types of construction work is as follows:

1 *Effective work,* or activities directly involved in the actual process of putting together or adding to a unit being constructed, such as necessary disassembly of a unit that must be modified and movements essential to the process that are carried out in the immediate area where the work is being done

2 *Essential contributory work,*or work not directly adding to but (through associated processes) essential to finishing the unit, such as handling material at the work station, cleanup, personal time, receiving instructions, reading plans, waiting while some other member of a balanced crew is doing productive work, and necessary movement outside the work station but within (say) a radius of 35 feet of it

3 *Not useful or idle,*or all other activities (see below)

Certain items that are in the work category when ratings are only between working or not working are often placed in the contributory category in productivity studies. An example is measuring. Measuring, which must be done in some manner, might be eliminated by using precut materials or a jig. To classify it as essential would direct attention away from that possibility. Another example is the category of studying drawings, which is time-consuming and might not be necessary if the foreman's instructions had been clear or if in preplanning the operation, a simple sketch showing the pertinent details had been provided. In both such instances, the contributory rating challenges supervision to look for a better way to do a particular task.

The difference between activities that are contributory and those that are nonessential and classed as ineffective or idle is sometimes small and must be carefully defined. For example, taking a drink of water might be considered to be a contributory allowance under personal time but having soft drinks should not be. Waiting while some other member of a crew is working would require strict interpretation. The person rated as contributory must be in the same crew and handling the same material; the crew must be properly sized for the task; and there must be absolutely nothing the person could do to use the time otherwise. Only one of the crew members standing by could be classified under contributory; the rest would be classified as being ineffective.

Ineffective work clearly includes being idle or doing something that is in no way necessary to complete the job. It might involve such activities as walking empty-handed, moving materials or oneself outside a radius of 35 feet from the work face, or doing tasks that would be classed as effective or contributory work had they not been done with the wrong procedure or tool. Rework of a job done incorrectly in the first place is also classified as ineffective.

The important concept in this analysis is that work is effective only when it directly adds to the completed product. The definition of the finished or end product is easily determined on most projects. One possibility is to class any unit of work for which a cost-accounting code number has been assigned as an end product. Thus, a cubic yard of excavation, a foot of pipe in place, or an electrical fixture would be called finished products. In concrete work, building or stripping forms, placing concrete, or finishing the surface all would be considered effective work, since they normally are reported separately to the cost accounts. Following this approach, carrying materials, erecting or disassembling scaffolds, or job cleanup would be essential contributory work. The argument that essential tasks such as these should be classified as effective work could be answered in part by the fact that they may be more susceptible to modification or even elimination than are those usually classified as effective work, so that expenditures for them should be brought to the attention of management.

Techniques for making field observations are the same as those discussed earlier under the topic of field ratings, so they will not be repeated here.

Over the range of construction activities, some of the items that might be put in the three categories are:

1 *Effective work:* painting a wall, placing bricks, attaching a valve to a pipe, nailing boards on a wall, hauling material from an excavation, or moving within 10 feet of the work position. Other essential items carried out away from the work station such as mixing mortar for bricks, threading pipe, and cutting boards before nailing can be classed as effective work, as long as they are done efficiently.

2 *Essential contributory work:* building a scaffold to serve as a work platform, measuring a piece of pipe or placing it in a machine preparatory to cutting and threading, returning an empty truck to be filled, or moving within the area extending from (say) 10 to 35 feet of the work position.

3 *Ineffective work:* walking empty-handed or carrying anything more than 35 feet from the work position, taking a coffee break, waiting for a truck, riding on a truck, correcting an error, going back to the shop for a tool or a part, or discussing last night's ball game.

Table 7-2 gives a sampling of productivity ratings developed by a large prestigious firm that has used them to assess relative productivity for many years. Its data clearly show the differences in ratings to be expected among craftsmen whose tasks are highly repetitive, i.e. bricklayers or painters, and others, such as carpenters and pipe fitters.

The classifications suggested here are not sacred. Individual activities should be placed in the category that for the task at hand best indicates whether or not they are effective, contributory, or ineffective. After all, the aim of the classification scheme is to focus attention on inefficiencies, and it may be desirable to adjust the focus under different sets of circumstances.

By examining how the various activities are distributed in the classifications given here, management can see items on which to focus for making improvements in the work plan. For example, rating workers as ineffective if they are more than 35 feet from the work position will make management think about proper material location and arrangement. Again, if a man working with the wrong tool is placed in the ineffective work classification, management will be reminded forcefully that the proper tool should be at hand.

The findings of the productivity rating can be used in a number of ways. At the foreman or even the superintendent level, it is valuable as a means of pointing out specific situations in which the work can be done more effectively

TABLE 7-2
PRODUCTIVITY RATINGS FOR SEVERAL CONSTRUCTION TRADES*

| Trade or craft | Percent of total time in category | | |
	Effective	Contributory	Not useful
Bricklayer	42	33	25
Carpenter	29	38	33
Cement finisher	37	41	22
Electrician	28	35	37
Instrument installer	30	30	40
Insulator	45	28	27
Ironworker	31	36	33
Laborer	44	26	30
Millwright	34	36	30
Equipment operator	38	22	40
Painter	46	26	28
Rigger	27	57	16
Sheetmetal	38	33	29
Pipefitter	27	36	37
Teamster	45	16	39
Average of above	36	33	31

*Data are from 2 years of ratings by a large construction firm which has used work sampling for many years. Ratings given represent good performance.

and at a lower cost. Such things as inefficient work layout, unhandy material placement and handling, poorly sized or balanced crews are among the candidates for improvement that may be found.

On some projects, the practice is to have the superintendent or foreman accompany the raters as they walk together around the job, with the raters explaining their decisions on the spot. In effect, this procedure provides job management with a set of eyes focused specifically on productivity, something that it is difficult for a busy manager to do.

At higher-management levels, the results of productivity ratings may be presented as numerical ratings, which are discussed below.

Obviously, productivity ratings if used improperly can create antagonisms and threaten job morale. As with preplans or schedules, they must be considered and presented as aids to line managers to make them more effective. Attempts to use the results of productivity ratings as a basis for criticism, discipline, or discharge are bound to defeat their main purpose. Many instances can be cited where workers or others have almost sabotaged the effort. As an example, workers were criticized for walking some distance to go to the toilet, because walking empty-handed classified them as not working. They changed their classification to contributory merely by carrying a piece of pipe or conduit when they made the same trip, thereby improving the rating.

The techniques for making productivity ratings are very simple, much like those described earlier for field ratings, except that a third counter is required on the clipboard. Details of individual observations of particular importance might be written out, possibly on a preprinted form adapted to the particular situation. On some projects, raters carry a portable tape recorder into which they dictate their findings and make pertinent comments and suggestions. From this, a written summary can be prepared for submission to higher-level management.

Productivity ratings can be carried out by several classes of personnel. One very large constructor has developed a professional rater classification. For such a career it is essential that the position satisfy the long-term expectations that these people set for themselves; otherwise, they will not stay permanently with the job. Technical institute graduates have performed well; it is to be anticipated that graduate engineers might not, since such a position would probably not match their career objectives. On the other hand, short assignments as raters can be an excellent training mechanism for them.

Labor-Utilization Factors

The principal goal of any work-improvement scheme is to increase the number of employees engaged in the effective-work category. But some essential contributory work is required in all jobs. Thus, making some allowance for essential contributory work when reporting overall performance is common. Some will argue that no such allowance should be made because the work involved is not adding to the final product. Others claim that substantial credit for the contributory work is justified. The labor-utilization factor takes into account this latter view.

One form of labor-utilization factor can be defined as the percentage obtained by summing the number of observations of effective work plus one-fourth the number of observations of contributory work and dividing this sum by the total number of observations. In formula form, this factor would be as follows:

$$\text{Labor-utilization factor} = \frac{\text{effective work} + \frac{1}{4}\text{ essential contributory work}}{\text{total observed}}$$

where total observed = effective + essential contributory + ineffective

Considered judgment as well as experience must be used in interpreting any labor-utilization factor. Acceptable values of it, however defined, vary with the type of work or trade. It is clear that a painter generally does more effective work and less contributory work than does a plumber, and that a labor crew doing cleanup around a project would rate low in effective work since very little of its time is spent directly adding to a finished product. For example, a crew of laborers might have a factor in the range of 40 to 50 percent; factors for electricians and pipe fitters might be 30 to 40 percent; and that for painters, 45 to 55 percent.

For a particular kind of work or a mixed crew, the factor would be different;

but once a backlog of past observations had developed an accepted base, current samplings could be relied on to give a fairly accurate picture of management effectiveness. These results could be used to make comparisons among projects or crews or to detect changes on a single project.

With good tools and a job to do, most workers will apply themselves if properly instructed and supervised. Under most circumstances, labor-utilization factors and any other form of activity sampling offer a numerical measure of the quality of the foremen, superintendents, and managers, not that of the workers.

The 5-Minute Rating Technique

The 5-minute rating technique is a quick and less-exact appraisal of activity than is that of the field-rating method. Even so, it is an effective method for making a general work evaluation. It is based on the summation of the observations made in a short study period, with the number of observations usually too small to offer the statistical reliability of work sampling.

The purposes of the 5-minute rating can be to (1) create awareness on the part of management of delay in a job and indicate its order of magnitude; (2) measure the effectiveness of a crew; and (3) indicate where more thorough, detailed observations or planning could result in savings.

The 5-minute-rating technique does not differentiate between delays which impede the progress of the job and those which do not affect progress but merely indicate higher cost. (For this reason, its findings must be used carefully.) To illustrate this distinction, consider a situation where management has deliberately assigned a crew of specialists to do a particular task essential to an operation which follows. Once the crew has finished, it will be idle until its services are needed again. Consequently, to include this idleness in the overall job rating would reflect adversely on the entire project. If the crew's rating is to be determined at all, the results should be reported separately and carefully explained.

To make a 5-minute rating, the observers with a watch and form for recording observations (Figs. 7-6 and 7-7) must place themselves in a position from which they can observe a whole crew without being conspicuous. In this way, crew members will not be aware of who is being observed and will not react to the observer's presence. For small crews working in close proximity to one another, all are observed at the same time. Large crews can be mentally divided into subgroups for ease of observation. Individuals in each group are then observed during consecutive blocks of time of from 30 seconds to several minutes, and the ratio of delay or nonwork to total observed time is noted. If the delay noted for an individual in any block of time exceeds 50 percent of the period of observation, then the rating for that individual is classified under delay. Conversely, if the delay is less than 50 percent, the appropriate block is classed as effective. The sum of effective times for each individual and for the crew divided by the total time of observation will then give an effectiveness ratio. When multiplied by 100, an effectiveness percentage for the whole crew is found.

The 5-minute rating technique is so named because of the rule that no crew

FIVE-MINUTE RATING

DATE _7-7-88_

JOB _Erecting precast panels_

CONTR. _N + E Corp._

SUPT _____ FOREMAN _____

TIME	IRONWORKER (1)	IRONWORKER (2)	CARPENTER (3)	CARPENTER (4)	CARPENTER (5)	WELDER (6)	Notes
START	1	2	3	4	5	6	
10:13	✓						Crew waiting for panel to be hoisted
:14	✓	✓	✓	✓	✓		Landing panel, welder waiting to
:15	✓	✓	✓	✓	✓		tack rebar
:16		✓	✓	✓	✓		Install upper bolts for braces
:17		✓		✓	✓		Install braces
:18			✓	✓	✓		Align panels
:19			✓	✓	✓		" "
:20			✓	✓	✓		" "
:21	✓	✓	✓				Unhook crane
:22	✓	✓	✓				Unhook crane
:23						✓	Welder tacks rebar, crew waits
:24						✓	for next panel to be hoisted
:25						✓	
	5	6	8	7	7	3	Effective unit totals
			36				

TOTAL MAN UNITS _78_   EFFECTIVE _36_   EFFECTIVENESS _46_ %

FIGURE 7-6
Example of the 5-minute rating technique, applied to precast panel erection—12-minute cycle.

FIVE-MINUTE RATING

DATE _8-88_

JOB _Pipe Manifold Revision_

CONTR. _Lupen Co._

SUPT _____ FOREMAN _____

TIME	CUTTER	WELDER					
8:35	✓						*Welder waits while cutter measures*
:40	✓						*delays between tacks and cuts*
:45	✓						*Welder waits while cutter marks cut*
:50	✓						" " " " " "
:55							*Delay while cutter leaves to*
9:00							*relieve pressure on lines to*
:05							*be cut*
:10	✓						*Welder waiting for cut to be*
:15	✓						*completed*
:20	✓						*Same*
:25	✓						*Same*
:30	✓	✓					
:35		✓					*Cutter waits as welder welds*
:40		✓					" " " " "
:45					1		*Stop for smoke*
:50							" " " "
:55							" " "
10:00							" " "
:05	✓	✓					
:10	✓	✓					
:15	✓						*Welder waits as cutter heats*
:20	✓						*and bends line*
:25		✓					*Cutter waits as welder welds*
	13	6					*Effective unit totals*

TOTAL MAN UNITS _46_   EFFECTIVE _19_   EFFECTIVENESS _41_ %

FIGURE 7-7
Example of the 5-minute rating technique applied to pipe welding—2-hour cycle.

should be observed for less than 5 minutes. A rule of thumb is that the minimum observation time, expressed in minutes, should be equal to the number of men in the crew. Thus, 12 men should be observed for at least 12 minutes. A longer period is sometimes necessary to satisfy observers that they have recorded the actual situation.

An adequate knowledge of crew effectiveness is usually achieved by making four separate 5-minute ratings in a day, two during the first half of the shift and two during the last half. Additional studies can increase the reliability of and confidence in the results. Grouping the values of the separate studies can be useful. Close agreement indicates validity; wide variation in results indicates that the work situation may have changed. An experienced observer, by analyzing the data, learns to judge whether longer or additional studies will materially affect ratings made for the minimum recommended times.

An example will illustrate the use of the 5-minute-rating technique. It involves hoisting precast concrete wall panels into place by a crane and having a crew that is landing and temporarily fastening them. The work sheet is shown in Fig. 7-6. Data at the tops of the columns designate the individuals being observed and their crafts. Descriptions of activities written in the spaces to the right help to make the ratings understandable. In this case, with only six men in the crew, the complete group was being observed at the same time. The observer divided the time into arbitrary blocks or units—such as 1-minute intervals—and evaluated each unit. Each person was rated during each minute of the study.

Rather than work on a minute-by-minute basis, the observer can wait until a specific task is completed and then rate the workers as to whether the majority of their time was working or delay, and the study sheet is so marked for the appropriate minute blocks. After the study is completed, the ratio of effective to total time is computed on a man minute basis as shown in the figure. More extensive studies can be undertaken using modifications of this technique. For studies lasting several hours, the time blocks can be 5 minutes in length. Each person is rated as before for each 5-minute block of time, and the final effective rating is computed as previously indicated (see the example in Fig. 7-7).

Any number of modifications to the various activity-rating methods outlined here are possible. As suggested earlier, eliminating certain classes of workers from the count may be appropriate if for some reason management has no control over their activities. Again, revising the working and not-working categories given earlier may be advisable. For example, with careful work planning, many operations such as carrying, holding, or supporting materials and much of the measuring, layout, and blueprint reading might be unnecessary. Such activities could then be placed in the not-working category.

Some rating schemes attempt to assess the level or pace of the various activities. For example, the working ratings might be reported in numerical terms. A fast-moving alert worker might be shown as 110 or 120 percent; one who moved slowly might be listed at 80 or 90 percent. The aim is to extend the activity rating so that it more accurately reflects job tone.

A word of caution regarding all rating schemes is appropriate here. Rating procedures evaluate groups, not individuals. Just as workers cannot be held responsible for following poor instructions, so the results of rating techniques must not be used to criticize or discipline individual workers. Rather, such examples of unproductive activity might be considered the fault of management.

RECORDING PRESENT WORK-FACE PRACTICES

When management has some indication, through hunches, informal discussions, cost reports, questionnaires, interviews, or sampling, that certain operations might be improved if given detailed attention (whether they are going well or not) the next logical step is to gather data on which to base decisions about changes. The kinds of data needed for dealing with the away-from-the-work-face or systems problems have already been described in Chaps. 4 and 5. In those instances, the follow-through involves reexamining and reworking the existing system or preplan. This section discusses several ways to gather data for studying and correcting problems at the work face. Suggested procedures for processing and presenting these findings are explored in Chap. 8.

Selecting the most promising problems to examine more carefully from among those found by preliminary studies may not be easy. One of the criteria might be that the operation in question is one that will continue for a period of time. As discussed before, the estimated net positive effects on costs, time, or quality must appear to be substantially greater than the costs for making the study, including the cost to make a change. Another consideration in undertaking a particular study might be the anticipated difficulty of implementing it if the solutions involve changing management practices or controls that are beyond the reach of those faced with the problem. Finally, time, energy, and support from higher management, along with their willingness to pay for the special personnel and equipment required to collect and analyze the data and present the findings, are essential.

Specialists in productivity analysis, particularly if brought onto projects that are in trouble, often encounter suspicion, resentment, and an unwillingness on the part of job management to cooperate. In such cases, it is often advantageous first to study operations that seem to be doing well, for example, those that are on time and under budget. Under these circumstances, job management will feel less threatened and will often participate willingly in data gathering and analysis. If given credit for the improvements that are almost certain to be found, they will then cooperate in later studies of the more troublesome situations.

The remainder of this chapter will discuss several of the more common data-gathering methods, under the headings of (1) records of physical and operational procedures, (2) still photographs, (3) time studies, and (4) time-lapse and video films, and explain how each procedure fulfills the need to provide easily understood facts.

Recording Physical and Operational Features

A first step in recording any construction operation is to portray clearly the physical layout as it currently exists and to have a detailed work plan. This commonly involves (1) taking measurements to show the positions of materials, machines that are fixed in place, and work stations, and (2) making a detailed flowchart of the operation. How accurately this is done depends on the circumstances; it can range from rough guesses to careful taping. Relative elevations should be determined when they affect the operation or any changes that might be made in it. Measurements, in turn, are sketched, often freehand but sometimes carefully drawn to scale. Drawings can be isometric or in plan, elevation, or cross section. The aim is to portray the situation in a manner that will be clearly understood by those who will be involved in studying it. Several examples of typical drawings are given in this book.

Knowing the number, craft or other category, positioning, and movements of the people and equipment carrying out the operation in question is essential. These should be recorded in a way that will give the required information. Examples showing how these records are kept and used are given in this book.

There are no rules which prescribe what information to record; rather, those doing the study must make their own rules before the data gathering is begun. This advance planning must be done carefully, for if insufficient data are gathered, the findings of the study will be of little value.

Acquiring Data for Productivity Improvement by Examining or Sampling Project Records

Records of past happenings can often supply valuable clues to those looking for ways to improve productivity, quality, or safety in on-site construction. At times, useful leads can be found in a single set of figures or oral or written reports. In other cases, substantial amounts of data might be processed to get a dependable picture of a given situation. In this latter case, it should always be kept in mind that conclusions drawn from sampling can be in error if the rules which establish statistical reliability, as described and applied to work sampling earlier in this chapter, are violated. The rules for sampling are no different here, although depending on the use to which the findings are to be put, a less rigorous criterion for accuracy may be satisfactory.

Projects do not have to become very large before higher management must rely on verbal reports or written records of past happenings to assess how the work is going and to plan future operations. It is also possible to use this information to search out ways for improving ongoing operations or for bettering productivity on tasks that are yet to be undertaken. Time, effort, and ability are required to use these many resources imaginatively. Fortunately, as discussed in more detail in Chap. 13, computers are making it much less difficult to gather, process, and sometimes abstract or rearrange this often voluminous information into more useful forms.

Only a few of the many ways in which records can show opportunities for productivity, quality, or safety improvement can be pointed out here, but the authors hope that this brief discussion will stimulate thinking about other means. Some suggested ways are:

1 Examination of costs and schedules showing past performance. The data on them can give good indications of the amounts of time, labor, materials, and equipment that are being used to do various activities, as compared with that of the estimate or budget. If these costs or the schedule depart substantially in either direction from the estimate or budget, the situation should be investigated in detail. Finding cases for study can be expedited if they are flagged in some manner on the periodic reports that management receives (or should receive) showing progress, payroll or unit costs, inventory or materials ordered, and equipment usage or rental charges. On the other hand, indiscriminate flagging can be a distraction rather than a help if it diverts attention away from large work items which are near or below their anticipated costs but which, if given careful study, might be susceptible to substantial improvement in productivity.

The problems involved in using the results of reports such as these, as discussed in Chap. 2, bear repeating here. Those preparing the reports learn very quickly how higher management will use them. If they are seen by the preparers as ways to improve management efforts and job performance and are to the preparers' personal and group's advantage, their dependability is usually good. However, if management attempts to use the reports as a tool for rewarding or punishing or as a way to find scapegoats to cover up its own failures, the data may be fudged to a degree that they are no help at all and may hinder attempts to find and implement possible improvements. (Fudging is not a figment of a bookwriter's imagination. Actual examples are common. To illustrate: recently on one very large project, a sampling study showed that the crews, sized as originally planned, were in fact performing a significant work item in a small fraction of the time allotted in the schedule, while the reported costs remained close to budget. Further investigation disclosed that the foremen were using this work account item as a "dump" for the costs of any task that did not have an account or for any work that was at or near budget. They had all learned that by using this approach they would not be reprimanded when costs on an item ran over budget, so it was an easy and logical thing to do. But sampling easily uncovered the problem, and a little questioning disclosed the reason for it. Far more important, it revealed the obvious fact that higher management had failed in its efforts at team and morale building, which would probably mean that any productivity improvement efforts would be fruitless.)

2 Studying safety reports. Safe and productive performances go hand in hand, so well-kept records on the past accident experience of supervision down through the foreman level can be revealing, not only with respect to safety but also indirectly with respect to the nature of the climate for productivity improvement. In addition, one form of productivity is to reduce costs, including accident costs, which can be substantial (see Chap. 12 for more detail). The evidence is abundant that studying safety records and doing something about unsafe supervision are very productive activities.

3 Examining records that measure the quality of the finished product. As with safety, attention to quality pays, and management must give it careful attention. Helpful records can include the cost and delays associated with rejected materials and with rework. Again, as with problems dealing with cost, schedule, and safety, the costs and delays associated with quality deficiencies and the attendant rework are morale destroyers. An example makes this case well. Again and again on atomic power plant construction sites where poor quality required tearing out and redoing, one would hear the complaint "Why do it? We will just have to tear it out and do it over again!"

Still Photographs

Still photographs, either prints or slides, preferably in color to provide better contrast and greater detail, can by themselves often provide enough information to satisfy the need for data to analyze less-complicated situations. A photograph or projected slide on which one can point out that there appear to be too many workers in this crew, a panel is misaligned, a platform on a particular scaffold is too high, or a truck is underloaded is often sufficient to generate discussion and thinking and provide a basis for productivity and safety improvements. Field supervision can often take the photographs. Some companies require that appropriate field personnel carry an instant print camera and film and take pictures of layouts or situations that they feel merit attention. This procedure has the added benefit of suggesting strongly to them and other managers that they are expected to be alert to opportunities for improving productivity or detecting unsafe practices. Moreover, such photographs can provide an excellent supplement to written diaries and serve as a basis for discussion and analysis at staff meetings or for work teams (quality circles) if they exist.

Modern procedures for reproducing the information recorded on still photographs offer opportunities not available in the past. For example, photographs taken with an instant-picture camera can be reproduced as slides if viewing by a larger group is desired. Conversely, slides can be made into prints in a variety of sizes. With either, a section of the original photograph can be enlarged to highlight a particular detail. Although such reproductions are not cheap, the benefits of using them under the right circumstances can far outweigh their costs.

A series of still photos can provide a basis for understanding the various steps and events that make up a construction cycle. Often the photographs will provide a suitable record. Photographs can also offer a very valuable sequencing of the process for further time or sampling studies and time-lapse or video filming.

To get a set of still photographs that depict the critical (necessary) steps of a complicated task or operation is often much more difficult than it sounds. However, once the set is made, it is easier to plan how to gather more detailed data. Also, still photographs are often helpful in organizing thoughts and discussions when an individual or group is looking for significant improvements. Again, the photographs permit an analyst to focus on details of the work face while away from its tension and confusion. Most people find concentrating there difficult because their attention is taken over by the problems of the job itself.

Still photographs are a most helpful means of finding good camera positions for subsequent time-lapse or video photography. Unless the operating area can be covered by a still photograph, it is a waste of time to attempt detailed recording with more sophisticated equipment.

Time Studies

The purpose of time studies, sometimes called stopwatch studies, is to record the incremental times of the various steps or tasks that make up an operation. Such a detailed record will show exactly how long each step of an operation took. Since there may be variability in the way a task is carried out or in the times for each step, it is often advisable to record a number of cycles. Such a carefully prepared record will reveal both the general procedure and variations within it. The results may be in conflict with how the job was planned, how the foreman or superintendent thought it was being done, how the plans and specifications required that it be done, or even how a reasonable person would have done it.

Time studies require a recorder for almost every person or machine being observed, which is almost prohibitively expensive. For this reason, except in special situations, manual observation and recording methods have been superseded by time-lapse or video procedures which permit bringing operations to the quiet of an office or study room for analysis. Steps now under way, which are discussed in Chap. 13, will bring computers into the analysis process. But the fundamental ways in which time studies attack problems are retained in these newer methods. For this reason they are given substantial attention here.

Time-studies were the fundamental approaches to productivity improvement employed by Frederick W. Taylor and Frank Gilbreth in the late nineteenth and early twentieth centuries. Later, in the 1910s, abuses of them by the so-called efficiency experts who used them to turn workers into productivity machines, brought strong reactions from the labor movement and government as well. The stopwatch became a symbol employed to identify oppressive management. One consequence was that time studies of federal employees were prohibited from 1913 until after World War II, in 1949. Today, however, they are often employed in manufacturing and other industries to establish performance standards and to serve as a measure for establishing bonuses in a variety of incentive-pay schemes.

Time studies require a minimum of equipment (a stopwatch, interval timer, or linked timers) and are a fast way to record a specified sequence of events involving at the most only a few units (workers or machines). Unfortunately the results are limited by the proficiency and training of observers and by their abilities to portray all aspects of the task.

One early and much-publicized construction time study was the TAMAP (Time and Methods Analysis Program) project, which used a team of industrial engineers with stopwatches to completely record the activities of each worker involved in building several houses. It had a tremendous impact on housing construction practices. Much can be learned by studying this project in detail.[5]

Time-Study Techniques   To make a time study, in addition to a stopwatch or interval timer, the observer needs a prepared form, clipboard, paper, and pencils. The aim is to record the times for the different tasks or elements of tasks that workers or machines perform. One observer can generally watch only one worker or machine at a time; efforts to cover more may lead to errors or incomplete records.

Data-recording forms must be prepared for each kind of observation and must include every item of information that will be required to make a subsequent analysis complete and meaningful.Those shown in Figs. 7-8 and 7-9 are two among many used in the late 1950s by engineers of the (then) Bureau of Public Roads for stopwatch studies of scrapers and drilling units on Bureau projects. They are reproduced to show the scope and limits of information that experienced work analysts can obtain from direct observations. The recorded data in several of the columns may be constant if the machines are of the same model or capacity. Other items of information may have to be estimated: for example, for scrapers, distances and sizes of loads; and for drills, sharpness.

When observing activities of short duration with a stopwatch, an appreciable error can result if the watch is stopped and read and started each time. It can be eliminated if the timer runs continually and the time of each observation is recorded and then the incremental times are calculated from the consecutive readings. This process is most easily carried out with an interval timer that has two hands, one of which runs continually and an auxiliary which travels with the other hand except when stopped for reading. Another and perhaps easier method is to use three stopwatches ganged together to a common touch bar so that depressing the touch bar will stop one watch, reset the second, and start

---

[5] See Jonathan Aley, "Results of Project TAMAP," *House & Home,* vol. 21, no. 1, January 1962; "How to Save at the Site," vol. 22, no. 3, September 1962; "Fresh Insights from TAMAP: On Site Study of Products," vol. 23, no. 4, April 1963.

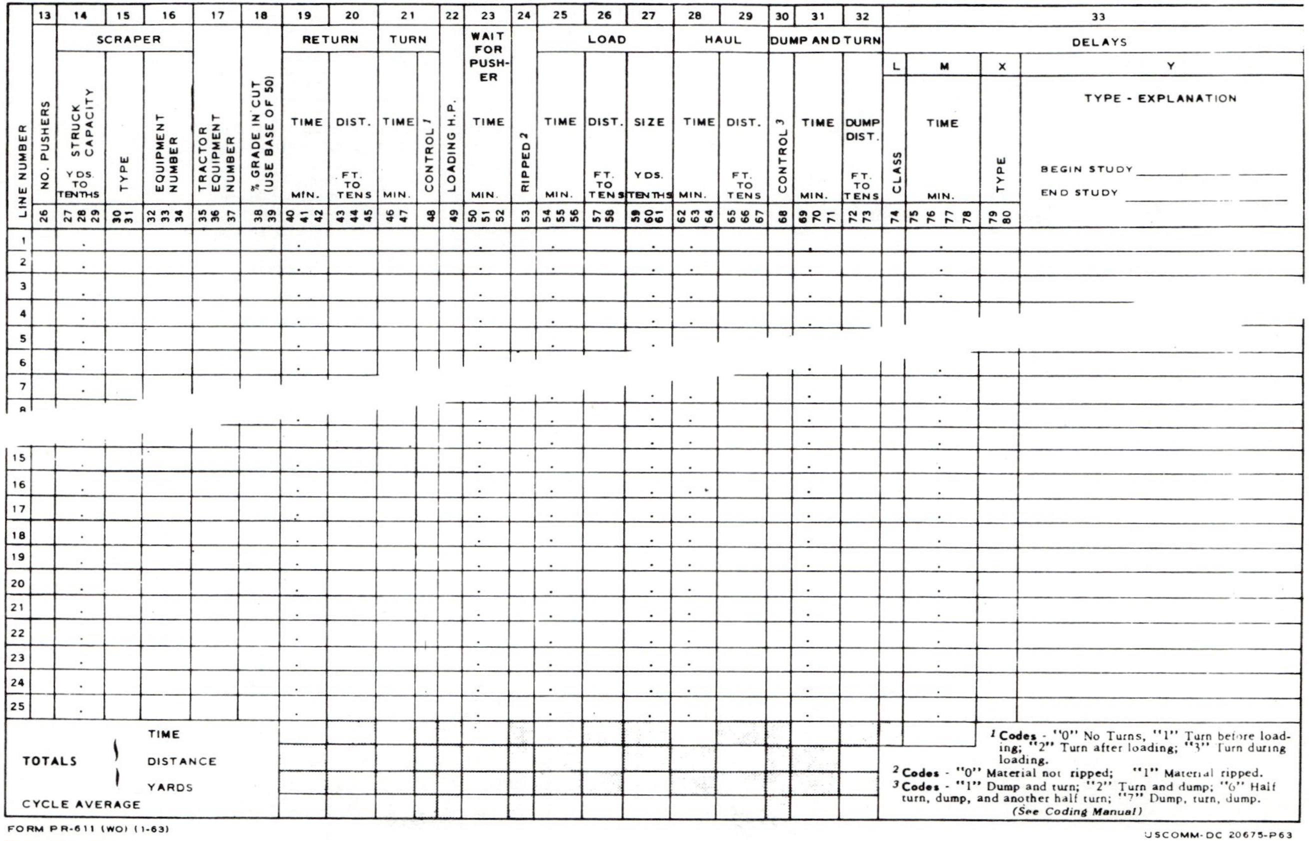

FIGURE 7-8
Form for recording time studies of scrapers.

FORM PR-611 (WO) (1-63)

USCOMM-DC 20675-P63

[1] **Codes** - "0" No Turns; "1" Turn before loading; "2" Turn after loading; "3" Turn during loading.

[2] **Codes** - "0" Material not ripped; "1" Material ripped.

[3] **Codes** - "1" Dump and turn; "2" Turn and dump; "6" Half turn, dump, and another half turn; "7" Dump, turn, dump. (See Coding Manual)

Column headers (rotated form):

23 24	25	26	27	28	29	30	31	32		33
DRILL BIT	EFFECTIVE SIZE IN. 1/8	DEPTH DRILLED — NEAREST BIT 0.5 FT. WITH EACH	DEPTH DRILLED — NEAREST HOLE 0.5 FT. COMPLETE	DRILLING TIME MIN.	CONTROL 2	REMOVE INSERT DRILL STEEL TIME MIN.	CLEAN OUT HOLE TIME MIN.	MOVE AND SET UP TIME MIN.	CLASS / TIME MIN. / TYPE	DELAYS
SHARP- NESS 1 — INITIAL / FINAL									L / M / X	Y

LINE NUMBER

Columns numbered: 49, 50, 51, 52, 53, 54, 55, 56, 57, 58, 59, 60, 61, 62, 63, 64, 65, 66, 67, 68, 69, 70, 71, 72, 73, 74, 75, 76, 77, 78, 79, 80

TYPE - EXPLANATION

BEGIN STUDY _____

END STUDY _____

Line rows: 1, 2, 3, 4, 5, 6 ... 28, 29, 30, 31, 32, 33

TOTALS

1 The following codes indicate relative sharpness of drill bits:
(See sketches at right):
Code 1 - Sharp
Code 2 - Dull
Code 3 - Very dull

2 The following codes indicate why and how drill steel was removed and inserted:
Code 1 - Remove section of drill steel from uncompleted hole and insert longer section.
Code 2 - Remove section of drill steel from completed hole and insert shorter section in new hole.
Code 3 - Add on section to increase length of drill steel in uncompleted hole.
Code 4 - Take off sections to remove drill steel from completed hole and insert section in new hole.

FORM PR-626 (WO) (10-62)

CODE 1 SHARP — 0" TO 1/8" FLAT
CODE 2 DULL — 1/8" TO 1/4" FLAT
CODE 3 VERY DULL — OVER 1/4" FLAT

USCOMM-DC 20391-P 62

FIGURE 7-9
Form for recording time studies of rock-drill units.

the third. This apparatus allows plenty of time to carefully read the dial and reduces the error so common in studies using a single stopwatch. It also saves the time otherwise required to make incremental calculations.

There are many niceties in stopwatch studies that will not be discussed here. Numerous references will be found in the industrial engineering literature.[6]

Limitations of Time (Stopwatch) Studies   Several constraints inherent in stopwatch studies in addition to the heavy demands for observer's time which limit their usefulness include the following:

1 The observer must decide instantly the point in time at which one phase or cycle stops and another begins, with no second guesses or chances for hindsight. When activities are not clearly separated and cycles are irregular in order and type, there can often be differences of opinion as to when one phase is completed and another starts. This type of error is of less concern when studies are made by a single person who has applied the same evaluation to a series of studies; however, it can be of real significance if several observers are comparing information that is based on different evaluations of the node points between phases of a cycle.

2 Because a substantial period of observation is involved, data collected in studies of larger groups of people and/or equipment may vary significantly over the different cycles. When a single observer is studying an operation that has 10 components (workers and/or machines), 10 cycles of observation will be required in order to time each component for one cycle. To get an average value of the times recorded for each component, it is advisable to study several cycles of work. With a 10-component crew, to observe 5 cycles per component would thus require 50 cycles to elapse during the time of observation. Repetition would not be necessary if all the cycles were duplicated exactly, but seldom are even two cycles of construction work done in exactly the same manner, as is shown by the discussion of learning curves in Chap. 6. It is thus inherently difficult for a single observer to cover activities accurately that involve many components. This can be accomplished only with several observers or by another method of recording.[7]

3 The information gathered by a stopwatch study is strictly limited to what is recorded plus facts that can be gleaned from the recorder's notes. This deficiency can be overcome only if the recorder is very observant about such factors as the interrelationships among components and is not completely oc-

---

[6] Among them are D. Karger and W. Hancock, *Advanced Work Measurement,* Industrial Press, New York, 1982; M. E. Mundel, *Motion and Time Study,* 6th ed., Prentice-Hall, Englewood, N.J., 1985; B. W. Niebel, *Motion and Time Study,* 7th ed., Irwin, Homewood, Ill., 1982; and "The MTM Journal," The International MTM Directorate of the MTM Association for Standards and Research, Fair Lawn, N.J.

[7] In the TAMAP study, each worker was under constant observation by a work-study analyst. See National Association of Home Builders Research Institute, "Report on TAMAP," *Journal of Homebuilding,* vol. 16, no. 6, 1962, pp. 85–108.

cupied in merely observing and recording times. As an example, a study of a scraper operation might show that on occasion, loading times are very long. The notes should thus record the cause of the delay, such as the scrapers were waiting for the pusher, the pusher needed mechanical attention, the scrapers were being overfilled, or a loading condition was unfavorable because it was uphill or along the edge or in a wet section of the cut. Only if the report included such information could management use the data to correct the situation. From this example it can be seen that stopwatch information alone may not be too helpful in evaluating some situations.

4 Stopwatch studies can suffer because of the physical limitations or biases of the observer. A study often involves recording a large amount of data in a short time. This taxes an observer's stamina and ability to maintain accuracy and objectiveness. Consequently, studies are often broken into blocks of observations with time for rest during which no recording is done. Biases can affect the observer's objectivity and with them will come a tendency to fudge the data. To avoid this natural tendency, the observer must strictly follow the rule that once the observation has been made, it is history. Reevaluation, hindsight, or second thoughts are not to be allowed.

Because of their very high costs, elaborate time studies of the sort described here are seldom employed today on construction sites in the United States. But the concepts employed are still applicable. And even now, stopwatch studies of limited scope can be an extremely useful tool, especially in instances where only one or perhaps a few elements or components are to be observed. The technique is inherently simple and requires only a small outlay for equipment. With nothing more than a wristwatch, data can be gathered that can lead to ways to improve productivity.

RECORDING AND VIEWING WORK-FACE ACTIVITIES WITH PHOTOGRAPHIC OR VIDEO METHODS

Photographic techniques using camera or video can carry out two functions: (1) permanently record certain field operations for later analysis and other uses, and (2) compress that record so that the viewing times involved in scanning operations or for sorting out operations to be studied in detail is minimized.

Pictures are chosen for analyzing construction operations because they offer the most effective and understandable way to record the activities of an individual or a crew or machine and the interactions among them. They provide all the advantages found in time studies, discussed earlier, without the disadvantage of extremely high data-gathering costs. Either film or video offer the added advantages of (1) being easily understood by any visually able person, (2) providing more detailed and dependable information, and (3) making possible review and study by analysts, management, or other groups away from the hustle and bustle of the work site. The principal limitations are those of any photograph or set of photographs: they portray only what can be recorded, given the light and other conditions; they can show only what is within range and view field and not obscured by objects in the foreground. Compared with techniques that generate only words or numbers, photography has the advantage of being understandable and believable to those who are not well-versed in studying written material or numerical data analysis or who question verbal or written reports. This applies particularly when the activities portrayed are not going as smoothly as they might, since it is difficult for a dissenter to argue successfully that the photographs do not portray the actual situation.

Real-Time Viewing of Photographic or Videotape Recordings

Either videotape or photographic procedures can be employed to record continuously the movements of each worker and machine and can be viewed again and again in real time for study in detail. A major disadvantage of such real-time recording and viewing on either film or videotape is that the period required for each viewing equals that of the initial operation. Seldom is that approach practical or cost-effective, especially when the cycles of an operation are long. An added problem with continuous photography on film is that it requires a substantial footage, which is costly. For example, Super 8-mm movie film costs over $150 per hour for film and processing. Hence, making motion picture films of work-face operations in real time is not usually considered to be a viable approach for gathering data for productivity-improvement studies, unless it is done with a very special purpose in mind. The use of videotape, however, is more cost-effective, since a cartridge costs less than $10 for 2 to 4 hours of recording.

How Photographic Time-Lapse Equipment Operates

In contrast to real-time photography, time-lapse filming has proved for many years to be a very useful means for recording work-face activities. Film taken in a time-lapse mode costs far less than than for continuous filming and yields benefits in reduced viewing time without diminishing understanding of the operation that has been recorded.

Time-lapse by filming involves taking still pictures at selected intervals of from 1 to 4 (or more) seconds with a specially adapted movie camera. The 1-second rate is employed when the aim is to record details such as hand movements, while the 4-second interval is usually satisfactory to show the overall body or equipment movements of interest in studying common construction operations. (Two short segments of time-lapse films of construction operations are given in Figs. 7-10 and 7-11.) In contrast to time-lapse, most moving pictures are taken and projected at rates of either 18 or 24 frames per second. At this rate, movements shown on the screen appear to flow smoothly, as in real life. Since individual frames of time-lapse are taken at

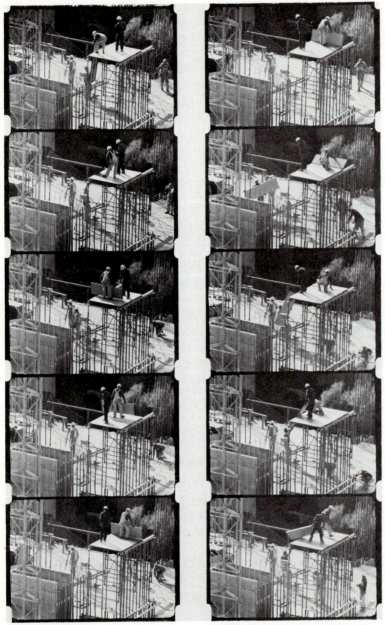

FIGURE 7-10
A time-lapse sequence of carpenters setting floor forms. The interval between pictures is 3 seconds.

FIGURE 7-11
A time-lapse sequence of a loader filling a truck. The interval between pictures is 3 seconds.

rates many times slower than the usual movie filming, viewing at the usual projector speed makes any significant movement that occurs between the frames of film become distinct and obvious to the viewer. Thus people or equipment appear to move in jerks rather than smoothly. On a first viewing, such movement seems amusing, but after a first introduction, it is neither annoying nor distracting; it has the advantage that depending on the rate of filming, observers can follow activities 24 to 60 times faster than the rate at which they were recorded. Even so, at these rates viewers are still able to follow and interpret the movements and relationships that the film portrays and to pick out segments to be studied in more detail.

Recording with Video Techniques

Video techniques are fast replacing photographic methods as the method of choice for recording construction operations. The method commonly chosen today combines portable video cameras and a special cassette player for projecting onto television-type screens. This equipment allows viewing the videotape by individual frames either backward or forward at specific speeds ranging up to 5 to 10 times faster than real time and with excellent picture quality. This equipment costs about $2000 to $3000 at early 1988 prices. There is also more expensive equipment for time-lapse recording on videotape. For this system, the camera records on the tape continuously at its usual rate, commonly $1/30$ of a second between frames. A special and very expensive ($20,000 in early 1988) editor permits selecting and recording individual frames from the tape at a stipulated rate, thereby producing the equivalent of time-lapse. This second recording can be run forward or backward or it can be stopped completely to give a still picture.

The third way to record the action using video equipment is to set the camera to take pictures at stated intervals, as with time-lapse. In this case also, the tape can be run forward or backward or stopped for viewing individual frames. The difficulties are that the pictures are far less clear than with full-time video recording and that the user has fewer choices of speeds at which the tapes can be viewed.[8]

Comparing Filming with Videotaping Procedures

While the pictures from video are often less clear than film, the advantages of instant replay, less-expensive equipment and recording costs, and more reliable equipment make full-time video the current choice.

[8]It would be pointless to attempt in a book such as this to describe the videotape equipment suitable for construction applications that will come onto the market in the next year, much less the next 5 years. Certainly it will become better and more versatile. Neither can the development of special applications for construction be predicted. A further dilemma is the probability that recording on discs rather than tape will become common. This approach will have a number of advantages, among them the ease in locating specific segments of information for viewing, whereas a tape must be reeled to the proper place before viewing can begin.

Among the reasons are:

1 Videotape cameras and the controls required to operate them are readily available and should become less expensive with time. On the other hand, time-lapse equipment is specialized and may become difficult to obtain.

2 Repair and maintenance of videotape equipment is available locally; this is not always the case with time-lapse equipment.

3 After time-lapse film is taken, the film must be processed. This often requires a wait of at least several days although overnight service can sometimes be arranged. Videotape can be played back immediately.

4 There is a strong possibility that the Super 8-film commonly employed for time-lapse may be difficult to obtain in time, while videotape is available from almost every television dealer.

Most film or video viewers can at the beginning follow the activities portrayed on film or videotape at the rate of several times the real-time rate of recording, and with a little experience, five to ten times real time. Experienced viewers often review records at rates up to 60 times real time, i.e., an hour of real-time viewing in a minute. However, this rapid viewing rate is only available with time-lapse film, as of early 1988.

Equipment Requirements for Photographic and Video Applications

As indicated above, the equipment currently available at reasonable cost for recording construction activities produces two different types of records: full-time video recording versus time-lapse photographic recording.

Many different makes and styles of video equipment are available, with more models coming on the market frequently. The recommended requirements are that the camera recorder be portable, have the ability to imprint frame numbers as well as time on each frame, have a continuously variable-focus camera lens, and have the ability to record under marginal light conditions. Playback should be in single frames as well as both forward and backward at variable speeds up to eight times real time with good picture resolution.

The equipment for taking and showing photographic time-lapse pictures consists of a specially modified movie camera coupled with a projector or viewing screen modified to compress viewing time. This can be done either by projecting single frames or by continuous viewing at an accelerated rate. Available film camera equipment can be programmed to take pictures at intervals from 0.01 to 99.99 seconds between pictures in steps of 0.01 second. For construction uses, the interval usually ranges from ½ second to 20 seconds between frames, with the majority between 1 second and 4 seconds. Film is currently available that gives good pictures under any light level at which construction can be carried out effectively.

Camera equipment for time-lapse applications should be very portable and capable of running unattended for periods of time ranging from hours to days, depending on the situation. Desirable camera features include variable-focal-length (zoom) lens that ranges between wide angle and long range, through-the-lens automatic exposure control with a manual override, a procedure for imprinting the time and date on each frame taken, and a battery-condition indicator. Capabilities of the projection equipment are as important as are those of the recording camera. While the camera, once set, continues at a fixed rate, projector speed is operator-selected. Choices can be made among various speeds ranging from viewing a single frame, through one frame each several seconds, to 30 frames per second, and run in either forward or reverse order.

The advantages of a continuously variable-focal-length lens for both video and camera recording cannot be overemphasized. With a zoom lens the camera operator can exactly fill the picture with the information desired, giving an optimum balance between broad coverage and detail. As the scene changes, the focal length can be modified without changing lenses, which is often a difficult operation under field conditions. The through-the-lens automatic exposure control ensures a properly lighted picture regardless of changes in light conditions. It also gives proper exposure for hours on end with an unattended camera. Imprinting the date and time on each frame allows observers to easily calculate elapsed time. Also, interrelating films from two locations is possible if separate operations that affect each other (e.g. a batch plant and a concrete placement operation) are being photographed.

Recording Techniques for Timelapse and Video Applications

Today, almost everyone is familiar with cameras, projectors, and television, so that people are readily available with the potential ability to operate the equipment and make the recordings. The principal problems lie in site conditions that cause difficulties in finding suitable camera positions. The general rule on placement is that the instrument should be considerably higher than the work being recorded in order to minimize masking subjects by foreground activities or other obstructions, to gain a better perspective, and to clarify the physical relationships. The term "higher than" is, of course, relative and can mean heights from a few to a hundred feet. The actual camera position depends on the operation that is being photographed, the setting, and, specifically, what is being viewed. Suitable locations often can be found on adjoining slopes, buildings, roof tops, vehicles, or on specially erected small scaffolds. On occasion, the camera has been clamped to the mast of a tower crane, to a pole, or to a light standard. When photographing general scenes, positioning the camera with the feet of the tripod on the same level as the work being observed should be avoided, since much of the action may be obscured by persons and/or equipment passing between the camera and the area where the activity is taking place. Since recording can almost always be carried out successfully from a height of 15 feet or more, arranging a suitable viewing position, even if this involves temporary scaffolding, is usually worth the cost, time, and effort.

The detail that can be recorded is a function of the proximity of the camera to the activity and the focusing distance for which the camera is set. If, for example, bench work is to be observed, the camera must be focused closely on the work. On the other hand, a camera distance of 100 feet or more is needed to cover a spread of equipment.

When using time-lapse photography for construction operations, the details of a method involving both workers and machines are most successfully recorded, as noted above, with frame spacings of 3 or 4 seconds. For bench work, 1 second or 2 seconds between pictures is better, since more pictures give greater detail. Those more interested in merely having a record of what was being done sometimes use rates of 8 and 10 seconds between views. This is sensible in part because with a roll of Super 8-mm film containing 3600 frames, a single roll will record a complete shift of 8 to 10 hours. With video, 4 hours is the usual maximum time per cassette.

The actual recording requires no greater skill than that required to take still pictures. Focus and exposure are important; artistic composition is not. The purpose is to record as much pertinent information as possible. Thus, particular attention must be paid to picture content and essential details. When an operator is monitoring the recording, events that affect the operation but that take place beyond the view of the camera can be tied into the visual record by notes correlating them with the frame or picture number or the time. A better procedure for recording this detail is to photograph any pertinent notes in their proper sequence at the time they are taken. Soft chalk and a small chalkboard or a pad of paper and a felt-tip pen are excellent for such note taking and photographing. One format for photographing notes is to fill the entire picture; another is to put the notes in a corner of the picture. The former method seems better for titles and for information that should be conveyed before a sequence is studied; the latter is usually more suitable for notes that are added when a film sequence is underway and should not be interrupted.

At the beginning of a recording, details such as camera data, project identification, names, locations, dates, construction-equipment characteristics (weight, size, model) are customarily recorded directly on the film or tape. Such information will be valuable for identifying the recording at the time it is played or when it is studied later. Special conditions and other information relevant to the study but outside the view of the camera will be incorporated directly into the photographic record, where it will be readily available.

When the camera equipment is not equipped to imprint each frame with a time, a clock can be placed in the foreground but still in focus so that it appears in the corner of the field of view. These on-film records involving clocks are not normally used as the basis for detailed time studies, although they can be if the timepiece can be read to seconds. This time record, however taken, can be invaluable to the person who does not have a feel for time compression when viewing at a fast rate.

Unless care is taken, photographic film titles, which occupy only a few frames, can be lost during film processing. For this reason, it is advisable to inquire about the way the processor crops the ends of rolls. For example, Eastman-Kodak's usual procedure is to crop both ends of all films from a Super 8-mm cartridge to get rid of the tails. This means that the first several exposures are sometimes lost, and, if retained, can be light-struck. To prevent losing a title, it is wise either to run several frames before photographing or to run extra frames showing the title.

Photographers with experience in taking motion pictures are acquainted with the problem of panning, which means moving the camera from one viewing direction to another. It is important to move the camera slowly so that the continuity of the sequence is not lost. In doing this panning with time-lapse equipment, it is usually more satisfactory to think of a time-lapse picture sequence as a series of still pictures than as a slow-motion movie, and to ensure that each picture overlaps the last.

To give orientation to a photographed sequence taken on a given project, it is helpful whenever possible to take several pictures of the area surrounding the location of immediate concern. For example, a study of carpenters installing cabinet work in an apartment building could be preceded by a view of the building, using job signs for titles. Again, a sequence showing a power shovel loading trucks could be preceded first by a pan of the overall pit layout, and, if possible, the haul situation for the trucks. Any other information that will help viewers orient themselves to the activity at hand is valuable, since being aware of the context in which the film was taken permits them to be mentally ready to understand it without having to fumble for orientation or to search for continuity.

Photographing large and complex crews carries with it the problem of identifying individuals. It helps if workers from the various crafts wear distinctive insignia such as hard hats of specified colors. Such a procedure often is already followed on large projects, where the contractor furnishes color-coded hats in order to make this distinction. For studies where special hats are not warranted, applying quick-drying spray paint or easily removable poster paint to the workers' hats can be effective. It is generally easier and more effective to mark hard hats than to provide variously colored vests or other articles of clothing. In photographing equipment, if several units look about the same, each one should be marked distinctly with a number or some other symbol that will show clearly in the pictures.

A major advantage of film or videotape as contrasted with on-site observation is that it can command the viewers' full attention, since they are not distracted by the hustle and bustle of the operation. Furthermore, collapsing the passage of time occasionally reveals questionable activities or interrelationships that may have gone undetected in real time. As noted, there is also the ability to stop, back up, rerun, and change speeds. This makes it possible to assess all the facets of an operation, for example, to follow and separately analyze what each worker does, and to do this in detail either by running the film or tape over and over again very slowly or by stopping it and, in effect, having a still photograph to examine.

In addition to providing an easily understandable record that workers, foremen, and supervisors can use to study and improve productivity, quality, and safety, time-lapse film or videotape are also invaluable as historical records, as teaching aids, and as a means for transferring knowledge from project to project. Furthermore, because such films are interesting and easily understood, crews and managers remember and discuss them and consider how improvements might be made.

Time-lapse is particularly effective for detailed studies of multicraft crews, since it is only a matter of rerunning the same film until the needed detail is obtained. In this way, the activities of all crew members and of each craft can be examined in their relationships with the others. This may disclose inefficiencies and unbalances, among which could be interferences among individuals or crafts, shortages in materials, equipment or tools, work habits that are customary but ineffective, minor changes in procedures as cycles are repeated, or work that will produce an unneeded refinement. Often these are not perceived and may pass unnoticed or may be considered immaterial when management observes an operation in the field. But by scanning the film or by a detailed analysis of it, patterns or inconsistencies may become apparent when they appear again and again. These will call attention to practices that can bring unnecessary costs and losses in productivity.

Informal Uses of Video and Time-lapse Recordings

Some users of photo recording focus its application on informal viewing by individuals or discussion groups to provide a medium for generating methods improvements and developing better on-the-job communications. The film or tape itself becomes the agenda for either formal or informal meetings. All that is required is that a group of interested people, be they managers, engineers, foremen, or craftsmen, commit time to the sessions.

In some organizations, this informal technique has been so successful that users see little advantage in undertaking more formal analysis. Although some problems can be solved and job teamwork and communication improved, other gains from the fine-tuning of techniques can often be best accomplished with more detailed and formalized studies, as outlined below and in Chap. 8.

A technique for making informal presentations more successful is to have a knowledgeable person preview the record, select activities that show promise for improvement and that will be of particular interest to the group, and gather detailed data on them. For example, on a concrete placement, such items as the layout and delivery system, number and craft of personnel, placement rates, and waiting and truck unloading times might be of particular interest. On an excavation project, the number of loads or yards per unit time or a summation and distribution of lost time might be of primary concern. Having the detailed data available before viewing by the group will save the time required to determine such detail and will permit actions triggered by the information. It is helpful if such information is presented in graphs, tabulations, or other easily understandable forms. Figure 7-12 shows such a graph of data collected from a film on a concrete tilt-up panel pouring operation, and Fig. 7-13 sums up the activities of a crew placing concrete for a floor.

Frequently, the supervisors who plan and carry out an operation will not

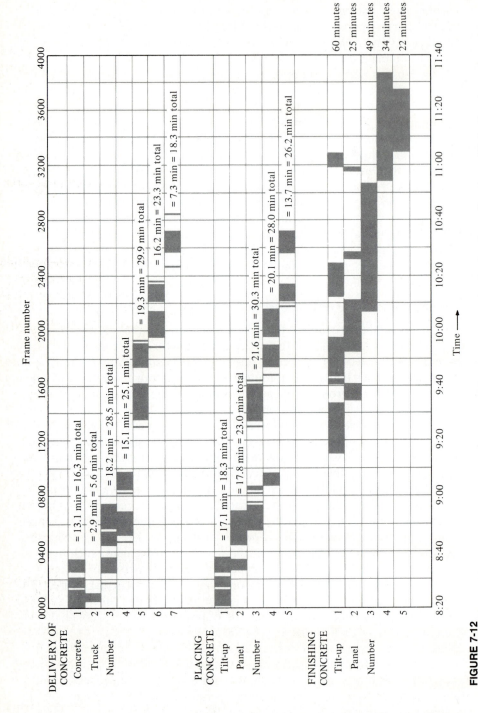

FIGURE 7-12

Analysis of times for placing concrete for tilt-up panels. This preconference analysis of a time-lapse film shows (1) the time to empty each concrete truck (and the time between trucks), (2) the time to place the concrete in each panel, and (3) the time required to finish each panel.

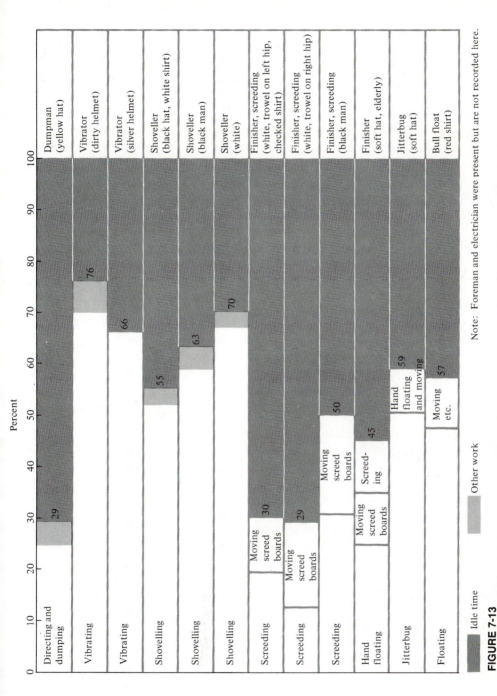

FIGURE 7-13
Times required for a concrete crew to place a floor slab, based on an analysis of time-lapse film. Data were taken from two 15-minute sections of film. The overall average working time (excluding foreman) was 52 percent.

be critical of it, unless there are glaring errors apparent to all. And they are often apt to be very defensive. It is sometimes good policy to let them view the film before it is shown either to higher-level management or to the crew. Fore-warned, they can have an explanation ready or, as often happens, can state flatly when others view the film that they have seen the situation and already have corrected it.

Having viewers unfamiliar with the operation at a presentation is often pro-ductive. They may see inefficiencies and inconsistencies not apparent to those who are close to the work. Also, because they do not have a proprietary in-terest, they can be more objective. From questions raised by these impartial observers, old hands often gain new insight into their practices and proce-dures. Of course these points must be made in a noncritical way so that the responsible managers do not take offense or feel threatened.

Group discussions of job problems are enhanced by time-lapse viewing be-cause it brings so many details into clear focus in an atmosphere where sharing is encouraged. Such discussions can produce creative solutions by combining the varied talents and ideas of all levels of management along with those of craftsmen. If the climate is right, all participants will contribute their special knowledge and strengths. Such an approach is often referred to as manage-ment by participation because cooperation is substituted for authority.

Detailed Viewing and Analysis

Once the method that is currently being used to carry out an operation is re-corded by either time study, film, or videotape, it is often advantageous to an-alyze carefully the data that have been collected. This can be done by methods which fit the various techniques for analysis. These will be presented in detail in Chap. 8; discussion here will be limited to techniques for detailed viewing of films or tapes.

Quite often managers, having employed time-lapse through the informal techniques just described, neglect the detailed-analysis step because it is time-consuming and its results are not immediately forthcoming. They do so at their peril, as illustrated by a typical example taken from a project involving a truck-loading operation. Casual group viewing of the film satisfied the managers that the operation of a 15-cubic-yard loader was acceptable since it was busy al-most all the time. To them, further study seemed pointless. However, detailed analysis of the time-lapse film revealed that all was not well. The clue was that the loader was making too many passes to load each truck. Further viewing showed that most of the time its bucket was only half full. Discussions with the operator followed, and they revealed that the loader had insufficient power to allow the bucket to be loaded more heavily. By making adjustments to the loader, full buckets were achieved and production was increased by 90 per-cent, with very little increase in cycle times.

One advantage of time-lapse films over the currently available video equip-ment is that more time compression is available with time-lapse. Time-lapse is not difficult for most people to watch. One's first impression may be that of watching an early silent picture, since the actions are jerky. However, if considered to be a connected series of still pictures, at a projection rate of two to four frames per second, comprehension of activities is not difficult. Time compression is from 6 to 16 times when films taken at 3- or 4-second intervals are run at two to four frames per second. Some observers can comfortably watch film when the ratio is 60 to 1. An advantage is that this time compression requires (and gets) full attention from the observer, which is not always the case with real-time recordings. Furthermore, the compression allows the viewer to interconnect dependent activities which at normal speed might appear not to be related.

Most viewers do not fully understand all the interrelationships of an activity when first observing a video or time-lapse film. Thus, the usual practice is to view the recording several times to bring out the subtleties, which are often very important. A common rule of thumb is that one must look at the film three to five times before the interferences, waste motion, lost time, duplicated effort, and a myriad of other factors become apparent.

People who have had extensive experience with analyzing time-lapse films and video recordings usually establish viewing patterns. For example, some start with a run-through at a fast rate of, say, 20 to 60 times real-time speed to give the viewer an overall understanding of the operation. After this, individual techniques vary; some go through the recording section by section, reviewing and restudying those that are of particular significance. Some run through the recording several more times at a continuous but slower rate. Others view the sequence backward before starting to study it in detail. Whatever the pattern, the general rule as noted above is that three to five repetitions are needed before full comprehension of the activity is achieved.

Are Video or Time-lapse Recordings Cost-Effective?

It costs money to implement a video or time-lapse recording program. The costs for buying cameras and related equipment and for paying for the time of those who take, view, and analyze the films are significant. The gains are often less obvious. Even when improvements are found, seldom will efforts be made to record the dollar savings. And many of the less-specific gains, such as improved job cooperation, communication, and morale, can probably not be quantified at all. It is unfortunate that only a relatively few cost-conscious constructors or those with jobs in trouble use such recording methods. Too many have found the technique a successful way to rescue a job, and yet let the equipment lie unused in a closet on the next job.

A few of the tangible and less-tangible gains that video and time-lapse recordings can bring can be summarized as follows:

1 Viewing film or videotape in contrast to on-site observation with its many interruptions and distractions can focus full attention on the task being stud-

ied, thereby leading to a better understanding of all aspects of an operation. For example, by collapsing the passage of time, questionable activities or interrelationships that may have gone undetected in real time are found. Again, the ability to stop, back up, rerun, and change speeds permits a much more critical examination of the activities of individuals or groups and makes it possible to assess all the facets of an operation.

2 Watching themselves work on the screen provides workers, foremen, and supervisors an easily understandable way to study an operation and suggest ways to improve productivity, quality, and safety.

3 Visual records are particularly effective when undertaking detailed studies of multicraft crews, since it is only a matter of rerunning the same film or tape until the needed detail is obtained. In this way, the activities of each crew member and each craft can be examined with respect to their relationships with the others. Inefficiencies and unbalances, among which could be interferences among individuals or crafts, shortages in materials, equipment or tools, work habits that are customary but ineffective, minor changes in procedures as cycles are repeated, or work that will produce an unneeded refinement, are disclosed. Often these practices are unplanned, but they may pass unnoticed or may be considered immaterial when management observes an operation in the field. On viewing the film or by doing a detailed analysis, patterns or inconsistencies may become apparent as they appear again and again. Both call attention to procedures that bring unnecessary costs and lower productivity.

4 Because visual recordings are interesting and easily understood, crews and managers remember and discuss the task and consider how improvements in the specific operation and others might be made. They gain particular satisfaction and knowledge from analyzing the work of other crews and proposing ways to improve them.

5 Film and videotape provide a detailed and indisputable record. This is advantageous in several ways, among which is:

a The records can clarify happenings or settle disputes among contractor or subcontractor personnel or with engineers or owner's representatives about what was done, what or who was responsible for delay or other difficulties, and support or disprove the accuracy of cost or other written or oral reports.

b Film and videotape can provide excellent support for or refutation of claims. Often it makes situations so clear that they are settled directly by the parties involved. If not, the film or tape is excellent support for witnesses in arbitration or court cases.

6 Time-lapse films and videotapes can be excellent teaching mechanisms. They are much better than lectures in creating analytical, questioning, and idea-sharing attitudes that make learning effective. Also, they can educate management and crews about good procedures developed on earlier jobs. Among the many examples that might be cited are handling, repairing, and overhauling equipment; a tricky forming or erection scheme; or the adjust-

ment and operation of a new machine or plant. The technically untrained can learn by watching the trained and thereby improve their skills. Providing a roll of film or tape is far cheaper than sending a foreman or mechanic a thousand miles away to school. A library of how-to tapes and films can be particularly useful by allowing individuals the opportunity to learn in a few hours the basic knowledge they need to do a complicated task, overhaul a piece of equipment with which they have no prior experience, or brush up their techniques by watching an expert perform the task. Many good mechanics or craftsmen not well-versed in technical drawings are uneasy when working from a parts book or written instructions. Giving them a sequential photographic record of the proper way to disassemble and reassemble a complicated piece of equipment can ensure that the work will be done in a shorter time and done correctly the first time.

Reactions of Workers and their Representatives to Films and Tapes

Over the past quarter century, during which time-lapse films have been taken of construction work, there have been some instances where craftsmen or their union representatives have objected. These reactions usually are holdovers from the past when, as reported earlier, unions and even government reacted strongly to stopwatch studies and the early managerial attempts to exploit workers by using them. In the past, labor union policies in some parts of the country, based on the supposition that on-the-job studies meant speed up and the elimination of jobs, have led to objections to time-lapse films. In general, however, in most areas, photographic methods are permitted and sometimes endorsed by unions, particularly where open-shop contractors are taking over a substantial part of the market. On open-shop projects there are usually no objections whatsoever.

In all situations, even those where no objections are expected, it is wise before undertaking filming to explain to both the workers and their representatives what is being done. The usual explanation is that management needs film records to improve or check on its own effectiveness. It should be made clear that filming will be done openly, that all employee questions will be honestly answered, and that in most cases workers will be invited to see themselves on film or tape. It should also be pointed out that better methods make the work easier and safer, as well as cheaper and faster. In short, it should be presented to workers as being an aid in doing exactly those things that a manager is charged with doing, so that a film or tape is really a way for managers to review their abilities. And experience has proved that when workers are shown the film, they are often anxious to help in achieving that goal.

It is sometimes charged that the workers will defeat the purposes for which film and videotape are taken by speeding up when being filmed but slacking off at other times. Actually, those who do the filming report that when photographing is first begun the worker's first reaction is curiosity. After the purpose of the study is explained and they have inspected the camera, they go back to work and soon are oblivious to its presence.

In summing up, worker resistance to time-lapse and video filming is seldom an obstacle if the reason for it is explained and the process is open. More often, managers who resist the use of video and film claim that workers will object, as a convenient screen for their own opposition.

SUMMARY

This chapter has described some of the practices that have been employed successfully to gather data useful to productivity improvement. At times it has suggested how particular findings can be employed almost directly to achieve that end. In many other instances the data gathering described here is only a step, with an analysis to follow, as discussed next in Chap. 8.

CHAPTER 8

PRESENTING AND IMPLEMENTING PRODUCTIVITY-IMPROVEMENT FINDINGS

INTRODUCTION

Data gathering on a particular work-face procedure or operation, as a preliminary step in productivity improvement, was discussed in Chap. 7. This chapter deals with techniques for organizing and presenting that data in forms that will be clear to managers and workers. The techniques must cover every aspect of the operation that is being considered and include information about both the setting and context in which the activity was going on, along with the details of how the task was being carried out.

As explained in Chap. 3, experience has clearly shown that organizing and presenting data and implementing the proposed solutions cannot be left solely to line management, because it is already fully committed to completing the tasks that are currently underway and to planning for those to come. The better practice is to provide time and resources so that a specialist or another designated staff person who collected the data can sort out the most promising activities among those that might be considered. This person can then organize and carry through a preliminary data analysis and prepare the groundwork for implementing the changes that are proposed. For this to happen there must be full and enthusiastic support and cooperation of line management at all levels, including consultation about the topics to be studied, provision of arrangements and a climate in which participation in analysis and decision making occurs, and assurances that implementation will be carried through. Probably the single most important reason that formal productivity-improvement efforts have failed is the assumption by management that the specialist or staff person can carry the ball without help. It just doesn't happen that way.

As indicated earlier in this book, solutions to many productivity-improvement problems are reached in a variety of formal or informal meetings. This chapter first offers guidelines for making such meetings effective. Following that, it deals with specific techniques for presenting and analyzing data to find productivity improvements, and it concludes with examples illustrating how these techniques have been employed successfully.

GUIDELINES FOR INFORMAL MEETINGS

Informal meetings to discuss the findings of work-improvement studies must not involve either too few or too many people. There are no magic numbers, but fewer than three or four or more than ten to twelve seem to be the limits. The idea is to involve enough people to get a spread of viewpoints, experience, and talents, but not so many that a free exchange of ideas becomes impossible. Those who should be invited might include the foreman, possibly a craftsman or two, engineers, and managers at appropriate levels. They must have a strong interest in the subject at hand. Also, as mentioned in Chap. 7, including someone who is knowledgeable but who is unfamiliar with the operation is often helpful, because this person has no commitment to or vested interest in present practices. If possible, persons who are generally antagonistic to others, who attempt to dominate every group or meeting they attend, or who are unreasonably defensive about their ideas and thus cannot be objective should be excluded. Even informal groups need a designated leader who can be impartial but maintain direction and order and when the situation is right, lead the group to a conclusion and recommendations. A well-accepted person, not necessarily the boss, and not the specialist who produced the data, should be in charge. Of necessity, specialists must be advocates and should not attempt to be otherwise. Moreover, presiding will take their attention away from presenting their information clearly and carefully.

Informal discussion of ways to use the gathered data to improve productivity can combine the talents, ideas, and experience of all levels and can often produce creative solutions. If the climate is right, all participants will bring their strengths and knowledge, ask questions, and contribute freely to the discussion and the development of solutions. Sometimes bosses are a hindrance rather than a help since they often find it difficult to shift from their custom of telling to that of listening. One stratagem is to have them attend, but under the condition that they listen and ask questions only to clarify situations. In all groups, it is a problem to get senior people to accept the fact that good ideas can come from those who lack authority or experience. One of the authors saw the payoff from getting this acceptance: In this case, the questions and ideas of a junior-high-school student visitor caused a subcontractor to change methods, thereby saving 20 days and thousands of dollars on the construction of a three-million-gallon underground tank.

If informal meetings involve viewing time-lapse, the film is usually first shown at a rate of four to six frames per second, as long as all participants are comfortable with that pace. Segments of the film that are of particular interest may be run several times, with the specialist supplying any data that group members request. Often giving a preview of the film to those in charge of the work being shown in the film is advisable. They can then be prepared to marshal facts and figures to defend the current methods if they so desire or, as often happens, to state that certain obvious problems have already been solved. Advance viewing at least precludes unpleasant surprises and lessens the chance of destroying morale or developing adversarial relationships. One way to avoid this difficulty when the film is shown to the entire group is first of all to point out good features of the operation and to emphasize that the aim of the session is to make a good operation even better. In any event, the possibility that such group sessions can be threatening to some must never be overlooked.

At informal meetings, plenty of time should be allowed for discussion, which might follow the brainstorming pattern described in Chap. 3. The specialist should record all ideas that are presented and any solutions and agreements that are reached, along with any assignments of responsibilities for implementing the agreed-upon solutions. These should be written up and copies distributed so that all have a record of what transpired and where responsibility for implementation, if any, was placed.

Often, these informal meetings develop solutions and the specialist can move on to studies of other problems. Alternatively, the group can conclude that more data must be collected or that detailed analyses such as those discussed further on in this chapter should be made and presented at a later time.

There are several benefits to be realized from informal group discussions. First, to release the creative ingenuity and inventiveness that are developed by such a group through mutually stimulating one another, pooling a broad base of experience, and generating constructive criticism. As ideas are tossed about, the process can rapidly transform vague ideas into tangible ways of improving methods. A second benefit has been described in various terms, but the phrase "psychology of participation" best describes the effect. By becoming intimately involved with the development of the plan for doing a job, individuals (supervisors or workers) become partners in carrying it out. Casting them in the role of consulting supervisors leads to an energetic, enthusiastic, and cooperative team effort.

Another benefit to an organization that can come from informal discussion sessions is that the channels of communication are opened in all directions—upward, downward, and laterally. For example, managers involved in each segment of a job can see one another's problems rather than operating as independent activities competing for limited resources of equipment and workers. Cooperation rather than overt or subconscious competition results. Furthermore, with full awareness and understanding of job-wide problems, managers can coordinate their work more closely with others. The experience of both managers and consultants has shown that this informal discussion technique becomes highly productive in a short period of time. Numerous examples can be cited where the free discussion and interplay developed in informal sessions aimed at productivity improvement have generated new approaches or methods that have brought major improvements in productivity and costs. Old-timers who have staunchly defended their practices have found their preconceptions unsupported by the film or other recorded data. On the other hand, technically educated experts learn that the old hands have ideas and tricks that can make their problems dissolve. What stumps one person is solved by another; together individuals can develop new and better ways of carrying out their mutual objective, which is to do the job in the most effective way possible. Some of the elements involved in making this communication process work more effectively are discussed in Chap. 10.

TECHNIQUES FOR DATA PRESENTATION

Careful preparation for data presentation, whether to an individual or group, is essential. A primary rule is first to think through the purpose of the presentation and the audience that will hear it, and then outline the elements that are important, arranging them in logical order. Only then should the decisions be made on which data should be presented and the exhibits that will be useful. Generally, these are in one or more of four forms. The first is photographs, still, video, or time-lapse, as discussed in Chap. 7. The other three are in words (text), numbers (digital), and graphics (analog). For most people, information in graphical (analog) form is most easily understood; a visual image is indeed worth a thousand words or numbers.

Graphical methods require drawings or sketches to show layouts and physical details and graphs and charts which present numerical data and the results obtained by observation. Depending on the situation being described, there are formats or methods for presenting the information that make them most easily understood by people with a variety of backgrounds. Those presented later in this chapter are suggested as representative of the usual practices. However, there are no set ways to display data graphically. Rather, one can and should feel free to select and modify displays to best fit the particular situation, as long as the methods are within guidelines, if any, set by the particular organization.

Today, there are many microcomputer software programs to assist in data preparation, analysis, and presentation that were not available even a few years ago. Among them are recorder-computer linkages through which raw data can be tabulated directly rather than by manual tallying. Other programs will easily do mathematical and statistical analyses and produce graphs and charts quickly and accurately. Also, pertinent information from a computer-based cost, material-control, or other record, if it exists, can be brought forward. Used properly, these computer capabilities can save time and greatly extend the preparer's capabilities.

It is essential that every member of a group be able to see the data being presented or any photographs, time-lapse, or similar exhibits used to augment them. If possible, each person should be provided with copies of essential items, so that the information can be referred to later in the discussion, or kept for future need. Only in this way, given the limitations of the human mind which prevent it retaining more than a few details (see Chap. 10 for further discussion), can discussions of presented materials be kept on track.

TECHNIQUES FOR DATA ANALYSIS

The specialist presenting the data collected for productivity-improvement studies or for their implementation usually prepares a critique of the methods, materials, tools, and equipment that were employed in carrying out the task being reported. In addition, the operation's interaction with others must be made clear. In sum, good data presentation allows those using it to answer six key questions. These were given earlier in Fig. 4-8 and were colorfully described years ago by Kipling, as follows:

I keep six honest serving men;
    They taught me all I knew.
Their names are What and Why and When
    And How and Where and Who.

In workday language these could be restated as:

- What is its purpose?
- Why is it necessary?
- When is the best time to do it?
- How is the best way to do it?
- Where is the best place to do it?
- Who is the best qualified to do it?

Using the six questions, and particularly the searching why, those evaluating the data must first examine every detail and question the suitability of the tools, equipment, materials, and place of work. Second, they must analyze the method in order to separate the necessary from the unnecessary or wasteful and the important from the unimportant. Following this will come a search for rational, simplified, and less-costly ways of carrying out the task. In this search for better approaches or new methods, managers must remember that the people who make up management sublevels and the work force must implement any proposed improvements. Their concerns about the physical requirements of the new approach and the relationships among people must be considered. Chapters 9 and 10 address these topics.

Detailed evaluation of tasks as proposed here is not a new concept. For three-quarters of a century industrial engineers and, more recently, athletic coaches, are among the many who have been using detailed time-and-motion, or methods, studies. They have demonstrated that better approaches and efficient teamwork are not the result of chance, good luck, or a well-intentioned supervisor and crew; rather, they are the planned outcome of study and efforts to reduce unnecessary work and wasted motion. However, the use of these approaches is not restricted to management. A motivated group of workers is a very important source of new ideas, and the workers are often in a better position to implement them. Many examples of worker participation in productivity improvement can be cited. To illustrate: on a recently completed project in the San Francisco Bay Area, the union was getting blamed by management for low production. To overcome this image, the foremen and crews became deeply involved with management in developing a microcomputer application which greatly improved management practices. After the scheme was implemented, productivity took a jump which greatly exceeded all expectations. It is interesting to note that the solution came through an upward flow of information about poor management practices, not from a speeding up at the work face.

From this brief discussion, it can be seen that productivity improvement of operations that are now being carried out begins by assembling data about them from all sources and then involving the appropriate parties, often including the workers, in developing improvements. Several ways of doing this are discussed in the remainder of this chapter.

ANALYZING WORK IN PROGRESS

The first step in improving work in progress is to understand and analyze how the work is currently being performed. Thus, the current method and the conditions surrounding it must be described in a way that makes clear what is being attempted and what is being accomplished. In simple situations, this can be done by verbal description, possibly accompanied by sketches. However, as shown in some of the examples given below and in App. C, presenting verbally all the relevant detail, even of simple operations, is difficult. One of the most satisfactory ways of overcoming this difficulty is to have recorded it previously by time-lapse, as already described in Chap. 7. A series of pictures offers details that are obvious and indisputable. Anyone can observe the work in progress by directly viewing the film. Furthermore, viewing can be carried out at a convenient time and place. For those who wish to gain added insights into an operation, as well as the specialist or staff person who wishes to analyze the operation in detail, the film may be run at slower rates. Particularly interesting sections may be viewed over and over; and if even greater detail is desired, individual frames may be studied one at a time.

Once the specialist has prepared the data, management must decide how the information will be presented and used. Under some circumstances an informal discussion with an individual or group of managers will bring implementation immediately. At other times the study findings and their use will provide the agenda for management meetings. In other instances, the data will be shown to the involved crews to gain their suggestions, interest, and support. And, where formally organized work-improvement groups or quality circles are operating (see Chap. 3), the data can provide a focus for the groups' efforts. Selling the findings of productivity-improvement studies requires the same creativity and thought as does the original data-gathering effort.

FORMAL TECHNIQUES FOR ANALYZING
WORK-IMPROVEMENT DATA

As indicated above, the term "formal" means that data analysis involves preparing arrays of numbers, tables, sketches, charts, and other displays, as opposed to examining the data as collected in its raw form. There are many formats for such formal reports, so a particular study may use one or several or may generate a composite that more clearly presents the information for a particular study. Photographic data are often an important part of the package.

Data collected by any of the methods outlined in Chap. 7 can when examined directly or in tabular form sometimes indicate where to look for improvements in productivity. Unfortunately, for reasons to be explained below, this approach works only under very simple conditions. To deal effectively with more complex problems requires analysis by some of the other methods outlined in this chapter or in Chap. 13.

Graphical presentations can serve several functions; among them a way of showing data, a method of analysis, and a means of communication. Of the many types and variations in use today, the two techniques described below, namely, crew balance and a combination of process charts and flow diagrams, seem to be the most useful ones, since one or the other or the two in combination usually cover most construction situations. There are other types and formats that can be employed; specialists must choose the one that they feel is best for the situation and for the data they wish to present. The particular format is not the end product, but only a tool to help marshal facts and data so that the way the particular task is being carried out can be shown and improvements in it be considered.

The crew-balance chart is used to depict the way all the individuals in a group use their time in carrying out a specific task at a particular work site. The process chart and flow diagram together describe processes where material is moving and being acted upon or processed by different machines and workers. The examples that follow will show how each can be used.

The most common application of these techniques is for operations that are repetitive. The cycle may be of relatively short duration (repeated within seconds or minutes), such as laying bricks and blocks or loading material with an excavator; or it can be of long duration, such as a multiday cycle, for example, raising a building a floor at a time. The advantage gained by studying repetitive cycles is that the benefits of any corrective action will accrue through succeeding repetitions. With such operations, a small improvement in output or a decrease in the time and effort required, when repeated many times, may create large savings.

Tasks that have less discernible cycles can also be excellent candidates for analysis and corrective action by one of these methods. A one-of-a-kind project is made up of many common everyday construction elements which involve materials and work skills. Excavation, concrete forms, concrete handling, steel assembly or erection, and almost all other tasks have repetitiveness, though the end result may be different. And, more important, observation and critical analysis of work methods, however done, provide techniques and ways of thinking that transfer readily to other tasks or to other operations, no matter how different. Good analogies are the skills of the skier and the golfer. The courses may be vastly different (as are construction projects), but the skills are easily transferable from one downhill run or golf course to another. Thus the skills learned by managers and workers from formalized techniques of analysis may be far more important than the improvements generated on that particular operation or project, because they can be carried on to other work in the future.

In certain situations, attempts to develop tabular data or crew balance or process charts and flow diagrams may show that there is no recurring cycle on what could be a repetitive task. This may be a clue indicating that there has been insufficient care in planning or managing the work. As pointed out in the discussion of learning curves in Chap. 6, there are often substantial savings from systematizing what have before been nonrepetitive procedures.

Examining Statistics or Tabular Data

There are situations in which an analyst or field manager can look at a table or a column of figures gathered in the field and from them (1) detect that something is awry, and (2) determine a credible if not optimal solution to the problem. The example which follows is typical.

Using Collected Data to Improve Concrete Delivery by Crane and Bucket A crane and 2-cubic-yard bucket were being employed to hoist concrete delivered in ready-mix trucks to the higher floors of a high-rise office building located on a restricted downtown site. There a crew placed concrete for the columns, beams, and slabs. Costs were far over budget because the placing crews were almost continually waiting for concrete. In desperation, the superintendent engaged a consultant to record concrete-delivery arrangements by time-lapse.

The accompanying table shows in the first column of figures the times recorded for each step in the concrete-transfer operation under the as-found condition. With it, only one truck at a time could be spotted under the crane and only one concrete bucket was used. The second column gives times after space was made available to spot two ready-mix trucks side by side, and a second bucket was provided. The third column reflects the change under the two-spot, two-bucket plan, with concrete deliveries greatly improved.

Steps in Cycle: **Average Duration in Seconds per Bucket**	**As found**	**Change 1**	**Change 2**
Waiting for concrete trucks	151	151	0
Delay while positioning trucks	24	0	0
Spot bucket	34	30	30
Switch hook to second bucket	0	18	18
Fill bucket	42	0	0
Hoist full bucket	30	30	30
Swing full bucket	30	30	30
Place concrete	59	59	59
Swing empty bucket	25	25	25
Lower empty bucket	30	30	30
Totals	425	374	222
Percent decrease from as found		12%	48%

This simple study illustrates that substantial savings can be found in actual situations by looking critically at the elements that make up an apparently straightforward operation. In this instance, time-lapse would not have been needed; a stopwatch would have sufficed. The improvements involve two elements. Change 1 was within the operation itself; the person supervising the concrete placement had failed to recognize the effects on output that the two-truck spot and second bucket could have made, a situation that is often seen on actual projects. The second and dramatic savings came by going outside the contractor's organization and looking at the system. In this instance, the ready-mix concrete supplier had grown so tired of having his trucks delayed on this site that delivery to it had a very low priority. When it was demonstrated that trucks would be unloaded promptly, it was to the supplier's advantage to make a strong effort to supply the project promptly.

This example has not examined many of the other effects that the change would have on the project. Two among many obvious but unanswered questions are:

• Can the increase in concrete delivery be accommodated by the existing forming, placing, and other crews, or are the savings imagined rather than real?

• What positive effects can the lessened demand for crane time for concrete placing have on other project activities?

Readers should not assume from this simple example that most construction improvements can be found by merely examining a row of figures and from them finding all the potential improvements. In most cases, decision making is far more complex, a topic discussed in some depth in Chap. 10. In particular, the seven-plus-or-minus-two limitation on human reasoning power makes judgments based on simple tables or rows of figures hazardous at best.

Crew-Balance Charts

Industrial engineers for many years have employed a man-machine chart, which has been very useful to them in analyzing the effectiveness of worker-machine combinations. They have been adapted to construction under the name of crew-balance charts and offer an effective way to show the interrelationships among the activities of individual members of crews along with the equipment they employ. Before the chart can be made, the time devoted to each element of the activity by each person and machine must be observed and recorded. As explained in Chap. 7, this can be done by timing and recording the activities of each in turn with a stopwatch or by taking time-lapse film. Ideally, the times should be taken for several cycles to validate their accuracy and the variation among cycles. Often much can be learned by studying the best among the cycles or examining the reasons for the variations. The obvious advantage of working from film or videotape as contrasted with stopwatch methods is that the times for each worker and machine can be obtained from a single cycle and verified by viewing several others. In contrast, with the stopwatch method employing a single recorder, times for each worker and machine must be taken from different cycles. In this case, if there is substantial variability between cycles or in the incremental times, the data are far less reliable and useful.

The crew-balance chart has vertical bars representing each person or machine element. The ordinate is time, shown either as recorded or as a percentage of the total cycle. Each bar is subdivided to show the times devoted to each of the various types and sequences of activities that make up the entire cycle, including idle, ineffective, or nonproductive time. Figures 8-1 and 8-2 are typical crew-balance charts. Since each element of time for the crew and equipment being observed is plotted to the same time scale, the interrelationship of the various elements of the activity can be seen by comparing them along any horizontal line on the chart.

A graph such as the crew-balance chart allows one to compare interrelationships among the tasks assigned to the various members of the crew and equipment and to appraise the amount of nonproductive or noneffective time of each. Very often one can find a pattern of noneffective time that moves across the full time cycle. By rearranging work assignments among various members of the crew, noneffective time can be reduced and productivity increased. Analyses such as these often suggest that the crew size could be modified.

It should be made clear that crew-balance plots such as those shown in Figs. 8-1 and 8-2 do not necessarily demonstrate the effectiveness or efficiency of an operation, since being busy is not synonymous with using a good method. Again, they do not necessarily reflect the work pace. However, where appropriate, the diagram can be coded to reveal these inefficiencies or greater or lesser levels of activity. Again, enforced idleness where an operation is machine-paced and a worker must wait (as in hoisting or lowering) might be coded to show that the idleness cannot be avoided unless the machine, process, or work assignment is changed. By using codings or other

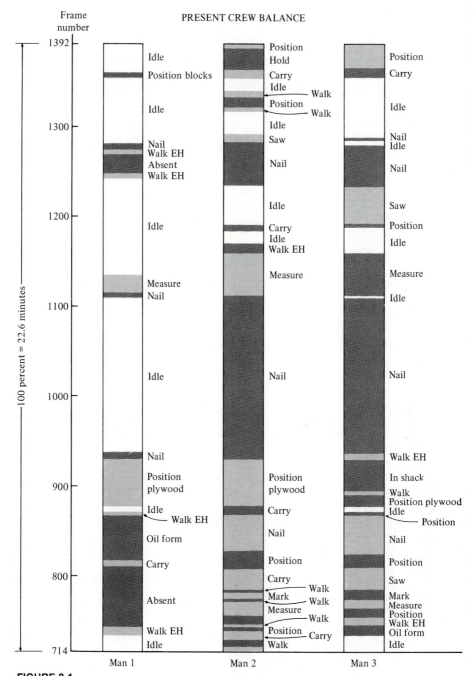

FIGURE 8-1
A crew-balance chart (taken from time-lapse film) of form panel construction showing the detail and sequence of each activity.

222

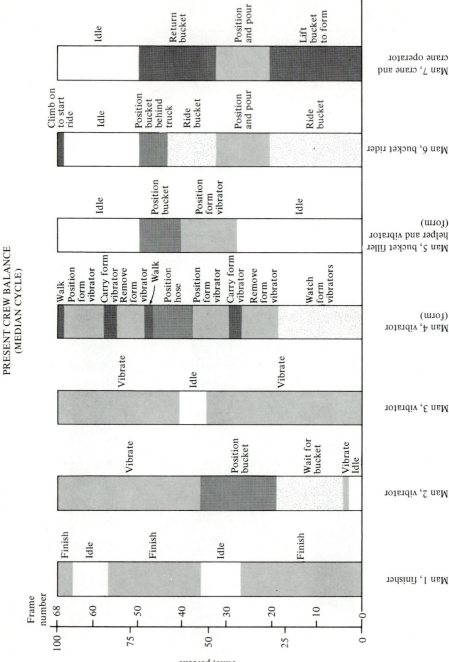

PRESENT CREW BALANCE
(MEDIAN CYCLE)

FIGURE 8-2
A crew-balance chart showing the sequence of activities of a crew in a typical cycle of placing concrete for a beam.

special markings, charts can then be drawn in a manner that portrays specific information of value in a particular analysis.

As indicated, specific forms of information can be clearly portrayed on the crew-balance chart by coding with different colors or with a variety of cross hatchings. There is no limit to the amount of information that can be shown. Some common and useful categories are:

- Working productively
- Working, but at less than normal capacity
- Doing work paced by controllable (throttled) machine speed
- Doing work paced by uncontrollable (governed) machine speed
- Standing by (necessary but unproductive)
- Holding (while someone else works)
- Transporting material
- Doing rework or repair
- Doing nothing (idle time)
- Waiting for instructions
- Waiting for another person to finish (space conflict)
- Waiting for material delivery
- Waiting for a tool or machine

By categorizing the activities of the various workers and machines directly on a crew-balance chart, it is often easier to see where resources are being wasted and where savings may be made. On the other hand, if the chart becomes too complex some of its value may be lost. The best procedure is to show things that are pertinent to the problem at hand and to omit details that are not. As with so many other uses of graphical techniques, the aim is to clarify situations under which all the details cannot be retained for consideration in the analysts' short-term memories. As noted earlier, this limitation is discussed in more detail in Chap. 10.

As already suggested, in addition to being applicable to work that is clearly cyclical in nature, the crew-balance chart can also be used to evaluate work that on first examination has no discernable overall cycle. As an example, refer back to Fig. 7-13, which charts the placing and finishing of a concrete floor slab. In this instance, the crew does two separable tasks. One group (laborers) does placing, shoveling, and vibrating; the other (cement finishers) does screeding and finishing. In this instance, it might be desirable to chart the operations of the two groups separately. On the other hand, by charting the entire crew together, savings from, for example, exchanging workers between them might offer ways to improve productivity.

The crew-balance charts shown in Fig. 8-3 are those for an actual case study made during construction of a high-rise building. That such a wasteful condition existed is testimony to the fact that someone was too busy to pay attention to the method adopted by a small crew that was doing a simple task assigned to it. It must be assumed that the boss was otherwise occupied and more concerned with other problems that were obviously in need of attention.

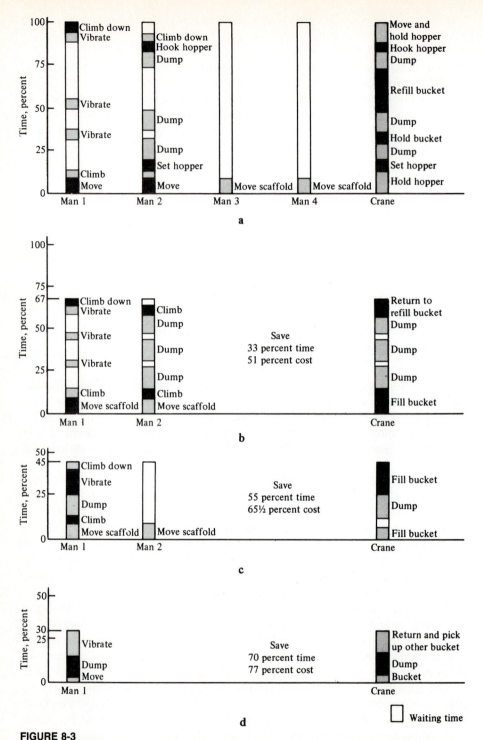

FIGURE 8-3
Crew-balance charts: analysis of concrete placement in a column: (*a*) present method; (*b*) first revision; (*c*) second revision; (*d*) third revision.

When this situation is examined in more detail, one sees a crew of four work-ers and a crane engaged in placing concrete in free-standing column forms. The concrete was supplied by a crane and bucket, and a funnel-shaped wooden hopper was used to direct the concrete from the bucket into the top of the form. Figure 8-3a shows the actual crew-balance chart as constructed after observing the time-lapse record for six cycles of work. Man 1 and man 2 were doing the actual placing and vibrating of the concrete from a small portable scaffold. Man 3 and man 4 moved the scaffold from one column to the next as each concrete placement was completed. Since the job-built wooden hopper was large and heavy and was not attached to the concrete bucket, it was moved by hooking it to the bucket and having the crane remove, transport, and place it. The ineffectiveness of this operation can be seen from the large amount of waiting time shown in the figure.

The six serving men described earlier can be called upon to analyze the method used by the contractor.[1] First there is the question, ''What is the crew doing?'' True, it is placing concrete, but this is not the basic task. From a larger point of view, the crew is constructing a column. With such a basic task determined, one can become imaginative. Alternative methods are first sug-gested and discarded only when it is apparent that they do not fit into the con-text of the overall job. Until that decision is reached, however, the task at hand must be approached with an open, inquiring attitude; for from a collec-tion of several ''impossible'' ideas may come a good one.

Given that the answer to ''What'' is constructing a column, the next inquiry is, ''Why do it this way?'' Here one might suggest other ways of accomplish-ing the basic task. If one can rule out columns constructed of wood or steel and limit the material to concrete, a good question might be, ''Why not precast the column?'' One might also ask, ''Why use the crane at all?'' This question might not apply here, but it raises a very basic point: namely, if equipment is available, workers will use it, even if the task could be done more quickly and efficiently by a method requiring human effort.

Changing the time of concrete placement would raise the question, ''When is the best time?'' If it is decided that the column has to be cast in place, the time of placement is generally limited to the period after the floor on which it stands is completed, and before the floor and column above are placed, but not necessarily before the forms for the floor above have been set. Thus there is again a broader question that could be asked, ''Why construct this column be-fore the next higher floor is formed or framed?''

''What is the best way to do it?'' The limitations set so far are that the col-umns must be cast in place before the next higher-floor concrete is placed. From an examination of the crew-balance chart, it becomes apparent that the effectiveness of the four man crew is very low. Only when the scaffold is being

---

[1] This example may seem almost insultingly simple to knowledgeable construction managers, but the reasoning process that is developed here is the same that is used in dealing with far more complex problems.

moved will more than one person be doing productive work; all are nonproductive while the crane returns the bucket for refilling. The most obvious change would be to have the two men who place concrete (man 1 and man 2) move the scaffold and put man 3 and man 4 onto some other task. This change could result in a saving of the labor costs of two persons but would not improve the effectiveness of the task as a whole, and it might tie up an expensive crane for a longer period of time (presuming that there are other uses for the crane). At this stage, some might suggest that a ladder rather than a rolling scaffold might be used. This alternative probably should be ruled out for safety reasons.

Another alternative would be to eliminate the waiting period while the crane returns the bucket for refilling with concrete. Since the crane must hold the bucket while the crew changes columns, perhaps these two operations, i.e., changing columns and refilling the bucket, could be performed concurrently. Although apparently by chance, the bucket held enough concrete to completely fill a single-column form; however, the crane was also being used to move the wooden hopper between columns. Here another possibility may have been overlooked. A common solution to placing concrete from a bucket into a small receiving area is to attach a rubber funnel to the bucket. These are available for most concrete buckets; they travel with the bucket and fold flat when the bucket is placed on the ground. In any event, by adopting the workable changes pointed out here and diagrammed in Fig. 8-3b, a 33 percent savings in time and a 51 percent savings in cost over the original method is possible.

There is always the possibility that further improvements can be made. A general rule is then always to generate several different improvement schemes, keeping in mind that finding successive changes becomes almost a game, thus making it easier to generate other ideas.

A change that might be considered relates to placing the concrete in lifts, as is common in wall pours. This procedure is usually dictated by the desire to reduce the concrete pressure in the forms. However, it costs very little more to design column forms to withstand the stresses of a full liquid head of concrete for their full height. They can then be filled all at once. Thus, given a concrete bucket of sufficient capacity, it is possible to fill the form completely and then to vibrate the entire height in one operation. The sequence would then be to (1) place the vibrator in the bottom of the column form; (2) completely fill it with concrete; and (3) start the vibrator and slowly withdraw it, thereby vibrating the concrete progressively from bottom to top. Such a change would make unnecessary the several successive cycles of dump and vibrate. Note that this solution still requires two people to move the scaffold. As can be seen in Fig. 8-3c this change would reduce the time by another 22 percent, with a resulting savings of another 14.5 percent, based on the original cost.

There are still more possibilities that might be considered for reducing the cost of this simple concrete-placing operation. If, for example, the concrete placement in the column were to be deferred until after the forms for the floor above were in place, the scaffold would not be required at all. This also would eliminate the need for the person who mounts and demounts and moves the scaffold. This approach, if adopted, would result in a crew-balance chart as shown in Fig. 8-3*d*. It would give an added savings of 15 percent in time and 11.5 percent in costs over the solution proposed in Fig. 8-3*c* and result in total savings of 70 percent in time and 77 percent in cost, as compared with the original method. However, there are trade-offs which might make this solution unattractive. Among them are the possibility of concrete spills which would have to be cleaned up, interferences with workers carrying out other tasks preparatory to placing concrete on the deck, and hazards associated with swinging concrete buckets over workers' heads. Even so, the alternative may be worth considering.

Doubtless the reader can propose other changes in this apparently simple concrete-placing operation which can bring additional savings. Among them might be an entirely different scheme for delivering concrete or for precasting the columns at a different site. In any event, this discussion of methods may suggest a step-by-step way of thinking that can be valuable in planning or revising any work-face construction operation.

In sum, our discussion has shown that the crew-balance chart portrays the productive and unproductive elements in work cycles and work assignments in a way that can be clearly understood. From this should come ideas for eliminating the unproductive elements and rearranging the relevant and necessary activities into a coherent and effective whole.

At this point, a doubter usually argues that any good construction manager could have seen all these changes and probably would have instituted them. Why, then, bother with the chart? The answer is that a good construction manager for a successful contractor did not see them. He was too busy doing other things. Furthermore, his only reporting system, that for unit costs, gave him no indication that anything was wrong or that costs might be reduced. And one might even say to a doubter, "Try a similar approach on your own job and see what happens."

One important point that many construction organizations are learning is that these techniques and ways of thinking should be taught to all levels of management and sometimes to the craftsmen as well. Then if participants are given information and time to study what it implies, improvements such as the one described here can be found by informal meetings, organized approaches such as the work teams described in Chap. 3, or suggestion schemes as proposed in Chap. 11. Management's job becomes primarily one of reviewing and implementing the proposed improvements and seeing that those who found them are rewarded.

The purpose of the simple illustration given here is to demonstrate the crew-balance chart and the way of thinking it can generate. Many far more complex applications could be cited. A few are given in App. C.

Flow Diagrams and Process Charts

Another commonly used graphical aid for the work-improvement analyst is the combination of flow diagrams and process charts. Industrial engineers use them to analyze situations in which materials are being processed in succeeding steps. Flow diagrams effectively show operations as diverse as routing invoices through an accounting department or moving individual parts in an automobile-assembly plant. Basically, the flow diagram is nothing more than a line sketch showing the movements of people or things.

The process chart which accompanies the flow diagram is a chronologically arranged listing of the various steps in an operation. Usually each item carries a symbol which classifies it in terms of its general nature. Figure 8-4 shows the five symbols which have been widely adopted, as follows: (1) a small circle: something is being done to the item, but only at a single location; (2) an arrow: the location of the item is being changed; (3) a square: the item is being checked or inspected; (4) a half circle: the item is temporarily stopped, delayed, or otherwise held up; and (5), a triangle: the item is in storage. Only one symbol can accompany a given listing in the process chart, thereby preventing the analyst from using too coarse a breakdown in charting the operation.

The two most important activities, a do operation and transportation, deserve further clarification. A do operation involves a physical or chemical change or a process whereby something is being created, modified, assembled, or disassembled. In work-sampling terms, productive work always falls into this category. Nonproductive work, such as required makeready and cleanup, can also be classified as do operations when they are carried out as distinct steps. Any movement of material integral to an operation such as pick up and lay down is usually included in the operation step.

Transportation covers movement between work stations, as well as movement during an operation that involves several steps to obtain or dispose of an item. Transportation is not included as a separate activity if no extra effort is expended in making the transition or if it is incidental movement between two operations or between an operation and some other activity. For example, no transportation is involved when a concrete mixer dumps into a hopper using a

FIGURE 8-4
Process chart symbols adopted by the American Society of Mechanical Engineers.

Symbol	Name	Result
●	Operation	Produces, changes
→	Transporation	Moves
■	Inspection	Verifies, checks
◗	Delay	Temporary storage, interference
▼	Storage	Keeps

short gravity chute, because no extra effort was expended. Again, rock blasted from a high bench which falls to the quarry floor to be loaded with a shovel involves no transportation subsequent to the blast. Here, even though the rock's location was changed, it moved by gravity in the blasting operation.

The detail into which an operation is broken down in making a flow diagram and process chart depends on the purpose for which it is used. It may or may not be logical to subdivide large operations into smaller operations or transportations. Clearly the operation "erect concrete forms" would involve many subdetails under the other classifications. But these will be of interest only when "erect forms" is the subject under study. It is important to the clarity of a study that the level of breakdown among the individual operations be fairly constant; for example, the operation of "hoist form into place and secure it" is not comparable with "insert form bolt."

After a sequence of existing steps in an activity has been properly sketched and analyzed and the flow diagram and process chart have been prepared, the next step is to look for ways to improve the operation. Here again, the analyst must question the basic reasons for doing each step, looking always at the broader phases first. Why is the step necessary? What is its purpose? Can substitute materials be used, or can a completely different method be employed? Where is the best place to do it? When is the best time to do it? Who is the best qualified to do it? After answers are suggested for these searching questions, a solution can be sought.

The most obvious savings found by analyzing flow diagrams and process charts generally involve excessive or duplicated transportation. Rearranging the positions of the separate "do" operations can often greatly reduce the need for ineffective movements. For example, work-station locations that come about almost by chance or are carried over from earlier operations sometimes are continued despite their inefficiency or uselessness. Because managers are too busy to think through the upcoming operation, they simply carry over earlier arrangements. Flow diagrams and process charts can be powerful tools for revealing such ineffectiveness. But as with all other methods for finding productivity improvements, one should not stop with the first but should develop at least two or three new ways of doing the task.

An Example of Separate Crew-Balance, Flow-Diagram, and Process-Chart Applications

This example of the use of flow diagrams and process charts is a simple one. Others are given in App. C. It involves installing short timber joists between steel floor beams in an industrial building (see detail, Fig. 8-5). With the present method, lumber to make the joists (1) was hand-carried as full-length 2- × 8-inch timbers from a stockpile to the second floor of the buildings; (2) was cut in a two-step process, into 5-foot lengths and notched, and (3) was installed and nailed into place. A flow diagram and process chart for this method are shown in Fig. 8-5a. A study of this figure shows that under

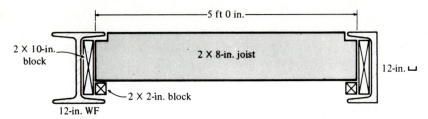

Detail of joist installation

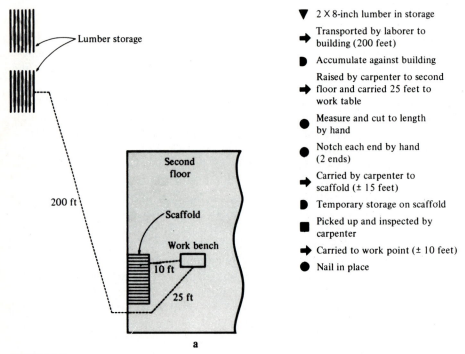

FIGURE 8-5
Flow diagrams and process charts: analysis of a lumber-cutting operation: (a) present operation: the raw material is delivered to the building where each piece is cut by hand and fitted into place; (b) revised method: the material is precut and delivered to the work site on wheeled pallets.

this method there were three do operations, four transportations, and one inspection.

There are many possibilities for improving the way this task can be done, with the final choice depending on a variety of factors. The first question might be, as discussed in Chap. 4, "Is this a system problem, or one that can be

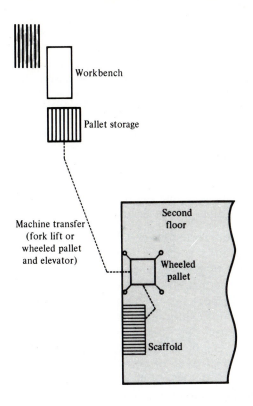

▼	2 × 8-in. lumber in storage
●	Jig cut to length
▶	Accumulate 10 joists in notching jig
●	Notch 10 joists
▼	Place on wheeled pallet
➡	Transport by fork lift or elevator to second floor
▶	Stored on pallet until needed
➡	Joists placed on scaffold
▶	Temporary storage on scaffold
●	Nail in place

SUMMARY

Step	Present	Proposed	Save
Operation	3	3	0
Transportation	4	2	2
Delay	2	3	−1
Storage	1	2	−1
Inspection	1	0	1
Distance	Manual 250 ft	Machine 200 ft	
		Hand 2 ft	

b

FIGURE 8-5 (Continued)

solved by the foreman or craftsmen?'' This question must be answered first in order to determine who will be involved and who will make the final decision. For example, the question, ''Can the shape of the joists be modified?'' is a system problem because it involves changing the design. If it cannot be changed, we must then work at delivering the same end product rather than looking for a solution of the more basic problem of side support for the steel beams, even though under different circumstances, this might be an excellent subject for study.

Using the outline given in Fig. 4-5 as a guide and assuming that system problems are to be considered, a few of the questions that can be raised are:

1 Why cut the joists on the job site? Wouldn't shop- or mill-cut joists save time and money?

2 Why move the material so far by hand? Couldn't the lumber be delivered to the second floor by hoist truck or forklift?

3 Why do the cutting by hand? Gang sawing with a power saw, as well as using jigs, could result in significant savings.

4 Is a power nailer used? If not available, could the toenails be started in the joists at the workbench, thus saving time for the installer?

One possible substitute for the present method of providing the joists is diagrammed in Fig. 8-5b. With it, the cutting operation is done at the lumber storage pile. Since it can be done at any time before installation, it may serve as a fill-in operation to use the spare time that often develops at the end of a work period. After cutting, the finished pieces are stacked on a wheeled pallet which is delivered to the second floor by elevator or forklift, again at a time when these facilities are not otherwise busy. The work platform, which should be designed for one-man manipulation, can be reloaded with joists from the wheeled pallet each time it is moved as the work progresses. The inspection was eliminated as unnecessary and expensive. Since the increases in storage and delay times do not cost anything and actually allow for independence in scheduling succeeding operations, they can actually represent savings, although unquantifiable ones. The laborer was replaced by the occasional use of a forklift or hoist; and the hand sawing and measuring with a tape[2] were replaced by the more efficient power saw and jig. This particular method, which involves changing both the system and work-face practices, is clearly only one of many changes that might be suggested, and it may not be the best.

An Example Combining Crew-Balance, Flow-Diagram, and Process-Chart Applications

Another example will demonstrate the use of the crew-balance chart combined with the flow diagram and process chart. In this case, the task is to cut tunnel blocking and cribbing which is used to fill the space above the steel ribs supporting the tunnel. A carpenter (power-saw operator) and usually three laborers (although the number could vary from two to four) have the task of cutting rough-sawn timber purchased in random lengths into shorter pieces. Although special lengths are sometimes prepared, almost all are cut from the several standard timber sizes into about 10 prescribed lengths. The crew's workday

---

[2] As demonstrated in Chap. 5, the unnecessary use of measuring tapes is one of the most wasteful work practices.

starts when a crane places several truckloads of lumber on the ground beside the power cutoff saw. The crew then feeds the long random lengths of lumber to the sawman and piles the assorted cut pieces into neat piles beside the railroad to the tunnel.

The first step in analyzing this job is to reduce the overall operation into its elements or tasks. We can list these tasks and the workers involved in each as follows:

1 Carpenter (sawman) and laborer 1: Move piece of lumber from pile to saw table.

2 Carpenter (sawman) and laborer 1: Cut lumber into designated lengths and push pieces off saw table.

3 Laborers 2 and 3: Carry cut pieces to pile and stack them.

Figure 8-6a shows the physical placement of the materials and equipment and the flow diagram for this simple operation.

The next step is to catalog the interrelationships of these activities. A crew-balance chart is convenient for doing so (see Fig. 8-6b). From this figure we can see the interrelationship of the activities of the various craftsmen and their principal piece of equipment. Although this represents their normal interrelationship, there are times, especially near the end of a day, when the remaining supply of lumber is relatively far away from the saw. Then laborer 3 helps laborer 1 bring lumber to the saw table. This change keeps the saw busy for short periods of time until the whole crew, including the sawman, must stop in order to stack the cut pieces. Operating procedures would be different if the

FIGURE 8-6a
Analysis of cutting operation—lumber for tunnel lagging, flow diagram for present method.

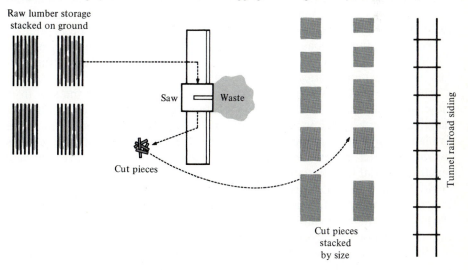

Raw lumber storage stacked on ground

Saw   Waste

Cut pieces

Cut pieces stacked by size

Tunnel railroad siding

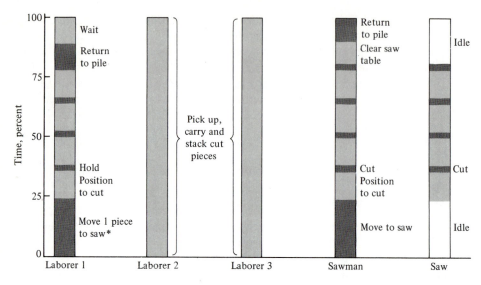

*Minimum time, any increase will decrease output

FIGURE 8-6*b*
Analysis of cutting operation—lumber for tunnel lagging, crew-balance chart for present operation.

crew had two or four laborers rather than the usual three. In cases where there are only two laborers, the operation is slowed down from time to time because all hands remove and stack the cut pieces. On the other hand, with four laborers, two of whom load the saw table, the extra person is not fully utilized inasmuch as only one person is required to help slide the material along the saw table. These special situations are merely noted but not addressed here.

A flow diagram such as that in Fig. 8-6*a* is usually accompanied by a more formal process chart using the standardized symbols for operations, transportations, delays, and storage. This diagram is given as Fig. 8-6*c*.

From Fig. 8-6*a* it can be noted that the lumber can be fed to the saw from either side, depending on the storage location of the uncut material. Furthermore, the storage place for cut pieces is not symmetrical. If more of them are available near a given end of the saw bench, the storage is greater there. This overall material and work-station arrangement results in an irregular and lengthy material flow line (see Fig. 8-6*a*), indicating that perhaps the material could flow in a shorter, more direct route. Note, for example, that because the benches run out from the saw, the material must be carried around the ends of the benches. Also, we can see from the crew-balance chart (Fig. 8-6*b*) that neither the saw nor the laborers are working full time.

The next step, as proposed earlier in Fig. 4-5, is to question each detail of the process. One of the variables not yet indicated is that since the demands in the tunnel for particular lengths and sizes of the cut lumber often change on very

Symbol	Name	Transportation, feet
▼	Lumber in storage	
➡	Carry piece to saw table (carpenter and laborer)	15 + 15
●	Cut to length	
➡	Push cut pieces to ground (slide)	
◗	Material temporarily stored on ground	
➡	Transport to piles by hand (2 men)	40 + 40
●	Stack pieces in piles	
▼	Materials stored in piles	———
	Total	110

FIGURE 8-6c
Analysis of cutting operation—lumber for tunnel lagging, process chart for present operation.

short notice, the answers to "when" would be "on a daily or even shorter basis" and "where" would be "at the specific location where that tunnel crew is working." Because of the union agreement, the answer to "who" is "the workers who are doing the job." Thus, what, where, and who are givens. In this case, when, what, and how are therefore the primary questions with which to analyze the operation. The answer to what ("What is the purpose of doing the job?") is to reduce random but long lengths of lumber to the stipulated lengths as required in a given segment of tunnel and to stack them for delivery there by the train. In answer to how ("How can this be done best?"), gang sawing was not considered to be the practical answer, since the sizes and lengths were changed frequently. The stock came in random lengths because timber in such lengths is cheaper. It was reasoned that by using a single cutoff saw, the sawman could schedule his cuts to reduce the waste. Thus the problem resolved into a most common one—making better use of the facilities and work force already available.

Two new methods were proposed for this job, each similar except for the physical layout where limited working space dictated the final choice between these two. The adopted solution involved rearranging the work area and substituting roller conveyors for hand carrying. Figure 8-6d shows the new area layout and flow diagram; Fig. 8-6e, the new crew balance chart; and Fig. 8-6f, the revised process chart and a comparison between the methods.

The new method offered the following favorable consequences:

1 It was possible to reduce crew size by one.
2 Although it did not change the number of operations, it made them flow more smoothly and reduced travel distances. The original method involved 110 worker-feet, the revised one 10 worker-feet.

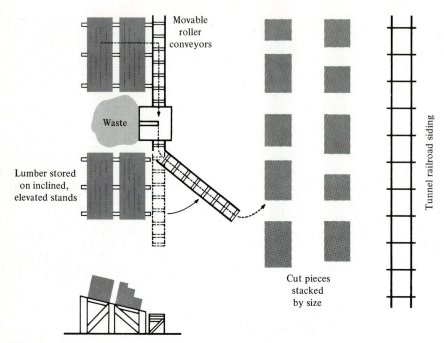

FIGURE 8-6d
Analysis of cutting operation—lumber for tunnel lagging, flow diagram for revised method.

FIGURE 8-6e
Analysis of cutting operation—lumber for tunnel lagging, crew balance for revised method.

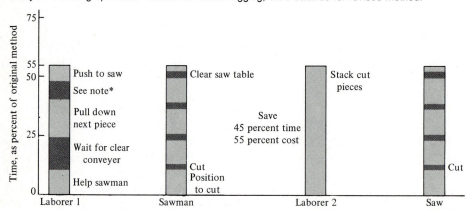

*Note: This time either waiting time or pulling down
next piece when extra time is required

▼   Lumber in storage on inclined racks

➡   Piece rolled to saw

●   Piece cut to length

➡   Cut pieces, roll to delivery point

◗   Intermittent storage

➡   Carry to pile (when necessary)

●   Stack in pile

▼   Store in piles

SUMMARY

Step	Present	Proposed	Save
Operation	2	2	0
Transportation	3	3	0
Delay	1	1	0
Storage	2	2	0
Distance	110 man-feet	0-10 man-feet 26-ft roller conveyor	100-110 man-feet

FIGURE 8-6f
Analysis of cutting operation—lumber for tunnel lagging, flow diagram for revised method, and summary.

3 Lifting, hand carrying, and handling heavy material, both very taxing and time-consuming chores, were replaced with gravity assistance, conveyors, and special tools, which were accomplished by rearranging the work area to give the material a better line of flow, by storing incoming lumber on elevated, inclined supports, and by replacing hand carrying with transporting on roller conveyors. Setting the incoming materials on the raised inclined bents rather than on the ground and providing roller conveyors made it possible for one laborer, assisted by gravity, to transfer the raw lumber from storage to the saw. This was done with special long- or short-handled tools with 1-inch pick points. With them, it was possible to reach the piles easily and to pull down individual timbers without leaving the work station. Also, with the conveyor, the saw operator could move the lumber along the table without assistance. Cutting time at the saw was reduced since positioning the timber was easier and the laborer could supply the next piece while the saw man was clearing the last piece to the delivery conveyor. Also, the roller conveyors were pivoted on the delivery side of the saw; thus the cut pieces could roll by gravity to the stacking location. This permitted handling of the stacking by one person instead of two.

The net result of these changes was a cost reduction of 55 percent per board foot cut and an output increase of 75 percent per day. Instead of working overtime to keep up with the demand, this crew was able to assist other crews in their work.

The other proposed method was to use a unidirectional flow of material (Fig. 8-6g), which would have made it easier to deliver material to the saw and to the person doing the stacking and could have utilized gravity to a larger degree for transportation. As stated before, space limitations precluded the use of this alternative method.

Added examples of the use of these techniques are given in App. C. All were developed using the approaches described here. As has been stated repeatedly, the techniques are merely an organized way of thinking that can pay large dividends. It is worth noting also that in each case, the physical effort has been reduced, making the work less fatiguing. Thus, with these techniques, crews work smarter, not harder. For this reason, workers seldom object to their use. Instead they are enthusiastic about them, often more so than management, which too often sees these techniques as a threat to its status and ego.

SUMMARY

This chapter has offered several techniques for processing and presenting data gathered as a first step toward improving operations at the work face. The

FIGURE 8-6g
Analysis of cutting operation—lumber for tunnel lagging, flow diagram for alternative revised proposal.

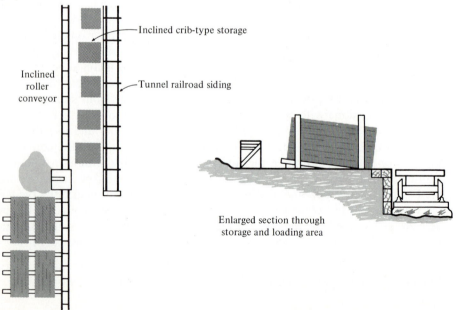

Inclined crib-type storage

Inclined roller conveyor

Tunnel railroad siding

Enlarged section through storage and loading area

techniques are simple—so simple that many construction managers reject them as unnecessary. But experience has demonstrated again and again that if they are incorporated into the project work plan, they can pay huge dividends. More advanced techniques which can also improve productivity are available or being developed by construction researchers. Some of these are presented in Chap. 13.

As indicated earlier, this book has been organized so that its users, particularly students, are introduced in this and the preceding chapter to hands-on techniques for gathering and presenting data useful in productivity studies. The chapters which follow expand on the brief references to people as workers and human beings, and on safety, both of which have been alluded to in this and earlier chapters.

CHAPTER 9

FACTORS AFFECTING HUMANS AS WORKERS IN CONSTRUCTION

INTRODUCTION

Construction is finally accomplished through the efforts of humans at the work face. First, of course, projects must be decided upon and site selection, financing, permits, and other necessary arrangements must be completed. This is followed by design, plans, specifications, development of owner and contractor organizations, and the assembly of construction materials, tools, and equipment. All these preliminaries finally lead to completion of work-face tasks by craftsmen supervised and directed by foremen. All too often, the physiological abilities and limitations and the psychological needs, desires, and aspirations of these key people and of those who direct them are largely ignored.

Although some employers still give relatively little attention to the needs of workers as human beings, management-employee relationships have changed dramatically during the past 80 to 100 years. Many management attitudes in that earlier time stemmed from the sweatshop, wherein workers were seen as beasts of burden. Employment involved long, hard hours of physical labor which brought little reward beyond food and shelter; with a few notable exceptions, there was little concern for the workers' physical well-being, and none for them as persons. Authoritarian, domineering management, with the threat of discharge and its consequences, was usual. This situation began to change in the late nineteenth and early twentieth centuries and today conditions are different. In construction in particular, the formation of unions and the advent of collective bargaining; legislation of several kinds at both federal and state levels; the institution of health, welfare, unemployment, and retirement plans; and sometimes a critical shortage of skilled craftsmen gave employees a stronger position. With the economic downswing and the increasing use of open-shop labor in the early 1980s, the power of construction unions has diminished, but this situation may or may not be permanent. Furthermore, the best craftsmen are always in short supply and have a choice among employers. In addition, some crafts, particularly those where employers are also union members, still set wages without effective management-union bargaining. In sum, management-employee relationships have changed from complete control by management to one in which labor may have equal and sometimes greater strength. Even so, many construction managers employ and craftsmen accept the old pattern of authoritarian top-down management, with its inherent weaknesses. Construction gets done, but often with difficulty, high costs, substandard quality, and time overruns, and with workers suffering unnecessary injuries or damage to their health. The question then is, can performance in construction be improved under different approaches?

This chapter examines construction personnel from two viewpoints: (1) as organisms, with the physical capacities and limitations of flesh and blood; and (2) as creatures subjected to unfavorable environmental conditions in the workplace. Chapter 10 considers them (1) as people with the needs, desires, and motivations (psychological factors) common to all humankind; (2) as decision makers, and (3) as communicators who transmit facts, ideas, directions, and feelings. These chapters draw on both research findings and on-the-job experience. An attempt is made to discover how recognition and possible mitigation of some of the adverse conditions in the workplace can gain cooperation and assistance from workers and supervision alike to improve productivity.

The concepts to be presented in this chapter have been developing since World War I under such titles as ergonomics, human factors, and industrial hygiene. Applications in industry, consumer products, space, and the military began in earnest after World War II. The aim here is to summarize the principles and applications that best apply to construction. The safety and environmental health aspects of the subject are discussed in Chap. 12.[1]

SOURCES AND CAPABILITIES OF CONSTRUCTION WORKERS

Workers enter the construction trades in ways far too numerous to recount here. Some do so through family tradition, where a son or daughter follows a parent into a particular craft. Others start with formal apprenticeship programs which, in time, lead to journeyman status. Still others get into the industry al-

---

[1] Recent books covering ergonomics and human factors include M. Helander (ed.), *Human Factors-Ergonomics for Building and Construction,* Wiley, New York, 1981; E. J. McCormick, *Human Factors in Engineering and Design,* McGraw-Hill, New York, 1976; and W. E. Woodson, *Human Factors Design Handbook,* McGraw-Hill, New York, 1981. Periodicals include *Ergonomics,* published by Taylor and Lloyd, London; *Human Factors,* the *Journal of the Human Factors Society,* Santa Monica, California (published through December 1978 by the Johns Hopkins University Press, Baltimore, Md.); and the *American Industrial Hygiene Association Journal,* Akron, Ohio.

most by chance, by working through employment agencies, following informal leads, or making inquires at union hiring halls or job sites. Among the compelling reasons that some individuals are interested in construction is the desire to work outdoors or with one's hands. Reasons for leaving construction are equally diverse. One is that construction jobs are not of long duration and new openings may be hard to find, particularly when the economy is depressed. Also in many areas, employment is seasonal. In addition, many find that construction jobs demand heavy physical effort which they find unacceptable.

In general, neither construction supervisors nor workers have thought seriously about humans as machines with special but finite physical capabilities. Seldom do they employ principles or measures such as those on fatigue and the need for rest from heavy tasks cited later in this chapter, nor on lifting ability, as outlined in Chap. 12; yet the evidence is strong that attention to such details may bring increases in worker productivity and reduce the potential for injury.

There are reasons other than productivity gains or accident reduction which may demand that construction managers pay greater attention to worker characteristics. Lawsuits and claims citing discrimination have become common where employers have denied jobs to individuals or to classes of people on the basis that they lacked the proper height or weight or were physically incapable of performing certain tasks. There is also the possibility of liability and increased worker's compensation premiums when workers are given assignments which overtax them. To avoid such problems, many employers now (1) as a part of the preemployment procedure, give standardized tests to qualify applicants on the basis of strength, lifting ability, agility, or similar characteristics, and (2) on the job, pay careful attention to work assignments to avoid excessive fatigue or overtaxing an individual's physical capabilities.

PHYSICAL FATIGUE

Human beings are organisms and as such have special capacities and limitations because they are made of flesh and blood. Superimposed on these characteristics are mental activities which affect their ability to perform work in a variety of ways. For the purposes of this book, all of these are considered to be physiological factors. Things that affect workers adversely are treated under the general heading of fatigue, which can be defined as "weariness from physical or mental activity." It is commonly subdivided into physical fatigue, mental fatigue, and boredom.

The measurement of fatigue and the need to set an allowance for it in productivity studies have occupied scientists, engineers, and doctors for decades. That a decrease in a human's ability to work takes place is accepted as fact, but quantifying this decrease and placing limits on its acceptable levels have not been agreed upon. This book does not propose to enter such arguments, but to bring to the reader's attention some of the results of studies in this field.

Physical fatigue, which is probably the best understood, largely results from an overuse of energy from short-term overexertion and the long-term requirements of the body, although other factors may come into play. It is discussed here under the subheadings of short-term, daily and weekly, diurnal (24-hour), and long-term fatigue associated with unscheduled or scheduled overtime. Mental fatigue refers to the inability to maintain high levels of concentration for extended periods of time. Boredom is the inability to maintain continuous attention to a task for reasons other than physical or mental fatigue. Environmental conditions, which also affect performance, are discussed later as a separate topic.

Short-Term Physical Fatigue

Humans As Machines   Humans and animals can be viewed as machines that develop and store energy from food, burn it with oxygen to sustain life, and, on demand, create movement and power through muscular contractions. As with other machines, there are limitations on the body's ability to perform, dictated by factors such as physical dimensions, weight, age, and condition. Also, being of flesh and blood, the muscles are self-renewing within limits, but subject to the limitations imposed by their composition.

Figure 9-1 diagrams in a simplified way the process by which humans maintain their bodily functions and produce or resist movement of the skeletal muscles. Foods, primarily carbohydrates and fats, are taken into the body through the stomach and intestines. By a series of complex chemical reactions,

FIGURE 9-1
The energy development process in humans and animals.

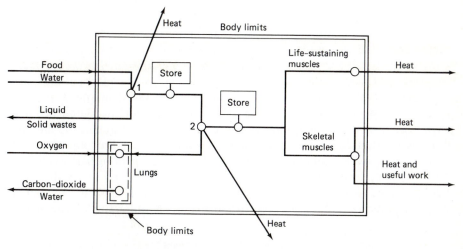

○ Points where age, skill, training, and other
factors can alter effects.

starches are transformed into glucose (sugar) and fats into glycerol. In this process roughly 5 percent of the energy in the food[2] is lost as heat. (point 1 in Fig. 9-1). If the body is at rest or activity is low, most of these substances are stored in the cells and blood for later use. In a subsequent reaction (point 2) these chemicals are altered to produce a high-energy substance called adenosine triphosphate (ATP). In this process, some 50 percent of the food energy is converted to heat. Then, by subsequent chemical reactions under involuntary or voluntary control by the nerves, the ATP reacts with oxygen and causes the muscles to shorten or to tense to resist elongation.

As noted on Fig. 9-1, cardiac and smooth muscles sustain the life functions; those attached to the skeleton cause or resist body movement. All muscular activities produce heat. Some also do work in the engineering sense by developing external forces.

The principal source of energy for producing muscular contraction or tensing comes from an aerobic reaction in which ATP is "burned" with oxygen. Less-strenuous activities are sustained by the small amount of oxygen retained by the muscles. This is renewed by the blood. Also, the blood carries away the lactic acid that is generated. Heavy muscular action also begins with a brief aerobic action. However, when the oxygen supplied by the cells and brought in by the blood is exhausted, anaerobic reaction (a chemical reaction without oxygen) begins. This is of very low efficiency: about 20 percent. The excess lactic acid remains in the cells of the muscles and can become concentrated enough to cause fatigue and pain and even prevent the muscles from functioning. During subsequent periods of low muscular activity or rest, the liver employs oxygen to neutralize the acid. In sports, athletes running sprints draw energy almost exclusively from anaerobic sources and afterward breathe heavily to quickly neutralize the lactic acid. Some claim that the second wind of those running longer distances occurs when the aerobic action fully supplies the muscles.

In discussing humans and animals as machines, it is important to distinguish between the energy consumed in contracting the muscles or holding them taut and "work" in the engineering sense, where work is computed as the product of force times distance. For example, muscular energy is consumed and dissipated as heat when a person walks on the level at constant speed or stands holding or supporting a weight. In neither instance is work accomplished in an engineering sense. In the first case, there is no external force in the direction of movement; in the second, there is no movement in the direction of the supporting force. On the other hand, some work in an engineering sense is accomplished when, for example, a person pushes or lifts a load or climbs stairs. Studies have found that of the energy made available in food, work in an engineering sense consumes between 10 and 30 percent of the total during normal

---

[2] By another complex series of chemical reactions, proteins are processed into forms in which they can replace body tissue. They supply little of the energy needed to activate the muscles.

working situations. The remaining energy is dissipated as heat within the body, as indicated in Fig. 9-1.

The human or animal machine has certain valuable features. To begin, it is largely self-renewing, replacing cells as needed. It converts fuel to energy at the relatively low temperature of 37 degrees centigrade (98.6 degrees Fahrenheit), far lower than other heat engines. It can react very rapidly, with less than a second between the apparent need to move and maximum muscular reactions. Furthermore, it can store energy in a quickly convertible form; normally, food is ingested only two or three times a day; and in a crisis, intake can be deferred for far longer periods of time. Also, although very little oxygen can be stored, a reasonable level of activity can be sustained for substantial periods of time by replenishing the supply from the air.

Three methods have been employed to measure energy demands and fatigue resulting from human activities. The first, that of Frederick W. Taylor, was to observe how much an individual could accomplish, given various ratios between work and rest times. For the second, the rate at which the heart beats during and after given tasks is recorded. The third involves measuring oxygen consumption. Each of these is discussed briefly in the paragraphs which follow.

Before 1900, Taylor and his coworkers, viewing human beings as machines, observed short-term energy consumption and fatigue. In one experiment,[3] they studied the work practices of several strong men who carried 92-pound cast-iron pigs from ground level up inclined ramps into gondola cars. Originally, output was 12.5 long tons per man per day. The experiment involved directing the men to work and rest different percentages of the time and observing the changes in production. Maximum output was raised to 47 long tons a day when the subjects worked 42 percent of the time and rested the remaining 58 percent. An extension of the experiment found that if the load were reduced to 46 pounds (half a pig), the optimum work-rest times were 52 and 48 percent, which resulted in a lower daily output.

Another of the Taylor studies is particularly applicable today. Around steel mills at the turn of the century, many materials were shoveled by hand. Some, such as iron ore, were very dense and heavy; others, such as ashes or rice coal, were porous and light. Through a series of experiments, it was found that maximum output was reached when shovels held 21 pounds regardless of the material being handled. To reach this target six or more sizes and configurations of shovels were stocked, and workers were told which of them to use.[4] The situation of many projects today is that only two kinds of shovels are stocked—square and round points with long handles—even though the cost of a shovel is less than half an hour's wages for a worker.

By the standards of his time, Taylor showed great consideration for the

---

[3] See Frederick W. Taylor, *Principles of Scientific Management,* Harper & Row, New York, 1911, pp. 40–60. This book was reprinted in 1967 by Norton, New York.
[4] See Taylor, op. cit., pp. 64–69.

well-being of workers, but his practices are often seen as dehumanizing and would not be acceptable in the labor climate in the United States today. On the one hand, he dictated to workers exactly how and how long to work and when they were to rest. Again, if the expected outputs were not reached after a trial period, they were reassigned to other tasks or possibly discharged if other jobs were not available. On the other hand, and far in advance of his time, Taylor advocated dealing with workers as individuals and providing pay ranging up to 50 percent over the usual levels for those who exceeded the expected minimum. One can only wonder if some of Taylor's approaches, modified to fit present situations, might not improve productivity today.

The other two approaches, namely, changes in pulse rate and oxygen intake, are employed in research and their findings are applied directly in work situations. These ergonomic concepts must be used as joint limitations on activity (that is, not as independent limits).[5] Each tends to establish a reliable, tolerable work limit for the average worker. One finding is that many previous work standards were set at unreasonably high levels.

Studies of pulse rates by E. A. Muller made over 30 years ago have shown that the pulse rate of a worker is a very sensitive and reliable indicator of overexertion and fatigue.[6] Since normal pulse rates vary widely, it is necessary to collect a set of data for each worker. Such a base can be established with a 20- to 30-minute test on a bicycle ergometer. Even without this test, however, pulse rate can provide a valuable criterion by which to measure fatigue. Muller reported that in general, an hourly increase of six pulses (heartbeats) per minute is a signal that a person is overexerting on a long-term basis. Further, he suggested an indicator (hourly or daily) of short-term fatigue that measured the ability of the body to recover from a temporary pulse-rate increase. It is called the recovery pulse sum—the accumulated number of pulses exceeding the resting-level pulse rate required before the resting-level rate is reestablished after work is stopped. If this sum exceeds 100, the work level is beyond a tolerable limit. For example, if the resting-level pulse rate is 80 per minute and the work rate is 120, then the total number of pulses accumulated between the rate of 120 and 80 after work stops would be the recovery pulse sum. To meet this criterion, and with a uniformly decreasing pulse rate, the minimum safe recovery period would be 5 minutes.

For various levels of activity, the oxygen-use approach to work and fatigue measures the oxygen that is converted to carbon dioxide and water in chemical reactions with food substances.[7] For every liter of oxygen used, 5 kilocalories of energy are produced. As indicated above, about 55 percent of the food en-

---

[5] See, for example, C. L. M. Kerkhoven, "Work Measurement Seen From an Ergonomic Viewpoint," *Proceedings of Seminar on Work Standards and Standard Allowances,* Syracuse University, Syracuse, N.Y., December 12, 1962, p. 13.

[6] E. A. Muller, "Ein Leistungs-Pulsindex als Mass Der Leistungsfahigkiet," *Arbeitphysiologie,* vol. 14, pp. 271–284, 1950.

[7] For a comparison of the three field methods for measuring oxygen consumption, see V. Loduhevaara et al., *Ergonomics,* vol. 28, no. 2, April 1985, pp. 463–470.

ergy is lost in the chemical-conversion processes. The remainder goes to maintain basic body functions such as breathing, temperature control, and blood circulation and, on demand, to contract or resist extension of the skeletal muscles. Thus, by observing the oxygen-conversion rate before, during, and after various activities, the demands on the human machine can be measured.

Using the kilocalorie[8], in the International System of Units (SI), researchers have found that the average young male adult—5 feet 8 inches (173 centimeters) tall, weighing 160 pounds (72.6 kilograms), in good physical condition— can develop energy over several hours at the average rate of about 5 kilocalories per minute. The energy required to sustain life processes (called basic metabolism) requires about 1 kilocalorie per minute, leaving a potential average excess capacity for muscular activity of about 4 kilocalories per minute. It follows that occupations with energy demands less than 5 kilocalories per minute (including 1 kilocalorie per minute for basic metabolism) can be carried on continually for a shift or more without being subject to human physical limitations. However, if tasks call for an energy expenditure greater than this, intermittent rest is required to recoup the energy taken from storage in the body. Recoupment is at a net rate of 3.5 kilocalories per minute—the 5 kilocalories per minute gross rate less about 1.5 kilocalories per minute demand as the individual rests on the job. For the average young adult female in good physical condition, all the values cited above are reduced by about 30 percent.

Energy Requirements for Construction Tasks   Approximate values for the total energy demands (including basic metabolism) for various construction activities are tabulated in Table 9-1.[9] The table shows that the energy demands of some less strenuous construction activities fall below the suggested control levels of 5.0 for males and 3.5 for females and therefore can continue without interruption throughout normal work periods. Other tasks, however, are more taxing so that they cannot be carried on continually; rest to allow recoupment will be required.

Data such as those in Table 9-1 are at best a guide in assessing human ability to carry out strenuous work tasks with or without intermittent rest. For example, among those of a given age, there are variations in overall body lean and fat weights, lung capacities, and blood-circulation rates. One rough rule of thumb is to adjust values directly in proportion to total body weight. Poor physical condition can result in a 20 percent reduction in performance and recovery rates. With age, heart-stroke volume and rate of beating decrease and

---

[8] 1 kilocalorie = 1000 calories, 3.97 British thermal units (BTU), 3086 foot-pounds, 1.162 watt-hours.

[9] Specific data sources include J. F. Parker and V. R. West (eds.), *Biometric Data Book,* 2d ed., National Aeronautics and Space Agency, available as No. AD-749877 from the National Technical Information Service, Springfield, Va., 22151; *Work Practice Guide for Manual Lifting,* National Institute for Occupational Safety and Health, March 1981; P. L. Burke, *Human Factors,* vol. 21, no. 6, December 1979, pp. 687–699; and Woodson, op. cit. Values for specific activities differ among sources but fall in the same general ranges.

TABLE 9-1

TYPICAL VALUES OF TOTAL ENERGY EXPENDITURES FOR VARIOUS TASKS (YOUNG
ADULT MALES AND FEMALES IN GOOD PHYSICAL CONDITION)

Activity	Energy consumption, Kcal/min*	
	Males	Females
Sleeping	1.1	0.9
Resting on job	1.5	1.1
Sitting	1.7	1.1
Standing	1.7	1.4
Doing office work	1.8	1.6
Driving car	2.8	2.1†
Walking, level, casual	3.0	2.2
Walking briskly, carrying 22 pounds (10 kilograms)	4.0	3.4
Driving a truck, local	3.6	2.5†
Bricklaying	2.5–4.0	1.8†–2.8†
Doing carpentry	4.0	2.9
Sawing with power saw	4.8	3.4†
Pushing wheelbarrow, loaded	5.5	3.9†
Doing average construction work	6.0	4.2†
Chopping wood	6.2–7.5	4.4†–5.3†
Shoveling sand	6.8–7.7	4.8†–5.4†
Doing heavy manual work (stonemason)	7.5	6.3
Sledge hammering	6.8–9.0	4.8†–6.3†
Digging, heavy activity	8.4–8.9	5.9–6.2
Continuous sawing and hammering	8.1	6.8
Running a marathon, swimming	10.0	8.8
Extremely heavy exertion	15–20	12–14

*All values include energy for basic metabolism.
†No specific data for women; set at 70 percent of values for men.
Source: Data from various references. Values are approximate at best and vary among sources.

the oxygen transport system operates less effectively, reducing the rate of oxygen intake by 20 to 25 percent between the ages of 25 and 60. For individuals
in these categories, either work pace will be slower or more rest will be required to keep up with demanding tasks. It should be noted, however, that
many older craftsmen who know how to "work smarter" and conserve energy, easily keep pace with or outdo younger workers who have not learned
the tricks of the trade.

Table 9-1 shows that on the average, women consume less energy in maintaining basic body functions and also that they devote less residual energy to
heavy tasks. Compared with males, the average female has less muscular
strength. Measurements have shown that on the average, women have 20 percent less body mass, 33 percent less lean body (muscle) mass, and 14 percent
more body fat. Body surface area for females is 18 percent less than that for
males. These factors, coupled with a lower blood circulation rate, explain why
women on the average have more difficulty than do men in carrying out heavy

tasks and why they are less able to adjust to heat and cold but have more endurance. Also, performance may be adversely affected for some women during the menstrual period. It must be recognized that figures such as these do not apply in individual cases. Many large, strong women in good physical condition can often outdo the average male. This could well be the case for women employed in some of the construction trades.

The data in Table 9-1 were computed for workers in the developed nations. Little quantitative information is available for those in the developing countries. The information in Table 9-1 would certainly require adjustment for differences in body weight and similar factors. Far more important, however, may be diet and food intake since, as pointed out earlier, fuel is necessary if the body is to develop the energy needed to do work. Thus behavior that has often been interpreted as lethargy or an unwillingness to work may actually stem from a meager diet, which, in turn, results in an energy deficiency that makes strenuous work impossible. Numerous instances have been cited where when regular meals or supplementary food were provided, such workers have been able to produce at high levels.

Performance requirements such as those cited in Table 9-1 can at best serve as guides in assessing the capability of workers to carry out construction tasks. At the same time, it shows that many work assignments cannot be carried on continuously, but require intermittent rest.

Figure 9-2 illustrates by an analogy with water storage how the relationship between work and rest times for an average worker can be quantified. There is a readily available reservoir of energy in anaerobic or aerobic form of about 25 kilocalories. Inflow into it is at the maximum rate of 5 kilocalories per minute.

FIGURE 9-2
Water-tank analogy of the human body's energy storage-replenishment capacity.

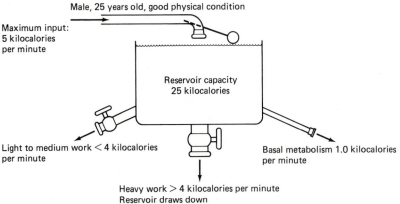

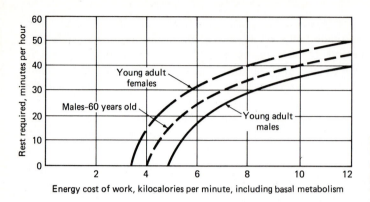

FIGURE 9-3
Rest required per hour for various levels of physical effort.

There is a steady outflow of about 1 kilocalorie per minute to sustain life, leaving up to 4 kilocalories per minute to supply the skeletal muscles. At energy demands lower than 4 kilocalories per minute, the energy reservoir remains full. At demands higher than this, the reservoir drains down. For an average construction task requiring 6 kilocalories per minute including basic metabolism, work at this pace could continue for no longer than 25 minutes before the worker became exhausted. An average male, sawing and hammering with an energy demand of 8.1 kilocalories per minute, must rest after about 8 minutes. In the extreme case of a drawdown of 15 kilocalories per minute, including basic metabolism, the stored energy would be gone in 2.3 minutes and muscular activity would stop. In all cases, recharge of the reservoir, which involves removing lactic acid from the body, is at the rate of 3.5 kilocalories per minute, since resting takes 1.5 kilocalories per minute of the 5 that are available. From exhaustion to full recharge takes 7.1 minutes.

The data on short-term fatigue indicate that for tasks too arduous to be continued through a shift, there is a constant ratio between work and rest times. These have been plotted in Fig. 9-3 in terms of rest required per hour for various energy demands, including basic metabolism. Data are for young adult males and females in good physical condition and for 60-year-old men. As indicated before, values are approximate at best. In using this information, one should keep in mind that performance is better if the work-rest periods are shorter than the maximum possible. Completely or nearly exhausting the body's store of energy, thereby increasing the amount of lactic acid in the body, can cause muscular cramps, pain, and other deleterious side effects.[10]

Techniques for Reducing Short-Term Fatigue   The supervisor making work assignments, knowing that heavy tasks cannot be performed continuously, should plan where possible to mix heavy and light work. In the light-work period, recuperation can take place, although less quickly than if the worker

[10] For verification of this "short-work, short rest" theory, see R. E. Janero and S. E. Bechtold, *Human Factors*, vol. 27, August 1985, pp. 459–466.

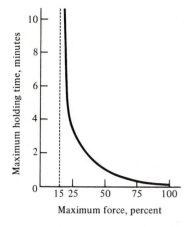

FIGURE 9-4
Rohmert's law for static work-muscle effort versus time. Graph shows the maximum length of time that a person can exert a muscular holding force as a variable of the person's strength. (*From C. L. M. Kerkhoven, "Rohmert's Law,"* Time and Motion Study, *July 1961, p. 29.*)

were resting. Another possibility is to schedule rest periods and use them for giving instructions, planning, or laying out subsequent assignments.[11]

It was noted earlier that the skeletal muscles require energy not only to contract but to resist movement. The time that this resistance can be sustained depends on the ratio between the required holding force and the maximum force that can be resisted. This characteristic of the muscles, known as Rohmert's law for static work, is illustrated graphically in Fig. 9-4. From this and supporting data it can be shown that a tensed muscle can sustain 15 percent of its maximum resistance indefinitely, 50 percent for a minute, but maximum force for only 6 to 7 seconds. Rohmert's law has important implications in task planning. First, it makes clear that fatigue results when muscles are under load, whether or not there is movement. Second, to prevent fatigue, tasks should be designed to avoid activities such as holding heavy loads or pushing hard against nonmoving objects. Hence tables, supports, props, jigs, fixtures, carefully designed tools, or other devices should be provided as substitutes for muscular effort. Here again, applying the engineering definition of work can provide a trap for the unwary.

Tool Selection to Reduce Short-Term Fatigue Construction workers employ a wide variety of tools and appliances to carry out their tasks. As will be discussed elsewhere in this book, failure to have tools of the right kind in sufficient quantity available at the workplace can be a major cause of lower productivity. Of concern here is the relationship between tools and short-term fatigue.

If a tool is used only occasionally and for very short periods of time, the main concern is how quickly, well, and safely the given chore can be done; this should be the principal control in selection. However, if the tool is employed for longer periods of time or its use calls for very strenuous muscular

---

[11] Casual observers or work samplers should recognize that those doing heavy tasks cannot work continuously and that periods of rest are essential. Otherwise, reporting will be unfair and inaccurate and lead to unwarranted criticism.

movement or tension or awkward positioning of the body, limbs, hands, or fingers, the user may have to rest at spaced intervals of time.

The design of tools for efficient use and reduction of fatigue has been studied intensively.[12] Among the fatigue factors that should be considered in matching tool to task and to worker are the following:

1 Characteristics of the tool
   - Weight, size, shape, temperature during use, vibration, presence of sharp edges or corners, protrusions, and openings
2 Requirements of the task
   - Feeling or seeing the point of application; the need to lift and lower, hold and carry, grasp, position, apply force, maintain posture, and reach
3 Worker considerations
   - Avoid heavy grasping, holding, or lifting or awkward or strained positions that tax the muscles
   - Reduce heavy contact pressures on body tissue or bones in the fingers, hands, arms, and body

A few examples of adaptations to standard tools to meet these criteria are:

- Pliers and wire cutters with gripping and cutting areas skewed from the handles to avoid cramping the wrist
- Cutting and welding torches and welding-rod holders aligned to reduce effort and make work more visible
- Sanders, grinders, drills, hacksaws, and similar tools with good weight balance and handgrips for easier handling and ways to apply force
- Wheelbarrows and buggies so designed that weights are balanced, thus requiring little lifting, and with pneumatic tires for ease in pushing and guiding

Most experienced construction managers are aware that tools such as these are available. The difficulty is that those doing the buying may not know of them and may let price or similar considerations control the purchasing decision. Again, those who control the tool supply may not have the goals and budgets, which, in turn, causes them to resist stocking special items. Finally, in the hurry to get tasks underway, the natural tendency is to use tools that are on hand or in the worker's kit or crew gang box, without considering the ultimate cost of doing so.

Daily and Weekly Fatigue

As has already been indicated, heavy exertion can produce fatigue during a work period that must be followed by a rest interval to allow the body to recoup. Observations in industrial situations have found that other influences

[12] See, for example, L. Greenberg and D. B. Chaffin, *Workers and Their Tools*, Pendall, Midland, Mich., 1977; S. Konz, *Work Design*, Grid, Columbus, Ohio, 1979; and Woodson, op. cit.

not related to heavy energy demands also affect productivity both during half-shifts and between morning and afternoon (see Fig. 9-5). These curves are for tasks involving heavy but not overtaxing efforts. For less-taxing assignments the curve shows similar but less pronounced variation.

Figure 9-5 shows that in industrial situations, the work pace is lower at the beginning of each work period, accelerates to a peak, and then slows down again before the work break. Also, morning productivity is usually somewhat higher than that in the afternoon. The increase from the beginning of the shift probably can be explained as getting up to pace. The later slowdown and lower afternoon outputs may well be associated with some form of overall physical and mental fatigue which dulls the senses and hampers the drive to produce. Although variations in performance cannot be fully explained, there is evidence that they occur in construction to varying degrees, except when the work pace is set by machine.

Generalized data such as those shown in Fig. 9-5 are of little use in looking for construction productivity improvements. Rather, management must examine specific situations. For example, output could be low when work begins because starts are late, work is not assigned in advance, or the necessary plans, materials, tools, and equipment have not been arranged for. It is a natural tendency for workers to slow down an almost completed activity near the end of a work period to make it last in order to appear busy, thereby avoiding criticism of foreman and crew. Also, when new tasks are undertaken, as is often the case in construction, a learning and coordination period may be required. All too often management accepts such falloffs in productivity as inevitable without examining the underlying reasons.

In many industrial and certain construction situations such as prefabrication, the work may be machine-paced, in which case the worker is expected to keep up with the machine. Thus, the variations shown in Fig. 9-5 would not be expected. Rather, the problem becomes one of setting an appropriate pace which will give satisfactory output without upsetting operations or causing work pileups upstream or downstream of a specific work station. Setting the appropriate pace becomes a human rather than an engineering matter and can become a focus of management-labor controversy. As an illustration, the speed of the assembly line in some automobile plants has been a negotiable item in the labor agreement. On the other hand, imaginative approaches have

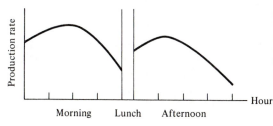

FIGURE 9-5
Typical daily production curves for industrial workers engaged in heavy work. Tasks involving less effort follow the same general shape but with lesser amounts of change.

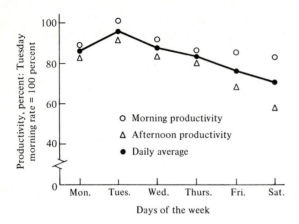

FIGURE 9-6
Typical production curves by days of the week for industrial workers engaged in heavy work.

been developed whereby through worker participation and agreement, arrangements are made to vary the pace at intervals and still meet an overall production target.

Figure 9-6 shows that in industrial situations productivity also varies with days of the workweek. Tuesday offers the highest output and Friday the lowest. Here also, some form of largely unexplained physical and mental fatigue appears to be at work.

In the United States, the 5-day 8-hour workweek has been traditional. More recently, four 10-hour days have been substituted. Research has indicated that the change has had little effect on overall fatigue.[13] How productivity is affected is less clear. While an extrapolation of Fig. 9-5 would indicate an additional productivity falloff at the end of each 10-hour day, Fig. 9-6 suggests that a low-productivity fifth day would be eliminated by a 4-day 10-hour plan. Also, travel to and from work on a fifth day is no longer necessary, so that another cause of fatigue and stress is eliminated. At present, considerations other than the marginal effects on productivity of five 8-hour vs. four 10-hour days are the controlling factors, such as scheduling consequences and union acceptance. Even more important is the recent change made by the U. S. Congress in the Davis-Bacon Act, which eliminates the requirement that time-and-a-half rates be paid for shifts longer than 8 hours which brought a substantial penalty in costs against the 4-day 10-hour workweek.

Diurnal-Nocturnal (24-hour) Work Behaviors—Shift Work

One of the basic differences among individuals is that alertness and mental capabilities peak at different times of the day. Some people are morning-oriented, others are afternoon- or night-oriented, while still others are able to perform well at almost any time. Recognition of the innate variability in the

[13] See M. Volle et al., *Ergonomics*, vol. 22, no. 9, September 1979, pp. 1001–1010.

basic 24-hour capability and assignment to jobs that more closely fit these cycles can do much toward increasing satisfaction, productivity, and safety.

Studies have indicated that approximately 25 percent of the work force classify themselves as day-oriented, 50 percent feel able to perform well at almost any time, and another 25 percent are night-oriented.[14]

For those who are day-oriented or who have no pronounced orientation, the normal day work shift is an appropriate work time. The night-oriented group is strongly off cycle to the normal day work and would be better matched to a night shift. Variations in the body functions during daytime and night and their impact on day vs. night work or by shifts has been studied in detail.[15] Research has shown, for example, that most body functions such as temperature, kidney activity, hormone level, and corticosteroid production follow a 24-hour (circadian) cycle. This is illustrated by Fig. 9-7, which for a certain group shows body temperature more than a degree higher in the afternoon or after work and an evening meal than while sleeping in the very early morning hours. These variations differ greatly among individuals, with lower variation associated with higher tolerations for shift work.[16]

Circadian variations in body activity have also been related directly to output. For example, production in heavy tasks over many hours decreased between night and day by 5 percent for well-trained individuals and by 10 percent for those with lower work capacity. Also, night work offered even more difficulty for those doing mental or repetitive tasks.[17]

In construction, it has been common practice on multishift projects to rotate workers among the shifts to be fair to them. In some cases the workday on swing and graveyard shifts has been shortened a half hour to compensate for the apparent inequity of working such shifts. Research findings do not support these practices. Rather, they show that the practice of rotating shifts is detrimental to alertness, ability, and safety.[18]

Adjustments in body rhythms to new work-sleep regimens are almost impossible for some people. Others require 7 to 12 days. Furthermore, reversion to the normal cycle can reset body functions in a single day or over a weekend, so that adjustment must start over again. Older individuals (over 45) have particular difficulty in adjusting to night shift work or to changes in shifts, primarily because sleep for them is more difficult.[19]

Where changes in shifts must be made, it is important to retain the relationship between work and sleep periods. But often it is difficult for night workers to retain the work-sleep relationship, since it calls for sleep at times when ex-

---

[14] See S. Folkard et al., *Ergonomics,* vol. 22, no. 1, January 1979, pp. 79–91.

[15] See, for example, W. P. Colquhoun and J. Rutenfranz, *Studies of Shiftwork,* Taylor and Francis, London, 1981. See also articles in many issues of *Ergonomics.*

[16] See A. Reinberg et al., *Ergonomics,* vol. 23, no. 1, January 1980, pp. 34–64.

[17] See "Physiological and Psychological Aspects of Night and Shift Work," National Institute for Occupational Safety and Health, Washington, D.C., December 1977.

[18] See W. P. Colquhoun et al., *Ergonomics,* vol. 11, no. 5, May 1968, pp. 527–546; and vol. 12, no. 6, June 1969, pp. 865–882.

[19] See T. A. Kersted and L. Torsvall, *Ergonomics,* vol. 24, no. 4, April 1981, pp. 265–273.

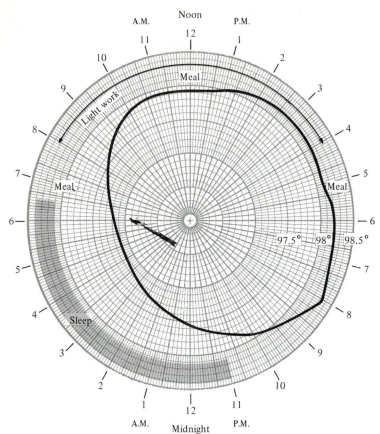

FIGURE 9-7
Mean body temperature of 59 workers taken 20 times a day. (*After W. P. Colquhoun, M. J. F. Blake, and R. S. Edwards, "Experimental Studies in Shiftwork,"* Ergonomics, *vol. II, no. 5, 1968.*)

ternal noise, family activities, and social life are at their peaks. The worker either must forego these or have lower efficiency on the job. Surprising as it seems, the fact that rest is in daylight rather than in dark does not seem to be an important influence.

There may be crisis situations where workers must sustain their efforts for long periods of time. One investigator found that for day workers, an 18 percent reduction in output occurred in the early morning after 18 hours of continuous work. This was at the low point in the circadian cycle. Output increased to 90 percent during the second day but fell to 67 percent the second night. Full recovery came with 24 hours off the job.[20]

To plan for higher productivity where shift work is necessary, management must recognize the following factors to the greatest extent possible:

[20] See B. B. Morgan Jr. et al., *Human Factors,* vol. 16, no. 4, December 1974, pp. 406–414.

1 Strongly oriented day and night people can recognize their cycle peaks. Such people should be assigned shifts that fit their feelings.

2 The normal daily cycle should be retained for those on the off shifts rather than revising the work-relax-sleep sequence. Persons working swing or night shifts should follow their usual day sequence, which is normally to arise from sleep 2 hours before work, have breakfast, work for 8 to 10 hours with a lunch period, followed by an after-work meal, several hours of relaxation, and sleep.

3 The work-sleep cycle referred to in 2 above should be followed on nonwork days in order to maintain (even enhance) the circadian rhythm shift.

4 When shift cycles are changed, management must recognize that the first several days are periods of change and that employees will be less alert, less accurate, and less cautious.

5 Night-shift assignments should be avoided for older individuals, for those with high temperature variation during the circadian cycle, and for those for whom daylight sleep is difficult to arrange (for example, those with small children in the home).

Long-Term Fatigue from Occasional Overtime

As already noted, a 40-hour workweek is generally accepted as the norm in the United States. There are, of course, exceptions set out in employer-employee agreements or under special conditions which reduce that period by a day or two, usually with the aim of spreading the work among all members of a union. In one instance, for this reason a carpenter's union restricted the workweek for its members to 36 hours, in which case, the unfortunate consequences were to limit the work period for all crafts, since no one could work if carpenters were not available when needed. With any of the shorter work periods, physical fatigue should not directly affect productivity unless the workers have taken second jobs and thereby worked a longer overall week. Of course, there might be other effects unrelated to fatigue, such as slowing down the work pace to stretch the job out.

At times the workday may be extended in order to complete a task, as for example to get ready for or complete a concrete placement. Again, work on a sixth or seventh day, possibly more than 8 hours, may be scheduled. Pay for such overtime is commonly at a premium rate, such as time-and-a-half or double time.

With overtime work, productivity drops such as those predicted in Figs. 9-5 and 9-6 are to be expected, although there are little data from which to predict the magnitude. Furthermore, there is usually the necessity to pay a premium wage, which tempts workers to slow down to gain the extra pay or to stay away on straight-time days to work the premium-pay ones. For such reasons management should carefully consider the desirability of even occasional overtime work. Prohibiting it entirely is probably unwise, since there may be occasions where working a strategic crew overtime may offer substantial overall job savings in cost or schedule.

In addition to the tangible losses or gains associated with introducing occasional overtime, complications develop for workers and management alike. For example, transportation arrangements to the job may be disrupted, and plans for family or other leisure time activities may be affected. Among the other consequences may be losses in motivation and possibly increases in absenteeism.

Since there are probably losses in productivity and certainly added labor costs, management should as a matter of policy closely monitor the use of occasional overtime to see that the gains in using it are substantial. Otherwise, it can become a substantial drain on job profits.

Long-Term Fatigue from Scheduled Overtime

There is strong evidence that at least 1½ days rest per week is needed to sustain normal output.[21] This and other findings gathered over the years support the contention that output per week is relatively constant over extended periods of time at 40 hours work or less per week.

Scheduled overtime has to do with situations where management deliberately sets up a workweek longer than the normal 40 hours on a regular basis. Commonly it develops when workers, or at least skilled workers, are in short supply and contractors use the longer hours and premium pay to attract them with the hope of rushing projects to completion.

In most instances, the introduction of scheduled overtime on even a single large project will disrupt the entire labor situation in an area. When it is introduced, workers shift to projects where they can gain its rewards. Often less skilled and other marginal workers may be attracted both locally and from outside the area, compounding the problem. Soon after, other employers feel forced to match these special arrangements, so that working hours and pay escalate on all projects that are competing for labor. Furthermore, it is seldom possible to confine the arrangements to the craft or crafts that are in short supply, since other workers will also demand them. Soon all work in an area is affected.

A very serious consequence of scheduled overtime is its apparently devastating effect on the productivity of workers and crews. Although the results are disputed, several studies claim to have shown this clearly. The findings most widely cited were developed and reported in the late 1960s for members of the National Constructors Association by Weldon McGlaun. These were republished by the Business Roundtable as Report C-3 in 1980. Figure 9-8 summarizes the principal conclusions, among which are the following:

1 There is a substantial loss in productivity during the first week of scheduled overtime (for example, 54 hours output for 60 hours worked). An expla-

---

[21] See for example, R. Schelling, "Prevention of Fatigue in Industry," *Modern Management*, vol. 6, January 1946, pp. 17–19; cited by L. J. Fogel, *BioTechnology Concepts and Applications*, Prentice-Hall, Englewood Cliff, N.J., 1963.

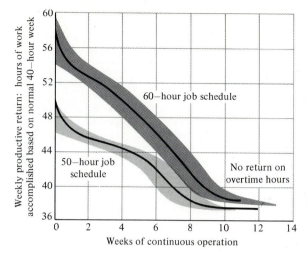

FIGURE 9-8
Effective return from working 50 or 60 hours a week for various numbers of weeks. (*Source:* Business Roundtable Cost Effectiveness Study Report C-3, *November 1980.*)

nation for this falloff may be that workers develop a more leisurely pace to reserve energy for the added hours.

2 Productivity in subsequent weeks continues to fall. After 5 weeks of 50 hours a week, productivity is about the same as is that for a 40-hour week. With a 60-hour week, the no-return for hours over 40 comes in about 9 weeks. Beyond these times, productivity for 60 hours is below the normal 40-hour output.

To illustrate the gains and losses from scheduled overtime, consider a situation involving a 60-hour workweek for 7 weeks with double-time pay for over 40 hours per week. Based on Fig. 9-8, work output per week during the 7 weeks would average about 49 hours, an increase of 22 percent. Pay, including overtime, would be the equivalent of 80 hours at the straight-time rate. This means that output, per unit of pay, compared with straight time, would equal 49 divided by 80, or 0.61 (61 percent). As another illustration, it can also be concluded from Fig. 9-8 that once scheduled overtime of 60 hours per week exceeds 9 weeks, productivity is less than that for a 40-hour arrangement, meaning that the ratio of output to unit pay is less than 0.50.

Results of an earlier (1964) study of efficiency losses in construction under overtime are given in Table 9-2. It shows for periods of 1 to 4 weeks the loss in efficiency (called percent inefficiency) for weekly hours ranges from 54 to 84. These data parallel the findings portrayed in Fig. 9-8, although stated somewhat differently.

From data such as those given in Fig. 9-8 and Table 9-2, it should be clear that there are only small gains in output from a work schedule that substantially exceeds forty hours per week. Furthermore, these gains rapidly diminish. Given that overtime usually requires premium pay, often double the straight-time rate and that usually entire projects are affected, scheduled overtime makes very little sense. There are recent instances where owners on cost-

TABLE 9-2
THE INFLUENCE OF OVERTIME ON EFFICIENCY IN CONSTRUCTION CREWS

Days	Hours	Weekly hours	Percent inefficiency			
			7 days	14 days	21 days	28 days
6	9	54	4–6	6–9	8–12	10–15
6	10	60	7–9	11–14	14–18	18–23
6	12	72	12–14	18–21	24–28	30–35
7	8	56	9–11	14–16	18–22	23–27
7	9	63	11–13	17–20	22–26	28–33
7	10	70	14–16	21–24	28–32	35–40
7	12	84	20–22	30–33	40–44	50–55

Source: Qualified contractor, March 1969

plus projects alarmed by substantial cost or schedule overruns have stipulated in their contracts that no scheduled overtime would be tolerated.

In addition to the inefficiencies and costs associated with scheduled overtime, it brings other undesirable consequences. Those identified include greater absenteeism and a higher accident rate.[22]

In sum, the experience with scheduled overtime has been very poor. It has brought high penalties in costs and has probably adversely affected morale.

MENTAL FATIGUE

The discussion so far has considered physical fatigue associated with heavy tasks or long work periods. In addition, there are mental tasks or stressful situations under which efficiency can decline and accident susceptibility increase.

Researchers have found it difficult to even define or measure mental work load as a first step toward assessing its effect on fatigue and productivity. One study[23] found that mental loads increased the respiration rate and heartbeat by 15 percent and that substantial decrements in efficiency occurred. It has also been reported that for work of a repetitive nature that requires a degree of attention, alertness, or vigilance, such as on assembly lines, there is a marked decrement in performance after 60 to 75 minutes. To optimize productivity and minimize errors requires a rest of 20 to 25 percent of the work period or, alternatively, a change in activity after 30 to 50 minutes. Another approach is to arrange work-rest cycles of 10 and 2 minutes.[24] Less repetitive tasks which

---

[22] Based on *Hours of Work and Output,* Department of Labor Bulletin 917, Washington, D.C., 1948.

[23] L. J. Fogel, op. cit.

[24] See K. F. H. Murrell, *Human Performance in Industry,* Van Nostrand, New York, 1965, chap. 17.

require concentration and attention are not as susceptible to errors or loss of productivity.[25]

Little is known about mental fatigue, except that it is a reality. There are, however, practical measures involving rest or changes in activity that can reduce its impact on mental effectiveness.

STRESS FATIGUE

There can be many stressful or emotionally frustrating situations in the construction environment. Managers often have deadlines or targets that demand high energy input, often for long periods of time and without sufficient rest. Again, they must make troublesome decisions about quality, safety, and crew performance or behavior. For workers, there may be cramped work space, noise, vibrations, glare, heat, or cold. Sometimes there are disturbing circumstances such as arguments with other workers or with supervisors or resentment about overly close supervision and possible censure. Research has shown that in such situations the heartbeat increases and those affected feel fatigued and become irritable.[26] As with other forms of fatigue, productivity can suffer under such conditions.

First and intermediate level supervisors are subject to all the stresses just described. They are often in addition faced with conflicting demands from workers on the one hand and from higher-level management on the other. Also, as discussed earlier, they are targets for criticism, whether deserved or not, when their crews fail to reach expected targets for schedule, cost, or quality. Many of these supervisors probably are what has been classified as type A managers who have high levels of drive, aggressiveness, competitiveness, and ambition coupled with a sense of urgency about time and a predisposition to select challenging jobs. All of these can contribute to a high level of stress, which in turn can lead to fatigue and possibly a proneness to health disorders, particularly those which may trigger heart attacks.[27]

Stresses produced away from the job (e.g., at home or through frustration in travel to and from work) have also been found to be related to fatigue and lower productivity.

BOREDOM

Boredom is a physiological condition sometimes defined as underexertion, as contrasted to fatigue, which is associated with physical or mental overexertion. It seems to be unrelated to physical exhaustion or discomfort. A

---

[25] For a recent inquiry and extensive bibliography on mental fatigue in the work place see N. Morey, *Human Factors,* vol. 24, no. 1, February 1982, pp. 25–40.

[26] For added discussion see J. K. Hennigan and A. W. Wortham, *Ergonomics,* vol. 18, no. 6, November 1975, pp. 675–681.

[27] See J. Sharit and G. Salvendy, *Human Factors,* vol. 24, no. 2, April 1982, pp. 129–162 for a detailed summary of the findings about stress in the workplace and an extensive bibliography.

bored worker can develop a feeling of dissatisfaction, emptiness, weariness, and detachment because the assigned task offers little challenge, is too monotonous or easy, lacks novelty, is repetitious, or requires little alertness. Psychologists often distinguish between boredom, a physical reaction, and alienation, a psychological response to feelings of depersonalization, isolation, or lack of power or meaning in the work situation.

Workers cope with boredom with such strategies as restlessness, daydreaming, and mental withdrawal. In effect, they are no longer mentally on the job. There are no guidelines or scales for measuring boredom. Nevertheless, this condition is of concern because a bored person may have lapses in attention and vigilance that can lead to mistakes or accidents. However, making jobs less boring or determining which people will not be bored is not an exact science. In general, studies indicate that extroverts and high achievers on mental tests are more likely to be bored by routine assignments. There is the classic example from precomputer days where an insurance company plagued by filing errors reduced them substantially by assigning low rather than high achievers to the task.

On-site construction tasks, as contrasted with those of manufacturing, are generally variable in location and content and of relatively short duration and have highly visible results so that boredom is a less serious problem. However, it can become one if assembly-line or highly repetitive operations are being employed.

Some correlation has been found between low productivity and boredom, attributed primarily to motivational factors.[28]

ENVIRONMENTAL ASPECTS OF ON-SITE CONSTRUCTION

Other things being equal, human beings perform relatively continuous physical or mental work most effectively when the temperature falls between 50 and 70 degrees Fahrenheit (10 and 21 degrees centigrade), at a relative humidity of 30 to 80 percent, under dry conditions, with the atmosphere clear of dust and other atmospheric pollutants and without excessive noise. Departures from these conditions have adverse effects on productivity, comfort, safety, and health.

Because much on-site construction is carried on in the open, productivity can be strongly affected by adverse environmental conditions. For example, it has been estimated that half of all construction operations are weather-sensitive, which means that productivity will suffer unless management devises a method for compensating for it or at least mitigating its effects. The traditional approach is to shut down work being done in the open when it is too hot, cold, rainy, or wet. Possibly the workday may be adjusted to avoid afternoon heat. Sometimes repetitive operations such as cutting and fabricating can be done under shelters or moved inside partially completed buildings or struc-

---

[28] For a review of research on boredom, see R. P. Smith, *Human Factors,* vol. 23, no. 3, June 1981, pp. 329–340.

tures. It is common practice in cold climates to follow the approach developed in Canada, where working areas are enclosed in a temporary structure, or closure of the exterior of buildings is rushed so that the work area can be heated.[29] In hot climates, air conditioning for the cabs of earth-moving and other equipment is available and is increasingly being provided. These few examples show that in many circumstances where the climate can adversely affect productivity, means are available to make working conditions more favorable, if management chooses to use them. The difficulty is that, as with so many other approaches to productivity improvement, firm dollar measures of the gains are difficult to establish but the costs are highly visible.

Even though means for providing environmentally desirable working conditions are often available, busy construction managers tend to accept less extreme but adverse conditions as givens. Some even assume the macho attitude that he-men are expected to put up with them. In either case, this approach may be shortsighted, since it overlooks the adverse effects of poor conditions on productivity, which are often substantial.

Primarily in the past decade, governments at national and state levels have enacted legislation and established monitoring and enforcement procedures to protect workers and the public from certain adverse environmental situations. These include regulations concerning conditions that may affect an individual's safety and health, such as unsafe workplaces or practices, air pollution, toxic agents, exposure to radiation, and noise. These topics are discussed later in Chap. 12. On the other hand, judgments about conditions not strictly construed under the law to endanger safety and health are largely unregulated, even though some of them may have adverse safety and health consequences. These are left to the constructor's discretion.

The sections which follow briefly present research findings on the effects of heat, humidity, cold, and noise on productivity.

The Effects of Heat and Relative Humidity on Productivity[30]

As stated above, humans operate effectively in a comfort zone where temperatures range from about 50 degrees to 70 degrees Fahrenheit under wide variations in relative humidity. In this comfort zone, body and skin temperatures remain near normal since heat is dissipated by evaporative effects in respiration and through the skin without much sweating. At higher temperatures, augmented by higher relative humidity, productivity declines.

---

[29] For detailed reports on the difficulties of and approaches to construction in cold weather and for extensive bibliographies, see F. L. Bennett, *Journal of the Construction Division,* ASCE, vol. 101, no. 4, December 1975, pp. 839–851, and vol. 103, no. 3, September 1977, pp. 437–451.

[30] Extensive research on the effects of heat and humidity and of cold on human performance and behavior has been carried out by the armed forces and space agencies. Findings are reported in numerous sources. See for example, *Prevention of Heat Injury,* Headquarters Department of the Army, Circular 40-82-3, July 1982; and J. F. Parker and V. R. West (eds.), *Biometric Data Book,* 2d ed., National Aeronautics and Space Agency, 1973.

Data on the impacts of heat, cold, and humidity on the productivity of four general classes of construction workers are graphed in Fig. 9-9. The curves for electricians show the effects on the productivity of two workers doing the repetitive construction task of installing electrical receptacles in premounted junction boxes. For this experiment, working conditions ranged from hot to bitter cold. Even for this task, which is not physically demanding, productivity at higher temperatures dropped off; for example, at 100 degrees Fahrenheit it fell by 15 percent at 30 percent relative humidity and by 40 percent when relative humidity reached 90 percent. Among the stated causes of this decrease in productivity at high temperatures and humidities were difficulties caused by perspiration, which resulted in steamed-up glasses and the necessity for workers to wipe their face and hands. In addition, the quality of workmanship declined. Observers also reported that at a 110-degree-Fahrenheit temperature and higher humidity, prolonged work could have resulted in serious damage to

FIGURE 9-9

Effects of temperature and relative humidity on the productivity of journeyman electricians and bricklayers and for equipment-operating and manual tasks.

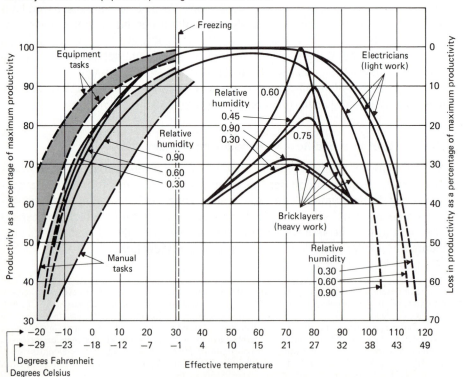

health. Other consequences included the necessity to take frequent cool-off breaks, difficulty in breathing, dizziness, and extreme instability.[31]

The group of curves in the upper right of Fig. 9-9 shows a few of the basic findings of controlled tests to measure the effects of temperature and relative humidity on the productivity of bricklayers. These were sponsored jointly by the U.S. Department of Housing and Urban Development, a group of brick manufacturers, and the Bricklayers, Masons, and Plasterers International Union; they were conducted at the University of Texas. As with the tests on electrical workers, the task was standardized; in all, some 283 identical brick wall panels were built under carefully regulated conditions.[32] Because bricklaying under the conditions of this test was heavier work, it was to be expected that productivity would be strongly influenced by both higher and lower temperatures and by relative humidity. The research results, depicted in Fig. 9-9, confirm this supposition.

The third set of values, developed by the U.S. Army Cold Regions Research and Engineering Laboratory, was reported in *Engineering News-Record,* Mar. 20, 1986. As shown by the crosshatching, it portrays ranges of values relating productivity to below-freezing temperatures for workers doing equipment-related tasks and those carrying out manual work. The spread in values and the difference in productivity between equipment and manual tasks are to be expected. This set of data is strong evidence of the difficulty faced in estimating productivity under adverse weather conditions.

The manner in which variations in productivity occur with changes in conditions, as shown in Fig. 9-9, is not completely understood. One fact remains clear: heat, cold, and relative humidity can all exert strong influences on productivity. Consequently, attention to and expenditures for creating near-optimum working conditions for craftsmen doing construction work seem justified.

There has been speculation about how productivity is affected by the worker's knowledge of temperature and humidity conditions. This situation has not been explored for construction.

Although humans are most comfortable and work most effectively at temperatures at or below 70 degrees Fahrenheit, they can tolerate and work at substantially higher temperatures without serious health consequences. The approximate upper limits of this neutral zone, which reflects (1) temperature, (2) relative humidity, (3) resting, and (4) working for unclimatized and climatized male adults are shown in Table 9-3. At temperatures and humidities lower than the limits shown, normal interior temperature is sustained as the body involuntarily adjusts such functions as blood-flow rate to internal and external locations, breathing, sweating, and salt retention. Above these temper-

---

[31] The data given here are based on the report *The Effect of Temperature on Productivity* distributed by the National Electrical Contractors Association, Inc., Washington, D.C. The task was standardized and carefully controlled.

[32] C. T. Grimm, *Journal of the Construction Division,* ASCE, vol. 100, no. C02, June 1974, pp. 171–185, and C. T. Grimm and N. K. Wagner, *ibid,* vol. 100, no. C03, September 1974, pp. 319–335, describe both bricklaying as a task and the testing procedure and its results that have been summarized here.

TABLE 9-3
MAXIMUM TEMPERATURES AND RELATIVE HUMIDITIES AT WHICH ADULT MALES
REACH THE NEUTRAL BOUNDARY CONDITION AND ARE RELATIVELY INDEPENDENT OF
THE ENVIRONMENT.

Physical condition	Relative humidity		
	10	50	100
Unclimatized			
At rest	120	100	88
Working	100	90	83
Climatized			
At rest	136	112	95
Working	120	100	88

atures, heat stress develops. It bears repeating here that as shown by Fig. 9-9, productivity can be seriously affected in the temperature and relative humidity ranges shown in Table 9-3, even though physical well-being is not threatened.

Heat stress is accompanied by a dulling of the senses, dizziness, and a reduction in efficiency, motivation, and morale. It has three forms of increasing severity, as follows:

• Heat cramps, which result primarily from excessive losses of salt. They are painful, sometimes severely so. They usually involve the muscles of the extremities and the abdomen. Body temperature remains normal. Seldom does death or permanent physical impairment result.

• Heat exhaustion, which arises from excessive loss of salt and water. Removing the patient to a cool environment, rest, and administration of a salt solution bring a prompt recovery. The mortality rate is low.

• Heat stroke, which is characterized by extremely high body temperature, usually accompanied by profound coma. It represents a breakdown of the body's heat-regulating mechanism. Heavy physical exertion, alcoholism, or diarrhea are contributors. Unacclimated persons are particularly susceptible. The mortality rate is high.

In situations where heat stress is likely, workers should be carefully observed and required to take frequent rests. It is particularly important that they replace water lost through perspiration, which on construction in warm weather has been measured to be 2½ to 3 quarts per day and under very hot conditions, 2 quarts per hour. Refreshment pauses should be at least hourly; some suggest that these intervals be spaced at 15 to 30 minutes. Unfortunately, thirst and a desire for water are unreliable indicators of dehydration; other positive and aggressive means, exercised by management and workers alike, must be employed to avoid dehydration.[33]

[33] See, for example, L. W. Stowlinski, "Occupational Health in the Construction Industry," *Technical Report 105,* The Construction Institute, Stanford University, Stanford, Calif., 1965.

As indicated, heat stress causes dehydration and salt loss in the body. Along with it are other consequences; for example, a loss of 1½ quarts of water will increase the pulse rate by 10 per minute and the same physiological stress as a temperature rise of 14 degrees Fahrenheit. Salt loss, which accompanies sweating, brings on salt deficiency. A study for the U.S. Public Health Service describes the problem as follows:[34]

> Small deficiencies can be replenished from stores in the tissues, but major deficiencies will lead to a reduction of the sodium chloride concentration in the body fluids. This, in turn, leads to increased loss of water in the urine, with still further embarrassment to the circulation, and in extreme cases to marked dysfunction of the tissue cells. Compensating mechanisms, such as the retention of sodium by the kidneys under response to reduced blood volume and lowered sodium concentration, are easily overwhelmed.

Heat stress and salt deficiencies were brought forcefully to the attention of the U.S. construction industry in the early 1930s during the construction of Hoover Dam in the southwestern desert. Under high humidity conditions, temperatures reached 130 degrees Fahrenheit and many workers collapsed. Among other measures taken to minimize the problem was a requirement that workers take salt tablets. Since that time, acclimatization of workers to be employed under such conditions has become common practice. The general aim is to induce modifications in body mechanisms in a manner that decreases internal heat production and lowers the salt content of sweat. Steps in acclimatization are roughly as follows:

1 At first exposure to the hot environment, even moderate amounts of work by unacclimatized individuals can produce discomfort, dizziness, nausea, and even collapse. Therefore, first-work assignments must be light.
2 Following the first exposure, adjustment is quite rapid and is well-developed in 4 to 7 days. The first work periods should be short and intermittent, totaling 2 to 4 hours daily. These can be increased quickly. Care should be taken to stay within the capacity of the individual until acclimatization is well advanced. Some work is essential, since inactivity confers only slight acclimatization. More detailed findings from research on acclimatization include:
   a Acclimatization to severe conditions facilitates performance at lesser conditions.
   b The general pattern of acclimatization is the same for short, severe exertion as for moderate work of longer duration.
3 Acclimatization to hot-dry climates increases performance ability in hot-wet climates, and vice versa.
4 Inadequate water and salt replacement can retard the acclimatization process. It has been shown that acclimatization will not take place if the salt

---

[34] See D. H. K. Lee, "Heat and Cold Effects and Their Control," *Public Health Monograph 72,* U.S. Public Health Service, 1964.

intake is less than 5 to 6 grams per day. Therefore, during the first 2 to 3 weeks of exposure to heat, workers should be required to supplement their salt intake in order to make up for the large amounts lost in sweating.

5 Acclimatization to heat is well-retained for about 2 weeks after the last exposure; thereafter, it is lost at a rate that varies among individuals. Most people lose a major portion of acclimatization in two months. Those who stay in good physical condition retain their acclimatization best.

6 Individuals who are convalescing from debilitating illness, have hangovers, or lack sufficient sleep lose significant amounts of their acclimatization.

The Effects of Cold on Productivity

When the temperature of human skin falls below the comfort zone, the body reacts involuntarily. The triggering mechanisms are cold receptors in the skin which send impulses to the hypothalamus in the brain. It, in turn, directs the smooth muscles in the skin to contract and reduce blood flow. Also, other muscles are activated and they alternately contract and relax to generate heat, as evidenced by shivering and chattering of the teeth. In more extreme situations, blood circulation to the limbs and the trunk is reduced. Finally, the entire heat-producing mechanism of the body can be disrupted. Unconsciousness develops if the body temperature falls to 70 degrees Fahrenheit, and tissue damage occurs at 39 degrees Fahrenheit. This process usually begins in the toes, because blood flow to them is most impaired and they are most exposed to external coolness. Fortunately, voluntary reactions to reduce exposure to cold are easier than is the case with heat, since it is usual to insulate the body from cold with clothing and head, foot, and hand protection.

Except in conditions of extreme exposure to cold, most of the disabilities probably result from ignorance or carelessness. For example, failure to protect the nose can lead to frostbite. Toes and fingers also are adversely affected, particularly if socks or gloves become wet. Merely touching metal colder than freezing will cause the fingers to adhere, with serious consequences. Heavy exertion and sweating inside protective clothing can lead to rapid body coolness, with serious results. Obviously, then, education to the hazards introduced by cold and instructions about sensible protection methods which still permit productive work is the key in cold-weather construction.

Air temperature is not the sole indicator of exposure to cold, since wind can greatly heighten the effect. This is illustrated in descriptive terms by Fig. 9-10. This chart, first offered in 1951, is still cited today. Its use is demonstrated by the dashed line, which indicates that at an air temperature of 30 degrees Fahrenheit and with a 10-mile-per-hour wind, exposed flesh will freeze in 1 minute. In recognition of the fact that severe injuries can result with exposure to cold under windy conditions, U.S. Army regulations call for protection at wind-chill factors over 1200. This would mean, for example, that special measures are taken at 32 degrees Fahrenheit in a 60-mile-per-hour gale. (See dotted line in Fig. 9-10.)

As with excessive heat, extreme cold decreases productivity (see the curves

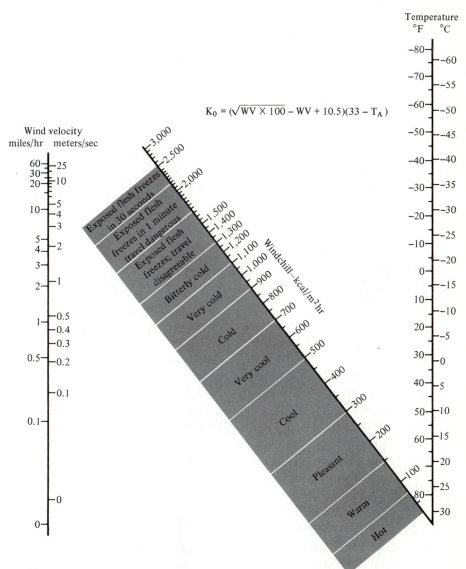

$$K_0 = (\sqrt{WV \times 100} - WV + 10.5)(33 - T_A)$$

FIGURE 9-10
The cooling power of moving air. The windchill factor is a measure of the effect of cool temperature and wind on the human body. Chill factor, K_0; velocity of wind in meters per second, WV; ambient temperature in degrees centigrade, TA. (*Quoted from C. F. Consolazio, R. E. Johnson, and E. Morek,* Metabolic Methods, *The C. V. Mosby Company, 1951.*)

for electricians and equipment and manual tasks in Fig. 9-9). To a large degree it results from the distraction and inconvenience of protecting against exposure, obtaining, getting into and out of, and working in cumbersome protective gear. One particularly troublesome cold-weather problem is that of having to do meticulous, careful work while wearing thick gloves.

Reports on the effects on productivity of temperatures above freezing but below the comfort zone do not agree. The study of bricklayers (see Fig. 9-9) shows substantial effects. However, studies of costs while working in cold weather in Ohio showed no effect at temperatures of 40 degrees Fahrenheit or above.[35] But at temperatures below that level, reported costs, compared with usual conditions, taken from the Ohio study, ranged as follows:

Contractor's work plan	Effective temperatures, degrees Fahrenheit			
	30–39	20–29	10–19	Below 10
Project planned for winter working percent cost increase	8	17	32	58
Project not planned for winter working percent cost increase	40	44	60	—

Data such as those for winter working make clear why, as mentioned earlier, contractors close in or otherwise shelter the workplace if they plan to operate in cold weather. Doing construction on projects not planned for occasional cold weather days can be costly. Furthermore, as Fig. 9-10 shows, wind chill can lower the effective temperature far below measured values and further decrease productivity.

Only in extreme cold should factors in the work environment itself such as motivation and morale suffer. On the other hand, because cold-weather work is often carried out in remote locations, other troublesome factors such as boredom, isolation, and inadequate housing may come into play and bring dissatisfaction and reduced productivity. To overcome such difficulties, it is common practice to give workers extended periods of released time so that they can leave the job completely.

Noise as a Factor in Productivity

Construction sites generally are noisy both in background levels and in peaks or bursts at specific locations or times. In either form, noise is measured in decibels (commonly denoted as dBA) which is set on a logarithmic scale. Background noise on construction sites probably falls into ranges paralleling industrial areas and commercial locations with heavy traffic (70 dBA). For areas around machinery, such as compressors, cranes, or loaders, the level is probably 90 or more. In the past, levels of 110 dBA or higher have been measured on the operating platforms of heavy equipment and around pneumatic machinery such as drills and jackhammers. These are sufficient to produce

---

[35] See E. Koehn and D. Meilhede, *Journal of the Construction Division*, ASCE, vol. 107, no. CO4, December 1981, pp. 585–595.

permanent hearing loss. Today OSHA regulations prohibit continuous noise levels higher than 90 dBA.[36]

Even the lower noise levels mentioned above are so high that communication becomes difficult. For example, at 65 dBA, conversation is difficult at 3 feet; at 75 dBA, a finger must be placed in the nonlistening ear to hear a telephone conversation. Since good communication is necessary to manage construction operations effectively, it follows that unless precautions are taken, noise can severely affect construction productivity and contribute to both stress and fatigue. In particular, high-pitched (high-frequency) noise, such as that of a whining saw or blowing air can be particularly stressful and tiring. However, little specific data have been developed. To date, productivity losses from noise have not been separated from those brought on by other operational, behavioral, environmental, and safety factors. It can be said with certainty that a failure to control unnecessary on-site noise, with the stresses and communication barriers that it can produce, is bound to be costly.

SUMMARY

This chapter has looked at the problems of using people as machines or organisms of flesh and blood as they work in the construction industry. It has examined the problem in terms of individuals and the effects of fatigue on their performance and at the environment in which they operate. Insofar as data are available, it has shown how these conditions affect productivity. It is hoped that those who study the material carefully will find ways for making their on-site construction operations more suitable for the workers as well as more productive and profitable.

---

[36] The safety and environmental health aspects of noise and the methods for controlling or correcting it are discussed in Chap. 12.

CHAPTER 10

HUMAN BEHAVIOR
AS A FACTOR
IN CONSTRUCTION
PRODUCTIVITY

INTRODUCTION

The last chapter discussed people as machines or organisms of flesh and blood which develop and supply the energy to get construction tasks done within the various environments where construction is carried out. This one looks at the behavior of people as humans blessed or encumbered with all the human traits—needs, expectations, prejudices, loves, hates—and with a brain which provides the ability to make decisions and to communicate.

Many construction managers in the past would have found the notion that they must understand people, be they bosses, equals, or subordinates, to be repulsive in the extreme. Somehow to them this smacked of babying grown men and women and did not fit their image of construction people as rugged and somewhat macho. But many managers today do not hold this attitude. They recognize that work gets done through people; they cannot do it themselves; thus they see that they must understand people if they are to gain their respect so that they can direct, lead, and if necessary, persuade them. Stated differently, it is becoming clear that in today's world, there may be better ways than the hard-fisted one of do what I tell you and don't talk back, so common in the past.

Introducing these notions into construction may be particularly difficult. Many managers came up in the old school or through the trades where authoritarian leadership has often been the rule. Furthermore, the attitude is prevalent that the gain from careful attention to people as individuals is not worth the effort, given the usual short duration of construction projects and the ingrained but often erroneous notion that construction workers and foremen are here today and gone tomorrow.

This chapter approaches the subject of human behavior under the following topics:

- The discipline of organizational behavior
- Factors that may affect individual behavior in the workplace
- Management-subordinate relationships that may affect workplace behaviors
- Application of human-relations concepts in the construction environment
- The decision-making process
- Communications

THE DISCIPLINE OF ORGANIZATIONAL BEHAVIOR

Organizational behavior deals with the human problems faced by people in the workplace and particularly with those confronting managers. It is primarily based on psychology, anthropology, and sociology, disciplines within the overall one called behavioral sciences. Since World War II it has received increasing attention as the problems have become increasingly complex, because people, both managers and workers, have higher expectations from life and are better-educated, more vocal, and less willing to accept authority blindly.

At the outset, it must be recognized that in contrast with the exactness often attributed to the physical sciences and often claimed for engineering and certain aspects of business such as accounting, many of the research findings in organizational behavior seem to conflict or at least to permit different interpretations. This is to be expected in a discipline that deals with a subject as complex as that of human beings. Even so, much of value has been learned and more is being discovered every year. Courses on the topic are standard fare in schools of business management and industrial engineering, and books and periodicals on the subject abound.

Frederic W. Taylor, mentioned earlier, is commonly credited with being the father of scientific management, a forerunner of organizational behavior, because he applied or attempted to apply rational rules to both operating procedures and human performance in the workplace. He has been accused of treating humans merely as machines with a "whiff of exploitation" and with taking an "us and them," adversarial viewpoint. Even so, Taylor's approach, which considered workers as individuals and advocated some incentive pay, was far ahead of the common practices of his time.

Emphasis on the human relations aspect of productivity in the workplace came later; it was first brought into sharp focus by the so-called Hawthorne Studies begun in 1924 at the Hawthorne plant of the Western Electric Company in Cicero, Illinois. These studies made clear that human beings are not mere machines, and that paying attention to them as individuals and groups, along with allowing participation in decision making, could dramatically influence productivity. From this beginning has developed the discipline now called organizational behavior.[1]

The sections which follow present specific topics selected and summarized from the extensive organizational behavior literature. They represent the findings that are most pertinent to productivity in construction. The aim is to extend the brief discussions given in Chaps. 1, 3, and 4 to provide an orderly means for thinking about and addressing the human behavior side of productivity.

FACTORS THAT MAY AFFECT INDIVIDUAL BEHAVIOR IN THE WORKPLACE

Almost all human beings have needs for the present and expectations for the future. These needs and expectations to a large degree dictate behavior by providing motivators or demotivators. They apply to all facets of life, whether at work, at home, or at leisure. Our discussion will focus on the workplace, although the same patterns, adjusted to the particular situation, will probably develop everywhere. What is reported here are the pertinent findings of more than 60 years of research, plus observations of construction people by the authors over many years. Specific application of these concepts to on-site construction will be discussed later in this chapter.

Individual Differences

Management's function at every level in an organization is to get work done by individual persons, whether in one-on-one or group situations. As will be discussed in more detail in the sections which follow, individuals, as humans, have certain common traits which cause them to react in similar manners. Thus an understanding of these traits, as developed in disciplines such as organizational behavior, is important. However, it is essential at the outset of this discussion to recognize that every person is an individual and comes to the workplace with "baggage" that has been accumulated while growing up, being educated, and working in a unique set of circumstances. It follows that although in given situations certain behaviors might be predicted from the findings about people in general, these must be tempered to fit each individual's reactions.

Examples of a few among the many influences acting on individuals that are particularly pertinent in considering their reactions to situations in the workplace include the following:

• *Culture and religious beliefs,* which affect many aspects of the work situation, are, to name a few: attitudes toward status; others, particularly the op-

---

[1] Two comprehensive accounts of the Hawthorne Studies are T. N. Whitehead, *Industrial Worker* (2 vols.), Harvard, Cambridge, Mass., 1938, and F. J. Roethlisberger and W. J. Dickson, *Management and the Worker,* Harvard, Cambridge, Mass., 1939. Less comprehensive summaries appear in almost all the standard textbooks on organizational behavior. For a fascinating look back at the Hawthorne experiments after 50 years by leaders in organizational behavior, see E. L. Cass and F. G. Zimmer (eds.) *Man and Work in Society,* Van Nostrand, New York, 1975.

posite sex; and responsibility. As an example of the latter, it has been reported that some, although far from all, of the Muslim employees on mid-East projects refused to take responsibility for errors. The response, if they were called to task, was that it was not their fault but rather it was Allah's will.

• *Upbringing,* which includes attitudes about and behavior toward parents, siblings, and the outside world, along with beliefs about honesty and fair play that developed as the individual was growing up. To illustrate the workings of attitudes about parents, consider individuals who are very bright in other ways but try to "buck" the boss. The fortunate ones will realize that they are bringing their attitudes about their parents to the job and discovering this will completely change their behavior. Regarding honesty, it is interesting to observe the differences in attitudes of individuals about, for example, expense accounts, as a reflection of their general ethical slants. Some feel it is proper to spend their employer's money freely and claim every item that can conceivably be charged; others are very frugal and pay for every questionable item from their own pockets. Both groups feel comfortable with their decisions; that's the way they were brought up to be fair about money.

• *Attitudes about work (sometimes called the work ethic).* For some, hard work is the norm; for others just getting by is sufficient. These attitudes develop both from an individual's upbringing and from past work experience. In this matter, a wide variation among individuals is to be expected.

• *Expectations about job availability, rewards, and future prospects.* Particularly in construction these can vary widely between tradesmen and professionals. Tradesmen usually expect to work only for the duration of a project or a phase of it, to be paid at a fixed rate, and, with rare exceptions, to see no long-range future with a particular firm. In contrast, professionals often judge employment prospects in terms of their continuation, prospects for advancement and increasing rewards, and sometimes a place in management and ownership of the business.

These and other subtleties, in addition to general patterns of behavior, explain why dealing with individuals is such a complex and challenging job.

Motivation vs. Commitment

Motivation has been defined in the organizational behavior literature as "inciting unconscious and subconscious forces in people to achieve particular behaviors by them." The early Greeks and later famous people such as Adam Smith and John Stuart Mills assumed that people were primarily motivated by desires to maximize comfort and pleasure and minimize discomfort and pain. Today in the world of work, as will be discussed later, motivating drives are said to include the need for food, to feel secure, for affiliations, and to achieve and have power and status. These drives, most of which cannot be measured directly, show up in the behavior of individuals or groups. Therefore, motivation involves satisfying these desires in a way that produces a desired behavior. As will be shown in the succeeding sections which discuss several theories of motivation, behavioral scientists and managers alike are wrestling to find answers.

The material on motivation presented in the sections which follow is based on the organizational-behavior literature, which seems to assume that motivation and commitment are synonymous terms. It is worthwhile repeating here a point made in earlier chapters: motivation is only part of the game; it involves creating a desire to be productive. But commitment implies a willingness to follow through with this desire and actually accomplish a task. In order to advance to this crucial second step, one must believe that the task can be carried out successfully. This calls for assurance that others will remove obstacles beyond the control of the motivated person or group. Readers who do not understand the difference will miss an essential point.

From the statement that management is getting things done through people it follows that not only must managers issue directions about what is to be done, but they must somehow get others to agree to carry them out. As indicated in the discussion of leadership styles later in this chapter, getting others to perform can seldom be done in today's world simply by commanding that the supervisor's wishes be followed. Rather, subordinates must want to carry out these directions. But that is not all. To restrict the use of strategies underlying motivation and commitment to dealings with subordinates is short-sighted, because these same strategies also apply when peers and even higher-ranking people in an organization are involved.

Needs as Motivators—A Hierarchy

A. H. Maslow[2] in the early 1950s advanced a theory of motivation called the hierarchy of needs. This is very useful for those who are trying to understand why people behave as they do, and it may suggest ways for motivating and getting commitment from them. Maslow proposed that man is a wanting animal and that these wants in turn become needs that man tries to satisfy. These needs are not only physical, to be satisfied by material things, but also psychological—that is, they act through the mind. Furthermore, these needs fall into a rank order (a hierarchy). Man first satisfies his most basic needs; once these are partially satisfied, they are in part replaced by new needs and he is motivated to attain these. Once largely satisfied, needs are no longer strong motivators, but there is always a new need to replace the old. These concepts are illustrated graphically in Fig. 10-1.

The most basic needs for food, clothing, and shelter are physiological; that is, they are the ones necessary to sustain life and provide physical comfort for the individual and any dependents. Only after these are nearly satisfied do humans seek to satisfy their needs for safety, belonging or love, esteem, and self-fulfillment, in that order. It is important to recognize that the basic physiolog-

---

[2] Abraham Maslow, *Motivation and Personality,* Harper & Row, New York, 1954.

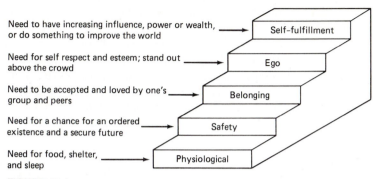

Need to have increasing influence, power or wealth, ——→ Self-fulfillment
or do something to improve the world

Need for self respect and esteem; stand out —— Ego
above the crowd

Need to be accepted and loved by one's—— Belonging
group and peers

Need for a chance for an ordered —— Safety
existence and a secure future

Need for food, shelter, ——→ Physiological
and sleep

FIGURE 10-1
Graphic representation of the hierachy of needs.

ical and safety needs are imposed by the external environment, but the higher needs, belonging, esteem, and self-fulfillment, are generated mentally as individuals think about themselves and their situations.

A corollary theory that can be derived from the hierarchy of needs is that the more a need is satisfied, the less it is a motivator that influences behavior. This notion has been portrayed graphically in Fig. 10-2; however, as noted, it merely illustrates a concept and is not to be taken too literally. The point of Fig. 10-2 is, for example, that people with sufficient food, clothing, or shelter have little desire for more, but they can become anxious about having them in the future, the assurance of which offers a feeling of safety. Once they feel reasonably secure about the future, worry largely disappears and no longer affects behavior substantially. Instead, individuals turn to actions that will satisfy the higher-level needs, one after another. Equally important is the fact that the motivations can be reversed. People concerned with the higher needs can through a change in financial circumstance or loss of stature or a position of importance find their security or even physiological needs threatened. All at

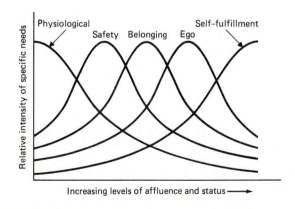

FIGURE 10-2
Graphic representation of how needs change with levels of affluence and status.

once the need for status and belonging becomes secondary. The situation for union construction craftsmen in some areas in the late 1970s and early 1980s vividly illustrates this reversal. When construction was booming and the unions controlled jobs, workers enjoyed high wages, status, and an apparently secure future. But the combination of a downturn in construction and non-union competition threw many of them out of work so that safety and even physiological needs took over. A turnaround in construction which would bring a high demand for skilled workers could produce another reversal.

Motivation by Job Enrichment

An extension of Maslow's hierarchy of needs is Herzberg's two-factor theory of work motivation.[3] This theory proposes that there are two distinct sets of factors in the work environment. One set, called motivators (job enrichers), includes achievement, recognition, the work itself, responsibility, and advancement. The other, labeled hygiene factors, includes company policy and administration, technical supervision, salary and fringe benefits, interpersonal relations, and working conditions. The theory is that if jobs are structured to provide motivators, workers will be challenged and possibly increase their productivity. On the other hand, although providing the hygiene factors will lessen dissatisfaction and probably keep the individual from seeking other employment, providing them will not bring the long-lasting satisfaction that will lead to motivation. Herzberg argues that all too often, management fails in its attempts to motivate employees because it puts all of the emphasis on removing dissatisfiers and neglects the satisfiers which bring motivation.

There is little challenge to the notion that job enrichment efforts carried out properly can improve worker satisfaction and the climate of the workplace. However, as will be discussed later, some will argue that job enrichment may not improve productivity, since a positive link between job satisfaction and productivity has not been firmly established.

Process and Expectancy Theories of Motivation[4]

Reports from job-enrichment programs show that they have apparently increased worker satisfaction on assembly lines or where tasks are unexciting and highly repetitive and workers seldom see any result from their labors. The positive effects on productivity have been more difficult to quantify. As will be discussed in more detail later, the job-enrichment approach to satisfaction may be of lesser importance to productivity in on-site construction than other fac-

---

[3] See Frederick Herzberg, *Work and the Nature of Man*, World, Cleveland, 1966; and "One More Time—How to Motivate Employees," *Harvard Business Review*, January-February 1968, pp. 53–62.

[4] For more detailed discussions and references see F. Luthens, *Organizational Behavior*, 2d ed., McGraw-Hill, New York, 1977, and T. R. Mitchell, *People in Organizations*, McGraw-Hill, 1982.

tors. However, construction managers who attempt to divide tasks too finely or introduce highly repetitive operations or assembly lines should be aware that they are taking away the variety and richness inherent in construction work. This may lead to monotony, boredom, and dissatisfaction and accompanying ills on the part of the workers.

Maslow's theory assumes that meeting people's needs will increase satisfaction; Herzberg holds that job content will do so. It is reasonable then to infer that these satisfactions will motivate people to strive for higher productivity. Since these theories were advanced, some behavioral scientists have challenged this inference on the basis that motivation is too complex to be explained so simply. Other factors that might be included are the presence or absence of a work ethic, job availability and the prospect of a layoff, and the pure pleasure of doing a job quickly and well. Process theories attempt to be more specific in defining these motivators and determining how they work.

A process approach, first advanced by Vroom, is expectancy theory, which relates behavior to an individual's expectations that certain behaviors will have predictable outcomes which satisfy organizational or individual goals. Organizational goals are measured in terms of quantity, quality, or timeliness of output; individual goals could include money, recognition, promotion, and security.

Porter and Lawler advance the notion that effort leads to performance that brings both intrinsic (originating in the mind) and extrinsic (external or tangible) rewards; the rewards, in turn, provide satisfaction. If the rewards are equitable and of sufficient value, the effort will be repeated. Lawler later added the concept that expectancies have two dimensions, that of performance (the job can be done) and that of reward. These theories have performance leading to satisfaction, whereas the Maslow and Herzberg theories have satisfaction leading to performance.

Smith and Cranny present a somewhat simplified model which, in effect, says that performance produces both rewards and satisfactions, and that the two effects are interrelated. Adams is credited with an equity theory, which proposes that motivation is related to an individual's perceptions of his or her rewards as compared with the inputs and outcomes and rewards of others. Research covering a few situations of overcompensated and undercompensated people showed that productivity changed up or down until performance and compensation were perceived to be in balance.

Laufer and Jenkins have proposed a quantification of expectancy theory as it applies to construction workers.[5] Their approach, shown graphically in Fig. 10-3, assigns numerical values to the subjective mental processes that go on in workers' minds when they decide on the level of effort they will exert toward accomplishing a specific task. This involves:

1 A subjective estimate by the worker of the probability that each among two (or more) levels of output can be reached. For example, a bricklayer might

---

[5] See A. Laufer and G. D. Jenkins, Jr., *Journal of the Construction Division*, ASCE, vol. 108, no. C04, December 1982, pp. 531–545. This paper includes an excellent bibliography.

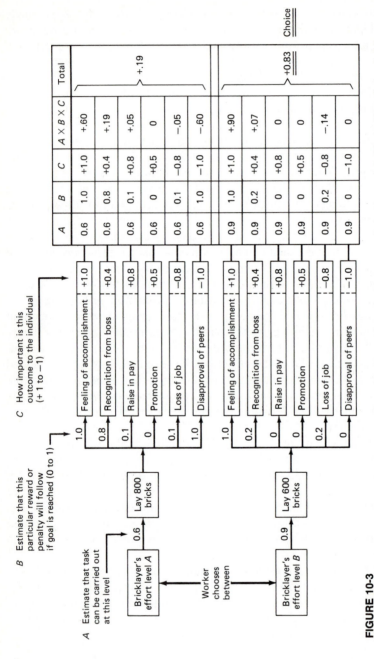

FIGURE 10-3
Expectancy model for a bricklayer selecting between levels of effort. (*Adapted from A. Laufer and G. D. Jenkins,* Journal of the Construction Division, *ASCE, vol. 108, no. 4, December 1982, pp. 531–545.*)

decide that the probability of laying 600 bricks a day is 0.9, but the probability of laying 800 bricks is only 0.6 (see item *A* in Fig. 10-3).

2 A subjective estimate by the worker of the probability that if a given output is reached, it will bring specific rewards or penalties in each of several areas. Estimated probabilities that each reward or punishment will occur range from 0 to 1 (see item *B* in Fig. 10-3). For the bricklayer example, proposed rewards include a feeling of accomplishment, recognition from the boss, a raise in pay, and a promotion. The penalties are loss of job and disapproval of peers. The rating that the worker assigns probably will be a combination of (a) intrinsic values, representing each individual's internal feelings, and (b) extrinsic values, imposed by external conditions. A different list would probably be appropriate for different classes of workers.

3 A subjective estimate of the importance of a particular outcome to the person involved. The importance of positive outcomes grades upward from 0 to 1; that of negative outcomes ranges downward from 0 to −1 (see item *C* in Fig. 10-3).

4 For each of the items listed, the values given under items *A, B,* and *C* are multiplied together and the results totaled, as shown in the figure. The greater total will predict the workers' subjective choices between the relative levels of effort to be expended. In this instance, the choice is to lay 600 bricks a day.

This example illustrates an attempt to express in numbers the mental process that goes on when an individual chooses the level of effort to exert in carrying out a particular task. A first reaction can well be that an attempt to quantify something as complex and subjective as a worker's behavior is an exercise in futility and a waste of time. With further reflection, however, it can be seen that reducing the process to numbers clarifies a very complex process that goes on in people's minds whenever a decision is made. Furthermore, making a list of rewards and penalties for a particular work situation can be helpful to managers as they think through and appraise how they deal with subordinates. One productive use of such a listing and numerical rating would be to have supervisors on the one hand and workers on the other make separate lists and appraisals. In most instances where questionnaires on subjects such as this have been employed, the differences have been startling.

Process and expectancy theories carry two important messages to construction managers. First, the perceptions of situations in the minds of the affected people are the motivators or demotivators. Managers cannot assume that their assessments of situations and/or of factors that motivate them personally will apply to others. Rather, they must develop empathy which crudely defined is "getting in the others' skins and looking out through their eyes" to understand how they think. Second, in every situation, there will be motivators and demotivators, and managers must sort out the ones that apply.

Money as a Motivator

Many people, including Frederick W. Taylor, have accepted as fact the notion that money will motivate people to work more productively. They cite the examples of piecework, prevalent in many industries often under sweatshop conditions, and of high output in construction, where agreed or informal (sometimes under the table) payments are tied to output. But instances such as these tell only part of the story; studies have clearly demonstrated that money isn't everything. Rather, how well money motivates depends on the circumstances.

It is helpful to relate the motivating power of money to the hierachy of needs. Unquestionably money or the fear of being deprived of it is the primary satisfier for physiological and safety needs; this has been demonstrated again and again during high-unemployment periods in construction. During such times employers could, within limits, pick and choose among workers and drop those who were less productive. Under such circumstances, money was the prime factor that motivated workers to produce more. Also they were generally more tractable. Money was also the primary concern in the early agreements between contractors and unions, focusing primarily on wages, which satisfied physiological needs. Later, monetary fringe benefits such as health, welfare, vacation pay, and provisions relating to job security have reflected a shift toward security needs. Union membership itself and provisions regarding seniority and jurisdiction, although not strictly financial matters, are primarily aimed at security needs. In general, satisfaction of the higher internally generated needs has not been a union target. In fact, unions have been charged with frustrating attempts by their members to satisfy these needs by, for example, demanding that all workers be paid equally except for a small hourly premium for foremen.

At the needs levels (often high) of today's skilled construction workers and supervisors the situation may be different. Consider first the belonging-level needs. Under one set of circumstances, high productivity is favored by the workers and monetary rewards can strengthen the bonds among individuals. On the other hand, there have been many instances where a high producer, seeking more pay or other similar rewards such as extended employment, may be ostracized as a rate buster. Unions and their individual members often exert such pressures. For example, the attitude behind the slogan a fair day's work for a fair day's pay, and the common requirement that members' paychecks be subject to examination by the business agent or fellow union members on demand, work against acceptance of extra-money incentives. In sum, money incentives even if available may or may not be effective as a motivator at the belonging level. At status and self-fulfillment levels in the hierarchy of needs, money is one motivator along with others, including title, recognition, and perquisites such as plush offices or company automobiles. It primarily functions to establish the relative worth of individuals as seen by company, associates, the community, and themselves. For many, this is a need as real and important as is food to the hungry. For such people money is primarily an attention-getter which points out the individual's importance. Its effects last only a short time; soon after such individuals have received raises, their thoughts turn to the next raise or to something else to satisfy the always unfulfilled need for status or self-fulfillment.

Four reasons why managers turn to money as a motivator, often without considering the consequences, are as follows:

1 Managers as a class are at the higher-needs levels and are achievement-oriented. For them, money is a valuable measure of satisfaction and status; they work hard to get more of it. Without thinking further, they assume that what motivates them also motivates others.

2 Money is the one motivator over which on-site managers have a large degree of control and which they can manipulate. At all levels from project manager to foreman, ways can be found to give monetary or closely related rewards to favored subordinates, regardless of the efforts of higher-level managers to prevent or control these practices.

3 Since measurements of the effect of money on productivity are undependable, the assumption is that it is a motivator.

4 For on-site managers the alternative approaches to motivation are often unknown, untried, tricky, or frightening. Who knows, money may work.

One of the authors of this book saw a classic example of how money as a motivator works differently depending on the circumstances. In Johannesburg, South Africa, in the late 1970s, building was booming and there was a critical need to increase bricklaying output. The white bricklayers were anxious to work overtime, but their black helpers refused on the grounds that the extra earnings were meaningless. Having no electricity and few other amenities in Soweto, the black reservation, they had little opportunity to spend the extra money, so money did not matter to them.

Regardless of these shortcomings of money as a motivator, all managers will continue to see money and its surrogates, or the withholding of it, as one of their most important management tools. They may fail to recognize that unwise or indiscriminate use of it to reward or punish can create serious difficulties. Only if it is used fairly and for clearly defined purposes based on fair and impartial measures will it help to improve productivity and bolster rather than damage job morale.[6]

Nonmonetary Rewards and Punishments as Motivators

It is common knowledge that praise and other recognition for work well done can be a valuable motivator—a continuing theme of this book. It must be emphasized here that all too often, busy managers fail to make the small effort necessary to capitalize on this factor. Rather they assume that it is enough if the individual recognizes that praise is due, whether given or not. Further-

---

[6] For an opinion survey among contractors, owners, academics, consultants, and union officials on the effectiveness of several financial incentive plans, see A. Laufer and J. D. Borcherding, *Journal of the Construction Division,* ASCE, vol. 107, no. C04, December 1981, pp. 745–756. W. F. Maloney, *Journal of Construction Engineering and Management,* vol. 109, no. C04, December 1983, pp. 475–477, directs attention to conditions peculiar to construction which limit the applicability of money incentives.

more, they do not realize that praise, in contrast to punishment, is most effective if given before others. One measure of a good manager is to whom and how he offers praise. Whenever possible it should be done before others and not in private. But unjustified praise is pointless and even destructive of morale.

Punishment in the minds of the unthinking is merely the opposite of praise and other nonmonetary rewards. However, it is far more than this and has many consequences that are largely unrecognized. Many managers see punishment as a way of correcting unsatisfactory behavior when in fact they are using it in an attempt to protect or restore their own status or battered egos. From the workers' point of view, punishment may not be seen as punishment at all. Rather it may be seen as a singling out, which can satisfy an individual's need for attention, since attention has not been otherwise given. Employees who mess up the same job again and again may not be stupid; rather they may have found a means of getting attention and are willing to accept censure and abuse to get it.

As a general rule, good managers feel that if punishment or censure are needed, they should administer it in private. This procedure has been found to be more effective, and supervisors are less likely to use it as an ego trip for themselves. An exception to this rule is when a worker commits an unsafe act. It has been found that foremen with excellent records in safety use a crew discussion of the act not to humiliate the worker but as an effective tool for teaching safe practices.

Overcoming Resistance to Change as a Demotivator

People, whether workers or supervisors, resist change of any kind which affects behavior patterns to which they are accustomed and with which they are relatively comfortable, or which make them feel vulnerable, because change:

1 Can create uncertainties in what may have been a relatively pleasant and secure world
2 May, in the eyes of the affected people, challenge their status
3 May appear to the affected people to be manipulation

For reasons such as these, it is probable that change will be resisted or resented, even if the steps to getting cooperation listed in Chaps. 1 and 3 are carried out. These, repeated here in slightly different form, are:

1 Workers and their supervisors must understand the change.
2 It must be physically possible to accomplish the change.
3 All those affected must mentally feel that the change can be carried out.
4 In the minds of those carrying out the change, it must be in the interest of the individual, group, or organization to which they owe their loyalty.
5 Individuals or groups carrying out the change must find it to their own personal advantage.

Resistance to change in industrial situations was studied soon after World War II, and the findings are relevant today. For example, Lawrence demonstrates that not all changes were resisted.[7] Many of them, such as the introduction of obvious labor-saving devices, were welcomed. He suggests that resistance is offered to changes that affect the "existing social structure." This term includes all changes that affect people's job satisfactions, as well as those that affect their interpersonal relationships on the job. On the basis of these ideas, his recommendations, restated somewhat, are:

1 Where possible, changes should be made through existing communication channels.

2 Workers should be given opportunities to be heard before decisions are made to change their jobs.

3 Changes should be explained fully and clearly to those who will be affected by them.

4 If a staff specialist rather than the regular supervisor introduces a change, management should make clear that it understands and supports the change.

5 There must be a follow-through until the change is in place and working.

Research into the characteristics of small groups proved to be another source of information on the problems of resistance to change. A series of experiments were conducted by Bavelas, Barret, Smith, and Leavitt to measure the effect of various communication patterns on the ability of a small group to solve problems.[8] They found that when people are placed in situations where they have to perform jobs while overloaded and under pressure, they become reluctant to accept new ideas. This was proven in experiments with groups who were able to communicate only along predetermined channels. In the tests, groups were directed to solve a simple problem, using the communications patterns shown in Fig. 10-4a and b. With the centralized network, the groups showed much less flexibility in developing new methods for performing assigned tasks and in their ability to adapt to changed conditions because this arrangement put almost all pressure and responsibility on the individuals in the center. In every experiment these people refused to accept suggestions from other members of the group about how to do their jobs better. They did, however, have high morale in contrast to the low morale of others in the group.

The situation on a construction site is closely analogous to the centralized network that was created in this early experiment. Both foremen and superintendents occupy centralized positions and are generally busy and under pressure. Often those in such positions tend to offer strong resistance to change. This general finding has been confirmed by field studies on construction sites. Here also it has generally been found that when foremen or superintendents

---

[7] See P. R. Lawrence, "How to Deal With Resistance to Change," *Harvard Business Review,* vol. 32, no. 3, 1954.

[8] See A. Bavelas and D. Barret, "An Experimental Approach to Organized Communication," *Personnel,* vol. 27, no. 5, 1951.

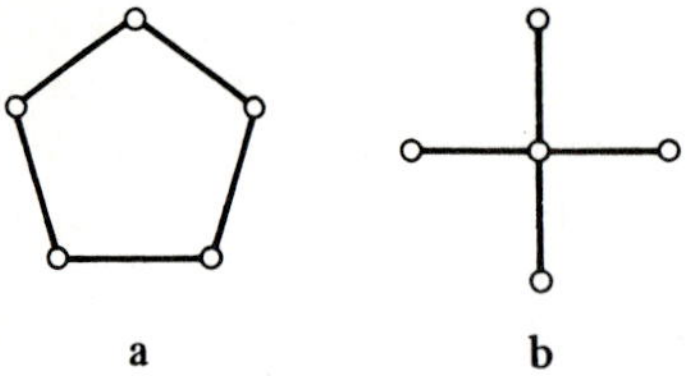

FIGURE 10-4
Experimental communication patterns: (a) circular network group interaction; (b) centralized network-dependence on central authority.

were busiest and under the greatest pressure, they were more reluctant to try any new method even though it was obviously one that would save money or effort in the long run. This behavior is to be expected; it has become a truism known as Greshams' law, which states that "people respond to problems that find them before they respond to problems they find themselves." Usually, an overabundance of problems developed by others find construction supervisors.

Yet another approach to the problem of resistance to change was investigated by Coch and French.[9] They devised an experiment based on the theory that employee participation in designing changes would bring ready acceptance of the changes. In a pajama factory, the experimenters introduced changes into separate but similar groups in three different ways. In the first group, they changed existing procedures by simply telling the operators what the changes would be. In the second, they introduced the change by having the operators participate in planning the change through representatives whom the operators had chosen. In the third, changes were made by having all the operators participate in designing the new jobs. The effectiveness of the three ways of introducing changes was measured by counting the pajamas processed each day by each group, both before and after the change. After a month, equally improved production was obtained from the two groups whose members participated in the change. However, the group that participated through representation took longer to implement the changes and to get their production up to the level of the group that had total participation. Meanwhile, the group that had not participated showed a decline in production. Even after 30 days of using the new method, this group was not producing as much as it had previously. This experiment indicates that the best way to introduce change is to allow employees themselves to be deeply involved in any changes in their methods. This is one feature of quality circles first used in Japan and is now in widespread use in the United States. Adaptations of this concept are called job teams or quality circles (see Chap. 3). They involve having a representative from each level in the hierarchy that has concerns and that can do something about a problem.

Changes are sometimes proposed that conflict with accepted status levels. An example would be where an equipment operator or carpenter is told to

---

[9] See L. Coch and J. R. P. French, Jr., "Overcoming Resistance to Change," in R. A. Sutermeister (ed.), *People and Productivity,* McGraw-Hill, New York, 1963, pp. 436–458.

work under a foreman from the laborers. Another example would be having a grade checker issue instructions to skilled scraper or blade operators. Status reversals such as these are almost certain to meet resistance, and managers who attempt to implement them can look for trouble.

It should not be presumed from this discussion that resistance to change is encountered only in craftsmen, foremen, or crews. The same factors apply at every level in an organization from owner through project manager, project superintendent, engineers, and the rest. It has been demonstrated that it can be overcome, but only with careful attention to human behavior at all levels from top to bottom.

Motivating by Overcoming Alienation, Apathy, and Frustration

Many derogatory things are being said today about workers by both behavioral scientists and managers. As an example, Lawrence has written that "some (blue-collar) workers, but not all, have been nurtured in subcultures that have deeply conditioned them to see work as a simple exchange of time and minimal energy without much involvement for fair pay and decent conditions. A decent living is seen as a right and not something to be earned."[10] They expect to "live away from the job," getting their satisfaction elsewhere. Whether or not this loss of the work ethic is a new phenomenon which accompanies a higher educational level and an accompanying individualism and willingness to speak out and to rebel against authority, or has only recently been discovered by researchers is not the point; the point is that it actually exists, or at least it exists in the minds of many managers in the United States today. Stated differently, it is certainly true that some workers are alienated from and are apathetic about their jobs and are merely putting in time to secure paychecks.

Behavioral scientists have associated alienation of and apathy in workers and lower-level managers with (1) higher management's being taken over by people concerned only with profits (Herzberg calls them MBA types) and (2) technology and the high degree of specialization and aloofness it creates between individual workers and workers and managers. Under such situations workers see little connection between their work and a completed product or some other accomplishment. They feel powerless and isolated, socially lonely, and without norms and rules of behavior. Work for the fun of working has little meaning. Such books as *The Organization Man* by William H. White, Jr., *The Lonely Crowd* by David Riesman, and *Escape From Freedom* by Erich Fromm described these feelings well. Today, discovering ways to combat alienation and apathy is a major concern of managers and behavioral scientists.

Traditional construction seems to have largely escaped the problems associated with alienation and apathy. Work assignments are varied, they generally involve crews rather than something done in isolation, and their results are apparent so that satisfaction comes with completing tasks. Despite impressions

---

[10] See P. R. Lawrence in *Cass and Zimmer,* op. cit.

to the contrary, foremen and workers have been much involved in decision making. Recently, however, and particularly on large projects, the situation seems to have changed. Such factors as deep management hierarchies and material and equipment shortages, along with design changes and incomplete planning, have led to delays and rework, which frustrates good workers and takes away the usual satisfactions. Workers and lower-level management feel powerless, isolated, and alienated and lose any desire to work. Absenteeism increases, and those with a strong work ethic quit. Foremen, usually highly dedicated in their assignments, feel they are without guidance and are reduced to being pushers.[11] Examples such as these indicate that construction is not necessarily immune from the destructive effects of boredom and frustration, especially very large projects and assignments on projects of any size involving merely putting in time, rework, or assembly-line or other highly repetitive or nonchallenging tasks.

Motivating Crews

Researchers have found that group behavior is as difficult to understand as is that of individuals. It has become a truism that "although assigned relatively straightforward tasks which call for simple sets of internal and external behaviors, crews somehow have a strong need to complicate their lives by developing attitudes, activities, and behavior that are not required and that have no connection with the assigned tasks." Knowing these peculiarities of group behavior, good managers work hard to create positive and head off negative forces that affect them.

When groups of any kind first form, they develop informal rules to which members must conform to be "regulars" in good standing. Among these are attitudes about and standards for approaching productivity improvement. For example, the "regulars" in a highly productive group will deny membership to laggards (deviants) after attempts first by persuasion and then by abuse fail to change their behavior. This form of crew self-discipline is particularly apparent in situations where a piecework rate or a bonus for high performance is paid to a crew rather than to individuals in it. In such groups, the laggards perform or are forced out. On the other hand, a group that is antimanagement and that fosters low productivity will isolate, abuse, and possibly drive off a hard worker as a rate buster.

Over time, groups establish a "pecking order" of leader and followers with the followers in a rank order. Often this informal leader is not the foreman des-

---

[11] See J. D. Borcherding, *Journal of the Construction Division,* ASCE, vol. 102, no. C04, December 1976, pp. 599–614, and ibid., vol. 103, no. 4, December 1977, pp. 567–575; N. M. Samelson and J. D. Borcherding, ibid., vol. 106, no. 1, March 1980, pp. 73–89; and J. D. Borcherding and D. F. Garner, ibid., vol. 107, no. C03, September 1981, pp. 443–453, for the results of surveys made on several large industrial construction sites. Also see, W. F. Maloney, ibid, vol. 106, no. C04, December 1980, pp. 618–622, which discusses these findings and critically evaluates the methodology.

ignated by management. When such is the case, it is the informal leader to which the crew looks for cues about their behavior. The group leaders, whether the designated foreman or someone informally selected by "regular" group members, maintain their standing as leaders by adhering to the standards set by the "regulars." In representing the group in dealings with managers or others, these leaders often find themselves caught by the "hero-traitor" dilemma. If they hang tough and follow the group's wishes, they are heroes to it and remain leaders. If, however, they compromise, they become traitors to the group and are rejected by it. Managers who push group leaders into making concessions may have won the battle but lost the war, for they may destroy the leader's credibility and influence with the group without changing the group's stance or attitude at all.

In dealing with groups, managers sometimes make changes that to them seem inconsequential but are then surprised when the changes produce violent group dissatisfaction and possibly lower production. A short list of such disruptive changes would include the following:

1 Reducing opportunities for group members to interact and move around. In addition to reduced productivity when interaction and moving around is restricted, an added consequence may be that horseplay develops, which can result in injury to workers or destruction of property.

2 Reducing the status of the group by playing down its importance.

3 Breaking up congenial groups or subgroups.

4 Attempting to impose stricter measures of performance.

5 Intensifying supervision and monitoring of the group's work, habits, and behavior.

Autocratic managers feel that as bosses they control situations such as those just listed and that they are within their authority in making such changes. However, they often learn to their dismay that it is the group that is in control, not they themselves. The exception is when workers find their basic physiological and (possibly) safety needs in jeopardy. Then the group, in effect, falls apart as the individual's belonging needs are superseded by the more basic ones, and each puts personal interests ahead of group membership.

It is sometimes argued that strong groups are undesirable in that they interfere with management's right to manage. Proponents of this view advocate rotating workers among crews or other stratagems to prevent groups from forming. H. J. Leavitt has argued, however, that under usual circumstances, there are advantages in building work situations around strong groups.[12] He lists the following:

1 There is support in times of stress and crisis.

2 It is a good tool for problem finding.

3 They make better decisions than do individuals by themselves.

---

[12] See H. J. Leavitt in *Cass and Zimmer*, op. cit.

4 They are better tools for implementation.

5 They can control and discipline individual members.

6 Through groups it is possible to fend off many of the negative aspects of large organizations.

7 If management creates groups it prevents their formation in adverse ways. They "sprout as flowers rather than weeds."

Skeptics may argue that the discussion here applies to industry but not to construction, but good field supervisors will disagree. They know that knowledgeable, dedicated crews, supportive of their supervisors and the contractor's aims, are a most valuable asset.

Job Satisfactions and Dissatisfactions as Motivators and Demotivators

In the discussion of motivation earlier in this chapter, the point was made that job satisfaction might not result in motivations leading to improved production nor dissatisfaction to demotivators and lower production. But there is strong evidence from data gathered on construction sites to indicate that these relationships hold. It can also be argued, as a practical matter, that there is a good chance that satisfied workers may be receptive to approaches which lead to increased productivity, whereas, even if dissatisfied workers accepted the approaches, they might leave the job before the approaches could be put into effect. Furthermore, if they stayed, productivity probably would decrease not only because of on-the-job actions but also because there would be high absenteeism and turnover.[13]

Because much job satisfaction and dissatisfaction may be different in construction than in other industries, a specific discussion of each is postponed until later in this chapter.

MANAGEMENT–SUBORDINATE RELATIONSHIPS THAT MAY AFFECT WORKPLACE BEHAVIOR

Seldom do individuals operate entirely alone in the workplace, completely isolated from others; they must deal with one another. One may be the boss and another the subordinate, the two may be relative equals in the job hierarchy or members of the same crew, or there may not be a clearly defined relationship; but interaction is necessary or desirable, for example, when a manager must deal with engineers, architects, or union representatives. In all these instances, it is essential that the principles of concern for and fair dealing with others be followed; there must be an understanding of the motivational forces acting on the parties involved, as discussed earlier in this chapter. The sections to follow look more specifically at the relationships between supervisor and supervised at every level in the job hierarchy.

---

[13] For a recent survey of the reasons for and effects of absenteeism and turnover, see "Absenteeism and Turnover", *Report No. C-8*, the Business Roundtable, June 1982.

Ways of Perceiving People—Theory X vs. Theory Y

In Chap. 4, which looks at construction as a system, there is a brief discussion of human resources as an element in generating and implementing commitments to productivity, with Table 4-2 being a primary focus of the discussion. It contrasted theory X vs. theory Y ways of looking at people and their effects on a worker's choice of workplace, behavior, and willingness to make commitments to productivity-improvement efforts. This discussion aims to flesh out and, to a degree, qualify the propositions advanced in that earlier discussion.

The basic premises about people summarized in Table 4-2 were advanced and later modified by Douglas McGregor.[14] He classified managers into the following two groups according to their underlying beliefs about people: theory X managers, who, as noted in Table 4-2, believe people "must be driven;" and theory Y managers, who feel that people are "mature, basically good and honest, and anxious to work."[15]

McGregor described the beliefs of theory X managers as follows:

1 The average human being has an inherent dislike of work and will avoid it if possible.

2 Because of this dislike for work, most people must be coerced, controlled, directed, or threatened with punishment to get them to put forth adequate effort toward the achievement of organizational objectives.

3 The average human being prefers to be directed, wishes to avoid responsibility, has relatively little ambition, and wants security above all.

In contrast, theory Y managers' beliefs were that:

1 The expenditure of physical and mental effort in work is as natural as play or rest.

2 External control and the threat of punishment are not the only means for bringing about effort toward organizational objectives. Human beings will exercise self-direction and self-control in the service of objectives to which they are committed, including those of organization achievement. The most significant of such rewards, e.g., the satisfaction of ego and self-actualization needs (discussed earlier in the chapter as the hierarchy of needs) can be direct products of effort directed toward organizational objectives.

3 The average human being learns under proper conditions not only to accept but to seek responsibility.

4 The capacity to exercise a relatively high degree of imagination, ingenuity, and creativity in the solution of problems is widely, not narrowly, distributed in the population.

---

[14] See *The Human Side of Enterprise*, McGraw-Hill, New York, 1960, and *Leadership and Motivation*, MIT Press, Cambridge, Mass., 1966.

[15] William Ouchi has written a popular book entitled *Theory Z*, Wesley, Stony Brook, N.Y., 1981. It treats the topic "How American Business Can Meet the Japanese Challenge." From its title, Ouchi's book might appear to be an extension of McGregor's theory. Rather, it describes the management systems of five successful U.S. companies that follow McGregor's theory Y practices and gives the origin and application in Japan of quality circles.

5 Under the conditions of modern industrial life, the intellectual potentialities of the average human being are only partially utilized, so that a valuable resource is often lost.

E. H. Shein,[16] in discussing these locked-in management beliefs and their effects, adds the following:

1 Theory Y managers have the ability to maneuver and devise strategies for gaining cooperation. Theory X managers have no such choices.
2 Theory X managers in today's society where authority is not absolute will find that:
  a Workers will undermine their authority.
  b Workers, not managers, will control the system.
  c Workers' responses will be illogical and not what is expected.

In discussing theory X and theory Y mental sets, a distinction has been made between actual and perceived, which deserves special attention. Actual describes how bosses see and measure people; perceived is what those affected conclude about the boss's beliefs from the messages they receive. This may well mean that as a result of early contacts, the mental sets of gruff, hard-nosed bosses will be perceived as theory X even though they are theory Y. In such instances, the initial reactions of workers will be to play it safe by following orders and keeping their mouths shut. In such situations, it may take longer for supervisors to develop creative participation from their crews.

In construction as in other enterprises there are many actual or perceived theory X supervisors; and higher-level management must decide what, if anything, is to be done about them. Unfortunately, there is little evidence that the ingrained beliefs and conduct of such supervisors can be changed either by training or by censure, so that the dilemma will haunt industry in the years to come.

Observers of management behavior in companies in terms of theory X and theory Y have discovered that attitudes at the top of an organization permeate it. In case after case it has been discovered that those lower in the hierarchy soon get the message about how they are to view and deal with their subordinates, and they either conform to higher-management's attitudes or quit or are fired.

In summing up, the evidence is clear that in the United States today, effective managers in most instances believe in and practice theory Y approaches. Many of them have never heard their behavior described in McGregor's terms, but they have somehow learned to follow the principles. Those who have not risk their survival.

The Management–Worker and Management–Foreman Dichotomy

Eric Hoffer, the longshoreman-philosopher years ago stated the management-labor dichotomy clearly.[17] He made the point that management's primary concern must be with the task and the results, whether the manager is a profit seeker, an

---

[16] See *Cass and Zimmer,* op. cit.
[17] See *The Ordeal of Change,* Harper & Row, New York, 1963.

idealist, a technician, or a bureacrat. Naturally, then, managers view workers as a means to an end and feel that they should try to get the utmost from the workers at the lowest possible cost. Thus, if circumstances permit managers to take workers for granted, they will do so and not worry about what the workers will say or do. Of course many management people will argue that when workers have the upper hand, they will trample roughshod over owners and managers alike, unless their interests are better served through cooperation. In sum, Hoffer's argument is that there is a cleavage in viewpoint which makes strain and strife between management and workers inevitable and a fact of life. The only protection from its consequences is to accept it as obvious and treat it in a matter-of-fact way. To him, the alternative of a millennial society where the "wolf and lamb shall dwell together" is stagnant and far less attractive. Hoffer also makes the point that the problem is not unique to capitalistic societies. It exists wherever there are managers and workers. He expresses a preference for the self-seeking capitalistic manager rather than for the ruthless idealist dedicated to a holy cause, be it communism, fascism, or religion.

The thinking just described draws the lines between the objectives and concerns of management and labor quite clearly. However, there are those in management, particularly at the foreman level, who have one foot in each camp. Usually they have been chosen from the ranks of labor and have strong emotional and possibly union ties to it. Certainly, labor claims them as its own. On the other hand, management sees them as supervisors and expects, if not demands, their loyalty and assurance that they will adopt management's interests as goals. The view through the foreman's eyes is that neither management nor labor recognizes his dilemma; he is caught in the middle and cannot please either side. He is cast in the classic "hero-traitor" role where if he takes one side he may become a hero (or be taken for granted) by that side, but will certainly be a traitor to the other. This problem is as old as industrial society and, like the management-labor one, will not go away.[18]

A thesis of this book is that if management is effective, and properly structures the work situation, the common interests of it and labor will offset the differences in basic objectives, so that clashes need not develop. But it would be naive to assume that the opportunity for conflicts to develop does not exist and fail to give attention to ways to head them off.

Authority and Power

In traditional approaches to management, it is often assumed that managers have two rights and privileges: authority and power. Each of these words has several meanings and, depending on individual attitudes, different connotations, some good and some bad. Behavioral scientists interpret them roughly as follows:

---

[18] A classic article titled "The Foreman—Master and Victim of Doubletalk," by F. J. Roethlisberger, *Harvard Business Review,* Spring 1945, pp. 283–298, states the situation in the late 1980s as if it had been written forty years later.

Authority   The word authority, as used here, refers to an assigned or assumed right to command, such as that of a general in the military. The divine right of kings, with its life-or-death control over subjects, is another example. Traditionally in organized business, government, or other enterprises also, authority has flowed from the top down. This form of authority has been classified under such headings as authoritative, legitimate, legal, or formal. It may be exercised coercively, dictatorially, or permissively, depending on circumstances. This traditional form of authority, sometimes described as "worn-on-the-shoulder" power, stems primarily from position in an organizational hierachy. Under it A is the boss and has the right to set the rules of the game and give directions for carrying them out, and B, the subordinate, follows these directions. In such situations, A bosses delegate authority to underlings as they see fit, establish the rules for accountability, provide rewards, and administer punishments.

Where managers have this kind of authority and operate in an authoritarian rather than a permissive way, it is alleged that certain advantages accrue, including the following:

• Efficiency, simplicity, speed, and orderliness
• Personal gratification and a feeling of superiority on the part of supervisors
• Orders carried out by poorly-qualified people or, as one skeptic put it, by bunglers or fools

Offsetting disadvantages that have been cited are:

• When the boss's activities interfere with the worker's desire to satisfy important needs, the employee may resist covertly or even openly.
• Censure for idleness or poor workmanship will soon bring such behavior as looking busy or covering up mistakes.
• Restrictions bring hostility and destroy any chance to gain feedback.
• The boss becomes the enemy to be fought aggressively with every tool the worker can command.

As indicated earlier, in recent years a number of forces have been generated, at least in the developed world, that challenge the viability of traditional authority. One of these forces is the change in needs levels of workers, which have been moving from the management-controlled ones for food, shelter, and security to the worker-generated ones for belonging, status, and self-fulfillment. In addition, forces generated by, among others, labor unions and the women's movement, along with many forms of social legislation, have limited the authority and ability to reward and punish that bosses once had. Today, some firms have or claim to have almost completely replaced their top-down authoritarian approach to management with schemes which share authority on many matters among management and workers. However, despite much talk, authoritarian approaches to management are very much alive in most American industry.

Power   Authority deals with the relationships between boss and subordinate; power has to do with all interpersonal relationships among individuals or groups under which one individual, A, tries to get another individual, B, (or a group) to do something, to refrain from doing something, or otherwise to change behavior. Power in some form is the moving force behind almost all interpersonal relationships and should be recognized as such. It is unfortunate that the word power has been so widely associated with abuse that it has a strong negative connotation which is widely held, as illustrated by a famous and oft-quoted statement attributed to Lord Acton in 1887, that "Power tends to corrupt; absolute power corrupts absolutely."

The resources on which power is based have been classified in a variety of ways, among which are:

Rewards, which cover a wide variety of actions that A can carry out which in A's mind will create incentives that will influence B's behavior in a positive way.

Punishment, which deals with the negative things that A feels he can do to influence B's behavior.

Information, and the person who controls it, can be powerful in whatever manner fits one's aspirations. There is a widely quoted phrase going around today that "he who has the key to the computer controls the company."

Legitimacy, which obtains when B feels that A's position, rank, or title gives A the right to request or demand a given action from B. Legitimate power corresponds to authority, as discussed earlier.

Expertise, which is knowledge that others lack and need. To be useful, these abilities must be established in the eyes of those the expert is trying to influence.

Referent power, which refers to cases where B looks up to A because he admires or respects him or because he sees him as a friend or ally. Because of this, B will often defer to A's requests whether or not he feels the proposed action is in the best interests of B or others that might be involved.

In the real world, some of or all these aspects of power are brought to bear in situations where one person wants to influence the behavior of another. Which specific tools are employed will depend on the situation and the judgment or feelings of the person attempting to assert power. For example, early military leaders such as Alexander, Caesar, or Napoleon would have relied particularly on reward and punishment, although some of the other forms would have been employed. Rewards for foot soldiers would have included food, shelter, loot, and women. Punishment for the smallest infraction would probably have meant death. Today's military leaders, at least in peace times, have few of these options at their disposal. Rather, they must find far more subtle means to get their orders carried out. And as indicated above, managers in the developed nations today must carefully tailor their use of the various forms of power to be effective.

Another dimension to the concept of power was introduced by D. C. McClelland in 1960 in a paper entitled "The Two Faces of Power."[19] He sees

---

[19] See D. C. McClelland, "The Two Faces of Power," in *Organizational Psychology*, D. A. Kolb, I. M. Rubin, and J. M. McIntyre, 2d ed., Prentice-Hall, Englewood Cliffs, N.J., 1974, pp. 163–178.

one face of power as negative; that of being powerful by forcing one's wishes on others and thereby becoming king of the hill. Whether correct or not, he associated behavior such as heavy drinking, exploitive sex, and an interest in shows portraying crime and violence with this negative power. The other and positive face of power is that of making others believe that they are important and can be powerful and accomplish the tasks which face them. Those who have positive power are often responsible for building successful enterprises and in helping others around them be successful. This form of power is less apparent and recognizable, but it is real. In contrast to negative power, it is an attribute of leadership, as discussed below.

It is important to recognize that power in some aspect is at the base of almost every relationship between individuals or with groups in the workplace. It therefore becomes a major factor in decision making, as discussed later in this chapter.

Leadership Practices and Styles

Ever since human beings began to organize society into small groups, there have been leaders who gave orders and followers who tried or were compelled to carry them out. This tradition of authoritarian leadership continued as groups assembled into tribes and then into nations. Only recently, beginning in the late eighteenth century, did the concept develop that leaders could not demand the obedience of subordinates. Today the right to protest against or to ignore higher authority is firmly established in democratic societies but not in totalitarian ones.

In the past, military forces usually operated and, some will argue that they must continue to operate, in the authoritarian mode with orders issued from the top to those at successively lower levels, who in turn are accountable to their superiors. Most industrial organizations and large governmental entities today are also organized around this long-established authoritarian pattern, on the assumption that it best fits their situations. A few enterprises have abandoned it or profess to have abandoned it for the less-rigid approach involving a collaborative style of shared decision making. Certainly the organization of most construction companies and individual projects operate in the authoritarian fashion. Everywhere else in society there are also leaders and followers. Examples are parents and children; and chairmen and committee members.

The absolute power of leaders began to be challenged in the United States and other developed nations about the turn of the century in a few isolated industries. This phenomenon accelerated rapidly after World War II. The level of dependence of workers on the employer and the relative effectiveness of authoritative-vs.-democratic leadership, as perceived by McGregor in 1960 and as it is to a substantial degree today, is illustrated by Fig. 10-5. Among the factors that brought this phenomenon about are economic growth which resulted in greater affluence, independence, and job choice for workers, along with a higher level of education and increased strength of labor organizations. Some see a swing back toward authoritative management today under the

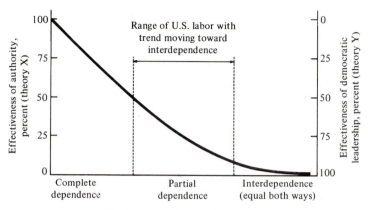

FIGURE 10-5
Level of dependence of workers on employer and the relative effectiveness
of authoritative and democratic leadership styles. (*After Douglas McGregor,
The Human Side of Enterprise, McGraw- Hill Book Company, 1960, p. 25.*)

combined influence of the difficulties faced by the manufacturing industries,
the movement of population to the sun belt, a loss in influence of labor unions
and a strong open-shop sector, particularly in construction, and a conservative
mood in the younger segment of the population. The extent and lasting influence of this swing is unknown and may be largely if not entirely offset by indications that management is, in some cases, adopting the thinking proposed
by McGregor and the present-day advocates of his approaches.

The ideas extracted from two outstanding presentations on the topic of
management styles are offered here as guides for those who must think
through the problem and decide how to proceed in a given circumstance. The
first is condensed into Fig. 10-6 and uses the decision-tree technique. The
reader is urged to study it carefully. The second draws from a classic article by
Tannenbaum and Schmidt and is summarized in the following paragraphs.[20]

Tannenbaum and Schmidt point out that there are a number of decision-
making styles among which managers may choose. They divide them into
seven categories that range from fully authoritative (boss-centered) to permissive (subordinate-centered), as follows: (1) making and announcing the decision; (2) selling the decision that has already been made; (3) presenting ideas
and inviting questions; (4) making tentative decisions subject to change; (5)
presenting problems and creating discussion but making decisions;(6) defining
limits and asking individuals or groups to make the decision; and (7) permitting
individuals or groups to make decisions within defined limits.

[20] See R. Tannenbaum and W. H. Schmidt, *Harvard Business Review,* March-April 1958, pp.
95–101, and republished as a retrospective in *Harvard Business Review,* May-June, 1973, pp.
162–180.

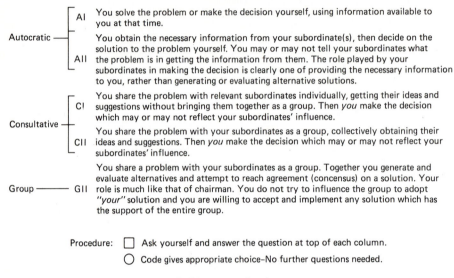

Types of management decision styles

Autocratic

- **AI** — You solve the problem or make the decision yourself, using information available to you at that time.
- **AII** — You obtain the necessary information from your subordinate(s), then decide on the solution to the problem yourself. You may or may not tell your subordinates what the problem is in getting the information from them. The role played by your subordinates in making the decision is clearly one of providing the necessary information to you, rather than generating or evaluating alternative solutions.

Consultative

- **CI** — You share the problem with relevant subordinates individually, getting their ideas and suggestions without bringing them together as a group. Then *you* make the decision which may or may not reflect your subordinates' influence.
- **CII** — You share the problem with your subordinates as a group, collectively obtaining their ideas and suggestions. Then *you* make the decision which may or may not reflect your subordinates' influence.

Group

- **GII** — You share a problem with your subordinates as a group. Together you generate and evaluate alternatives and attempt to reach agreement (concensus) on a solution. Your role is much like that of chairman. You do not try to influence the group to adopt *"your"* solution and you are willing to accept and implement any solution which has the support of the entire group.

Procedure: ☐ Ask yourself and answer the question at top of each column.

○ Code gives appropriate choice–No further questions needed.

Decision process flowchart

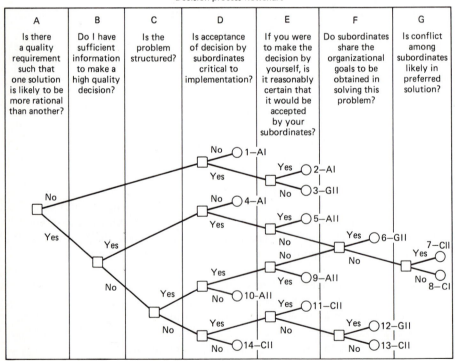

A	B	C	D	E	F	G
Is there a quality requirement such that one solution is likely to be more rational than another?	Do I have sufficient information to make a high quality decision?	Is the problem structured?	Is acceptance of decision by subordinates critical to implementation?	If you were to make the decision by yourself, is it reasonably certain that it would be accepted by your subordinates?	Do subordinates share the organizational goals to be obtained in solving this problem?	Is conflict among subordinates likely in preferred solution?

1–AI 2–AI 3–GII 4–AI 5–AII 6–GII 7–CII 8–CI 9–AII 10–AII 11–CII 12–GII 13–CII 14–CII

FIGURE 10-6

Choice of management style, based on decision-tree analysis. (*Adapted from Victor H. Vroom, in Organizational Dynamics, Spring 1973.*)

In selecting among these seven leadership styles, one must take into account the forces on the manager and the subordinates arising from the particular situation. First of all, individual managers have their own value systems, leadership inclinations, patterns established by past practices, and feelings of security and confidence in themselves and their subordinates. Second, subordinates may or may not have the ability, knowledge, or experience to cope with the problem or be interested or ready, willing, and accustomed to participating in making the decision. Furthermore, they may not share common objectives with the manager. Finally, situational effects such as the practices of the organization or of higher-level management, the nature of the problem, and the time available to solve it must be considered.

Tannenbaum and Schmidt also insist that the managers must, at the outset, make clear to subordinates (1) what the selected approach is and, (2) the fact that the boss assumes responsibility for the decision, however made.

The preceding discussion involves strategies for making individual decisions. The approach selected will not only bring about decisions but will also have long-range effects on job operation and worker attitude and possibly on motivation and productivity. If an authoritative approach is dictated by higher management either by its directives or actions and is followed consistently by managers at any or all levels, the long-range effects on subordinates, which often go unrecognized, may well lead to the following effects:

1 Subordinates will have, and realize that they have, little control over their workday world.

2 Subordinates will become passive and dependent.

3 Subordinates will develop a short-time perspective, that is, they will be concerned only about today but not tomorrow.

4 Subordinates will learn and execute only shallow simple tasks. They will not attempt or be prepared to deal with ambiguous situations.

5 Subordinates will not expect to get psychological rewards or to satisfy higher level needs from the job.

On the other hand, participative or group decision making, effectively done, can have opposite results. But employees will soon see through pretensions of participative methods to cloak authoritative approaches, and bad situations can be made even worse.

Attributes of Proficient, Effective Leaders

There is an old adage that maintains that leaders are born, not made. This led early researchers to look, unsuccessfully, for certain physical or mental characteristics and behaviors which distinguished successful leaders from others. That thesis has since been abandoned; it is clear that there are many kinds of successful leaders. In a discussion of leadership, John Gardner has stated that leaders "come in all shapes and sizes, with strikingly diverse styles and strengths and weaknesses.'' What we do know about leaders and leadership is that almost every individual can learn the techniques of good leadership, and those who are already good can become even better. But this does not come about easily; it requires study and application of concepts such as those outlined in this book and in many other sources. Even more difficult, to improve one's leadership abilities requires examination of one's behavior and a willingness to admit failure and to try again.

Leadership in today's complicated world involves working with others. It is not the popular vision of the individual with charismatic personal magnetism which arouses support and enthusiasm for a cause, but it does involve behavior that develops trust and a can-do attitude.

Some of the characteristics of the climate in which leaders must operate today and which must be recognized and planned for are as follows:

* Leadership is, of necessity, based on the judicious use of power of a positive nature.
* Leadership is not confined to the top levels; it is needed and must be dispersed through all levels of an organization.
* Effective leadership involves individuals acting as a team. However, effective leaders do not confine their attention to those below them in the hierarchy. Their leadership moves in all directions.
* Unless care is taken, leaders and followers can be separated by intervening systems. Two-way communication through all levels is essential.
* Those designated as leaders should not mix the concept of assigned authority with leadership. They are never as much in charge nor are followers as submissive and powerless as may seem to be the case. Exercising or attempting to exercise authority is one thing; winning a following is something else.
* Leaders serve a group well when they help followers develop their own initiative, strengthen them in the use of their own judgment, and enable them to become better contributors.
* Although leading and managing are not the same thing, most leading must have a distinct management ingredient. Successful leaders are not bunglers.
* The ability to motivate and get commitment from people is probably the most important attribute of a good leader.

Certain behavioral patterns of outstanding leaders operating in one-on-one situations or in dealing with small groups have been noted by students of organizational behavior. These will be found in many writings by authorities on the subject. Table 10-1 represents an incomplete consolidation of several of their ideas. It is clear that they represent theory Y behavior, applied in practical terms.

As mentioned above, to become a leader rather than merely a manager is not easy. But those who accomplish it find the result rewarding both through success in the workplace and in personal satisfaction.

TABLE 10-1
A COMPARISON BETWEEN TYPICAL BEHAVIORS OF LEADERS AND NONLEADERS

Leaders	Nonleaders
Thinks of ways to make people more productive, more focused on company goals, and more aware of how to provide rewards	Thinks of personal rewards, status, and outside appearances
A common touch with all, bosses and workers alike	Strained with others, particularly subordinates
A good listener	A good talker
Available to all at all times	Hard to reach from below
Tolerant of open disagreements and willing to accept suggestions	Unwilling to accept suggestions or criticisms
Seeks anonymity for self and attention for others	Wants spotlight on self
Takes blame for failures	Looks for scapegoat
Sees mistakes of self and others as learning experiences	Sees mistakes of others as punishable offenses
Handles unpleasant tasks, e.g., firings	Ducks unpleasant tasks or sluffs them off onto others
Honest, even under pressure	Evasive, equivocates
Looks for controls to abolish	Seeks new controls over others
Consistent	Unpredictable
Wants no special privileges, e.g., special parking, private toilet	Expects privileges
Arrives early, stays late	Arrives late, but expects others to be on time
Says we or us	Says I or me

APPLICATION OF HUMAN-BEHAVIOR CONCEPTS IN THE CONSTRUCTION ENVIRONMENT

How Construction Differs from Other Industries

The concepts of human behavior discussed above apply generally to people in present-day work situations, including that of construction. However, there are peculiarities of construction people and the environment in which they operate that make construction unique. Many of these peculiarities have been discussed in Chapter 2. A few others which deserve special mention because of their strong human-behavior implications are as follows:

1 Construction projects are generally of short duration, which leads to frequent reconstitution of the organization and changes in personnel. At least at the craftsman level, hiring procedures and dismissal policies, often handled at the foreman level, are less stringent. At all levels, there may be greater difficulty in applying motivational techniques because of the diversity in the work ethic between crafts or at different project locations, job availability and the prospect of layoff, and the relative pleasure or discomfort of the work environment.

2 Construction has a greater division of authority and responsibility among many parties at various levels (see Chap. 2), which means that human relationships become very complex. For example, one study found that to carry out a medium-sized building project involved many interchanges among parties at or associated with the project site. These included:

- 40 subcontractors and 20 materials fabricators
- 500 submittals and approvals of drawings, samples, etc., of 20 different kinds.
- Between 50 and 150 design changes.
- 1000 materials deliveries.
- An untold number of exchanges of information, often with several people involved.

Each of these and many other complexities introduce special "people" problems, three of which—job satisfaction and motivation, special demands on management, and the construction foreman's situation—are discussed here.

Needs Satisfaction on Construction Sites

The On-Site Construction Climate: As already mentioned, most but not all construction workers, including foremen and superintendents, expect to change employers from time to time, particularly workers who follow large industrial and heavy work from project to project. Of course, there are some who have worked for local-area contractors for many years. A few key people are kept on during any winter or rainy off season. Even so, one might generalize that in normal times, most skilled construction workers and foremen and many unskilled workers as well will have relatively continuous employment with one contractor or another and foresee a reasonable chance that it will continue into the future. Also, hourly pay for construction workers has been higher than that in other industries, based on the argument that employment is not continuous and the higher wages are often offset because of (1) loss of pay during inclement weather, (2) shifts from job to job which involve interruptions in pay and disruption of plans, and (3) the possible necessity of moving one's home, with the accompanying personal and family disruptions.

There are some construction workers, particularly skilled ones, who are affluent. They own homes and have other resources and are not threatened financially. Almost all workers have some cushion from unemployment insurance which covers a period of several months. In addition, union hiring halls or other contractor or government services notify them of available work; many unions spread the work by having first in—first out referral practices. In sum, the conditions of employment and the financial security of construction workers and foremen vary greatly but in normal times are relatively good. It is within this changeable climate that individual construction workers and fore-men arrive at a feeling about their positions in the hierarchy of needs.

The next few pages will deal with those needs of a construction worker that fall into two categories: (1) needs, particularly higher-level ones, that may be partially or wholly satisfied by the work itself and (2) needs that will be influenced by the introduction of methods improvement. The complex role of money as a motivator has already been discussed.

Physiological and Safety Needs Craftsmen who see themselves at their physiological- and safety-needs levels are concerned about being employed. They may feel threatened in one or more of three ways if productivity improvement programs are proposed. First, that they would be displaced from a particular operation and from a crew or job; second, that the project will be finished more quickly so that work must be sought elsewhere; finally, that there will be fewer jobs available in the workers' trade or in the industry at large.

If production-improvement studies are undertaken after a project is under-way, the severity of the reaction of the workers to the threat of immediate lay-offs depends largely on how management presents the program to them. If management assures them that no layoffs will follow the study and actually absorbs all workers into other parts of the job, there may be little resistance. And in almost all instances, a little forethought can avoid such layoffs. For example, on a project on which a time-lapse study was conducted, almost half the crew was eliminated on concrete placements. However, the superintendent chose to place the concrete twice as fast, so that two crews were needed. On the other hand, if workers had believed that some of the jobs were threatened immediately, a strong resistance to the new method would develop. In this particular instance the second threat, that the project would be finished more quickly, was not a concern, since other area contractors were in need of workers. If this had not been the case, resistance to the new method could have been strong. The third problem, the threat of fewer jobs available in the industry or to union members can be much more serious to the union representatives than to the craftsmen, whose thinking is largely oriented to the near future. The economists' argument that better methods mean lower costs, which mean more construction, is of little consequence to union officials if there are few jobs to be had. Strong evidence that this concern exists was the continual argument among construction unions over jurisdiction, with each attempting to retain its share or gain more of the available jobs. Also, next to wages, job security as measured by equipment manning and journeyman-apprentice ratios has been the area where unions have put their most strenuous efforts in collective bargaining with contractors. Often this has meant that the climate was unfavorable for introducing and implementing productivity-improvement schemes since they threatened jobs which were the primary satisfiers of the physiological and safety needs of both craftsmen and unions.

The recent emergence of open shop in construction, which has taken many jobs from the unions, represents another threat to needs at the physiological and safety levels. This has brought a change in attitude, so that today unions are encouraging and often participating in productivity-improvement efforts as a means of making union contractor's costs competitive with open shop.

A strong argument for preplanning operations before a job begins is that it may head off some of the difficulties just described. For example, new concrete forming or placing arrangements introduced at the beginning of a project will be accepted without argument as long as they conform to established work rules. The story could be different if the changes were made after on-site productivity studies had been made.

Belonging Needs   It may sound ludicrous to suggest that construction people try to satisfy their belonging (love) needs in their work environment. Yet it is an undeniable fact that many construction workers make serious efforts to work with their buddies. Some groups of friends have been together for many years, traveling from job to job.

Soon after World War II a very interesting experiment was conducted to measure the effect on a construction site's production costs when friends worked together.[21] It involved carpenters and bricklayers on a housing project. After the job had been going on for some months and the work pace had been established, the craftsmen were asked to nominate, in order of preference, the three men with whom they would most like to work. Work teams were then formed, based as far as possible on the tradesmen's preferences. Inasmuch as the groups of houses were identical and no changes occurred in either management practices or weather, it was possible to measure accurately the changes in productivity resulting from the restructured work teams. After the change was made, cost records based on 11 months' production showed a marked decrease in both labor and material costs. The labor cost index (the total labor cost in dollars divided by a factor to preserve the anonymity of the company's figures) dropped from 36.7 to 32.2 percent. The material cost index dropped from 33.0 to 31.0 percent. The total savings in production costs to the contractor was about 5 percent. There was also a substantial reduction in turnover, and the men reported that they were happier with their work. One of the men summed up the change by stating: "Seems as though everything flows a lot smoother. It makes you feel more comfortable working—and I don't waste any time bickering about who's going to do what and how. We just seem to go ahead and do it. The work's a lot more interesting when you've got your buddy with you."

The results of this experiment suggest that work-improvement techniques that enhance the chances for compatible groups to work together will be accepted favorably. Supervisors who break up such groups will be resisted and their projects may well have lower productivity; certainly there is little chance that it will improve.

One effect of belonging needs is that they may complicate the promotion of workers to foremen. Individuals could well see the camaraderie with their bud-

[21] See R. H. Van Zelst, *Personnel Psychology,* vol. 4, no. 3, 1952.

dies threatened or lost. These would not be offset by the few rewards of join-ing the management team. Among the possible alternatives are to rotate the foreman's job or to permit the crew to operate without a foreman within the limits set by schedule, cost, or union work rules.

Ego Needs   Workers, supervisors, and crews often find satisfaction for their ego needs in competition, praise, or status. There are many instances of performance beyond the minimum acceptable standard and where extra efforts have been made, not in a spirit of rate busting, but because of the satisfaction from the knowledge that the person or group could produce more than others.

Construction managers have long exploited ego needs that are satisfied by competition and belonging. They know that individuals compete, often without added pay, to outdo fellow workers and that piping, tunnel, and earthmoving, and other crews doing repetitive measurable tasks vie to beat records set by themselves or by other shifts. Often the competition reaches such levels that quality is threatened and safe practices are ignored. There is, of course, the danger that if workers suspect that they are being exploited, they may react by slowing down the work pace even below normal.

Another use of competition and belonging is to set target dates for complet-ing projects or units within them and to give praise and sometimes special re-wards when they are met. As with other approaches involving some form of competition, targets must be realistic, possibly set a little on the high side. To raise them beyond these limits usually results in the worker or workers giving up, with a resulting drop in productivity, dissatisfaction that accompanies fail-ure, and anger at management because it made unreasonable demands.

As mentioned earlier, praise by superiors and fellow workers is one satisfier of ego needs. Of course, it would never be a motivator if the concept that all should be treated equally (a fair day's work for a fair day's pay) were rigidly applied. In this case, praise would be showered equally and equal praise is al-most meaningless.

In an early study of railroad maintenance-of-way workers, the Michigan University Survey Research Center found that the supervisors of high-production gangs were more willing to give praise and encouragement than were supervisors of low-production gangs.[22] Also, it is common knowledge that superintendents and foremen are well aware that a few appropriate words of approval can make significant changes in an individual's performance.

Status is often one of the most important motivators. It is closely related to satisfaction and dissatisfaction (which appears as a separate topic below). Sta-tus motives are often difficult for a supervisor to comprehend, since status can be conferred in so many ways.

The previously cited studies at the Hawthorne Plant of the Western Electric Company are the classic example of how an unintentional granting of status

---

[22] See D. Katz et al., *Production Supervision and Morale Among Railroad Workers,* Univer-sity of Michigan, Survey Research Center, Ann Arbor, 1950.

can become a motivating influence. In this instance, studies were conducted with female factory workers to determine which working conditions were most favorable to high production. Rest breaks, light intensity, and other physical conditions were varied; in each instance, productivity increased. Finally, the physical conditions were returned to those before the experiment was started, and productivity went still higher. It was concluded that the status conferred on the group of women who were chosen as a test group and given special attention by higher management proved to be more significant than any of the variables that were changed. If construction managers at all levels are willing to make an effort, they can achieve similar results. As one example, the president of one of the largest construction firms in the United States with projects all over the country always asked for the names and safety records of every foreman when he first went on a site. On the job tour, he then shook hands with each one, addressed them by their first names, asked some personal but nonthreatening questions, and talked about the importance of safety. This company has always had a team spirit and an excellent safety record, in part from the president's action and also because lower-level managers adopted the practice.

Involving workers and crews in productivity studies or having them fill out questionnaires (given that there is a meaningful follow-up) are other ways of giving status. As one instance, a particularly skilled mechanic was chosen to be photographed on time-lapse film (in this case at the rate of one frame per second to show hand movements) as he rebuilt a clutch. He did the job in 6 rather than the standard 10 hours! Many other cases will be cited elsewhere in this book showing productivity improvements where status is the primary motivator.

Just as status can be a motivator, the loss of it or a perceived threat of losing it can be a demotivator and adversely affect productivity. For example, skilled craftsmen invariably will resist assembly-line or other approaches which reduce the importance and status of their trades. This problem is not confined to union members who must follow jurisdictional rules; skilled craftsmen on non-union jobs have responded in the same way when attempts were made to downgrade their work by assigning menial tasks.

Superintendents and foremen as well as craftsmen will resent new approaches that in their eyes infringe on their authority and status. For example, in many instances methods-study specialists have encountered resistance to suggested new methods or tools or reductions in size in crews or reassignment of tasks. Success in introducing such changes began by recognizing these apparent threats and defusing them. One approach was to involve the affected supervisors in the study from the beginning and to be sure that their contributions were recognized and acknowledged. It was also important to let them institute the changes, sometimes gradually, so that loss of face did not occur.

Any number of examples of how status plays a part in construction productivity can be cited. From them it can be seen that people at the ego level in the hierarchy of needs can not for long be coerced or tricked into being cooperative or productive. Furthermore, autocratic managers, even theory Y ones, will encounter difficulties, and theory X managers will be ineffective in using status as a motivator.

Self-Fulfillment Needs   The highest need in Maslow's hierachy is self-fulfillment. The concept of what constitutes self-fulfillment is hazy; among the definitions are ''the need to realize one's potentialities'' or ''the need to be what one is.''

Seldom are self-fulfillment needs fully satisfied. Many rich men continually strive to become richer; some contractors try to do larger contracts or increase the size and prestige of their firms; many officials work to extend their power and influence. Others may direct their attention into new areas such as political activities or social causes. Self-fulfillment for skilled craftsmen, as with artists, might be the sense of pride accompanying a task done beautifully. In all cases, there will remain a goal to be reached.

As with belonging and ego needs, self-fulfillment goals can be directed into productivity improvement. This involves having a clearly defined measurable and long-term objective that if reached will do far more than satisfy a person's ego.[23]

Needs Satisfaction as Affected by Position in the Hierarchy

The discussion of satisfiers and dissatisfiers at various levels in the on-site hierarchy which follows is based largely on the pioneering studies of J. D. Borcherding who through in-depth open-ended interviews developed a comprehensive list of satisfiers and dissatisfiers for construction workers and supervisors.[24] His findings are very briefly summarized in Table 10-2, which is simply a list of satisfiers and dissatisfiers. The source references also list the factors involved and some tentative recommendations for improving each situation. Although readers may feel that other satisfiers and dissatisfiers than those listed in Table 10-2 are more important, or that the rank orders of those listed should be changed, the rank orders serve a useful function in pinpointing the situations at various levels in the hierarchy of construction projects.

Among the many conclusions that can be drawn from Table 10-2 are:

1 Foremen and craftsmen alike get satisfaction from being productive and doing good work. Their strongest dissatisfactions are with conditions or with other workers who prevent that happening.

2 Higher-level people are satisfied when management functions are executed well and dissatisfied when they go awry.

---

[23] For added discussion of both McGregor's theory X and Y and the hierarchy of needs as they apply to construction, see C. G. Hazeltine, *Journal of the Construction Division*, ASCE, vol. 102, no. C03, September 1976, pp. 497–509.

[24] See J. D. Borcherding and C. H. Oglesby, *Journal of the Construction Division*, ASCE, vol. 100, no. CO 2, September 1974, pp. 413–431; and vol. 101, no. CO2, June 1975, pp. 415–430.

TABLE 10-2

THE MOST IMPORTANT SATISFIERS AND DISSATISFIERS FOR PEOPLE AT VARIOUS
LEVELS IN CONSTRUCTION (IN RANK ORDER OF IMPORTANCE)

Position	Satisfiers*	Dissatisfiers†
Owners	1. Job makes a profit 2. Customer satisfied 3. Job completed on schedule 4. Tangible physical structure (something built) 5. Quality work	1. Problems with unions 2. Company or personnel mistakes 3. Customer dissatisfied; collecting overdue accounts 4. Dealing with unqualified individuals such as field inspectors or members of city planning commissions.
Project manager (home office)	1. Job makes a profit 2. Quality work 3. Customer and owner satisfied 4. Good work relationships with and within crews 5. Tangible physical structure	1. Personal or company mistakes 2. Inability to maintain job schedule 3. Poor engineering (design) 4. Inadequate field supervision
Superintendent (top field position)	1. Maintain the job schedule 2. Meeting a challenge 3. Quality work 4. Job costs below estimates 5. Good work relationship with and within crew	1. Lack of necessary coordination by supervisors to maintain job schedule 2. Poor engineering (design) 3. Union workers not cooperating 4. Personal mistakes 5. Continual criticism by home-office management
Foreman	1. Challenge of running the work 2. Maintaining the job schedule 3. Quality work 4. Tangible physical structure (something built) 5. Good work relationships within crew and among trades	1. Uncooperative workers 2. Lack of incentive in workers 3. Lack of management support by not having engineering information, materials, and equipment 4. Union problems

Journeymen and apprentices	1. Complete tasks
	  a. Quality work
	  b. Productive day which often ends in physical exhaustion
	2. Tangible physical structure (something built)
	3. Social work relations
		1. Poor interpersonal relations
		2. Unproductive workers in their crews
		3. Poor quality work by their crew
		4. Unfair job assignments
		5. The work itself

*J. D. Borcherding and C. H. Oglesby, *Journal of the Construction Division,* ASCE, vol. 100, no. CO2, September 1974, pp. 413–431.
†Ibid., vol. 101, no. CO2, June 1975, pp. 415–434.

3 The lists of highest satisfiers and dissatisfiers change as people's levels in the hierarchy change. This is important in activities aimed at improving productivity, since to be effective the operations selected for study must fit the satisfactions and dissatisfactions of those at the target level. For example, the table shows costs to be a principal concern of higher-level management but not of foremen and workers. To get foremen's and craftsmen's interest, the focus should not be on costs, rather it should concentrate on targets such as challenges in the work, the excitement in building a structure or plant of high quality, or finishing tasks on time.

W. F. Maloney and J. M. McFillen have developed a numerical measure of job satisfaction and motivation using the contingency theory approach discussed earlier in this chapter.[25] Their method is shown in Fig. 10-7, which outlines the relationships among certain core job characteristics, psychological states (feelings), and outcomes in terms of motivation, growth satisfaction and general satisfaction, and work effectiveness (productivity). The resulting rating scheme was applied to data collected on questionnaires administered to members of union construction crafts. Their findings are reported in the next section of this chapter. It is important to note that the approach could be applied to all those involved in on-site construction, whether managers, engineers, other staff, or workers. Both it and the Borcherding study reported above represent attempts to measure subordinate reaction to management's attitudes and styles.

As indicated by Table 10-2 and Fig. 10-7, neither Borcherding nor Maloney and McFillen specifically include as dissatisfiers certain job-site physical conditions. They recognize that managers and craftsmen alike seem willing to accept rain, heat, cold, and other adverse site conditions as givens of the job that are beyond management's control. On the other hand, they resent and react negatively to neglect of things which management can do something about. Many examples could be cited, among them a lack of good drinking water, shortage of or dirty toilets, and poor arrangements for or discriminatory treatment in job access or parking.

Large projects offer particular difficulties to management if it tries to capitalize on the satisfactions and feeling of belonging that can develop on smaller jobs. On very large projects, only top managers have a sense of what is happening and even those in lower-level staff and line positions feel alienated. Foremen see their role only as that of pusher and not as a part of the team, and workers find few satisfactions. Among the strategies that progressive companies have employed to overcome these problems with the aim of improving job relations and productivity are to set up suggestion systems, provide information through newsletters and other sources, and conduct job tours by means of which workers and possibly their families can identify with the project (see Chap. 11 for added detail).

[25] See *Journal of Construction Engineering and Management*, ASCE, vol. 112, no. 1, March 1986, pp. 137–151.

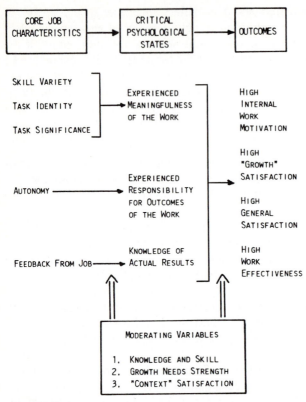

FIGURE 10-7
Relationships among core characteristics, critical psychological
states, and outcomes in kinds of motivation, satisfaction, and
work effectiveness. (*Source: W. F. Maloney and J. M. McFillen,*
Journal of Construction Engineering and Management, *ASCE,*
vol. 112, no. 1, Mar. 1986, pp. 137—151.)

Do Construction Managers Satisfy Workers' Needs and Expectations?

It is a given in construction that some construction people, be they project managers, superintendents, or foremen, have reputations for running profitable and safe operations on which morale, satisfaction, and productivity are at high levels. Generally, the performances are measured after the fact through the cost-reporting system and by comparison with the performance of others. Reports about these managers indicate that they are very people-oriented and have outstanding qualities of leadership, including a theory Y attitude toward others. Only recently, however, have attempts such as those described above been made to actually measure these qualities on jobs in progress. For this reason, very little is known about the general ratings that subordinates give their superiors.

Recently, Maloney and Fillen surveyed employee attitudes about construction employment and managers on union and open-shop projects in the middle west.[26] Some of the findings indicate that, in the eyes of workers, management practices could be improved. A few of the items reported included the following:

- The workers were confident that they had the skills necessary to do their jobs well, even if this was not recognized by management.
- The working environment contributed little to motivation and satisfaction. The tasks did not contain high levels of intrinsically rewarding elements.
- Rewards came from the job itself or from coworkers but not from management.
- Higher performers did not reap higher rewards from management. Instead, penalties for not performing were more likely.
- Coworkers were more likely to punish than reward high performance.
- Management support for field operations often failed, leading to an average delay of 1.3 hours per day.

Based on the findings of their studies, Maloney and McFillen indicate the following five areas to which construction managers should (but usually do not) direct attention if they wish to improve productivity and worker satisfaction:

1 Provide stable employment so that crews exist as an entity long enough to develop positive norms and goals
2 Build effective work crews by assigning individuals to them who like and are liked by the other members
3 Develop a program to train these crews in the group processes needed for team building
4 Given a cohesive and trained group, encourage the development of appropriate performance goals
5 Develop and implement a reward system that recognizes achievement of the goals

The parallels between these recommendations aimed at construction crews and the more general ones in the section of Chap. 3 on "Tracks Which Management Must Follow to Improve Productivity" is striking.

Given the findings of these studies of present industry practices, it seems logical to conclude that construction management is generally either (1) unaware of these approaches or (2) not interested in the opportunities they offer to enhance crew motivation and satisfaction and thereby improve productivity.

---

[26] See W. F. Maloney and J. M. McFillen, op. cit.; and *Journal of Construction Engineering and Management,* ASCE, vol. 113, no. 2, June 1987, pp. 208–221.

Difficulties to be Overcome in Motivating Foremen[27]

In construction, as in other industry, foremen occupy a crucial position. It is through their direction of individual workers or crews that drawings, specifications, and instructions provided by higher management become completed structures or plants. All too often, the difficulties foremen face and must overcome go unrecognized by higher management on the one hand and workers on the other.

In dealing with job personnel and, in particular, with foremen, higher-level management often fails to recognize that its failure to perform and ill-considered or thoughtless acts on its part can diminish, if not destroy, earlier commitments to productivity that foremen expect to make or have made. Examples of all kinds can be found, but a summary of one cited by a consultant is typical.

> Higher-level management had been getting complaints from the warehouse, accounting department, and equipment managers. The warehouse now had excess materials because foremen, when items essential to job operations were not available on site, picked them up locally without purchase orders and directed that the bills be sent to accounting, and this fouled up accounting's records. Also, those in charge of equipment had complained that the foremen rented the needed machines from local dealers while company-owned equipment sat idle in the yard. The general superintendent, under pressure from the managers of these service groups and without consulting the field staff, issued a directive that said, "In the future, no materials shall be purchased or equipment rented locally without advance permission from this office."
>
> Soon, through the job grapevine, reports began to circulate that foreman morale was low and that some of them were threatening to quit. The consultant, after running down the rumor, quoted a carpenter foreman as saying, "We have been getting all this talk about improving productivity. To me it's a lot of hogwash. I go to the warehouse for material to do the job scheduled for my crew and they say it's on backorder and will be in next week, or at least they hope so. To top it all off, the general foreman came around yesterday and started to bawl me out because my unit costs had gone up in the last couple of weeks. I told him that because of that stupid directive I spent all my time trying to find materials because I can no longer buy them locally. His response was that the general superintendent has said that he wanted the foremen to 'shape up' the way they handled materials and equipment." The foreman concluded by saying that, "If things don't get better I am quitting. I don't want to get a black eye for the stupid things management is doing here. On the one hand it talks about how important the productivity of my crew is; and how a ten percent saving on the $400,000 in annual labor costs that I control is big money. At the same time management makes it impossible for me to get the work done efficiently by setting up a bunch of silly procedures that tie my hands. On top of that they lay the blame on me." Joe, the sewer foreman, told a similar story. He said that he had been renting pumps because those the company owns never work and his entire crew stands around an hour or two while one man tries to get the engine to start and the pump to work.

---

[27] For the results of a study of the relationship between productivity and decision-making styles of construction foremen, see J. Hinze and K. Kuechenmeister, *Journal of the Construction Division*, ASCE, vol. 107, no. C04, December 1981, pp. 627–639.

The incidents just related are all too common on construction projects. Foremen in particular seem to be victims of mixed-up and inconsistent thinking by higher management such as that just described. As a result, foremen, on the one hand, see themselves as carrying the responsibility for the efficient use of very costly resources, particularly labor. On the other, they feel encumbered by rules forced on them by managers who do not know or care about their problems, whose performance is measured differently, and who therefore have different goals and motivations.

In planning crew activities on work of any complexity, foremen must rely on others to see that preceding operations have been carried out; that drawings, materials, tools, and equipment have been furnished; and, finally, that workspace has been provided. Failures in any of these areas makes execution of assigned tasks impossible. Rather than acting, foremen must react to the failures of others. There are ways, for example, by careful preplanning and monitoring the performance of staff, that management can sort out these conflicts and correct them. If it does not do so, foremen rightly see management as uncaring, possibly stupid, and uninterested in improving productivity. Any earlier commitments those foreman may have made will vanish.

Management's expectations of foremen often include a knowledge of (1) costs, materials, equipment, and quality-control systems; (2) efficient and safe work methods; and (3) company and, often, union work rules and policies. In addition they are expected to (1) induct, instruct, and train workers; (2) maintain discipline in the crew; (3) head off or handle worker grievances; (4) cooperate with higher-level management, engineers, and owner representatives; (5) deal with business agents and shop stewards on union work; and above all, (6) complete tasks under budget, on time, and with acceptable quality. In all these, they are expected to be evenhanded and fair and never lose their tempers. And, even while management is making all these demands on foremen, all too often it sees them (particularly if they are union members) as a part of a casual labor force to be hired when a particular need develops and dropped when that task is completed.

Workers also make demands of foremen. First of all they are expected to look after worker interests and front for them with higher management. Often their role is that of a father figure, counselor, and teacher. On union projects they are expected to see that labor agreements are followed to the letter.

As pointed out earlier in this chapter, students of management have long discussed and often decried the fact that the foreman is caught in the middle. But it is only recently that a few contractors have recognized this problem and developed and conducted training programs to help their foremen cope in their difficult role. All these programs emphasize the human side of foremanship.[28]

---

[28] As noted in Chap. 11, the Associated General Contractors has produced two comprehensive sets of manuals for in-service training of superintendents and foremen under its STP program. Each has a section devoted to personnel management.

Workable Strategies for Gaining Productivity Improvement[29]

The objective of motivating and gaining commitment from construction people is to induce them to work more effectively and safely and to produce a product of suitable quality. These objectives are far more than getting workers to exert more physical effort and managers and staff to put in longer hours. Rather, it means applying all of a person's mental and physical abilities and talents to think, plan, and execute tasks. It also means developing interest and involvement, which in turn reduces absenteeism and turnover. It bears repeating that in order to turn motivation into commitment, there must be a positive mental attitude and a strong commitment from people at all levels in the hierarchy.

The purpose of this chapter so far has been to sort out from the vast number of research findings and its literature in the field of organizational behavior the topics of most importance to construction managers. In so doing it first explored several aspects of motivation that apply to almost all persons as individuals as they are involved in work situations. The motivators were needs (a hierarchy), job enrichment, expectancies, money, nonmonetary rewards and punishments, the overcoming of alienation and frustrations, and individual and crew satisfactions and dissatisfactions. It then examined factors that affect interactions between supervisors and subordinates under the headings of ways of perceiving people (theory X vs. theory Y), the management-worker and management-foreman dichotomy, authority and power, and leadership styles and effectiveness. Finally, it focused more specifically on construction: how it differs from other industry and the specific topics of needs satisfaction in construction, how needs change with position in the hierarchy, how well managers now satisfy those needs, and certain specifics of the problems that foremen are heir to. This is a very long list of items, each difficult in its own right, that all construction people must consider in dealing with or making decisions about individuals or groups, whether they are bosses, equals, subordinates, or anyone among the many outsiders with whom they must deal.

Developing a Plan of Action   In each of the areas just listed, guiding principles have been pointed out which if followed may provide means for keeping construction people at all levels from becoming adversaries and inducing them to become more cooperative and productive. In the opinions of the authors of this book, successful efforts must:

1 Start with a belief in and concern for people (call it a Theory Y attitude)

---

[29] "First and Second Level Supervisory Training," *Business Roundtable Report A-4,* May 1982, indicates that only 13 percent of the contractors who responded to its survey give any supervisory training to individuals before they assume management jobs. Most of the training programs that were in use were developed in-house by individual companies; one was proprietary.

2 Determine which among the concerns listed above must be taken into account and how that can be done[30]

3 Regardless of what parties are involved (subordinate, equal, boss, or other), select an appropriate leadership strategy that takes into account the relative abilities and knowledge of those involved, adjusted to fit the existing or projected authority and power relationships

4 Follow through on a sensible decision-making and implementation sequence, as outlined below

5 Consider and establish a communications scheme, taking into account the factors that lead to effective communication, including listening, as outlined below

To turn these principles into actions is difficult indeed. In the simplest terms, it calls for, in each situation, the recognition that people, although different in some regards, all have needs to satisfy and a desire to participate, demonstrate their abilities, and share in or at least understand decisions that affect them. Furthermore, those wishing to persuade must demonstrate and communicate honesty and sincerity. (See the discussion about communications later in this chapter.) Actions construed as manipulation can destroy credibility and make motivation and commitment difficult if not impossible to achieve. Developing such positive attitudes and loyalties is particularly difficult in situations where long-time adversarial (''them and us'') traditions exist between workers (and possibly foremen) and managers, or in relationships among owners, engineers, and contractors.

In construction, adversarial relationships between company and employees may be less prevalent for small local contractors where the work force knows top management personally, is almost continuously employed, feels loyalty to the company, and can be kept informed of happenings by word-of-mouth. However, larger local companies have found word-of-mouth communication to be both inaccurate and ineffective where there are several levels of hierarchy between top managers and workers. Some have successfully employed newsletters or other handouts or fact sheets (see Chap. 11) which highlight information about individuals and their families as well as the company to create this feeling of belonging. They also provide a mechanism for dispelling rumors by providing facts, even when they might be unpleasant ones. Company events such as picnics which involve families also can be effective, if planned and executed properly.

Large projects which must be staffed with a casual temporary work force drawn from a large area have particularly difficult motivational problems. With them, workers and even foremen often see few physical results from their ef-

---

[30] There is a close parallel here with the procedures that are sometimes employed to get hard-nosed field supervisors to accept advance planning as a substitute for reacting to problems after they happen. It has been found that these individuals will attack people problems, as they will planning, if they think that it is a way to ''get those guys to work.'' They will reject it if it is seen as a management-imposed behavior change that they are expected to make to become better managers.

forts so that this powerful motivator is lost. Also, rumors, often erroneous, seem to spread like wildfire. Several successful strategies for overcoming this feeling of alienation are discussed in Chap. 11.

Gaining Participation Meaningful participation in the job itself can be a strong motivator. Questionnaires or interviews which allow workers, foremen, and even superintendents to point out deficiencies of all kinds offer one means. Examples of such questionnaires and interview forms and how they are applied to productivity studies are given in Chap. 7.

In some bigger companies and on large projects a variety of formal schemes for recognizing, rewarding, and publicizing long service and high individual or crew productivity, or for developing new or better solutions to problems have been instituted. These generally follow the patterns in long use for rewarding exceptional performance in safety. There is little consensus among companies or consultants about the appropriateness of types and values of awards. Some involve money in substantial amounts, or gifts ranging from gold watches to company belt buckles, others rely solely on publicity and recognition by higher-level managers.

In recent years, work-team (quality-circle) or crew-level approaches have been employed to gain the participation of workers and supervisors in solving job problems and creating a more favorable job climate. These were described in some detail in Chap. 3. Participation in them, or in suggestion systems (see Chap. 11), has often proved to be a strong motivator.

It should be clear from this brief discussion that motivating and getting commitments from construction people does not come about merely by half-heartedly employing staff specialists or consultants or otherwise attempting some of the approaches described here. Without exception, motivational programs have been successful only when there is a strong commitment of time, energy, and money, and continuous follow-up by top managers. Only they can remove the barriers that often frustrate those at the level or levels at which particular programs are aimed.

THE DECISION-MAKING PROCESS

Earlier chapters in this book discussed the climate of the construction industry and the technical and operational aspects of productivity enhancement. Chapter 9 examined people as organisms that perform in construction's physical environment; earlier sections of this chapter considered them as individuals or groups that respond as humans in complex ways. The rest of this chapter discusses (1) how, given all these inputs, sensible and imaginative decisions can be reached and (2) how these decisions can be communicated to those who are to carry them out.

To be most effective, decision making involves three distinct steps, each of which calls for somewhat different skills: (1) problem finding or identifying, (2) problem solving or seeking out viable alternatives, and (3) choosing the most suitable course of action. It is often an iterative process, which involves going through the successive steps more than once as new ideas or concepts develop and mature.

Before examining the three steps in decision making, it is appropriate briefly to consider three aspects of human behavior and thought processes that can improve or detract from creative decision making: (1) distractions and interruptions as barriers to thinking, (2) quirks in the way humans record, recall, and process information, and (3) the left-side vs. right-side division of the human brain, all of which affect its working as a decisionmaking instrument.

Removing Distractions and Interruptions

Some of the roles that managers must fill are discussed in Chap. 2 under the topic of "Complexities in the Construction-Management Situation." If the complexities discussed there and other pressures are allowed to dominate a manager's attention, decision making will suffer. Managing one's time effectively is one of the most important techniques for freeing the mind for decision making. Much has been written on this topic, all of which cannot be treated in detail here. One effective strategy is to delegate responsibilities and accountability within stipulated time limits, at the same time being sure the subordinate or other person involved clearly understands that the "monkey is on his back."[31] Another difficulty which traps the unwary when trying to conserve time is the clearly established human tendency to respond to demands for attention from others by giving their problems precedence over working on one's own. Yet another is an unrestricted open-door policy. Being readily accessible is essential but having uninterrupted privacy at designated times is equally important. With thought, ways for heading off intrusions by telephone without being rude can be developed and instituted.[32] Where groups are involved, a strategem relatively new on construction sites is the regularly scheduled conference or staff meeting, with (almost) compulsory attendance and a previously announced agenda at the set time that will be least disruptive to other duties. All those who can affect or are affected by decisions should be notified in advance, since better decisions result when the participants have time to prepare before the meeting itself.

Some managers tend to rely too much on routine or traditional practices rather than searching out new approaches. An often-found example is an obsession discussed (in Chaps. 2 and 6) for using cost reports as almost the sole search method. This can easily develop if such a technique is stressed by higher management. The trap is that signals from the cost system may not give the best clues and often come too late to be of much value.

---

[31] An amusing but most helpful article on this topic is W. Onchen, Jr.and D. L. Wass, "Managing Time—Who's Got the Monkey?", *Harvard Business Review,* November-December 1974, pp. 75–80.

[32] Traditionally, secretaries or assistants have been expected to handle this situation. Today, recorders or computer terminals on which messages can be left are available. Unless planned and executed skillfully however, such schemes can create animosities and ill will.

Quirks in the Way Humans Record, Recall, and Process Information that Can Impede Creative Decision Making

The human brain has a remarkable ability to record and store in long-term memory information that reaches it through the senses, to recall that information into its short-term memory, and to process it there to reach decisions about possible courses of action. However, research has shown that there are certain quirks or built-in peculiarities in each step in this data-handling sequence, any one of which can lead to less-than-optimum or even faulty decisions. Each of these processes is discussed briefly in the paragraphs which follow.

Information Recording into Long-Term Memory   All humans, in every circumstance are bombarded with impressions brought in from the surrounding environment. A selection process must occur, and the impressions which survive the sorting are recorded, after a rehearsal time of 5 to 10 seconds, into the individual's long-term memory, which has an almost limitless capacity. Recording is in chunks, as simple as a number, letter, or word or as elaborate as a picture or landscape, or the physical appearance, facial expressions, and voice of a person. One specific example of complex chunks is the chess expert's ability to record the positions of 20 to 25 pieces on each of many chessboards. Apparently, chunks based on vision are recorded pictorially; the eye collects images by moving from point to point in about a one-degree radius, at a rate of four to five fixations per second. These are collected and form the image that is recorded.

It is impossible for all the impressions projected by the environment to be recorded into long-term memory, so that there must be selective attention given and selective recording done. As a result, there can be a failure, often unconscious, to perceive some messages. For example, while a particular mechanic may be highly conscious of engine noise from a machine, many others may not be conscious of it at all nor record it in long-term memory. Possibly more important, messages that are uncongenial or unpleasant are frequently repressed and may reach long-term memory in garbled form, if at all. Furthermore, there is a strong tendency not to record new or oddball ideas nor for individuals to reconsider or modify their beliefs or philosophies. In effect, selection of items to go into long-term memory provides a mechanism for psychological reassurance about one's beliefs and prejudices and a defense against threats to them.

In considering the strengths and weaknesses of long-term memory as a factor in decision making, it seems clear that although a valuable resource, it can be unreliable or flawed.

Information Recall into Short-Term Memory   The mental activity which leads to decisions takes place in the short-term memory, based on information brought up from the long-term memory and from other sources at a rate of about 2 seconds per chunk. There appears to be a heuristic process by which long-term memory is searched for items relating to a particular problem or decision, with the more probable ones brought forward first. As with recording into long-term memory, recall is also selective. Research has demonstrated that the mind has a tendency to bring forward items that are psychologically pleasing and is less likely to recall those that are upsetting or that challenge preconceived beliefs, prejudices, or accepted solutions. Here also, reliance on memory alone may bring incomplete or distorted images to the thought process.

Information Processing in Short-Term Memory   As indicated above, information must be in the short-term memory if it is to be assimilated and thought about and to be the means by which a decision is reached about the problem or opportunity it presents. Holding time in short-term memory is brief, lasting about 30 seconds, after which it must be refreshed.

In using information in the the short-term memory to find solutions, the mind again follows a heuristic pattern. It searches among possible solutions and organizes them in a fashion that seems most likely to lead to the best answer. Favored approaches are explored first; less likely solutions may be neglected or even ignored. The process makes it difficult to think creatively, since creative thinking involves, as its first two steps (1) preparation, that is, assembling and organizing all known information pertaining to the problem, and (2) incubation, or mulling over possibilities, shifting from one to another, free of rigid, rational, or logical preconceptions and constraints.

Another barrier to finding creative solutions is what has been called the "seven-plus-or-minus two" concept—a weakness in the way that humans handle large amounts of information which has been established by psychological research. The concept maintains that the human mind can hold between five and nine (with an average of about seven) chunks of information in it's short-term memory where reasoning and decision making takes place. As noted above, these chunks can be simple (a single number) or fairly complex (a graph or picture). In any event, when problems involve more than about seven variables, the brain will exclude as many of them from the reasoning process as is necessary to get the number down to this manageable level. Consequently, in making decisions on complex matters entirely in one's head, some elements important to the outcome may not even be considered.

There are ways to overcome or partially overcome the seven-plus-or-minus-two difficulty. One is to make lists or otherwise put the various considerations into a form where they do not get lost. Another strategem is to increase the complexity of individual chunks. Ways for doing this include showing physical features and details on drawings or displaying tabular data in graphs rather than tables. Several of the techniques given in this book for information recording and analysis illustrate how the seven-plus-or-minus-two limitation can be mitigated.

In sum, the point to be made in this brief discussion of quirks in human reasoning and decision-making processes is that it is very difficult to make rational decisions about complex matters in one's head. Those who attempt it may well encounter difficulties that could be avoided.

Left-Side-vs.-Right-Side Thinking Findings from medical and psychological research on left-side-vs.-right-side thinking also have strong implications for decision making about all but the simplest productivity problems. Briefly, the research shows that given the same information, the left and right sides of the brain process it differently and may reach conflicting conclusions. A simple illustration of this dichotomy is the statement that many of us make: "His facts seem to be correct" (left-side analysis) but "I don't trust the conclusions he drew" (right-side synthesis).

With either left-side or right-side thinking, information taken in through the senses is available by cross connections to both sides of the brain. But they use it differently. The thought processes in the left side can be described by such terms as linear, convergent, step by step, rational, logical, and time-conscious. Verbal and mathematical skills and problem-solving and decision-making abilities are centered there. The left side is the reasoning side. In contrast, the right side is not verbal and does not reason. Rather it is holistic (sees things and puts them together as a whole), intuitive (makes leaps of insight), tentative (does not decide), divergent, playful, and without a sense of time. What is needed for productivity improvement is to capture the values from both ways of thinking.

It can be generalized that most people in construction (managers, engineers, field supervisors, and craftsmen alike) have through education and experience developed left-side thinking patterns. However, a little thought will show that problem finding and solving and implementation of a solution require both forms of thinking.

To bring about right-side thinking requires that left-side thinkers augment their skills, which involves what is called a cognitive shift of the thought process from the left to the right side of the brain. Specialists claim that this ability is not inborn but it can be developed with practice and observant individuals can learn to recognize it when it is achieved. It is also known that cognitive shift takes place when a person is not working on a problem but is occupied with routine, nonstressful tasks which permit thoughts to wander. Occasions where this might happen include driving on a freeway or the open road, doing home chores such as gardening, or bathing or shaving. For some, these shifts happen when sleep has been interrupted, which lends credibility to the statement that one should sleep on a problem before deciding. Construction people, being action-oriented, may distrust and even ridicule the value of taking the time necessary to develop and employ right-side thinking, but they may therefore be overlooking a valuable means of solving problems.

Problem Finding

Problem finding, where productivity is the goal, can be simply stated as finding problems worth working on and where something good can be made to happen. On most construction projects this is easy. It has been said that "finding problems on construction sites is like looking for the water when you are swimming."

Problem finding usually begins with clues that indicate either an opportunity or a problem. For the alert, thoughts about where problems exist and what they are can be set in motion almost spontaneously in a bull session or conference, during a discussion of a task, by a chance remark made by a craftsman or foreman, by the formation of a mental image of something being amiss, or when in the act of reading a magazine article. Again, problems may emerge from data analysis, comments on questionnaires or interviews, or productivity studies, or progress and cost reports. There is no theoretical explanation for how these clues lead searchers to discover problems. For this reason, there seems to be no specific way for teaching how it is done. Some level the charge that focusing on problem solving, which can be taught, actually diminishes an individual's abilities for problem finding.

Although the mental processes that underlie problem finding are obscure, much study has been given to approaches to, techniques for, and barriers that inhibit problem finding. A few of the many ideas, summarized here, are as follows:

1 There must be a climate which encourages problem finding. Management at all levels, but particularly at the top, must demonstrate strong interest in and encourage and reward efforts to explore and question existing procedures and practices.

2 An attitude that there are better solutions, and probably more than one, to be found.

3 Ideas, particularly unconventional ones, from craftsmen as well as managers and particularly from mavericks in the organization must be given consideration.

4 Physical arrangements and sufficient time to permit right-side, intuitive, thinking by individuals or brainstorming and similar activities by groups (see Chap. 3) are essential if tunnel vision and quick fixes are to be avoided.

The list just given, and other ideas as well, indicate ways for developing both a climate for and broader thinking about looking for problems. But certain guidelines are helpful in keeping the search on track and within limits, such as:

1 Know about and think through the data available or that can be gathered by techniques such as those offered in this book in order to find the best focus on the problem.
2 Limit the focus by questions such as:
  a What drew attention to the problem?
  b Does the problem involve several activities or can it be pinpointed in one?
  c Does the problem lie away from or at the work face, or at both?
  d Is the problem:

(1) Related to the system now being followed or to activities now under way?

(2) Outside any system or procedure because people do not know what they are doing or should be doing?

Problems falling under (1) should be tackled by examining structure and procedure; problems under (2) should first consider people and their actions, attitudes, and behaviors.

e With what person and at what level in the organization does responsibility lie for creating or solving the problem?

An actual situation in high-rise building construction provides an example of problem finding involving all the items listed above. It concerns the operations involved in hoisting, holding in place, and securing panel forms for cast-in-place concrete walls. The problem first came to management's notice when the work fell behind schedule and costs exceeded the budget. The first data collected by a productivity specialist was by work sampling. This showed that there was a low percentage of working time and a high percentage of idle time for the crew. Of course the first and easiest explanation was that the craftsmen were loafing; this was dismissed after earlier work-sampling results which showed good ratings for the same crew were examined. The specialist then made a time-lapse film of the crew in action to look for possible reasons for the excessive time consumption. Also, through a foreman's questionnaire, it was learned from the foreman that there were long waits for the crane which hoisted and held the forms until they were fastened in place. This crane was under the control of the foreman in charge of deck forms, including reinforcing steel. That foreman's questionnaire response complained about lack of crane time for his crews because when it was released to the wall-panel crew, the members of that crew kept it tied up for long periods of time.

In summing up, it can be seen that there were problems involving several activities, some away from and some at the work face. Also, the concerns and well being of several individuals must be considered, each with objectives in mind that were in conflict. Some of the difficulties were related to the system and some to single activities. Finally, responsibilities were frequently ill-defined and many were assigned at the wrong level. The solutions, which may appear obvious but which management had not seen before, are discussed below under problem solving.

There can be legitimate concerns about the problem-finding procedure just outlined. It is based on data collected after the fact, as a result of the discovery of a bad existing situation. One can ask, "Why were the operations being so badly managed? They should not have existed at all if preplanning of schedules and the operations themselves had been done properly and in sufficient detail." Even so, the studies provided a means for raising important questions for job management to consider. Furthermore, since this operation would be repeated many times in the future, straightening it out would produce substantial savings on this project and possibly on future ones.

As already indicated, there are no specific guidelines for teaching problem finding, but the skills needed for doing so can be developed by practice and by repeatedly asking questions, even those which seem ridiculous. Moreover, one must consider not only the technical and procedural questions but also those at the human-behavior level, such as for example, "Which person or persons are involved in and might contribute to the problem" or, if personally involved, even the more devastating one "Am I the problem?"

Problem Solving

Problem solving is a supposedly orderly process which (1) defines the problem, (2) sorts out the information available in a given situation in order to retain the relevant and discard the irrelevant, (3) arranges the information logically to reflect the relative importance of the various factors, and (4) weighs the various effects and, if possible, estimates the cost of and establishes the relative merit of each reasonable alternative. The overall aim is to sort out clearly all important possible consequences of each feasible solution, including that of doing nothing.

Problem solving cannot be neatly separated from problem finding; in fact, attempts at problem solving may redefine the original problem. A simple real-life illustration is that of an analysis of the cost of providing and setting forms for concrete curb and gutter. The problem was first defined in terms of setting optimum crew size and task assignments when using form boards, steel stakes, and clamps. The analysis involved the conventional process-chart and crew-balance techniques. However, before recommending a change in procedures, the analyst, looking for an alternative solution, contacted a form-hardware supplier and discovered that steel support assemblies for the header boards were available that would cut the required labor about in half. Hence, there was now a new problem to solve; that of developing a revised crew size and task assignments. (Of course, today, this second method has been replaced by a slip-form machine which has made conventional curb and gutter forming outmoded on all but the smallest or most complicated projects.)

As with problem finding, imaginative problem solving draws on the appropriate segments of the technical and analytical skills and requires an understanding of human behavior. Choosing the more suitable among the possible alternatives is the first step. Then the left-side (analytical) thinking process becomes the key to sorting out the important and less important factors properly and evaluating the important ones systematically so that the alternatives are clearly defined. In doing so, it is important to remember the seven-plus-or-minus-two limitations on human ability to think logically. This means that in some manner, the differences among the alternatives must be boiled down to not more than about seven variables. These will provide the factual but not total input for choosing a course of action.

With respect to the concrete-wall-forming example given under problem finding, two separate but interrelated solutions emerged. The first, which involved the system, required more care in scheduling crane use. It was clear that with more than one crew dependent on the crane, control of its scheduling should not be with one of the foremen who quite naturally would put his interests first. The solution was to designate a scheduler for the crane so that operations were improved rather than adversely affected. By a combination of shortening demands for crane time and developing and following a specific schedule of crane use, all crews were served satisfactorily. Also, in this instance, each foreman was forced to enlarge the scope of his planning to fit his schedule to the allotted crane time.

As noted in the description in the panel-setting example, cost and schedule data and information from questionnaires indicated that setting the form panels took too long. To investigate this difficulty the company's productivity specialist carefully examined the time-lapse films of the operation. These were viewed and analyzed by the specialist to find the segments most suitable for later viewing and analysis by job management and the foremen and their crews. Based on these discussions, the specialist developed and presented revised methods for implementation in the field. Among several time-consuming operations that were analyzed and revised was that of engaging the opposing sections of the she bolt form ties. This was done by grinding the male ends of the connecting bolts to points so that they could be more easily inserted in the female socket by a worker who had to operate in a cramped position. In addition to saving substantial sums of money on this project, these efforts have been publicized throughout the company and have sensitized (1) personnel at all levels to think about planning, including more careful attention to assigning responsibilities and (2) supervisors and crews to look critically at apparently simple operations where small changes in procedures or supplies can effect cost reductions and, at the same time, make the work easier.

Choosing a Course of Action

The productivity-improvement scheme just cited implied that only one solution fitted each of the two problems. The reality is that behind these solutions lay analyses to define the problems and narrow possible choices on both the physical and people sides. This section lists and briefly discusses some of the factors that may be important in making such final choices. It is hoped that such a listing will provide a framework for making more imaginative and logical choices.

In choosing a course of action, decision makers should recognize the seven-plus-or-minus-two limitations which can inhibit their own thinking. Otherwise, in all but the simplest cases, some important considerations may be overlooked, such as putting details of the alternatives in writing, preferably under headings such as those listed below or others that may be devised.

Basically, there are three elements to consider in making choices among alternatives. The first two, if addressed first in the form of questions, can sometimes save effort:

1 Why do it at all?
2 Why do it now?

Sometimes it can be discovered quickly that change made merely for the sake of change does not make good sense, given the physical and human costs that may result. Possibly, also, the time and energy can be better spent in exploring other problems. In a similar vein, it may be that for a variety of reasons the time is not right to change a procedure or an activity. For example, it may be that control over certain steps in an operation requires commitments from others whose resistance must be overcome. Here it may be helpful to review the section on resistance to change given earlier in this chapter, which lists five requirements for getting change and five suggestions for implementing it.

The third and most difficult question in choosing is, "Why do it this way?" Here, answers from the problem-solving phase are needed. It is helpful to segregate them into physical and operational on the one hand and people considerations on the other, as follows:

1 Physical and operational factors
  • What will this particular scheme cost in money and effort and what will be the savings?
  • Does this scheme recognize critical conditions and constraints?
  • Is it possible and at what cost can these conditions or constraints be changed to make the scheme work?
  • Can time deadlines be met or modified?
  • What is the probability that this scheme will fail and what are the consequences of failure?
  • Is this scheme dangerous and if so, what precautions must be taken?

2 People considerations
  • What are the authority and power aspects of the situation?
  • Who makes and implements the choice? Is it higher management, affected supervision, crew, or specialist? Will there be leadership?
  • What individuals or groups were involved in developing the scheme?
  • Will these individuals or groups support its implementation?
  • Is higher management strongly in favor, supportive, not supportive, or hostile to it?
  • Is commitment assured, or can it be obtained from those who will carry the scheme through?
  • Are there individuals, groups, or organizations who will oppose the scheme because it threatens their jobs, authority, status, preconceived notions, or self image?
  • In situations where schemes are new and therefore on trial, is this alternative the most suitable?

Some readers may feel that this long discussion of choosing a course of action attempts to complicate what is basically a simple process. But a thoughtful analysis of the decisions they make will show that all these and probably other factors that often are treated subjectively or discarded as unimportant should be taken into account. In any event, because choices involve human behavior and thought processes, they are bound to be complicated, and every aid in simplifying them should be welcome.

COMMUNICATION

Communication in some form is necessary to implement choices made about particular courses of action. It involves the flow of material, information, perceptions, and understanding through a variety of media, as was shown earlier in Fig. 2-2. Having these processes work well is essential to maintaining or improving productivity.

The following sections look briefly at communications as a process and the influences that make it effective or ineffective.

How People Communicate

Fig. 10-8 diagrams the communications process between two or more people and indicates some of the complicating factors that can make it ineffective. In simple straightforward situations almost all the impediments shown in the figure may become unimportant and communication will be easy, simple, and direct. As an example, when two long-time field supervisors who know each other well exchange simple messages such as, "Hi, Bill, when will you be ready for me to make that concrete placement?" and the response, "Glad you telephoned, Joe, I'll have it ready for you this evening." In this case there are no encumbrances and information along with a simple gesture of friendship is transmitted. However, in another case, of that of a brief meeting between two ambitious and antagonistic area superintendents who are competing for favor with the project manager, this same exchange of words might be delivered in a way that reflects all the underlying jealousy, animosity, and bitterness. Communication between these two will leave much to be desired. Consider also a joint meeting of these individuals with the project manager. In such a situation, the messages each sends will be carefully planned to reflect their appraisals of the effect on their expectations and those of others, their personal situations, and most of the other elements shown on Fig. 10-8. Interpretations by the others present will be screened through their filters. Often the resulting statements and responses will provide little true communication, or the message will be so garbled as to be meaningless.

The implications on communication of the factors shown on Fig. 10-8 may seem obvious. Even so, it may be helpful to take a closer look at some of them and to indicate a few of the difficulties they create.

Expectations　　Expectations are the results the sender anticipates or hopes will come from the communication. Some are simple and explicit and can be

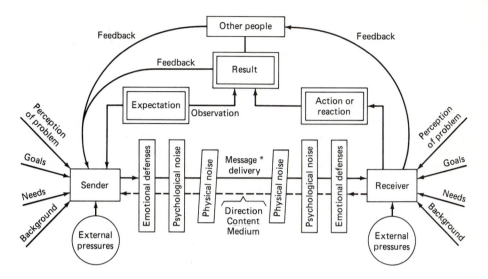

*Messages are delivered through a variety of media including:

Spoken word. Variables include clarity or limits of vocabulary, word choice, loudness, accents, tones, and inflection.

Body language. Variables include facial expressions, eye contact or lack of it, gestures, tenseness, posture, nearness to or touching listerer, body odor.

Written word. Variables include format, tone, available vocabulary, word choice, personal touches, and way of signing and delivery.

Pictorial representations. Variables include use of rough sketches, drawings, photos, physical models, computer printouts, and displays.

FIGURE 10-8
Factors that are involved in communications between individuals.

easily expressed; where relationships between the parties are clearly established, once expectations are stated by the sender, the receiver will carry out the desired action. Many routine superior-subordinate communications or those among persons who know each other well follow this pattern. A far-different situation exists when expectations are unclear to the sender who then fails to make them understandable to the receiver. Again, messages may be expressed vaguely rather than explicitly and are merely intended to feel out the receiver on a problem or issue.

Confusion and resistance can also result when expressed expectations are a cover for other desires. For example, instructions may direct that certain acts be carried out with the intention of forcing receivers to give in on other issues, to demonstrate that they are subservient, or to boost the sender's ego. In such instances, by clouding their real expectations, senders may well alert receivers so that they bring their defenses into play or reject the message.

Goals, Backgrounds, Needs, Perceptions, and External Pressures   In some situations senders and receivers will have substantially similar goals, backgrounds, needs, and perceptions, and feel somewhat the same external pressures. An example would be a well-coordinated, experienced management team on a construction project. In this situation, communications on at least routine matters will flow smoothly. But all too often the concerns will not be the same for the two parties. The case of the old-time superintendent and the young engineer is typical. In this instance, all the five influences impinging on sender and receiver, as shown in Fig. 10-8, are dissimilar. It is apparent that before meaningful interchanges, clearly understood by both parties, can take place, these differences must be taken into account. Often first encounters will be stormy and there will be many misunderstandings. Over time, both parties will come to recognize that these differences exist and are to be expected. Then understandings can be reached, even mutual trust and respect may be established. But if either party refuses to recognize the other's situation, there will be no real communication.

Emotional Defenses   All human beings erect defenses to protect their psychological health. In addition, it is common to play down, distort, or withhold information that may appear to senders to be damaging to their personal well-being. In turn, receivers will screen or purposefully or unknowingly garble messages in order to protect their egos or well-being. It can be expected, then, that in situations where sender and receiver have conflicting objectives or one appears to be a threat to the other, there will be little actual communication.

Psychological Noise   Psychological noise refers to the mental impediments that individuals face in sending or receiving messages, such as distractions introduced by other concurrent activities or demands for the individual's attention which overtax long-term memory, so that the intended messages are not recorded or are incomplete or garbled. Also there are the confusions and mental preoccupations associated with job or nonjob personal problems and concerns. In other words, psychological noise is the condition wherein, for a variety of reasons, senders or receivers are mentally absent or partially absent from the communications process.

Message Delivery   Some of the common ways of delivering communications and a few variables in each are shown on Fig. 10-8. Of particular concern here are those that occur face-to-face where the spoken word, body language, and other nonverbal devices come into play. Most of us fail to realize how complex and sometimes devious this means of communicating which we continually carry on can be. For example, it has been reported that from 65 to 93 percent of oral communication is in nonverbal form. Again, many common words have several meanings and senders and receivers may interpret them differently—and sometimes to their personal advantage. This is not a trivial item since on the average, there are 28 meanings for each of the 500 most common words.

Those sending messages often try hard to make them simple and sensible and, in most cases, attempt to please the receiver. In doing so, some of the content or intended emphasis may be lost. Other problems can develop when the speaker's intent is misinterpreted, as when a suggestion or ironic remark is taken as an order to be executed immediately.

Readers will gain more if they use the remaining items listed on Fig 10-8 as a checklist to reconstruct and analyze a few recent conversations than read more discussion about them. One of these might be a pleasant nonstressful interchange with a friend; the other could deal with a disagreement with someone with whom there is little common background and viewpoint. The result will, in all probability, be a real learning experience.

As with speaking, written messages can also be framed to deliver a variety of very subtle messages. One among many examples would be the form of salutation and signature on a memorandum or letter. The informal salutation "Dear Bill" and signature "Hal" convey a message of personal friendship and trust. Another communication, with the salutation "Mr. William T. Smith, Project Superintendent," and signature "Harold D. Jones, Project Engineer," shows neither friendship nor trust, and may convey a veiled threat. Of course, the refusal of a sender to put an important communication in writing is bound to raise suspicions and doubts about the sender's intentions and motives.

In engineering and construction, pictorial representations such as drawings and sketches are a main basis for communication. They convey most of the intentions of the designer from concept to minutiae. One great value of employing drawings is that they make it possible to focus on such divergent aspects of the problem as general layout or on a specific detail and still not exceed the short-term memory limit of seven-plus-or-minus-two chunks, with each of these chunks containing substantial detail. Another is the clear permanent record that is available for future reference.

Messages whether clear or unclear that have survived after passing through more noises and defenses, will be evaluated in terms of the receiver's own values and pressures. If there is two-way communication (see discussion below), responses will be directly to the sender, but through the impediments already discussed (the dotted arrow right to left in Fig. 10-8). Then the process, beginning with either clarification or revised expectations, will be repeated, possibly several times. If the communication is one-way only, the receiver will respond through some action or reaction, which of course may mean merely paying lip service, defying, or ignoring the message. The climate which will control the choice among these will be discussed subsequently. Following communication, there will be a result, such as action or inaction. Sometimes it will meet the sender's expectations, but often it will not.

Feedback   Feedback is a term, originally from electronics, which in the context of this discussion describes messages returned to the sender or given to others (see Fig. 10-8). As indicated above, if there is two-way communication, feedback from receiver to sender is direct. But there are numerous other forms. For example, the receiver may communicate directly with other people, either to inform them of a coming action or to solicit their aid in reinforcing, modifying, or frustrating a directive from the sender in a variety of ways. One means is through an informal organization. To illustrate: if engineers at all levels in a management hierarchy report officially to line management but have strong contacts with and ties to engineers at higher levels, they may attempt to get the higher-level engineers to intercede with higher levels in line management. Sometimes this form of feedback goes outside the organization. For example, a safety engineer for a construction company, directed by management to ignore certain hazardous conditions, might surreptitiously alert OSHA inspectors or union officials.

Action (or inaction), of course, provides feedback to the sender, either directly or indirectly, based on the sender's observations and analysis of what happens; or, as shown in Fig. 10-8, it can come from other people who observe or are affected by the results or lack of them.

Giving feedback on other than factual matters is difficult at best, particularly when it involves assessing the behavior or performance of another. In such instances, to focus on the factual and descriptive is better than being personal or judgmental. For example, if the message is that in a given situation a person talked too much, it is better to say, "I kept track and you talked 80 percent of the time" rather than "you certainly were a loudmouth." Again, the theme should be on sharing ideas and information and exploring new approaches, not on shouting recriminations or giving advice. And, of course, the discussion should center on the problems or needs of the receiver and not on those of the sender.

This analysis of some factors affecting communications is not intended to be exhaustive. Rather, it is hoped that readers will use Fig. 10-8 and this discussion as a tool for diagnosing their approaches when communicating with others, thereby improving their effectiveness.

Individual Differences or Behaviors that Affect Communication

The communication process outlined in Fig. 10-8 and discussed above is further complicated by the individual differences among people. It is common knowledge that even brothers or sisters born to the same parents and reared under the same conditions act, react, and communicate very differently. In the work environment of a construction site, disparities among people are further complicated by differences in such things as position in the job hierachy, background, education, profession, experience, career objectives, and how individuals were socialized.[33]

Incompatibilities among any of these can lead, at least in the beginning, to feelings of superiority or inferiority, create other conditions that breed distrust and disharmony and, as noted earlier, bring emotional and psychological defenses into play.

---

[33] Socialized is a behavioral-science term that encompasses the effects on standards of behavior and ways of looking at the world that an individual gains in growing up under a particular set of family, racial, national, regional, economic, political, educational, and financial conditions.

Communications can be one-way or two-way. From a manager's point of view, at least in the short run, each approach offers advantages and disadvantages depending on the particular situation. Attributes of the two approaches, solely from a communications point of view, are listed in Table 10-3. Traditionally, construction managers have leaned toward one-way communication. It should be remembered, however, that one-way communication is often associated with theory X attitudes toward people and an authoritarian management style, either of which can result in serious and often adverse consequences.

The character of the verbal or written exchanges between individuals has a

TABLE 10-3
ONE-WAY AND TWO-WAY COMMUNICATION COMPARED

Situation	Attributes of approach	
	One-way	Two-way
Time required for action	Less—no waiting for feedback	More—feedback process takes time
Accuracy	Satisfactory on routine or often-repeated problems; error-prone on complex ones	Better on new, complex, or unexplored problems; with routine problems or repetition of the same ones, communication shifts to one-way
Planning requirements	Communications must be clear, orderly, and systematic	Can be less exact, since process permits trial-and-error approaches
Time required from managers	Less for simple matters; few demands after original communication; may be more with complex issues if original direction is unclear or incomplete	More; soliciting and responding to feedback takes time
Pressures on managers	Fewer, since it follows traditional authoritarian management patterns so that they do not have to defend communications; requires less attention to people problems	More, since management pattern is less orthodox; more attention needed to people problems; often must defend and possibly reconsider original action
Difficulty in carrying out communications process	Easier, particularly for inexperienced managers who are beginning to learn the skills required to deal with people	More difficult; requires a higher degree of management skill
Threat to managerial power and status	Less; remain the boss and can more easily cover up mistakes	More; procedure can dilute or challenge authority and lead to loss of control; also brings mistakes out into the open

marked effect on the chance that they will communicate effectively. Choice of words weighs heavily. An early but often quoted article by Jack R. Gibb titled "Defensive Communication" provides excellent thinking on this subject.[34] He categorizes actual or perceived behaviors of individuals toward other individuals or groups, as measured by the language used, as defensive or supportive. Defensive means making listeners defensive; supportive involves giving them the feeling that they can make a meaningful contribution. Table 10-4 summarizes Gibbs's thinking. It should be apparent that when the parties have roughly equal status and power, using the defensive stratagem is almost certain to make communication ineffective, whereas the supportive approach can at least generate a useful discussion.

It goes almost without saying that with a defensive communication strategy there will be a strong effort to dominate verbal discussions. In addition, when not talking, the listeners' attention is primarily on what they will say next or how they will rebut an argument presented or that may be presented by the other parties. With the supportive stratagem there will be much more actual listening.

Another concern affecting both communication and leadership styles distinguishes between being assertive and being aggressive. Assertive persons state their positions with self-confidence to make their presence and influence felt. Aggressive individuals go further; their attitudes are often militant, combative, unbending, and overtly assertive. It is possible for a person to be assertive and self-confident and still develop a supportive climate with others; seldom, however, will this be possible when one becomes aggressive and even abusive. This distinction is more than a mere play on words, since it reflects basic attitudes and behaviors. Recently, assertiveness is being put forward as a virtue to be cultivated, with training programs being offered under titles such as "Success through Assertiveness."

Listening—An Important Aspect of Communication

Figure 10-8 indicates that in every communication there is supposedly a receiver (listener) involved. Unfortunately, in today's western cultures, being dominant and assertive is emphasized and listening is downgraded. Furthermore, in our educational approaches far more attention is given to speaking than to listening. Thus this important part of the communications process is neglected.

Listening is a distinct form of behavior that is learned almost from birth. It is selective; listeners choose a few cues from the many that bombard their senses. They tend to hear favorable, pleasant things and screen out the unfavorable and unpleasant. They listen to some people and not others. Attention

---

[34] See *Journal of Communication*, vol. XI, no. 3, September 1961, pp. 141–148. Reprinted in H. J. Leavitt and L. R. Pondy (eds.), *Reading in Managerial Psychology*, 2d ed., The University of Chicago Press, Chicago, 1973.

TABLE 10-4
LANGUAGE AS A FACTOR IN DEVELOPING DEFENSIVE OR SUPPORTIVE CLIMATES
FOR COMMUNICATION

Item	Climate Created	
	Defensive	Supportive
1. Approach to discussion	Strategic: attempts to maneuver to speaker's advantage	Spontaneous: encourages free interchange
2. View of the situation	Evaluative: states facts and opinions from speaker's point of view	Descriptive: states facts only and those objectively
3. Conduct during discussion	Controlling: attempts to dominate and mastermind	Problem-oriented: attempts to get others involved in getting situation clarified
4. Attitude toward other participants	Neutral: impersonal, deals at arm's length	Empathetic: attempts to see other's viewpoint (get inside listener's skin and look out through listener's eyes)
5. Ways of expressing thoughts	Superior: makes pronounce-ments as if from superior knowledge; prone to criticize destructively	Equal: projects feeling that other party has worthwhile things to say; criticizes constructively if at all
6. Approaches to conclusions	Certain: speaker's views are correct	Provisional: reserves judgment until all issues have been resolved

is not continuous; even the best listener cannot be alert all the time, so it follows that only a small portion of the signals that reach the ear are recorded in long-term memory for recall as needed. And, as indicated earlier, listening involves collecting cues that come through the eyes and other senses as well as the ears. It follows that to listen well involves keeping all the senses alert.

Senders can help listeners. For example, they can be sure they have the listener's full attention, including eye contact. If attention has lagged, the sender can, without expressing annoyance, go over the message again. A particularly helpful strategy is to encourage listeners to repeat the message in their own words.

Among the stratagems for making listening a more valuable communication tool are the following:

1 Talk less and listen more.

2 Listen through body language. This means among other things, keeping eye contact, noting facial expressions that indicate interest, leaning toward the speaker, and keeping the arms down. These give the speaker a feeling of importance. An overly relaxed attitude or apparent inattention indicates a lack of interest, and standing or sitting with the arms across the chest may suggest belligerence.

3 Encourage speakers to express their feeling with appropriate statements and body language.

4 Draw speakers out and help them express themselves better or clarify their ideas with appropriate nonthreatening comments or questions.

5 Do not make premature evaluations. Something about a message or the person delivering it may turn you on or turn you off almost at once, such as very positive vs. quiet speech; aggressive vs. submissive posture; or making or avoiding eye contact. All such cues can mean different things in different situations and to different people. For example, in western culture looking one straight in the eye creates a positive impression and is taken as a sign of honesty. In certain other cultures, to look directly at a person of higher rank is interpreted as an act of disrespect.

6 Outline in your mind the message that was intended. It may be that you did not hear or remember every important thing that had been said. As indicated, one cannot be continually fully alert and you may have been distracted. If you feel that you lack information, ask questions or otherwise get the item repeated.

7 If possible, rephrase the message in your own words to be sure that it is clear and complete. As noted, many words have more than one meaning and you should be sure which is intended. Again, intentions or concepts may not be clearly or completely defined, and misunderstandings should be cleared up.

8 Be patient. This is the hardest but most important task of all.

According to experts, listening is a most important element in the overall communications process, because by doing so:

1 You learn and develop a valuable fund of knowledge. Furthermore, it can whet your appetite to learn more.

2 You record messages or instructions more completely and accurately.

3 You can help others in several ways, by
   a Thinking out loud and more fully developing concepts and ideas.
   b Providing a way for persons to get things off their chests or take care of their hidden agendas. This is essential if an individual is to recover the mental attitudes necessary to express their ideas or concerns.

4 You are thereby able to size the speaker up. There is the truism that by listening to Peter talk about Paul, you learn more about Peter than about Paul!

5 A person who is unfamiliar with a foreign language is following a better approach for learning it by listening rather than by speaking.

6 By showing concern, you can develop friendships and earn respect and co-operation.

Given that listening is a learned behavior or habit, to become a good listener requires having a desire to listen, observing one's present behavior, and finally developing and practicing better approaches. To do these takes a conscious effort to place listening among the top priorities in allocating an individual's time and energy.

Oral Communication in On-Site Construction

On-site construction managers knowingly or otherwise contend with the complex communication processes described above when they deal with workers, bosses, or those of equal standing. In addition, the fact that construction involves fast-moving, ever-changing work assignments rather than repetitive ones brings many pressures on supervisors that are less common in industrial situations where changes come less rapidly. On shorter projects the work force is less stable, so that it is more difficult to establish and maintain set procedures. Also, as noted earlier, discussions often take place where noise levels and weather conditions make hearing and listening difficult. Finally, most on-site construction people are action-oriented, and taking the time from what is considered more useful work to talk seems to them to be wasteful. To have good oral communication on construction sites requires particular attention, always recognizing the complexities that have been outlined here.

Nonverbal Communication on Construction Sites

Nonverbal communication is a given on small projects where managers and craftsmen know each other and work closely together and patterns of responsibility and behavior are established. On such projects, most people can read the nonverbal signs of approval or displeasure from others. On larger projects the nonverbal signals become far more complex, since there is almost no personal contact between management at higher levels and the work force.

On all sites, management by its actions sends many messages on a variety of topics. Safety is a good example. If management insists on a clear, safe workplace and devotes time and spends money on safety equipment, the message is clear: safety is important. Again, having enough clean toilets mean that management is concerned about the physical welfare of the workers. Providing appropriate and well-maintained equipment and tools says that productivity is a major concern. The list of nonverbal signals can be long. What must be recognized is that both action and inaction send nonverbal messages to everyone on the project.

Written Communications on Construction Sites

Plans, schedules, delivery timing, and arrangements for materials, change orders, and cost and progress reports are essential elements in managing all but the smallest construction projects. It is from such records that supervisors at all levels control and monitor the work. Also, diaries, directives, and letters or memorandums to owners about on-site and materials contracts and others are essential to the business side of construction.

Other sections of this book dwell at length on written procedures for (1) finding, investigating, and solving away-from-the-work-force problems of all kinds and (2) approaching methods improvement systematically. For success in both these areas it is essential that plans for and thoughts about situations be recorded in graphs, tables, printed descriptions, and other written forms. Only in this way can rational approaches be developed and tested, because of the limits on how much the human mind can retain in short-term memory.

In general, on-site construction managers at all levels resist paperwork of any kind. They see it as an obstacle that keeps them away from "being out where the dirt is flying." Yet, since this form of communication is vital to highly productive work, it is essential that it be high up on the list of items that construction managers must do promptly and accurately. There is, of course, written information in several forms provided to craftsmen. Some, such as job rules regarding award schemes, safety, and alcohol or drug use are discussed at appropriate places in this book.

Communications In and With Crews

A substantial part of the communication on construction sites involves groups. The most obvious examples are foremen giving instructions to crews, superintendents developing schedules or other plans with foremen, and project managers conducting briefing and planning sessions with line managers, project engineers, and supervisors of ancillary functions. As with written communications, many on-site supervisors see group meetings as barriers to getting on with the job. Today, however, and particularly on complex projects, it is common to insist on regularly-scheduled meetings during working hours as often as once a day for each of the supervisory groups. These meetings provide a mechanism by means of which common problems can be discussed and previously set tentative schedules can be altered, with all those concerned furnishing information and being informed of decisions made about the job on hand. Also, on most projects, foremen are required to hold safety meetings and are often encouraged to hold other meetings to orient crew members as a group when a task is undertaken. Specific items of discussion for such meetings are outlined elsewhere in this book.

The success or failure of any group meeting rests primarily with the chair, who may be the senior individual in the group, although sometimes it is better if someone else is designated for the task. To be effective, chairs must take charge without taking over. A listing of functions and procedures would include the following: (1) planning carefully in advance and providing an agenda and timetable, if appropriate, (2) starting the meeting promptly at the designated time and ending it on time, (3) introducing topics and discussing them or designating someone else to do so, (4) stimulating discussion using techniques such as brainstorming, but at the same time keeping it on the track, (5) handling problem members, (6) summarizing conclusions and delegating responsibilities for action, and (7) preparing minutes, if appropriate. Items 4 and 5 in this list are often complicated by what has been called a hidden agenda, whereby certain ones in the group have concerns that are apparently unrelated to the topics to be discussed but try to bring them up anyway. Seldom will the group make progress on the scheduled topics unless the hidden agenda is brought into the open. It should be clear that chairs, be they foreman, superintendent, or project manager, must develop skills for planning and conducting meetings and for dealing with the people problems presented at the meetings.

Participants in a meeting can also contribute greatly to its success or failure. Talking only when a contribution can be made and paying attention while others are speaking is essential. Furthermore, participants who feel that their contribution is being ignored tend to try again and again. They are usually better off if they wait for a better climate at a later time. Participants also should be aware that they can assist by doing things that the chair cannot do and still remain impartial. For example, there is the common situation when a consensus is near but the chair is not in the position to say so; another member of the group can. Frequently, also, participants can help the chair by taking over the discussion from those who talk too much. The chair treads on dangerous ground when attempting to silence such people, but another group member can do so with a seemingly innocuous remark such as "Didn't you say that before?" or "Now that Joe has had his say, it's Mary's turn."

This discussion of improving communications in groups assumes that the leaders want participation and ideas from members. If such is the case, they should make it clear. If bosses chair meetings, they can take over by being authoritative and projecting aggressive body language, thereby signaling to the other participants that they are at the meeting to receive instructions and to listen but not to talk. It is particularly difficult for the bosses to avoid leaving this impression if they chair the meeting. A better approach often is to designate another chair with the boss staying silent on the sidelines, except for possibly summarizing the conclusions and making assignments for the future.

Learning about Skills in Communication

This discussion has done little more than indicate the ways that communication takes place in on-site construction, and has briefly suggested means for improving it. To implement these suggestions, requires a wider recognition of the importance of effective communication by highest-level construction managers and the buyers of construction, followed by a determined action to do something about communications on their projects. Ways for doing so are becoming increasingly available. For example, the AGC superintendent and foremen training programs (see Chap. 11) both include communication skills among their topics. Some large construction companies and buyers of construction have proprietary in-service programs for their employees. Also, the topic is an element in packages available through consultants. To date, however, hardly a beginning has been made. Most of the highly fragmented construction industry operates on the basis that any individual who is a supervisor at any level is an expert in communications. Unfortunately, there is substantial evidence that such is not the case.[35]

SUMMARY

This chapter has examined (1) psychological factors in humans as thinking beings, (2) the roles of authority, power, and leadership, (3) how these factors in 1 and 2 apply to on-site construction, (4) decision making, and (5) communications, or how information, ideas, and feelings are transferred among individuals orally and in writing and within groups. Most good construction managers somehow know instinctively or have learned by experience many of the concepts offered here. For them, the ideas may be useful mainly in providing a framework for thinking about and appraising behaviors that they have observed. There are others, new and old-time construction hands alike, who are either ignorant of or antagonistic to the concepts offered here. One example is the person who after 20 hours in a human-behavior seminar for construction managers burst out, "All this is a lot of bunk—forget it! If you want them to do something tell them to do it and don't let them bother you with questions or back talk." If individuals really feel that way, they are giving away much by failing to tap the resources available in human beings. For they are far more than machines. And, in today's world, construction managers who fail to recognize this fact do so at their peril.

Put another way, there is a question that all construction managers must answer about how to deal with people. It is "How much difference would it make if workers on the job cared more, knew more, and had more to say about their work?"

The aim of this chapter has been to provide tools that might make that difference happen.

---

[35] For a more detailed discussion of communication in construction see H. W. Parker, *Issues In Engineering,* ASCE, vol. 106, no. E13, 1980, pp. 173–180.

CHAPTER 11

ORGANIZED APPROACHES FOR IMPROVING PROJECT RELATIONSHIPS AND THE ABILITIES OF MANAGEMENT AND THE WORK FORCE

INTRODUCTION

Chapter 1 dealt with a prescription for improving productivity in on-site construction and proposed that project goals be developed to which all the parties involved in a project should subscribe. Chapter 3 explored what management can do to improve productivity, discussed strategies for putting together productivity-improvement programs, and gave examples of some of the approaches that have proved to be effective in gaining management and worker support. This chapter goes further, detailing other elements that have been employed on a number of large projects to gain management and worker participation in productivity-improvement efforts. These, if altered to fit the differing circumstances, might be applied to medium-sized or even smaller jobs as well.

As has been stated repeatedly in this book, any scheme such as those to be described here will work only where a unified approach involving the following elements is followed:

1 A belief that if all the parties work together in good faith to be productive and complete their tasks safely in a timely manner and to suitable standards of quality, all those involved, both organizations and individuals, will gain

2 A willingness to devote time, energy, and money to carry through on this belief

3 An assurance that the agreements, systems, procedures, personnel, training, and reward systems required to carry through this effort will be put in place and implemented

4 A follow-through to see that the entire organization knows about, agrees with, or at least accepts the plan and is implementing it

5 Being alert for telltale signs that the plan is or is not working well and making modifications to get it on track if it is not

This prescription is broad and fits any on-site construction operation from the smallest to the largest. However, strategies for implementation will differ with project size, duration, and other variables. For very small operations, top management can establish its program informally and almost single-handedly. Communication is largely by word of mouth among the owner's and designer's representatives, the contractor, and individual foremen, craftsmen, or crews. In contrast, a project costing hundreds of millions of dollars and of long duration involves many individuals employed by the owner, designer, contractor, subcontractors, and labor representatives. The program will then become the responsibility of the oversight team, as described in Chap. 3. On site, a full-time staff of several people may be required to carry out the details of such a comprehensive effort.

Most of the programs described here have been carried out on projects of considerable magnitude, long duration, and with a large work force that remained in place for a considerable time. A brief description of certain aspects of one such program is given in Exhibit 11-1. As in other people-related activities, procedures that are instituted will require modification to fit the specific situation and to correct problems that are bound to develop.

The remainder of this chapter describes several activities that have been successfully employed to improve productivity by developing understanding and support for goals set by higher management. It also examines briefly a few longer-range problems, including the situations in education, research, and training for the industry. The topics are:

- Setting and maintaining standards for good job relations
- Presenting a favorable project image
- Using employment techniques that create a favorable climate
- Telling what is happening on the project
- Dealing with complaints
- Granting awards and prizes
- Furnishing recreational and related activities
- Providing site visits and other informational activities for families of workers
- Establishing schemes for stimulating and handling suggestions
- Offering formal educational and research programs for construction
- Instituting training for craftsmen and field supervision

SETTING AND MAINTAINING GOALS AND STANDARDS FOR GOOD JOB RELATIONSHIPS

It may seem ridiculous to most construction people to set down, agree to, and publicize formal written objectives for job relations that will be followed by on-site personnel of owner, designer, contractor (or construction manager), and subcontractors. At the beginning of most projects some such objectives

EXHIBIT 11-1

THE RICHMOND LUBE OIL WORK-IMPROVEMENT PROGRAM

A $550,000,000 project in Richmond, California, involved the construction of an entirely new plant to produce lubricating oils. The owner was Chevron Oil Company; the designer and constructor was Bechtel Petroleum. The constructor's staff included a three-person employee relations and work-improvement team, augmented by a consultant skilled in work-improvement and human-relations techniques. Costs of this group and its activities were paid for by the owner.

The overall program operated under the assumption that people are willing to commit to project goals and will work to achieve them if invited to do so and are given adequate resources and strong management support.

Six of the basic elements of the program were:

1 An employee-relations program. Among other features were:
  • A video presentation to introduce new employees to the project and to management's goals and philosophies. At this time, baseball caps carrying the project's logo were issued.
  • A monthly newsletter which discussed project activities, progress, and difficulties without a management bias. It featured articles and photographs of individuals or groups and publicized awards from the suggestion system.
  • Attention to worker complaints. Typical issues included inadequate parking and difficulties of access to the site.
2 A suggestion program which brought substantial savings and created a strong interest in methods improvement.
3 Regular meetings of supervision and foremen to keep them informed about job progress and problems.
4 Training programs for job supervisors to familiarize them with the constructor's planning, operating, and reporting procedures.
5 Short courses for craftsmen to teach project-related skills.
6 A productivity-improvement program.

These activities were planned and directed by a work-improvement committee. In addition to the program administrator and consultant, membership included representatives of the owner, the constructor, and the building trades unions. The group met monthly to review the program, plan new activities, and discuss areas of concern.

The productivity-improvement program was highly participative. Among its elements were:

1 Development and adoption of a more rigorous work-planning procedure.
2 Formation of task forces to improve the functioning of various planning and resource-supply systems. For example, one task force developed a piping materials information system on a small computer which dramatically reduced the time required to assess material availability and identify available tasks.
3 Field studies of work methods leading to improvements and substantial cost reductions. These involved collecting data on delays using craftsman and foreman surveys and other appropriate methods. Efforts were directed not only to work-face activities but were extended to nonmanual and administrative operations.

The focus of the work-face studies was on highly repetitive operations involving large quantities: for example, piping, conduit, and wire and cable, totaling, respectively, 517,000, 730,000, and 2,900,000 lineal feet. Details of two of the piping and conduit studies are given in App. C.

Results of the effort were impressive. Examples include:

1 Higher project morale for both management and workers
2 A reduction of $20,000,000 in the labor budget
3 Suggestions leading to awards of $30,000 which saved the project $500,000
4 A doubling of output in conduit installation

may be in the minds of some or all of the parties. Seldom would they have been discussed, much less put in writing and agreed to. The result is that the good intentions about such standards often disappear as the job progresses and changes and other unforeseen circumstances affect the status and well-being of the parties or their employers. It is when there is evidence that things may be about to go awry that the agreed-upon standards can be brought into play to head off adversarial relationships that can otherwise develop.

To put the problem of goal setting in perspective, it is helpful to summarize earlier discussions of the motivations of owners, designers, contractors, and the work force at the project level.

- *Owners or owners' representatives* want the project completed on time, at or below owners' budget, and to suitable quality; may or may not have concerns about safety.
- *Designers or their representatives* want a profitable and satisfying job; costs, quality, and timeliness that will satisfy owner and enhance firm's reputation; may or may not have concerns about safety.
- *Construction managers (if on the project)* want the same that the designers (above) and contractors (below) want.
- *Contractors and their project management* want to make a profit; are concerned about quality and timeliness if it affects profit or ability to get work in the future; will be concerned about safety if they appreciate the fact that safety will affect profits or prospect of getting work in the future, and, possibly, have humanitarian feelings about the matter.
- *Subcontractors and their management teams* want the same things that contractors want.
- *The labor force* want to gain and maintain employment at a suitable wage, under favorable working and safety conditions.

Given these motivations for the various parties, it is easy to see why each party will put its concerns first unless there are clearly defined and understood goals and standards that will override these important self-interests. It becomes essential then, at the start of a project, to make clear the behaviors that are to be expected. This is best done by establishing formal standards and by backing them up on the job by behavior that abides by these standards. The procedure is particularly important on construction projects where the parties may not know each other, management may change often, and the all-too-prevalent adversarial approach and win-lose attitude set the tone. In such cases, representatives of each of the parties will, unless they are clearly committed to the plan, act under the assumption that what is best for their interests should be the primary control over their behaviors.

It would be an exciting and unusual experience to find prominently displayed in the project offices of owner, designer, contractor, subcontractors, and labor representatives a statement such as the following:

PROJECT OBJECTIVES

The Common Objectives of Everyone on This Project Are that

1 This project will be completed on time, of quality that meets the owner's stipulated requirements, with all tasks completed safely.

2 The project will be financially rewarding to and enhance the reputations of all.

3 Working on this project will be a personally rewarding experience for managers and craftsmen alike.

4 There will be fair dealing and open communication among all the parties.

It is obvious that agreeing to and posting a statement such as this would be only a first step. Moreover, every specific action by individuals or groups would have to be examined to see that it matched the criteria. This would not be an easy task, given that human beings are involved, that the rules fly in the face of traditional management attitudes, and that the reward systems of the various parties probably do not match these broader goals.

Among the factors that make it difficult to get project personnel to accept and strive for these objectives is that usually they do not know each other well enough to establish mutual trust and respect. As already mentioned, most projects are short and personnel changes often. All too often, the first meeting of representatives of the various parties is when a job crisis has developed and their primary concerns are to protect their own, their employer's, or their workers' interests. There are instances where to mitigate this situation, carefully planned team-building meetings, sometimes 2 or 3 days long, are held before the project begins. In some instances, shorter meetings are held at appropriate intervals throughout the length of the project. These should be far more than mere cocktail parties or ground-breaking ceremonies. Such meetings could start by all of the parties telling something about themselves. Games and problem-solving topics involving teams made up of representatives of the different parties also might be employed. Before the meeting is over, when the group has established a degree of friendship and trust, it could hammer out and agree on a statement like that given above. The activities could be augmented with social events involving the personnel's families. Done right, such a program could create an atmosphere of cooperation and trust that would continue throughout the project.

A common reaction to such a plan as this is that it costs too much. This argument overlooks the real but unquantifiable gains that will come when the parties on the project work together to foresee and overcome problems rather than having them sour the job atmosphere.

Many examples can be cited to illustrate how traditional on-site construction management practices conflict with the broad set of objectives outlined here. Handling claims for changed conditions is one. In such cases the common practice is to start with the assumption common to legal practice that the parties—owner and contractor, for example—are adversaries. Each party then looks to its own interests and collects data and develops arguments favoring its side and suppresses information and withholds or hides data favorable to the other. Given this kind of behavior, trust among the parties is destroyed not only with respect to this issue but with that of all aspects of the work.

It is not argued here that changed conditions and differing opinions over who should bear the costs do not or should not exist in construction; they are to be expected. But they can be handled in ways that avoid confrontation and charges of unfairness or even dishonesty which will bring a disruptive job climate. The alternative approach is to be open and honest. When either party sees a situation where a claim might arise, the other should be notified immediately. If possible, factual data should be collected and kept jointly by the parties. If possible, settlements could be made at the project level based on the data and the contract terms. If resolution cannot be reached on site, the matter could be referred to higher levels, while work on the project proceeds under amicable conditions. One way to bring this about is to keep site personnel out of activities that deal with pending claims.

As already indicated, even the best-planned scheme for developing and maintaining trust and good relationships among the parties on a construction project can fall apart as work progresses and pressures of many sorts develop. For this reason, it is important that all involved look for early signs of trouble which can be trustbusters and bring about adversarial situations. Among the indications that trouble is brewing are the use of phrases such as, ''it's your problem'' or ''it's in the specifications.'' Memos begin to replace informal information exchanges and formal letters replace the earlier informal notes. Also, the parties begin to withhold information. Each party to a telephone conversation may insist that a second party from the firm listen in or that the conversation be taped. When these incidents occur, perceptive people will examine both their own and the other party's position and the pressures that bring on the behavior. Then it may be possible to defuse the situation at this stage.

To summarize this discussion: It is a fact that projects go better when all parties on a job site work toward the common goals of profitability, safety, timeliness, quality, and satisfaction for all. It is unusual in construction that an agreement for achieving them is spelled out explicitly as has been proposed here. But a strong argument can be made that by formally setting and adopting objectives and making them clear to everyone on the project and by taking heed of the signs, it may be possible to head off trouble before it starts.

PRESENTING A FAVORABLE PROJECT IMAGE

First impressions make a difference. When construction hands—craftsmen, supervisors, or staff—approach and enter a site for the first time, they see and sense things that produce good or bad impressions of the project and its management. Changing these images, once formed, can be a slow process, and can influence attitudes and behavior for a considerable period of time. For example, good feelings can be generated at the very beginning by a well-planned and maintained sign identifying the project's owner, designer, and contractor. The good feelings are reinforced if there is a nearby well-signed place to park along with directions and access to an attractive project office. Again, a well-maintained signboard making clear that this is a hard-hat job and showing the project's accident record indicates clearly that attention is paid to safety, which means that management is concerned about people. Being pleasantly received in the project office also indicates that people matter to management. Opposites of these examples which create unfavorable impressions are all too common. They may begin with an absence of directions by which to find the project office; the lack of, or a disorganized, rough, or muddy parking lot; and difficult or cluttered access to an unattractive job shack which is in shambles, and with no one around to answer questions. The visible portion of the site has disorderly piles of scrap lumber and other debris. Craftsmen are working without shirts or hard hats, and openings and roof edges are unguarded. Scaffolding appears to be unsafe and certainly violates safety requirements. Any observant potential employee will at this point conclude that job management is incompetent and slovenly and cares little about the safety of employees. Unless work is tight in the area, competent people may not ask for a job but will look elsewhere. If they stay on, it may be with an uncaring, listless, if not pugnacious, attitude.

In addition to the importance of creating a favorable project image, attitudes and actions of owners and designers also can play a significant role. For example, owners that are concerned with having a good image in a community might include local requirements on, for example, provisions for safe pedestrian passage by the site in the project specifications and follow through to see that they are carried out. Furthermore, owners can make clear that the contractor's performance in establishing and maintaining good community relations and an attractive workplace and its attitudes about and precautions taken for accident prevention can strongly influence eligibility for future work. An examination of construction projects underway today shows that jobs done for large private owners, particularly if negotiated rather than bid, as a rule have an excellent appearance. This is less likely to be the case on projects such as public works where awards go to the lowest bidder and the owner has little control over the award of future work.

Doubters will ask what job image has to do with improved job productivity, something that cannot be proved statistically. But a project's image is one way that its management's attitudes, procedures, and overall competence is judged. Those coming onto a neat, clean, and well-ordered job expect to be part of an effective rather than a losing team.

EMPLOYMENT TECHNIQUES THAT CREATE A FAVORABLE JOB CLIMATE

Craftsmen and foremen come onto construction sites from a variety of sources. Often the contractor directs them there from another project. If they are new hires on union projects, they are commonly referred to the site by the hiring hall of their craft. For nonunion projects or when union or nonunion sources cannot supply people, hiring is done at the job site, in which case knowledge that jobs are available may come by word of mouth or through advertisements. Regardless of how workers reach the site, the contractor must have a mechanism for hiring them individually and assigning them to their supervisors. How this is done deserves careful attention and planning if new hires are to become part of productive work teams. Again, there is strong evidence that good hiring procedures not only create a climate conducive to productivity but also reduce absenteeism and turnover and the disruptions they bring.

Procedures for hiring craftsmen and putting them to work will of course differ because of many factors, including the size and length of the project and the number and duration of the need for certain skills. On small rather short jobs, the superintendent or foreman may handle all the details of employment, firings, and layoffs, including post-employment interviews, if they are appropriate. Large projects may necessitate a full-time staff. In any event, there are good and bad methods of carrying out these tasks.

Regardless of how hiring and induction is done, the employer must obtain certain personal data from each craftsman in order to pay them and to provide information to certain agencies of government, sometimes to the union, and often to those who administer a variety of fringe-benefit accounts.

How an employer proceeds in gathering personal information about individuals and in introducing them to the job and to their supervisor and crew can have a lasting effect on their attitudes toward job management and to the work task itself. The aim should be to make new employees feel that they are needed as a part of the job team and that they are important as individuals in the eyes of management. There is no right way to develop these feelings, since the situation on every job and even among crafts and crews may be different. Some large successful contractors have developed procedures such as the following:

1 The personal data form is filled out jointly by the craftsman and the person doing the hiring. Every effort is made to have a friendly, relaxed atmosphere in a quiet, comfortable setting.
2 The craftsmen, whether or not they have worked for the company before, are asked to view a video tape or slide presentation in which a top executive of the company welcomes them to the company and to the job. This presentation places strong emphasis on safe practices and what such practices mean to the company and to the individual. From the company's point of view, emphasis on safety with both new and old employees is important. As reported in Chap. 12, statistics show that new workers on a project, regardless of past employment with a company, have more accidents, but safety orientation can reduce that risk.
3 New workers are given a small packet of printed material to take with them. The purpose of each item should be explained by the person doing the hiring

and the employee should be encouraged to study it further. In the packet would be such items as:

a Safety rules for the project. (Some employers insist that the employees read the safety rules and sign a slip saying they have done so before actually going onto the project.)

b A written description of the project and of any formal programs that are underway, some of which are described below. On very large projects, these formal programs can be quite elaborate. Often they go under designations (LOGOS) which indicate their content or in some other way identify with the project. A few examples are LMPT (labor-management participation teams); PRIDE (people respecting integrity, dedication, and excellence); RLOP (Richmond Lube Oil Project); TIP [tripartite improvement program (a joint effort of owner, contractor, and labor unions)]; CTP (communications, training, planning).

4 Craftsmen are directed or escorted to the foreman or other supervisor under whom they will work. These managers have been instructed to take time to show the new hires around the job, taking particular care to point out the hazards, to introduce them to fellow workers, and, if possible, to team them up with an experienced craftsman who has been on the job for a while. They also make sure that the new workers have the proper attire, including hard hats and hard-toed shoes. Once the workers have begun their assigned tasks, their work patterns are observed not only to see if they are competent but also to help them develop proficiency.

5 After a craftsman has gone to work, supervisors make sure that personal concerns, if any, are heard. Within the limits of good sense, they should be acted upon. This does not mean acceptance of poor performance or rule breaking but simply fair treatment.

6 When workers must be terminated, every effort is made to do so fairly and impartially and in a way that preserves the company's image of fair dealing. Termination for cause is handled with particular care. For example, in some companies, this is not included in the foreman's responsibilities. Rather the superintendent reviews the foreman's recommendation and makes the final decision, thereby minimizing the chance that personal dislike rather than poor performance underlies the termination. Often the superintendent can straighten out the difficulty or retain good workers by reassigning them.

This very brief discussion of employment techniques aimed at creating a favorable job climate merely scratches the surface of a complex problem. How it is done varies from project to project and situation to situation. The purpose here has been to raise the issue by giving examples of a few among the many ways that such problems have been handled well.

TELLING WHAT IS HAPPENING ON THE PROJECT

Individuals on a project must know what is going on if their interest and commitment are to be developed and retained. This means that top management must develop strategies for doing so, fitted to each particular project. On small jobs, information can be passed on largely by word of mouth, but in a carefully planned and executed manner rather than leaving it to chance or rumor, which often distorts or introduces error. In so doing, management must be prepared to answer questions and gripes in a frank, open, and nonbelligerent way. As projects become larger, the effort to provide information must be more formally organized. Here again, top management must be involved but can delegate details to others. On the so-called super projects lasting several years and employing hundreds or even thousands of people, supplying information will be among the tasks assigned to a separate staff or added to the duties of those who deal with personnel, suggestions, complaints, awards, or possibly methods studies or safety. It must be recognized at the beginning that an activity such as this requires a substantial investment in management time and in money and should not be undertaken without a clear understanding of this fact on the part of those paying the bills.

The approaches to information development and dissemination described in the paragraphs which follow are typical of those employed on very large but well-managed projects. They are offered here with the recognition that the suggested scopes may not be suitable for shorter or smaller jobs. Even so, knowing of them may stimulate thinking, offer ideas, and reinforce the concept that getting the word out is important.

Some form of bulletin or newsletter is probably the most common means for getting information to all those who are involved with the project. In its simplest form, which may fit a small job, the release might be a duplicated project fact sheet that briefly describes the parties involved, project size and duration, job progress, and similar sorts of data. It might be put out as a supplement to a company magazine, if one exists. For large projects, elaborate newsletters several pages long, printed rather than duplicated, and carrying photographs and other interest-getting features are common. Through them information about matters such as suggestion and complaint handling or publicity about recreational or other project-sponsored activities and site visits are made known. Special attention is given to individual or group awards for productivity-improvement and safety.

The aim of such newsletters or bulletins is to create and maintain a favorable job climate. To gain attention, they must carry material of interest to all levels on the job, but particularly to craftsmen, since they constitute the largest number of potential readers. A willingness to spend money on an attractive publication creates the feeling that management feels its people are important and not merely cogs in a machine. All the information must be unbiased; readers will be suspicious of any statements that to them seem to be one-sided or manipulative. In general, creating the feeling that views are objective and impartial can best be accomplished by dealing almost entirely with factual matters and by avoiding topics that represent opinions. If a controversy or misunderstanding is discussed at all, both sides must be represented.

Material presented must be written with the readers in mind. In most instances, long, involved presentations are to be avoided, because the interest levels of most of the potential readers may differ from those of management. Also, because turnover on most construction projects is frequent, some of the basic information that employees need should be repeated from time to time; hence the writer must be ingenious in presenting the same information in different ways so that repeat readers are not turned off.

On very large projects, newsletters might be issued once a month. The method of distribution should be fitted to the job situation. For example, if paychecks are distributed at a central location or at stations around the project, copies of the newsletter could be made available there. But merely having the copies piled about is not enough; some positive means of getting them into the reader's hands is essential. On some projects, the newsletter has been mailed to the employee's home. This has the advantage of gaining family as well as worker interest. In format, the newsletter is usually four to six pages long. It emphasizes the job logo or symbol to associate the letter with the project. Photographs or write-ups of individuals or groups that are the subject of special attention are valuable.

The newsletter can provide a means for announcing, publicizing, and explaining the initiation, progress, and results of any special programs. Some of these are discussed individually below or in other chapters of this book. A sampling of other topics that could appear in one or more editions of a newsletter is as follows:

1 Messages from owner, contractor, and union people welcoming newcomers to the project and inviting participation in the available activities.

2 A description of the project. This could list the parties involved, project magnitude, duration, and milestones; give a site layout which emphasizes items of particular interest such as work areas, offices, and parking; and current and anticipated employment levels by craft. Some of these topics would be updated as appropriate.

3 Information on various matters affecting employees, such as routine announcements about such matters as holidays or changes in working or shift arrangements or discussions of situations that annoy the work force. Examples of topics that might be covered are:

- Security arrangements such as searches or locked gates or buildings. Whatever the reason for them, these are offensive to and time-consuming for the work force. Even more important, unless justified, they indicate a lack of trust, which can turn workers against management.
- Project rules about alcohol and drugs. These can be justified on safety grounds and should be cast in that light.
- A description of the safety program. Each newsletter would give accident statistics, including rates and trends on the project. It could explain any serious accidents that had occurred and what had been done about them. As a specific and actual example of the value of such an explanation: An ironworker fell from a high beam and wound up sprawled over another beam well above ground. He was not immediately rushed to the hospital; fellow workers were instructed to hold him there until the paramedics ar-

rived and a stretcher could be lifted by a crane to carry him down. The job rumor mill painted a picture of inhuman treatment but the rumor was defused by a newsletter article that explained that potentially more serious injuries could have resulted if the injured man had been lifted down improperly.

- Locations of various emergency centers such as doctors' and nurses' offices or ambulance, police, and fire stations.
- Locations of offices where information about payroll or other matters can be obtained.
- Discussion about traffic and parking inconveniences. Often these are unavoidable, but explanations about why they are can help alleviate the workers' annoyance. As a specific example: on a particular site there were long delays in leaving the project because of a traffic signal favoring a major highway. Intercession with local officials got some relief, but did not completely solve the problem. Even so, knowing that management had tried helped to remove some of the resentment.
- A description of arrangements for such services as buses, car pools, and lost and found.

This list merely suggests topics that might fit newsletters or other means of keeping job personnel informed. Every job will require its own list. The trick is to search for pertinent topics and discuss them fully.

To sum up this brief discussion of telling what is going on on projects: Without a reliable source of information, workers who want to know what is going on with respect to what affects their well-being will fall back on rumor and other sources that are incomplete, inaccurate, and often biased or distorted. Management can, through appropriate media, tell the true story, good or bad. Also, through these media it can give information about activities of interest to workers and thereby demonstrate that the workers are important to the job and as people. Knowledge is a prerequisite to developing interest, cooperation, and commitment.

SCHEMES FOR INITIATING AND HANDLING SUGGESTIONS

Part of all good construction supervisors' jobs, be they project manager, engineer, superintendent, or foreman, is to encourage and utilize suggestions from every possible source. If done poorly, or left to chance, many opportunities may be lost. To be sure that suggestions are encouraged and put to use where appropriate, many industrial firms including a few involved with construction have undertaken formal suggestion programs. Done right, these efforts have paid off by reducing costs, improving quality, and creating better morale. This section summarizes some of these organized efforts. It bears repeating that as with the other procedures discussed here for gaining employee commitment, suggestion systems must have the strong and continuing support of top management, which must see to it that involvement with it is incorporated into the reward system for all levels of the management team.

Data on the value of suggestion systems in construction are fragmentary.

However, the experience of one nonunion firm, Daniel International, is enlightening. It has reported that its profit-sharing suggestion plan, which gives the workers 80 percent of budget savings, increased productivity on its projects by 43 percent.

The first suggestion system on record was established in 1880 at the Yale and Towne Lock Company. Another, set up at Kodak in 1898, continues today. Currently in the United States some 7000 companies have formal programs and there are an equal number of informal ones. An analysis of the 1983 activities of about 700 companies showed savings estimated at $800 million.

A National Association of Suggestion Systems, Chicago, Illinois, founded in 1942, has about 1000 firms as members. Among its other activities, the Association produces a comprehensive guide called the "Key Program," a publication titled "Profile of a Suggestion System Administrator," and another called "Legal Guidelines." These are available for purchase to both Association members and others.

Suggestion systems are not always successful. It has been reported that 90 percent of the attempts to start them have failed, usually because they lacked the continued support of top management.

Figure 11-1 is a flow diagram showing a formal procedure for handling a suggestion made by an employee. Underlying the flowchart is the assumption that there is a suggestion system, it is known to employees, and employees feel the system is impartial and fair and that it is to their advantage to use it. The steps outlined in the figure have been developed to gain the advantages that suggestions can give while mitigating the adverse effects that a poor approach can have. Among the features are:

1 The procedure must be open and free of bias and prejudice.

2 Before suggestions will be considered, they must have substance and be developed to a point that they demonstrate merit and appear to be workable.

3 The procedure must encourage employee and supervisor cooperation but must not permit supervisors to veto submittals because they feel that they are unworkable or because they threaten the supervisor's judgment, status, or ego.

4 Responsible persons must be involved who will handle proposals systematically so that they do not get lost or ignored. They will also see to it that submitters know that their proposals are being considered.

5 Proposals must receive a fair evaluation by a knowledgeable and impartial review group.

6 Regardless of the findings of the review groups, management retains the right to implement or reject proposals. At the same time, management must notify the employee that the proposal has been accepted, modified, or rejected, and explain its actions.

7 There must be a clearly understood, uniform, and acceptable award and recognition plan.

Among the questions to be answered before a suggestion system is undertaken are:

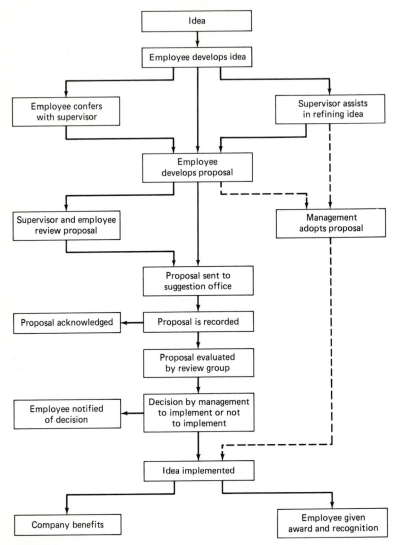

FIGURE 11-1
Flow diagram for handling employee suggestions.

1 Who should set up the program? Should it be undertaken in house or should a consultant be brought in?

2 Who should administer the program? On large projects it probably would be the individuals or groups involved with productivity improvement, news dissemination, or both. Again, it must be emphasized that top management involvement is essential.

3 What will be the composition of the group or groups that evaluate the suggestions? It or they should cut across management levels and include labor representatives. In each instance, there must be demonstrated knowledge and freedom from bias.

4 Who in management should choose which suggestions to implement, notify all concerned parties of the decision, and prepare needed explanations?

5 Who should establish criteria for and choose the levels and kinds of awards? On major projects, these may be a part of some larger scheme which also involves safety, work teams, and other contributions.

Setting the level of monetary awards is particularly difficult. If they are too low or merely token amounts, suggesters and others may feel that management is getting something for nothing and may be turned off. Management or others interested in profits will object if they seem to be too high. One way that has been suggested to overcome this difficulty is to set awards at a percentage of estimated savings, if such an estimate can be made. Surveys reported through the National Association of Suggestion Systems show awards ranging from 10 to 30 percent of evaluated first-year savings, with an average of 17 percent. Another approach is to set a sliding scale which decreases as the estimated savings increase. Whatever the arrangement, it should be settled and publicized when the system is instituted to avoid argument and controversy later.

There is divided opinion about whether supervisors should be eligible to participate in the suggestion system and the awards that may result. The reported majority opinion is that they should. The caution is that this right should be clearly spelled out when procedures are established so that supervisors do not profit from suggestions made by their subordinates and do not exhibit any favoritism.

Formal suggestion programs are more often associated with large construction projects. Even so, on smaller jobs a less formal effort can help to create a favorable job climate and be profitable as well. In planning and executing even simple programs, it is important to follow the steps shown in Fig. 11-1 and to consider the suggestions and cautions given above.

In proposing that carefully planned suggestion programs are valuable, the authors are not discounting the informal approaches that exist on every well-run construction project. This approach has been suggested by the shortcut procedure outlined on the right side of Fig. 11-1. Skilled foremen always draw on the experience and ingenuity of their individual craftsmen and crews in planning and carrying out tasks. Similarly, capable higher-level managers enlist the skills and interests of everyone on the job. Seldom is financial reward expected or offered, since contributing ideas is considered to be a part of the team effort that gets projects built. Here, recognition and thanks, and inclusion of such informal suggestions in the formal award system, if there is one, is necessary to keep the process going. But it is unwise to overlook the value of a carefully planned suggestion system. It seems to be the best way to focus attention on finding means for improvement rather than leaving their finding to chance.

DEALING WITH COMPLAINTS

Traditionally, construction managers and workers alike have dealt with complaints about working conditions and other matters in a direct face-to-face manner. This approach stemmed from the tradition of informal and often casual hiring, layoff, quitting, or firing that has been prevalent in the industry, at least until recently. If managers failed to listen and do something about complaints, workers would quit and move to another project if work were available there. Otherwise, they would tolerate the condition but be as uncooperative as possible, up to the point of risking being laid off. It is probable that as theory X thinkers (see discussion in Chap. 10), managers who failed to respond to justified complaints were also insensitive to the fact that dissatisfied workers were probably less productive or that it was costly to hire and train replacements if they quit. Today, employers are required by law or by union agreements if they exist, to rectify complaints in a number of areas. Among the recourses available to workers is to report to the authorities or union any unsafe or unsanitary conditions or failure on the part of the employer to supply certain amenities such as cooled drinking water or suitable toilets. In other instances, charges about neglect of seniority or other union rules or of discrimination in hiring or layoff because of race or sex may be brought. Failure to deal sensibly with such complaints, even if management considers them trivial or without foundation, can consume precious time and energy, divert attention from important management tasks, and often create adversarial relations and a poor job climate.

On projects where good relations exist between management and workers, many complaints can be handled by the immediate supervisor. If not, it is probably wise to move them further up the management ladder informally. This does not necessarily mean that higher management will overrule the immediate supervisor. Far from it. Rather, particularly if tempers are high or saving face is involved, a cooler head may be able to reconcile the differences. An example is the sensible provision mentioned earlier that superintendents rather than foremen handle layoffs or firings.

Today in some other industries and occasionally on construction projects, the suggestion system includes a mechanism for registering general complaints that are not dealt with by informal actions. Such a procedure would roughly follow the steps outlined above for handling suggestions. As with suggestions, the aim is to have the issue considered openly and fairly in a factual way by impartial parties and with built-in feedback to those concerned.

AWARDS AND PRIZES

Strategies for getting commitments to be productive and to join in team efforts are discussed in detail in Chap. 10. That discussion emphasizes that construction work when well planned and executed can offer tremendous satisfactions which in turn can bring higher levels of performance. There are other motivators also, among them money (see Chap. 10) and recognition, which if used properly can be effective in creating a job climate conducive to productivity-improvement.

Some awards may be distributed freely as visual reminders. These, usually of nominal value, permit employees to identify with the project and to realize that they are a part of it. Included are items such as hard-hat stickers, pocket protectors, caps with the project logo on them, and special stickers to go on hard hats to advertise some achievement such as an excellent safety record. They may serve a useful purpose as an additional morale builder when employee relations are good but are probably of little value alone.

Relatively inexpensive attention-getting awards are sometimes included in a company's employee-relations program. The range of choices among them is enormous. Common items are watches, pocket or hunting knives, shirts carrying the project or company logo, and a variety of sports items. Compiling lists of possible awards is not difficult; there are several firms that will furnish catalogs showing a wide variety of choices. The trick is to find those that are particularly appropriate.

Many construction companies give awards in money or its equivalent such as savings bonds to recognize special contributions or achievements. The reasons for such awards can cover a wide spectrum including acknowledgement of helpful suggestions, good productivity team performance, or an excellent safety record. Awards can be given either to individuals or groups. No industry consensus exists about the amount of such awards. One approach is to give token amounts such as from $25 to $100. These serve primarily as attention-getters. The difficulty is that if savings from the idea are substantial, the award may create bad rather than good feelings. As indicated in the discussion of suggestion systems, awards based on savings might fall in the range of from 10 to 30 percent of estimated first-year savings, as set by an unbiased individual or group. The more usual practice is to give set amounts determined by top project or home-office management. Records show that other industries have made cash awards as high as $100,000 while the top amounts reported for construction were in the $5000 range.

Nonmonetary awards can run the gamut from items of high value to those that are largely symbolic. Expensive awards might include merchandise such as television sets, radios, or other electronic gear or items for recreational or home use. Food certificates or trading stamps are often given in nontrivial amounts. Always the aim is to select items that will be desirable and get attention. For foremen and superintendents, in particular, some firms organize dinner parties that include spouses at a first-rate restaurant to celebrate significant events such as project milestones or achievements of some other kind. Expense-paid travel is another possibility. It is significant that many of these awards are designed to enlist the interest and good will of spouses as well as the individual.

For either monetary or nonmonetary awards, a policy for setting the kinds and sizes of awards, the purpose they are to serve, and a distribution plan and mechanism must be established. This must be viewed as workable and fair by line management and the work force and must be established in advance. Furthermore, once begun, such programs must be continued. If these conditions present difficulties, it may be better to look for other ways to create and maintain a good job climate.

Awards are intended not only to reward people for work well done or other past accomplishments but also to encourage the recipients and others to adopt the same attitudes and behavior. Therefore, awards must be publicized through whatever media are available. The project or company newsletter, if one exists, offers an excellent means but should not be the only one employed. Personal presentations by someone in higher management, accompanied by publicity, is one approach. Under other circumstances, the superintendent or foreman could make the award before an assembled crew or crews. The latter approach is sometimes more sensible, because it makes the awardee's peers and buddies aware of the recognition. As in so many other approaches discussed in this chapter, there is no one right way. The person who selects the awards and arranges for their presentation must be sensitive to the particular situation, choose a course of action after consulting with others, observe the results, and, if necessary, make appropriate changes.

RECREATIONAL AND RELATED ACTIVITIES

Employee leisure-time activities on large and long or continuing construction projects have much in common with those in manufacturing or other industrial enterprises. Under these circumstances, it is common as a part of the employee-relations function to develop a variety of recreational activities such as softball teams or bowling leagues. On multishift jobs, these sometimes can be scheduled to fit the most-effective work, leisure, and rest cycle, as described in Chap. 9.

No attempt will be made here to describe these activities in detail. However, the fact that they are being done in construction, with equipment, scheduling, and publicity arranged by the contractor's staff, should be noted.

SITE VISITS AND OTHER INFORMATIONAL OR
FAMILY ACTIVITIES

On very large construction projects, workers often have no concept of what the project is all about, since they are hired on and assigned to crews that are working on some small segment of the job. Under such circumstances, they often develop the feeling that they are isolated and merely cogs in a wheel, and that their efforts produce no tangible results. This is particularly true when for whatever reason, they must appear to be busy or when they must tear out old work and redo it. At times the workers take these feelings home with them, and their families sense their dissatisfaction and are also disturbed. Under such circumstances it is to be expected that a worker's attitude toward the job and toward the contractor will be adversely affected.

As discussed above, information about the project is often provided through newsletters and similar outlets. A very useful addition to or substitute for these has been to arrange site visits for the workers and their families. These are often coupled with employer-sponsored picnics or other activities. However carried out, they can make the project come alive for both the workers and their families and can do much to overcome the feeling of alienation and a sense of being neglected that workers can develop.

A different approach may be appropriate for the local contractor, large or small, who maintains a relatively stable work force. In such instances, picnics or other activities for families or social events for workers and spouses sponsored by the contractor can be valuable morale and interest builders.

Examples can be cited where on long projects in remote locations, the contractor has established trailer parks and even schools so that employees may have a normal family life. These do much to provide a stable and productive work force and head off the Monday morning hangovers and losses in efficiency common on such projects in the past. In addition, they eliminate the hazards associated with the long treks from site to town and return.

There is no one right way for contractors either on large projects or in localized situations to promote a feeling of belonging in the worker and his family. Strategies will be different in each case. But firms that have planned and executed such activities have found them to be well worth the effort.

FORMAL EDUCATION AND RESEARCH PROGRAMS FOR CONSTRUCTION[1]

Before World War II, most owners and managers of construction firms and their field supervisors came up through the trades. There were, of course, a few in these roles with engineering, business, or other professional backgrounds. Although there was some talk about establishing formal programs in construction management in engineering or architecture at the college level, only a handful developed. Rather, it was expected that the necessary skills would be learned on the job while working under the direction of already proficient managers. This approach was in stark contrast to preparation for careers in design, which usually called for a professional degree in engineering or architecture and, beginning in the 1920s, professional registration.

Serious attention to construction management as a professional career began to develop at the college level in the late 1940s. Today, about 200 schools have programs in place pointing their graduates toward construction management as a career. These programs are diverse in content and depth of attention to construction

---

[1] Underlying data for this brief discussion, modified to fit the current situation, come primarily from *Business Roundtable Cost Effectiveness Study Report A-5,* June 1982; C. H. Oglesby, *Journal of the Construction Division,* ASCE, vol. 108, no. CO 4, December 1982, pp. 605–616; A. Warszawski, *Journal of Construction Engineering and Management,* ASCE, vol. 110, no. 3, September 1984, pp. 297–310; C. E. Haltenhoff, ibid., vol. 112, no. 2, June 1986, pp. 153–162; and G. D. Oberlender and R. K. Hughes, ibid., vol. 113, no. 1, March 1987, pp. 17–26. B. S. Ledbetter, ibid., vol. 111, no. 1, March 1985, pp. 41–52, reviews the development of construction education soon after World War II.

topics. Each has developed from and is fitted to the school's particular situation. However, almost all fall roughly into five categories, as follows:

- Graduate-level programs in civil engineering departments, leading to an advanced degree, usually master of science. A relatively few doctoral (or engineer) degrees are also given. No accreditation procedure exists. Professional affiliation is usually with the American Society of Civil Engineers (ASCE).
- Four-year programs leading to the bachelor of science degree in civil engineering. Most of these are accredited by the Accrediting Board for Engineering and Technology (ABET). Professional affiliation is with ASCE.
- Four-year programs in construction technology. Some of these are accredited by the Accrediting Board for Engineering and Technology (ABET).
- Four-year programs in construction management, distributed among a variety of university schools or departments, such as, architecture, civil engineering, and business. Some of these schools offer advanced degrees. Accreditation is granted by the American Council for Construction Education. Professional affiliation for these schools is with the Associated Schools of Construction.
- Two-year programs, primarily in community colleges and technical schools offering courses dealing with construction topics. Many of their graduates enter the construction industry.

No attempt will be made here to explore the differences among and merits of the various educational approaches to construction management. Each has found its niche and its academic and industry proponents. The key point is that construction management, as is the case with engineering, law, and medicine, among other fields, is now a profession and is recognized as such.

Industry liaison with construction educators can take several forms, among them:

- Committees of professional societies such as ASCE, in which academics who teach construction often play a major role.
- Committees of trade organizations such as the Associated General Contractors, often with academics as members. Some of these associations sponsor and support student chapters and provide financial aid.

As a whole, financial support from industry for construction education and research has been small. Reported scholastic aid is about $30 per enrolled student per year. Industry-financed research on construction topics is about $1000 per faculty member per year. The entire industry contribution to research has been estimated to be about 0.1 percent of the industry's annual volume, as contrasted, for example, with 5 percent from the electronics industry to schools serving it. In sum, until recently the construction industry's support for education and research has been minimal when contrasted with other industries.

There are signs that industry's support for university education and research in construction management is increasing, partially as the result of the

Business Roundtable's Cost Effectiveness Study. The formation and support by the industry of the Construction Industry Institute, which has developed and now administers a program of university-based research, offers strong support for this new trend.

The construction industry is also becoming aware that in-service education for today's construction managers is essential. It is taking several forms. One is company- or industry-sponsored seminars on special topics. These are supported by a variety of publications, books, pamphlets, slide and cassette presentations, and videotapes suitable for groups or individual study. Another approach involves university-conducted full-time short courses several weeks in length, patterned after the executive training programs so popular in other industries. The Construction Industry Institute is setting up or cooperating in seminars where its researchers present their findings to groups of construction managers, owner's representatives, and academics. There are even instances of simultaneous nationwide presentations sponsored by trade associations employing communication by satellite. In sum, one can expect a continuing expansion of activities aimed at bringing modern management methods and research findings directly to construction industry managers. All these activities provide added evidence that the industry is changing and becoming more professional.

TRAINING FOR CRAFTSMEN AND FIELD SUPERVISION

From earliest times, craftsmen have learned their trades by working under their seniors. More recently, usually through the joint sponsorship of government, industry associations, and unions of the various construction crafts, formal apprenticeship programs have been established. These combine on-the-job work experience with classroom instruction. Beginning wages are generally substantially lower than those of journeymen, but increase with time. Candidates become journeymen on completion of the prescribed program of study and work.

Craftsman training programs are currently in a state of flux. The craft unions continue the apprentice-training approaches which has served well in the past. Open-shop contractors are not as fortunate, but are working through their associations to establish a government-contractor framework for such programs. In addition, some large open-shop firms have undertaken their own in-house training. Others rely on finding union members who will work non-union. Still others are attempting to diminish the importance of craftsmanship by designing work assignments which limit the level of skill needed. Given these developments, the future of craftsman training is uncertain. Overlying all this is the concern that the supply of skilled craftsmen may diminish, and the entire industry will suffer in the long haul.

In the past few years, the construction industry has recognized that being a construction supervisor calls for skills not learned in the traditional apprenticeship programs or on the job. To correct this deficiency, several of the trade associations and some large owners or constructors have developed supervisory and foreman training programs and materials to support them, fitted to the crafts they employ. For example, the Associated General Contractors has developed and now markets instructional materials for both a supervisory training program (labeled STP) and a shorter overview of the same material called the foreman course.

The series of short courses that make up the AGC supervisory training program (STP) provide for field supervisors courses and material to support them on a number of topics. These are directed to the management level. Topics covered are: (1) leadership and motivation, (2) oral and written communication, (3) problem solving and decision making, (4) using contract documents, (5) planning and scheduling, (6) cost awareness and production control, (7) project safety and loss prevention, (8) project management, (9) construction law: changes, claims, and negotiations, and (10) productivity-improvement.

Being aware that materials such as these and many others for training field supervision are available is important. Beyond that, the fact that they are available offers conclusive evidence that industry leaders recognize that construction supervisors at the field level need training in subjects other than their crafts.

SUMMARY

This chapter has briefly described several stratagems that have been employed on projects to develop understanding and acceptance of productivity as a goal for everyone involved on construction projects. Some of these approaches, as outlined, may seem to be appropriate only for very large projects of long duration. Even so, the techniques contain underlying principles that are sound and that ingenious construction people who will make the effort can adapt to their projects. In addition, the chapter describes briefly the formal educational efforts that have been developed or are developing to improve the abilities of both managers and field personnel.

CHAPTER 12

SAFETY AND ENVIRONMENTAL HEALTH IN THE CONSTRUCTION INDUSTRY

INTRODUCTION[1]

Accident statistics (see Tables 12-1 and 12-2) show that on the average, construction is among our most hazardous industries. In 1983, some 2000 workers were killed and 200,000 suffered disabling injuries. The hazards in construction work are also demonstrated by the fact that the industry's work force, representing 5 percent of the nation's total, account for 18 percent of the deaths and 11 percent of the disabling injuries for all industry. Also, construction's fatality rate is 3.4 times that of the all-industry average; only agriculture, lumbering, and mining have more recordable cases per employee. Furthermore, construction accidents tend to be more severe, as indicated by the disabling injury rate, which is double that of the all-industry average.

It is easy to cite reasons why construction appears to be so dangerous. Among them are that it may involve large, heavy materials and equipment and often requires work at heights, in excavations, underground, and in other high-hazard locations. Furthermore, activities and membership are ever-changing for crews performing the work. Generally the workplace is outdoors and variable, so that means of avoiding or protecting against hazards must change con-

---

[1] R. E. Levitt and N. M. Samelson, *Construction Safety Management,* McGraw-Hill, New York, 1987, addresses the topic of construction safety in terms of what management can and should do about it, and the gains that will result. The underlying concepts and statistics given in that book also appear here, but are treated here in less detail and from a somewhat different perspective. The Levitt and Samelson book should be required reading for construction managers at all levels.

TABLE 12-1

INDUSTRIAL WORK ACCIDENTS IN THE UNITED STATES, 1983

Industry group	Number of workers, 000	Deaths		Disabling injuries*	
		Number	Rate†	Number	Rate†
Construction	5,400	2,000	37	200,000	3,700
All industries	100,100	11,300	11	1,900,000	1,900
Trade	22,700	1,200	5	300,000	1,300
Service	27,600	1,800	7	350,000	1,300
Manufacturing	19,000	1,200	6	330,000	2,800
Government	15,700	1,500	10	280,000	1,800
Transportation and public utilities	5,300	1,300	25	140,000	2,600
Agriculture	3,400	1,800	52	180,000	5,300
Mining, quarrying	1,000	500	50	40,000	4,000

*A disabling injury prevents a worker from returning to work on the next scheduled work period.
†Annual rate of death or disabling injury per 100,000 workers.
Source: Accident Facts, National Safety Council, Chicago, Ill., 1984 Edition.

TABLE 12-2

RECORDABLE OCCUPATIONAL INJURY, ILLNESS, AND LOST-WORKDAY INCIDENCE
RATES FOR ALL PRIVATE INDUSTRY AND FOR CONSTRUCTION IN THE UNITED STATES

Industry division	Incidence Rates*		
	Total recordable cases†	Lost workday cases	Lost workdays
Private sector, 1984, all	8.0	3.7	63.4
Private sector, 1981, all	8.3	3.8	61.7
Construction			
All, 1984	15.5	6.9	128.1
All, 1981	14.9	6.3	112.1
General building, 1984	15.4	6.9	121.3
Heavy, 1984	14.9	6.4	131.7
Special trades, 1984	15.8	7.1	130.1
Roofing and sheet metal, 1981	20.7	—	—
Mining, 1984, all	9.7	5.3	160.2

*Incident rate is per 200,000 worker-hours (100 full-time employees working 40 hours per week for 50 weeks).
†Recordable cases are those involving an occupational injury or illness, including death. They do *not* include first-aid cases which involve one-time treatment and subsequent observation of minor cuts, scratches, burns, or splinters which do not require medical care, even though such treatment is provided by physicians or registered medical personnel.
Source: Monthly Labor Review, U.S. Department of Labor, March 1986.

stantly, precluding the development of set patterns. Under such circumstances, it is easy to understand why a mental attitude has developed that tends to take injuries and deaths on construction sites for granted and assumes that they are inevitable. This view was typified in the past by such statements as construction takes a life for each million dollars or each mile of tunnel.

The justifications for construction's high average accident rates as cited above and shown in Tables 12-1 and 12-2 can and should be challenged. There are data from certain companies which show conclusively that work can be carried out largely free of accidents and impairments to health. However, this requires that accident prevention become a major focus of attention at all levels of owner and contractor management and of workers as well. Data supporting this statement and procedures and practices by which this can be done are the subjects of this chapter.

ACCIDENTS DEFINED

In dealing with the problems of safety and environmental health, there is a tendency to look only at situations in which humans suffer physical injuries or occupational diseases because they become the immediate focus of attention and are recorded and counted. However, the overall impact of accidents on productivity extends much further. It helps then to define an accident as "any avoidable action by personnel or any failure of equipment, tools, or other devices that interrupts production and has the potential of injuring people or damaging property." Under this broad definition, accidents include about ten times as many instances of work interruption as are reported in accident statistics. Unfortunately, the near misses escape the attention of all but the most observant and are seldom, if ever, reported.

Accidents as defined here include not only direct physical injury to persons or damage to property, but also the short- and long-term effects of other exposures on construction sites that affect a worker's health or physical well-being. Examples of such short-term effects might include lung, skin, and eye irritation from exposure to fumes and noxious agents or illnesses associated with these exposures. Long-term consequences include maladies resulting from exposure over time to, for example, asbestos or radiation, or hearing loss from working in or around noisy equipment or in noisy environments. Certainly, happenings like these are as much an industry concern as are injuries and must be considered as such.[2]

AN APPROACH TO THE CONSTRUCTION
ACCIDENT PROBLEM[3]

Many books, pamphlets, and handouts as well as laws and regulations on the site-specific side of construction safety and health are readily available. Some

[2] The distinction between safety and environmental health has grown out of differences in the backgrounds of the professionals in the two fields: safety is from engineering; environmental health from the medical sciences. As far as construction is concerned, the distinction is unimportant; all job-related injuries and illnesses, both long- and short-term, are costly and result in other undesirable effects that are to be avoided.

[3] For another view of the factors that bear on construction accidents, see A. Laufer, *Construction Management and Economics,* vol. 5, no. 1, Spring 1987, pp. 73–90.

of these will be discussed and others given as references in this chapter. However, the emphasis here will be on the management and human aspects—on what owner and contractor managers at all levels and the workers themselves can and should do to carry on construction operations without the unplanned and unexpected events called accidents. This approach is based on the premise that accidents have no single cause but result from a chain of circumstances or events and that in more than 90 percent of the cases, the chain includes human error or human negligence. It follows that giving attention to human contributions to the accident chain should offer a primary but not the sole approach to preventing accidents.

This chapter treats accidents and occupational health in construction under the following topics:

- The chain-of-events and fail-safe approaches to accidents
- Statistics on industry-wide, off-site, and safe contractor accident records
- Direct and indirect contractor accident costs
- Construction accident costs to others
- The effects of attitudes and behavior of management and workers on construction safety
- Site-specific aspects of construction safety
- Environmental-health hazards in construction
- Workable approaches to construction safety
- The union's role in construction safety
- Contractor trade-association's role in construction safety
- Government's role in construction safety
- Owners' roles in construction safety
- Engineers' and architects' roles in construction safety.

THE CHAIN-OF-EVENTS AND FAIL-SAFE CONCEPTS

As mentioned, incidents that result in job interruptions, injuries, or health hazards to workers in construction are the culmination of a series of events in sequence. If one of these events can be prevented or its effects mitigated, the outcome will be different. This way of analyzing accidents or of coming up with ways to prevent them is very useful but too often overlooked. The following real-life example is cited to illustrate this point. Every construction worker and manager can offer others.

The newspaper report of a construction fatality was as follows:

40 Foot Fall Kills Bridge Worker

A 50-year old carpenter fell 40 feet to his death yesterday as he worked on the _____ Bridge. The victim was moving supplies on a scaffold on a bridge pier when his glove caught on a nail and he was thrown off balance. He is survived by his wife and three children.

This unnecessary death resulted in a deprived and desolated family, disrupted job operations, lowered morale on the part of workers, a loss of money to the contractor, and the loss of a skilled worker to society. It could have been avoided.

The tabulation which follows is a very simple chain-of-events analysis of the accident, listing some questions that if considered by construction managers and workers could prevent similar happenings.

Event	Questions
Work located 40 feet high	Were there provisions for safe access? Was worker wearing safety belt that was tied off? Did scaffold have suitable railings and a toe board? Was there a safety net below? Had someone in management alerted him to the hazards?
Worker wearing gloves	Were gloves the right kind, in good condition, and should he have been wearing them?
Glove caught on nail	Why a protruding nail?
Worker thrown off balance and fell	Why, when in a hazardous situation, did worker make an impulsive move that threw him off balance? Why didn't handrail or toe board stop his fall?
After-injury care	Were first-aid, medical care, or ambulance available in case worker survived fall?

This brief and incomplete account tries to make the chain concept clear. Also it indicates events where the chain leading to the accident might have been broken had either management or the worker observed one among several sensible practices. This does not mean that reliance should be placed on only one safety measure. In both planning and executing safety precautions, every step should be taken to make things safe as well as productive.

Setting up fail-safe conditions is another way to prevent accidents. If something goes wrong, all actions and movements stop. A classic example is the dead-man's throttle on railroad locomotives. Pressure is required to keep the throttle open; if that pressure is relaxed, power is cut off and the locomotive stops and the brakes set. An opposite and very unsafe situation is the case with automobiles. If the driver's foot slips from the brake, it hits the accelerator and the vehicle lunges forward. Unfortunately, this situation will probably not be corrected, since it would involve breaking the long-established habits of over 100 million drivers in the United States alone.

Certain construction safety rules violate the fail-safe principle. For example, automatic backup horns on trucks and other equipment are not fail-safe, because they do not alert the driver if there is a person in the vehicle's path. Rather, they usually lead the operators to assume that the way will be cleared behind them. In this instance, the fail-safe approach would require positive evidence that the path is clear. Usually this is done by a signaler, but it might also be accomplished with some kind of electronic detector which scans the area. Signalers should use the standard form of hand signal, which indicates to the truck operator that all is clear by continuing hand movements. A nonmoving hand says to stop. This moving-hand approach is employed in other recommended signaling practices such as that directing crane operators. The moving hand requires positive action on the part of the signaler. A nonmoving hand can be not only a signal to stop, but it can also be a way to force a stop when the signaler's attention is directed elsewhere, so that it is fail-safe.

STATISTICS ON INDUSTRY-WIDE, ON-SITE, OFF-SITE, AND SAFE-CONTRACTOR ACCIDENT RECORDS

Industry-wide Data

Workers in all industries, including construction, get hurt on the job or suffer work-related illnesses. Table 12-3 classifies these, for California, by how they occurred. It gives data for all industry, for construction in general, and for five from the numerous specific categories of construction. Under each of these types it gives the percent of the total that was found. Of particular note, based on Table 12-3, are the following:

• Injury patterns for construction are consistent with those for industry in general. The principal exception is that of falls from elevations, where exposure is high in many forms of construction. These patterns also have been consistent over time. Data for earlier years, not given here, show almost the same distributions among the various physical factors.

• Four out of ten disabilities are associated with sprains, strains, and other injuries associated with stressing the body beyond sensible limits.

• Almost one-third of the occurrences involve injuries to the back or trunk of the body. This targets an area of accident prevention so important that it is given detailed attention later in this chapter.

• Workers' compensation claims for occupational injuries were roughly 10 times as frequent as claims for occupational illnesses. This should not be taken as a cause for neglecting illnesses. Rather, as demonstrated by the recently published data on long-term effects of exposure to asbestos, it should be recognized that claims for occupational illnesses may be sleeping giants.

Generalized data such as that in Table 12-3 may not apply directly to specific activities on individual construction sites, but they can indicate that efforts to improve safety must apply to the types of occurrences peculiar to the task being undertaken.

Although Table 12-3 shows that the kinds of industrial injuries workers suffer fall into a common pattern, it must not be forgotten that the rates at which these injuries occur vary substantially among industries. This is demonstrated by Table 12-1, which gives comparisons based on two common measures of accident frequency, namely, deaths and injury rates. These data demonstrate clearly that compared with other industries, construction ranks high in both.

TABLE 12-3
PHYSICAL FACTORS CONTRIBUTING TO DISABLING WORK INJURIES AND ILLNESSES UNDER WORKERS' COMPENSATION IN CALIFORNIA 1985

Industry	Total number*†	Over exertion	Struck by or against	Falls		Bodily reaction‡	Caught in or between	Rubbed or abraded	Contact with radiations, caustics, etc.	Highway motor vehicle accidents	Non-highway motor vehicle and equipment	Other or not stated
				On same level	From elevation							
All industry	406,683	28	26	9	6	8	5	3	4	3	1	7
All construction	46,565	24	32	6	13	7	3	5	3	2	1	4
General building contractors	12,152	22	39	6	14	6	3	4	2	1	0	3
Heavy construction contractors	5,760	23	30	6	9	7	6	4	5	3	2	5
All special trade contractors	28,653	24	30	6	15	7	3	5	3	2	0	5
Plumbing, heating, air conditioning	5,686	24	31	6	13	7	4	6	3	2	0	4
Roofing and sheet metal	3,090	21	24	6	20	6	3	2	2	2	0	14§

Percent of total

*Of this total 83 percent were classed as occupational injuries, 7 percent as occupational illnesses, and 10 percent were not classified.
†Of these 43 percent involved sprains and strains. Body parts affected were: back and spine, 23 percent; trunk, 10 percent; upper extremities, 25 percent; and lower extremities, 20 percent.
‡Involves slips, walking, running, reaching, turning
§Almost all these were from burns and scalds.
Source: California Work Injuries and Illnesses, California Department of Industrial Relations, 1986.

Table 12-2 offers another set of accident data stated in terms of incident rates, which apply to all industries, all construction and its main subdivisions, and other high-hazard industries.[4] It supports the conclusions based on Table 12-1 that construction as an industry has high accident frequency and severity. The extremely high rate for roofing and sheet metal explains why the Occupational Safety and Health Administration (OSHA) targeted them for attention in its New Directions program.

Data such as those in Tables 12-1 and 12-2 indicate that work in construction is dangerous compared with other occupations. Even so, some ask, "Is the record so bad?" and use the statistics to support a rare-event myth and attitude which makes accident prevention difficult. Supporters of this myth argue that only 1 in 2700 construction workers is killed each year or that only 1 in 27 has a lost-time injury. The follow-up is then that such rare events will not happen to me or my crew or crews, so why worry!

That the rare-event myth affects behavior is demonstrated well by the reluctance of motorists to buckle their seat belts. Even though it requires little effort, before seat-belt laws were enacted (and only in some states), only about 11 percent of the motorists in the United States buckled up, even though most motorists knew that when belts were worn, fatalities in crashes were reduced by one-half. Somehow, such drivers assumed that they were immune; accidents happened infrequently and only to others.

To overcome this rare-events thinking and keep attention focused on accident prevention, the statistics must be looked at differently. First of all, for the average construction worker, given the incidence rate of 15.5 injuries per 100 workers (see Table 12-2), there is one chance in seven that during a year a construction worker may suffer an injury which can in certain cases and largely by chance result in permanent injury or even death. For a foreman, supervising seven workers, there may be one such happening per year; a superintendent supervising twelve such crews might have one injury each month. Looking at this in terms of interruptions that upset the work, the foreman could have at least one incident a month and the superintendent one every two working days. None of these can be classed as rare events. But as will be pointed out in more detail later, if work is carefully planned and executed, with safety and productivity being a principal concern, very few of these unhappy and largely unforeseen things need happen.

Far more meaningful to contractors and buyers of construction are statistics showing the rates that the industry pays for several forms of insurance that they carry to pay for anticipated accidents. The most costly of these is workers' compensation, although as discussed below, liability (third-party) insurance is becoming an increasingly large factor. A sampling from recent manual

[4] Accident data such as those cited here are routinely gathered and distributed by the Bureau of Labor Statistics of the U.S. Department of Labor. The National Safety Council also gathers and publicizes accident data. Reporting agencies may employ different data bases, use somewhat different definitions, and report in different units. However, all are good indicators of the accident situation.

TABLE 12-4
MANUAL (BOOK) COMPENSATION RATES FOR SEVERAL STATES AND CLASSES OF CONSTRUCTION WORK*

Classification of work	California	Connecticut	Florida	Kansas	Texas
Carpentry, one- and two-family dwellings	12.91	11.16	13.30	7.15	11.23
Carpentry, general	13.91	14.44	15.72	6.93	10.85
Concrete work, bridges, culverts	15.91	21.29	16.91	6.42	—
Concrete work, dwellings, one- and two-family	6.50	12.74	9.63	6.16	—
Electric wiring, inside	5.26	4.05	6.83	3.05	5.39
Excavation, earth, and not otherwise classified	6.33	6.72	12.15	4.07	—
Lathing	8.64	9.30	10.27	6.65	4.89
Masonry	11.66	16.89	12.02	5.70	7.89
Painting and decorating	11.75	10.88	10.97	6.78	7.73
Pile driving	23.33	19.75	22.18	16.58	17.02
Plumbing	7.59	6.75	8.57	3.77	5.90
Roofing	30.33	31.69	30.77	11.49	23.25
Steel erection, structural	14.11	42.88	22.84	8.81	18.43
Steel erection, not otherwise classified	19.44	37.61	21.69	16.35	11.87
Timekeepers, watchmen	7.82	6.05	5.65	2.96	6.46
Wrecking	23.33	19.95	22.18	16.58	17.02

*Rates are for $100 of straight-time payroll.
Source: Abstracted from *Engineering News-Record*, Sept. 18, 1986.

(book) rates for workers' compensation for construction activities is shown in Table 12-4. These rates are industry-wide weighted averages determined by representative states for specific trades or work classifications. These are compiled annually by special rating agencies.[5] Manual rates are stated in dollars per $100 of payroll, usually excluding premium pay for overtime. They include the overhead costs of administering the program, which roughly equals benefits paid out. Workers' compensation pays for medical,

[5] Manual rates and experience modification ratings (to be discussed later) for 40 states are determined by the National Council on Compensation Insurance. Each among four states (California, Delaware, New Jersey, and Pennsylvania) has its own rating bureau. In these states, private insurance carriers usually compete with state-run companies. In the remaining six states (Nevada, North Dakota, Ohio, Washington, West Virginia, and Wyoming) workers' compensation insurance must be carried with a monopolistic fund of the state government. Each of these likewise has its own rating system.

TABLE 12-5
COST OF WORKERS' COMPENSATION INSURANCE FOR "TYPICAL" INDUSTRIAL PROJECTS
(per $100 million of total project cost, using California WC type-of-work rates)

Type of facility	Contractor position				
	Lower decile, $	Lower quartile, $	Mean, $	Higher quartile, $	Higher decile, $
Paint plant	480,000	620,000	760,000	920,000	1,060,000
Paper mill	530,000	680,000	835,000	1,010,000	1,116,000
Chemical plant	575,000	747,500	920,000	1,115,000	1,276,500
Power plant (coal)	1,320,000	1,700,000	2,100,000	2,540,000	2,920,000
Power plant (nuclear)	1,360,000	1,750,000	2,160,000	2,610,000	3,000,000

Source: "Improving Construction Safety Performance," *Business Roundtable Cost Effectiveness Study Report A-3,* New York, N.Y., 1982.

surgical, and hospital care for injured workers, weekly or permanent disability allowances, and death benefits when workers are killed.

The manual rates given for workers' compensation shown in Table 12-4 clearly reflect the hazardous nature of construction. Furthermore, they demonstrate the difference in hazard levels among the numerous tasks that make up the construction process. Also, since they are based on claims paid earlier, they reflect the state-by-state differences in the level of benefits paid to workers or their survivors (see below).

An examination of the rates for workers' compensation shown in Table 12-4 makes it abundantly clear that injuries to workers add substantially to the costs of construction. For example, it can be seen that in California there is an add-on of 14 percent to carpenters' wages. Even in Kansas, accidents add over 7 percent. In the California case, assuming that a carpenter's wages make up one-third of the total cost of construction, accidents add another 5 percent to project costs. For the more hazardous tasks, these costs loom large; that of roofing is an example. In California, labor costs are increased by one-third to pay the costs of anticipated injuries to roofers.

Table 12-5 gives data on the average and spread in workers' compensation costs for typical industrial projects.[6] For average situations, at the time the data were gathered (about 1980), these ranged from 1 to 2 percent of total project costs and are thereby in the range of 3 to 6 percent of labor costs for this kind of work.

In addition to the legally required workers' compensation insurance, all public contracts and many among private parties require the contractor to carry bodily injury and property-damage insurance. Whereas workers' compensation is for the benefit of the contractor's employees, these coverages pro-

[6] For the research underlying the report, see R. E. Levitt, H. W. Parker, and N. M. Samelson, *Technical Report No. 260,* Department of Civil Engineering, Stanford University, Stanford, Calif., August 1981.

tect the contractor and indirectly the owner against their possible liability to third parties. In some states, subcontractors' employees may also file claims against the general contractor or owner. Suggested rates for a few typical construction activities are shown in Table 12-6. It can be seen that these rates clearly reflect potential exposure to hazards.

Construction, along with other industries and private individuals as well, has been caught up in recent years in a rapid escalation of the cost of third-party insurances. For example, the bodily-injury rate for general carpentry shown in Table 12-6 is three times as great as it was in 1982; the property-damage rate has increased five times over. In a few cases the current rates are ten times what they were in 1982. The result is that the cost of these insurances is no longer a minor but a major element in the cost of construction.

As noted in Table 12-6, the rates given are for a coverage of $1 million and the costs of lower coverages are proportionally less. The dilemma is that $1 million is considered by many to be a minimum sensible protection level. There are many instances today where small contractors and designers, if they have a choice, may forego these coverages entirely or, in the language of the trade, "go bare."

Also, as noted in Table 12-6, rates for third-party insurances are negotiable with the insurance carrier. Contractors who can show a good record in the areas covered by the insurance may be able to get more favorable terms. Thus, there may be a savings in these costs attributable to a contractor's safety record. Unfortunately, these savings will be difficult to quantify.

TABLE 12-6
SUGGESTED RATES FOR BODILY INJURY AND PROPERTY DAMAGE INSURANCE FOR 1987 FOR CERTAIN CONSTRUCTION CLASSIFICATIONS*

Work classification	Bodily injury	Property damage
Carpentry—two family residence	4.277†	4.380†
Carpentry, general	6.580	4.330
Concrete construction, forms, scaffolds, etc.	9.541	4.672
Electrical wiring, inside	3.290	3.796
Excavation, not otherwise classified	7.896	4.672
Lathing	1.645	1.110
Masonry	3.619	2.336
Painting and decorating	2.632	8.760
Pile driving for foundations	11.515	7.884
Plumbing and piping	3.948	11.096
Roofing	7.896	33.580
Steel erection, bridges	25.620	15.470
Timekeepers	9.541	3.504
Wrecking, buildings or structures	44.835	‡

*Rates are for $100 of straight-time payroll. The limit of coverage is $1,000,000. Rates are negotiable between contractor and insurance carrier.
†For a limit of coverage of $25,000, the bodily injury rate is 1.300 and that for property damage is 1.500.
‡Rates are set for each project following a risk appraisal.

Contractors may be required or choose to carry several other kinds of insurance to protect owners and themselves. Since most of these are less directly related to accidents, they are not discussed here.

Whether construction accidents are evaluated by comparison with other industries (Tables 12-1 to 12-3) or in terms of their costs (Table 12-4 to 12-6), the situation is clear; construction as usually carried out is a high-hazard industry and the costs of the resulting accidents are high. But, as will be demonstrated below, this need not be the case.

Off-site Accidents to Construction Workers

Primary attention to accidents for the construction industry has been focused on on-site situations. However, off-site accidents should be a cause for serious concern. The statistics are startling: for example, in 1983 there were 91,000 accidental deaths in the United States, distributed among motor vehicles, 44,600; work, 11,300; home, 20,000; and public activities, 19,500. On-site industrial accidents accounted for 11,300, or 12.5 percent of the total. Of these, construction contributed 2000, or 18 percent of the industrial total. Of the 8,800,000 disabling injuries, motor vehicles accounted for 1,600,000, work 1,900,000, home 3,000,000, and public activities 2,500,000. Included in the work total is construction's 200,000.

If only motor-vehicle and home accidents are spread uniformly over the population, and construction workers (excluding their families) constitute 2.5 percent of the total population, one can surmise that their off-site accidental deaths would total about 1500, including automobile and home-accident totals of 1000 and 500, respectively. For them also, disabling injuries would total about 115,000, distributed as 40,000 for automobile accidents and 75,000 at home, respectively. Summing up, it can be seen that off-site deaths to construction workers from these two causes occur about 75 percent as frequently as do on-site ones, and off-site disabling injuries come to about 60 percent of on-site ones.

These statistics make clear that effective accident-prevention programs for construction must reach beyond the job site and focus attention on other accidents, particularly those that involve motor vehicles and home incidents. Regardless of where the accident occurs, the workers are lost to the project. Although contractors do not pay directly for these accidents, they suffer many indirect penalties, including the short-run dislocations that accompany the workers' absence and the temporary or permanent loss of the workers' abilities and skills.

Records from Safer Construction Operations

As has been indicated, Tables 12-1, 12-2, and 12-4 to 12-6 portray average accident experiences or costs for typical situations in the construction industry.

However, averages such as these are deceiving because they lump together the records of firms of all sizes, all kinds of construction, and safe and less-safe contractors.

A commonly used measure of the disparities among individual companies is the experience modification factor. This is a multiplier based on past-accident experience and weighted by the volume of work done (see below for more detail) that sets the actual workers' compensation payments that individual contractors must make. Figure 12-1 is a plot of the distribution of experience modification factors (EMF) among two groupings of contractors, small and large. The factors range from about 0.5 to something over 1.9, indicating that the least-safe contractor pays a workers' compensation bill that is four times higher than that of the most-safe contractor. For reasons to be explained later, the lower scale on Fig. 12-1 (which shows the experience modification factor minus 0.5) is a better indicator of the actual spread in accident experience among contractors.

Other statistics offer far more dramatic illustrations of the difference in construction accident experience. The extreme is a record set by E. I. DuPont de Nemours on a construction site in Deepwater, New Jersey. In this instance 9.3 million hours were worked without a lost-time accident. Similarly, a group of contractors on the Satsop nuclear power plant project in Washington accumulated 4 million hours without a lost-time occurrence. By comparison, on the basis of data from Table 12-2, the industry average expectation would have shown 300 lost-workday accidents for the DuPont project and 130 for the Satsop project. More recently, *Engineering News-Record,* June 27, 1985, reported that on the Cape Fear plant in North Carolina (DuPont, owner; Daniel Construction Company, contractor) the lost-workday record was 0.4, as con-

FIGURE 12-1
Frequency distribution of California Experience-Modification Factors (EMF) for construction companies. 1983–1985 data from California Rating Bureau. Ordinates show percentages accumulated in the 0.05 EMF range below the plotted point.

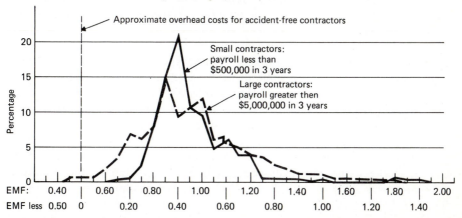

trasted with the industry average of 6.8. These three are not isolated instances; the DuPont safety record has been almost equally outstanding on many sites over many years as have those of many other companies. All of them use some or all of the procedures detailed later in this chapter at all levels in their organizations. Smaller companies with exceptionally low experience modification factors such as 0.5 (see Fig. 12-1) will testify to the attention they give to safety, with programs fitted to their particular situations. These examples make clear that most construction accidents need not happen. The fact that they persist indicates that for a variety of reasons, construction companies, their management, and their workers have not included accident prevention among their priorities.

Statistics on accidents introduce size of work force as another variable in safer-versus-less-safe performance. The pattern is the same for all industry as is that for construction. In 1981, firms having 19 or fewer employees had an incident rate (total recordable cases) of 8.8 per 200,000 worker-hours as contrasted with 14.9 for the entire industry. This rate rose to 16.6 for 20 to 49 employees, reached a peak of 21.2 for 100 to 249 employees, and fell to 7.8 for more than 2500 employees. Probably there are several factors that account for this variation. From research findings, to be discussed later, a supposition could be that, with small firms, there is a buddy system among long-time employees. With very large firms, there is a strong likelihood that safety receives strong management support. In contrast, moderate-sized firms may have neither.

CONTRACTOR DIRECT AND INDIRECT ACCIDENT COSTS

Measurable (Direct) Costs

As already indicated, individual contractors are required by workers' compensation laws to pay the costs of their anticipated accidents along with overhead costs and profit for their insurance carriers. Also, specifications may require or contractors may elect to carry other selected insurances. It is essential that contractors and buyers of construction understand the method for computing these projected costs of future accidents; for this reason, the preceding discussion will be expanded here.

Workers' Compensation Insurance Workers' compensation is social legislation intended to distribute some of the costs of accidents between workers and employers. Before it was enacted, the level of compensation given to injured workers was set by employers. If workers were dissatisfied they could sue under common law for damages for negligence. The usual claim charged an unsafe workplace or inadequate supervision. In turn, employers could defend themselves by claiming that workers knew their jobs were hazardous, that the workers were negligent, or that fellow workers and not the employers were responsible for the accidents. Under workers' compensation all this is changed. In return for settlements stipulated by law (see below), workers are deprived of their right to sue employers and employers cannot block these awards with the legal defenses available to them earlier.

The rules under which workers' compensation is administered and the levels of compensation are set by the legislatures of each individual state. In some states, benefits to a worker may be reduced from those usually stipulated if a reviewing agency finds the individual guilty of contributory negligence. Benefits may be withheld entirely if the injury is self-inflicted. Also, the reviewing agency may assess added charges against employers who are found to be guilty of negligence or serious and willful misconduct.

Workers' compensation operates generally as follows: Each year, insurance companies are required to provide the rating bureaus (mentioned earlier) with data on the total straight-time payroll and workers' compensation payments made for medical and hospital care and for disabilities and death benefits for each work classification. To this is added an allowance for insurance company expenses and profit which is roughly equal to the payments to cover the total direct outlays. The ratio between the sum of these costs and allowances and the straight-time payroll for each classification gives its manual (book) rate. This rate is stated in terms of dollars per $100 of straight-time payroll. Typical examples for a few of the common work classifications for five states distributed over the country are given in Table 12-4. An examination of these rates will show that they vary widely both with respect to geographic location and the hazards of the various classes of work, as measured by claims paid in the past. In all cases, they show that workers' compensation is a substantial element in the cost of construction.

Workers' compensation rates have increased somewhat over time, as the costs of medical and nursing care and disability benefits have gone up. But these increases are far less dramatic than those of the third-party insurances discussed later.

Rating bureaus also collect data on worker compensation payments made for individual contractors by their insurance companies. These provide the fundamental basis for setting individual experience modification ratings. This basic rating is adjusted by a formula which attempts to equitably recognize the rare-event aspect of accidents. The first adjustment is to average the contractor's record for the past 3 years (excluding the latest year) rather than take a single year's record. The second is to establish a risk pool which spreads the payments for high-cost or catastrophic accidents among employers rather than assigning all of them to the firm which incurred them. This distributive effect, which applies primarily to smaller employers, is illustrated by Fig. 12-1, which shows a bunching of the experience modification rate for the low-payroll employers. In contrast, the broader spread in experience modification rates for larger contractors shows that they receive far less cushioning against costly accidents through the risk pool.

The experience modification factor for contractors starting in business is set at 1.0. This is adjusted as the firm accumulates an accident record until after 4 years, the factor is based on an average for 3 years, with the past immediate year excluded. Firms with very bad records and high experience modification factors sometimes reorganize and start over under a different name, which allows them to return to the 1.0 factor.[7]

To show the effects on costs of a contractor's experience modification factor, consider a common work situation, carpentry, with a manual rate, say, of 10 percent. Assume further that contractor A has had an excellent safety program and thus an experience modification factor of 0.65. Contractor B, however, pays little attention to accident prevention and has an experience modification factor of 1.3. It follows that contractor A's workers' compensation premiums are (0.10)(0.65), or 6.5 percent of payroll, while contractor B's are (0.10)(1.3), or 13 percent of payroll. If labor is one-third of total job costs, contractor B is at a disadvantage of 2.2 percent of total job costs. This is often more than the profit margin on highly competitive projects. Furthermore, these statistics do not take into account the very considerable hidden costs of accidents, to be discussed later. For higher-hazard classifications, such as pile driving, roofing, steel erection, and wrecking, these differences can be much more; for others, such as electrical contractors, they are less. In any event, all contractors can easily compute their competitive edge or penalty from data on their experience modification factor obtainable from their insurance broker or carrier.

Table 12-5 summarizes the costs of workers' compensation insurance per $100 million of project cost for different kinds of industrial projects for a range of experience modification factors. It shows what workers' compensation costs are, and also illustrates how project costs are affected by the contractor's safety record.

Subcontractors' employees who are injured receive workers' compensation benefits through the subcontractors' insurance coverage. However, in some states they are permitted to sue the general contractor, owner, architect, or engineer, or all of them, for negligence and collect damages from them if they win their case. In other states, the courts have denied employees the right to sue on the basis that workers' compensation satisfies all claims. If such claims are allowed, general contractors who supply scaffolding, hoists, and services such as cleanup to subcontractors are particularly vulnerable.

Bodily Injury and Property Damage (Third-Party) Insurance   Third parties not employed by contractors or subcontractors who are injured or incur property damage through acts of contractors or their employees are also legally entitled to compensation. In contrast to workers' compensation where the level of compensation for injuries is set by administrative or legislative action, third par-

---

[7] This description of the costs of workers' compensation insurance although oversimplified does report its essence. In some states, large employers who can prove their ability to pay all reasonable claims are permitted to self-insure. They usually carry the costs of smaller incidents but often take out insurance against catastrophes. In all such cases, however, workers have the same level of protection offered by workers' compensation and lose their right to sue the employer for negligence.

ties must prove that the injury or damage resulted from negligence of the contractors or their employees. Settlements for such claims have no limits; the amounts are set by agreement between the parties or in the courts. Often the settlements are extremely large.[8]

The bodily injury and property-damage insurances (see Table 12-5 for a few typical rates) and certain other insurance policies are designed to protect contractors, their employees, and sometimes owners and designers against these claims. Until recently, the costs of these third-party insurances were relatively low, but under the insurance crisis of the past several years, they have skyrocketed and have become a substantial part of a contractor's costs. For example, rates for bodily injury coverage for carpentry on two-family residences increased by 300 percent between 1983 and 1987; those for property damage for the same classification grew by roughly 400 percent. These increases were not uniform among classifications. Some went up as much as 1000 percent. There is a feeling that these rates will probably remain stable at the present level but probably will not go down.

It is sometimes argued that the cost of these third-party insurances are unaffected by a contractor's safety record. However, this is not necessarily the case. As indicated in Table 12-6, the rates are negotiable between the contractor and the insurance carrier, and an excellent safety record will argue strongly for rate reduction.

Other Direct Costs of Accidents   Among other direct accident costs are damage to plant, equipment, or the uncompleted facility. Damage may result from weather, floods, or other natural occurrences or from carelessness of workers. Contractors often carry insurance to protect them against these and other potential losses; thus they are a part of the accident picture and must be included in summing up accident costs. Some of them are obvious and their costs easily determined as when, for example, a crane boom buckles or the blade on a pusher tractor damages a scraper tire and repairs must be made. Others are less obvious, such as the results from someone's failure to properly lubricate equipment. The damage costs usually differ from costs incurred

---

[8] These subcontractor-employee and third-party lawsuits stem from the obligation imposed by common law and federal and state statutes and regulations (OSHA, in particular, for construction) that employees and third parties be protected from hazards. To institute such suits, the parties alleging the injuries must prove tort liability to the satisfaction of the appropriate court. This involves demonstrating that (1) the defendant had a duty to protect the injured party; (2) there was an obligation to perform that duty; (3) an injury occurred; and (4) there was a causal connection between the failure to perform the duty and the injury.

In recent years, courts have increasingly permitted such lawsuits to be filed. Furthermore, they have reduced the importance that the defense can attach to contributory negligence on the part of the employee. Once a lawsuit is permitted, it is to be expected that trials will be held before juries and that the juries are likely to make generous awards. Defendants or their insurance companies, of course, have the option of making out-of-court settlements rather than face the expense of trials and the potentially very large court settlements. Often they take this recourse, considering it as the easier way out.

when workers are injured in that damage costs are reflected in a contractor's profits or losses far more quickly and if insured against are paid for under different policies.

Contractor Indirect Costs of Construction Accidents

Premiums that contractors pay for workers' compensation, bodily injury and property damage to third parties, and other insurances represent direct money outlays to pay for accidents. However, these costs although large are but a small portion of the total, since the indirect, unquantifiable, or hidden costs or ripple effects have been found to be far higher. These can vary greatly from accident to accident. Off-the-cuff estimates have often been cited which place them at 4 to 7 times the direct costs. In a study for the Business Roundtable, Levitt, Parker, and Samelson (op. cit.) queried a small number of large safety-conscious owners and contractors, asking them to evaluate as best they could the indirect and hidden costs of accidents for no-lost-time and lost-time categories for a range of direct costs. Average results from that survey are given in Table 12-7. It can be seen that for the average no-lost-time and less costly lost-time accidents, the indirect-to-direct cost ratio is 4.1 to 1 or higher. For more severe accidents, the average ratio is much lower. However, one instance was reported where the ratio was 24 to 1.

Among the indirect cost items considered in preparing these data were:

1 Transportation of injured worker (if not in direct costs)
2 Wages paid to injured workers and others while work was interrupted

TABLE 12-7
RATIOS BETWEEN INDIRECT AND DIRECT ACCIDENT COSTS

Range of benefits paid	Number of cases	Average benefits paid (direct costs)	Average indirect cost*	Average ratio indirect cost: benefits paid
No lost time:				
$0 to 199	13	$ 125	$ 530	4.2
200 to 399	7	250	1,275	5.1
400 plus	4	940	4,740	5.0
Lost time:				
$0 to 2999	9	869	3,600	4.1
3000 to 4999	8	3,947	6,100	1.6
5000 to 9999	4	6,602	7,900	1.2
10000 plus	4	17,137	19,640	1.1

*The indirect costs reported here are for a limited set of factors. In most instances, there are others that are not included.

Source: "Improving Construction Safety Performance," *Business Roundtable Cost Effectiveness Study Report A-3,* New York, N.Y., 1982.

3 Overtime costs necessitated by accident

4 Costs of losses in crew efficiency

5 Costs to break in or teach a replacement worker

6 Extra wage costs to rehabilitate injured workers or to allow them to perform at reduced efficiency

7 Costs to clean up, repair, or replace damage

8 Costs involved in delaying and rescheduling the work

9 Wages of supervisors while attention is given to the accident

10 Costs of safety and clerical personnel in recording and investigating the accident

Since only the costs of these items were tabulated in deriving the data for Table 12-7, it is probable that many indirect cost items were omitted. For example, adverse publicity and attitudes of workers, supervisors, and the public often result.[9] Increasingly, supervisors are called into time-consuming investigations and must spend many hours in the courts. In numerous cases, managers or supervisors have been faced with misdemeanor or even manslaughter charges. Even if cleared of such criminal accusations, their time is consumed and their attention is diverted from managing the work.

The ratios of indirect to direct costs stated here were developed for situations where a worker was injured and the accident reported. It must not be forgotten that as noted earlier, for every time that a worker is injured, there are possibly 10 work interruptions caused by human carelessness or an avoidable failure of a worker or machine. If the costs of these were added in, the ratios of indirect to direct accident costs would be far higher than those cited here.

Total Accident Costs to Contractors

The average cost of a construction accident leading to a disabling injury in 1985 has been reported to be $18,650—an alarming amount. But this single number says little about the charges for accidents that individual contractors can anticipate in the years ahead. Rather, they must be estimated. This can be done with reasonable accuracy by combining data given in Tables 12-4 and 12-6 with a contractor's experience modification factor and estimates of the indirect-to-direct accident cost ratio. These are substantial, even for the safest contractors, and high for the less safe ones. It is therefore difficult to understand why many contractors are ignorant of or are indifferent to their magnitude. Explanations, to be discussed in more detail later, include a lack of understanding of insurance and how it works, and the fact that in contrast to other costs, those for accidents are not immediately apparent, but are time bombs with delayed fuses.

---

[9] Interviews with construction superintendents, foremen, and crews (see below) revealed repeatedly that an employer's good or bad reputation and behavior with respect to safety was a principal reason for a worker's selecting on the one hand or leaving a given employer on the other.

Recently, in an attempt to bring the costs of individual construction accidents into clearer focus, M. R. Robinson collected data on the average direct costs of different classes of accidents.[10] These were obtained from insurance company records and verified with contractor safety administrators. His findings are summarized in Table 12-8. Results, stated in labor hours, are separated into the part of the body injured and the form of injury and whether it was a no-lost-time or lost-time accident.[11] In developing these costs Robinson tripled the directly measurable labor hours charged against an accident. The two items added, each in equal amount, were (1) insurance company overhead and profit required to make the workers' compensation system work and (2) indirect (hidden) costs. As explained earlier, there is substantial evidence that the amount calculated is too low.

The great value of having data such as those in Table 12-8 is that it makes possible an assessment of probable cost immediately after an accident occurs. Without this information, costs come to light, if at all, only as they are charged 2, 3, and 4 years later against contractors' experience modification factors and finally their workers' compensation insurance bills. With the tabulated information, management has a powerful and immediate tool for bringing accident costs and accident prevention forcibly to the attention of all. Some companies immediately charge the estimated costs against the individual project and the management involved. To do so, Table 12-8 can be used as follows: Assume that a worker fractures an arm. If the individual is able to return to work the next shift, which is very doubtful, the probable cost of the injury is 75 labor hours. At $20 per hour, including fringe benefits, this amounts to $1500. If the return to work is delayed, the probable cost is $20 times 450 hours, or $9000. As another example, the probable cost of a sprained back, without lost time would be ($20)(150), or $3000; with lost time it would be ($20)(750), or $15,000. For a contractor with a 5 percent profit margin, $300,000 worth of profitable work will be necessary to make up the loss from this single occurrence.

The tabular values of Table 12-8 may not be suitable for all contractors. However, they have a solid base, having been compiled from the records of 12,000 accidents that occurred over a 3-year period. For the company that wishes to introduce accidents directly into its cost system, the table provides a starting point which can be changed as a company accumulates its own records.

Contractors who have studied the data in Table 12-8 and compared them with their costs usually find those in the table to be too low. This results in part because in recent years, direct medical costs have risen far faster than inflation. Their records also show that the multiplier of 3 to incorporate all insurance company and indirect costs is also too low. This difficulty can be easily rectified. If, for example, a contractor or owner feels that indirect costs are four

---

[10] See "Accident Cost Accounting as a Means of Improving Construction Safety," *Technical Report 242,* Department of Civil Engineering, Stanford University, Stanford, Calif., August 1979.

[11] Lost-time accidents are ones where workers are not able to return to work on their next regularly scheduled shift.

TABLE 12-8
COSTS OF TYPICAL NO-LOST-TIME AND LOST-TIME CONSTRUCTION ACCIDENTS*

Body part	Amputation	Strain, sprain, crush, mash, smash	Fracture	Cut, puncture, laceration	Burn	Bruise abrasion	Other
Head, Face	NA	NA	50 600	20 220	25 550	20 75	25 450
Eye(s)	3,300 (1) 18,000 (2)	NA	NA	20 220	15 380	20 75	20 380
Neck and shoulder	NA	25 520	110 600	20 220	25 380	20 150	20 520
Arm(s) and elbow(s)	14,000 (1) 18,000 (2)	25 300	75 450	20 220	20 380	20 220	20 450
Wrist(s) and hand	3,800 (1) 18,000 (2)	20 190	50 650	20 220	25 380	20 300	25 450
Thumb(s) and finger(s)	600 ea. up to 2,800	20 190	25 380	20 220	15 380	15 220	15 380
Back	NA	150 750	NA 7,400	20 220	25 550	25 380	25 750
Chest and lower trunk	NA	35 300	NA	20 600	25 380	20 220	20 680
Ribs	NA	25 75	35 300	NA	25 380	25 220	20 680
Hip	NA	NA 260	35 900	15 220	25 380	25 380	35 300
Leg(s) and knee(s)	6,600 (1) 21,000 (2)	30 300	35 1,100	20 220	25 380	20 220	20 600
Foot (feet), ankle(s)	3,300 (1) 6,600 (2)	20 190	35 650	15 190	20 220	20 75	25 150
Toe(s)	520 ea. up to 3,000	20 110	15 190	20 220	25 150	15 75	20 150
Hernia rupture							15 600
Heart attack							2,200
Hearing loss							750
Death							6,600

*For further information, see M. R. Robinson, "Accident Cost Accounting as a Means of Improving Construction Safety," *Technical Report 242*, Department of Civil Engineering, Stanford University, Stanford, Calif., 1979.

Numbers on the left side are for no-lost-time accidents; numbers to right are for lost-time accidents. NA = not applicable. All costs are in equivalent labor-hours; to obtain a dollar value multiply by job labor rate including fringe benefits.

These values are set at three times that of the actual claims paid. The total is the sum of (1) claims paid, (2) overhead costs of the insurance carrier, set equal to claims paid, and (3) estimated indirect costs to the contractor, set equal to claims paid.

times direct ones, a multiplier of 6/3 can be applied to the tabular values. Then the cost of a lost-time back injury, cited above as $15,000, would increase to $30,000.

If one computes the costs of injuries to workers using workers' compensation manual rates adjusted by an experience modification factor and adds in estimates of indirect costs directly from data such as those in Table 12-8, one thing is clear. It costs money to have workers hurt. Not to be overlooked also are costs incurred from damages to machinery, equipment, and facilities, which are not included in the figures given above.

CONSTRUCTION ACCIDENT COSTS TO OTHERS

The earlier discussion has indicated how certain direct costs of construction accidents are charged to contractors, mainly though insurance. Also it has demonstrated that contractors also pay substantial indirect costs. In turn, some or all of these costs are passed on to owners in the prices they pay for completed projects. But this is not all; workers and their dependents and society in general also carry a share of the costs of construction accidents.

Distribution of costs among all those affected by construction accidents depends to a degree on legislation, such as that for workers' compensation and in part on judgments by administrative agencies and the courts. But seldom do any of those affected come out ahead. In other words, accidents are a lose-lose game.

All too often, it is forgotten that workers and their families and dependents carry a substantial share of the costs of accidents. Workers' compensation usually covers most of the costs of medical, surgical, and hospital care. However, the weekly disability and injury compensation and survivor benefits as set by the laws of each state (see Table 12-9) generally are far less than would be received if the individual continued to work. Also, the period of employment has often not been long enough to receive the maximum, low as it is.

Comparisons between wages foregone and the maximum benefits given in Table 12-9 show that the weekly compensation which an idled worker receives or the death benefits for the family are less than even laborers would be paid if they worked 2 days a week. And, as mentioned, in many instances actual benefits can be far less than the maximum and fall to almost nothing. Furthermore, money in any sum cannot measure the many other personal and family consequences of injury or death. Among these are the pain, shock, and stress at the time of and after the injury and the short- and long-term disruptions and deprivations in the lives of the injured workers and their families. Workers who appreciate these facts try to avoid getting hurt, because they know that the personal costs to them are very great. But all too often, job confusions and pressures or personal problems take over or obscure a worker's consciousness of job hazards.

It is sometimes charged that workers feign injuries in order to draw workers' compensation benefits when they are faced with layoffs at the conclusion

TABLE 12-9
A SAMPLING OF BENEFITS FOR INJURED WORKERS OR THEIR SURVIVORS UNDER
WORKERS' COMPENSATION

Recipients of benefits	Benefits—by State				
	California	Connecticut	Florida	Kansas	Texas
Injured worker					
Maximum benefit for total disability, $ per week	224	397 to 595.50	315	239	217
Minimum benefit for total disability, $ per week	112	79.40	20	25	37
Maximum limit to benefit, as percent of wage	66.67	66.67	66.67	66.67	66.67
Survivors, if worker is killed					
Spouse, maximum benefit, $ per week	224	397	315	239	217
Spouse, minimum benefit, $ per week	112	79.40	20	25	37
Spouse plus minor child, maximum $ per week	224	397	315	339	217
Spouse, maximum, as percent of wages	66.67	66.67	50	66.67	66.67
Compensation for permanent impairment					
Loss of hand	43540	100044	*	35850	32550
Loss of thumb	9595	37715	*	14340	13020
Loss of foot	33740	74636	*	29875	27125
Loss of one eye	21105	93295	*	21680	21700
Loss of hearing, both ears	43540	61932	*	26290	32550

*No specific schedule; based on an assessment of degree of impairment from accident and loss of earnings.
Source: Abstracted and simplified from *Analysis of Workers' Compensation,* U.S. Chamber of Commerce, 1986.

of the job or near the onset of winter. Back injuries at such times are particularly suspect because they are difficult to disprove. The frequency of such behavior is unknown but may well be overstated. Even so, contractors are on the lookout for such malingerers and avoid hiring them if possible. There may even be a grapevine which possibly illegally passes the information to others. Furthermore, certain criteria and rules have been developed to discourage this practice. For example, in some states, benefits are reduced if it can be shown that the worker's negligence contributed to the accident, and benefits may be denied if the injury is self-inflicted.

Families and relatives along with society in general absorbs many of the costs of construction accidents. First of all, any costs of medical care or caring for disabled workers and their families not paid for by employers or workers are passed on to the families or to society in general through various public agencies and charities. In addition, society is temporarily deprived of the services of workers while they are recovering from injuries or impaired health and permanently loses them altogether from workers who are killed or do not recover their full capabilities. The costs to society are largely unmeasured but must be substantial.

THE EFFECTS OF ATTITUDES AND BEHAVIORS OF MANAGEMENT AND WORKERS ON CONSTRUCTION SAFETY

Research studies based on interviews have been conducted to determine the effect of the attitudes and behaviors of three levels of managers and of crews on accidents. Along with the interviews, information about the accident records of companies, supervisors, or crews was obtained. Findings were tabulated and analyzed by computer and only results that were statistically significant were reported. In sum, the studies clearly show that attention to safety reduces accidents significantly. Further details of the findings of each of these studies follow.

Earlier in this chapter the point was made that jobsite accidents are often viewed as rare events, which makes it difficult to keep the attention of managers and workers continually on safe practices. One common feature in all the successful approaches described below is that they maintain a constant and persistent focus on accident prevention.

Top Management and Safety

It has been proven conclusively that companies whose top management (1) knows in detail about the safety records of their company's field supervisors, (2) talks about these records individually with them, and (3) considers safety records when making salary and promotional decisions have far better accident records than those who do not. The research findings supporting this conclusion are summarized in Fig. 12-2. Data for it came from in-depth interviews with top-level construction managers, with the interviewer probing to find their attitudes and practices regarding safety. The experience modification factor minus 50, a reasonably accurate measure of accident experience, is the method used for comparisons. This deduction of 50 was selected since even accident-free companies must pay about 50 percent of the manual rate for insurance coverage.

Looking in detail at Fig. 12-2, beginning at the top, one can see that management in 23 companies was involved and that their accident records were

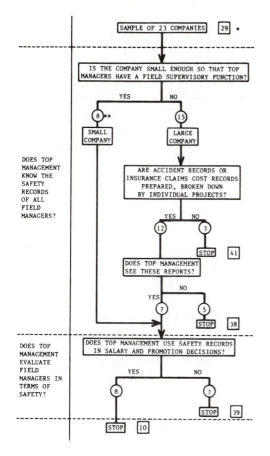

FIGURE 12-2
Effects of top management's knowledge
and evaluations on accident costs.
(*Source: R. E. Levitt and H. W. Parker,*
Journal of the Construction Division, *ASCE,*
vol. 102, no. CO3, September 1976).

* Figures in squares represent the accident claims cost
portion of the experience modification rate (i.e., the
rate less 50 points) averaged for all companies at that
branch on the flow chart.

** Figure in circles represent the number of companies on
each branch.

considerably better than average as shown by an overall experience modification factor of 79 (the 29 in the box in Fig. 12-2 plus 50). In the middle of the
figure, for large companies, are answers to the question, "Does top management know the safety records of field managers?" This is approached in two
steps as follows: (1) "Are accident records prepared and broken down by individual projects" and (2) "Does top management see these records?" Negative answers to one or both of these questions brought the higher-than-average
experience modification factors of 91 (41 + 50) and 88 (38 + 50). For the remaining large and all the small companies, the key question was, "Does top
management use safety records in salary and promotion decisions?" If the answer is yes, the experience modification factor drops to 60 (10 + 50); if no, the

factor is 89 (39 + 50). The message is clear; knowing accident records is not enough; a strong follow-through and reward system that is obvious to all lower-level management is essential.

Data from Fig. 12-2 can also be used to give a rough measure of overall accident experience. It appears that the safer group, with an experience modification rate of 60 (10 + 50), had about one-fourth as many accidents as the other group for which experience modification is 90 (40 + 50). Remembering that accidents lead to many indirect costs as well as to the direct ones measured by workers' compensation, the reductions gained can be impressive indeed.

Given that by their attitudes and actions top managers have done much to ensure safer construction practices on their projects, what specific steps did the research find that, based on statistical analysis, have improved accident records?

A short listing is as follows:

1 Know the accident records of individual supervisors and make it clear that those records play an important part in determining salaries and promotions.

2 Talk safety along with cost and schedules in every contact with field personnel at all levels. Also, expect them to respond with comments that show their knowledge of safety problems.

3 Incorporate the cost of accidents into the cost-reporting system.

4 Pay all costs of the safety program and safety equipment and supplies through a company-wide overhead account. (Spending money is a sure way to make clear that safety is important.)

5 Demand that safety be an integral part in planning every operation.

6 Establish a program to train both managers and workers, and particularly new workers, in safe practices.

The findings summarized here demonstrate conclusively that top managers' attitudes and actions can dramatically affect their company's safety records. But this comes about only through strong and continuous attention.[12]

Middle Management and Safety

Middle managers are individuals who act as top company representatives resident on construction job sites. They may hold job titles such as project manager, construction manager, or job superintendent. Regardless of their title, middle managers provide a vital link in the communications between the home office and the job site and very often have broad management responsibility

---

[12] This discussion is based on a paper by R. E. Levitt and H. W. Parker, *Journal of the Construction Division of ASCE,* vol. 102, no. C03, September 1976, which in turn is abstracted from R. E. Levitt, "The Effect of Top Management on Safety in Construction," *Technical Report 196,* Department of Civil Engineering, Stanford University, Stanford, Calif., July 1975.

for projects. Because of the unique nature of the position, middle managers play a significant role in safety on the jobs for which they have management responsibility.

Research to distinguish among middle managers whose projects were safer or less safe paralleled the interview method employed in looking at top managers and foremen (see below). Answers were solicited from about 90 middle managers to a series of questions about how each interviewee managed the work, without focusing on safety as such. Because few companies had kept accident records by individual managers, it was necessary to rely on the managers' replies about their safety records, an approach which proved to be satisfactory. Higher-level company officials provided subjective ratings on the managers' abilities to meet costs and schedules, to run safe jobs, and to work under extreme pressures. Added appraisals covered administrative ability and independence (acting outside company guidelines for the good of the project). The assembled data were analyzed and a series of correlations was developed; from these, certain characteristics of safer middle managers were recognized, such as that safer managers:

1 Rate higher with their superiors in their ability to meet costs and schedules.
2 Do not apply unrealistically high cost and schedule pressures on foremen and workers. They avoid crisis situations on the job through effective preplanning.
3 Actively support job safety by including safety as a part of job planning and by giving positive support for tool-box meetings.
4 Accept responsibility for eliminating unsafe conditions and activities from the job.
5 Work hard on safety even if they have little support from higher levels of management.
6 Establish a good working relationship with foremen and workers on the job by adopting the following measures, among others:
   a Provide foremen some latitude in the selection of workers in their crews. Crews made up of workers who get along with the foremen and one another are safer and more productive.
   b Insist that foremen orient new workers to the job and introduce them to other personnel.
   c Are aware of worker-foreman conflicts and take steps to ensure that workers are treated fairly and are not dismissed capriciously. Workers preoccupied by a conflict with a foreman are much more susceptible to accidents.

It is particularly difficult for many construction executives to accept the notion that the better and safer middle managers are people-oriented rather than being hard-driving and hard-nosed autocrats with an "I give the orders, you follow them" attitude. Furthermore, they simply cannot accept the premise that putting strong pressures for production on foremen and crews has been found to be both ineffective and unsafe. But the research on middle management proves that they have the wrong outlook.[13]

Foremen and Safety

As indicated earlier, foremen directly supervise individual workers and crews and are responsible for turning plans and the directives of higher management into completed structures and plants. One important aspect of good foremanship is careful attention to safety. Research confirmed by field observations has made it clear that more-safety-conscious foremen operate differently from less-safety-conscious ones.

The research relied on in-depth interviews with a number of foremen and on data provided by their superintendent's and from their company's records. Statistically significant differences in foreman practices were found which suggested that they could be grouped as follows:

1 High safety, high productivity
2 Low safety, high productivity
3 High safety, low productivity
4 Low safety, low productivity

It is the mode of operation of the first group and how it differs from that of those in the other groups that is of interest here. It includes the following approaches:

1 Assess all workers new to the crew by asking about previous work experience, and—without prying—learn something about their personal situations. Then orient them by showing them around the job and making them feel at ease. How this is done will depend on the work assignment, but the rule is always keep a close watch on the new workers. For example, if they are to work closely with others, be sure that they are integrated into and accepted by the group. If, on the other hand, they are to work more or less alone, as with scraper operators in a spread, bring them into the operation only after carefully testing their skills and knowing that they understand all procedures thoroughly. That orientation of and attention to new workers is important, particularly in construction, is supported by other research, as shown by Fig. 12-3. From data illustrated by this figure it can be seen that for construction:

- 24 percent of the total accidents to workers occur in the first month of employment
- 36 percent occur in the first 3 months of employment
- 46 percent occur in the first 6 months of employment

2 Make instructional and tool-box safety meetings specific and relevant to the task at hand. Use these meetings to give and discuss assignments and their

---

[13] For details of this study see Jimmie Hinze, "The Effects of Middle Management on Safety in Construction," *Technical Report 209,* Department of Civil Engineering, Stanford University, Stanford, Calif., June 1976.

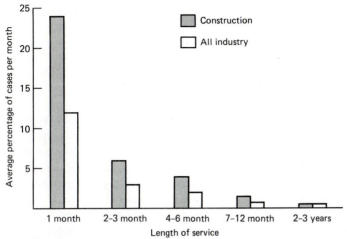

FIGURE 12-3
Work injuries and illnesses per month by length of service—for all
industries and for construction. [*Source: N. Root and M. Hoffer,* Labor
Monthly Review, *January 1979. (Data from the State of Maryland, 1976.)*]

safety implications. Hold additional meetings whenever a crew's task or exposure changes.

3 Take steps immediately to eliminate unsafe work conditions, practices, or
behaviors of the crew. Use these instances as opportunities to stress the
foreman's personal safety goals and to make clear that unsafe practices will
not be tolerated.

4 Keep unnecessary pressures for production off the crew and its individual
members. When there are productivity problems, discuss the difficulties that
curtail production and look for solutions rather than pushing crew members to
work faster—an action which leads to unsafe acts.

In addition to these findings, which identify the behavior of safe-productive
foremen as a class, the interviews uncovered facts about individuals who in
many years as foremen had had no accidents in their crews. A few of their
practices can be phrased as the following suggestions:

1 Know your workers as individuals and keep an eye on them if they do not
seem up to par. In such cases, watch them carefully and alert the crew to do
likewise. In particularly bad situations, send them home or even dismiss them
after consulting with the superintendent.

2 Keep your cool rather than lash out at individual workers or the group in
general if performance is unsatisfactory. Attack the problem, not the worker.

3 Seek out or be available to employers who emphasize safety.

4 Think and talk "us" rather than "me" on matters concerning productivity and safety.

5 Keep an eye out for freeloaders who may feign injury to get easy work or to qualify for workers' compensation.

6 Tone down or get rid of the macho workers who see taking risks as manly or smart.[14]

Crews and Safety

Many construction operations involve crews rather than individuals working alone. Some crews may involve only two people, such as a carpenter, pipefitter, or welder and a helper, who receive only occasional attention from a foreman. On the other hand, there may be a dozen or more workers in a crew, closely directed by a working or nonworking foreman. Regardless of crew size or attention from supervisors, research on crews doing work closely paralleling construction has found that there are good correlations between their attitudes, behavior, and cohesiveness and their accident records.

In attempting to study how crews themselves can influence safety, researchers met an impasse; there were no accident records for construction crews of similar characteristics doing comparable work over substantial lengths of time. This paucity of information is common in research on construction; few records are available. Moreover, it is argued that the personnel of construction crews changes so often that crew influences on safety can sometimes not be separated out as a variable. This argument may be partially true, but many instances can be cited of crews whose personnel changed little over substantial periods or where key personnel remained. In any event, the researchers sought and found accident records in a situation which closely paralleled construction, that of crews doing heavy repetitive work in an industrial environment.

In most respects the crews covered by the research were very similar to those in construction. The work was dangerous but so highly variable that it could not be standardized, which precluded the use of standard safety procedures or the installation of permanent barriers or guards. Often it was in elevated locations or confined quarters and involved handling heavy materials and large equipment. Often several crews were working at once in the same area. The principal differences from construction were, first, the work was for a large organization in a single facility; and second, the first line supervisors did not belong to the workers' union. In evaluating the results, these differences should be weighed but should not greatly diminish the importance of the findings.

Data about crew attitudes and behavior were obtained through freewheeling group interviews with 18 individual crews. These were held on company time, after first gaining approval from company safety managers and union representatives. Supervisors were not present during the interviews. It was also agreed in advance that individual and crew responses would be kept confidential and that neither the company nor the departments involved would be identified.

---

[14] For added detail, see N. M. Samelson, "The Effect of Foremen on Safety in Construction," *Technical Report 219,* Department of Civil Engineering, Stanford University, Stanford, Calif., June 1977.

Fifty specific questions dealing with the work situation, each of which called for open-ended answers from group members, were employed in the interviews. Subjects covered were contact and help among crew members; methods for handling new members; work pace and pressures; job interest, attention level, and feelings of accomplishment; support from managers in providing tools, materials, and instructions; routes for upward mobility for workers; relations with foremen; and the importance of safety to foremen and managers. Group members responded freely and openly and without tendencies to promote a company line.

The company provided data on total lost-time accidents, doctor visits, and first aid cases for the previous year. From this was computed the number of occurrences per crew member per month. Stated as a percentage, these ranged from 10.7 down to 0.

To evaluate crew safety factors in terms of interview results, paired comparisons were made. In each instance, type of work, skill level, and exposure to risk between the paired crews were the same but the accident records were very different. A descriptive criterion was chosen to determine whether answers to a particular group of questions were correlated with safety. Specifically, if the answers of three-quarters of the crew members (e.g., 9 out of 12) showed a particular tendency, this was considered to be strong evidence that the demonstrated attitude or behavior pattern influenced safety.

The following four factors relating to crew attitudes in addition to those identified in the top- or middle-management or foreman studies were found to be strongly correlated with their accident records:

1 Safer crews had and used upward influence through channels of communication in addition to the formal one through the foreman. These channels provided another means for correcting work or safety situations that the crews did not like. One of these channels was that of conferences with higher management. Another was through the union representative. Less safe crews did not feel that such communication channels were available to them.

2 In safer crews, there was an anchor person other than the foreman who provided continuity and know-how to new members.

3 Safer crews found that their work was less boring, offering variety, learning experiences, and a feeling of accomplishment. One way this was accomplished was to rotate tasks among crew members. Although not verified by the research, it may well be that boring jobs have a stupifying effect which brings indifference to accident potential.

4 Members of safer crews had friendships and social contacts with each other off the job. From these seemed to grow a buddy system where crew members cared about and looked out for each other. Thus, managers who attempt to prevent friends from working together because they might waste time in talking may be forfeiting important gains.

The research on crews further verified some of the findings in the earlier construction safety studies, such as:

- The importance of orienting new workers
- The importance of higher-management's commitment to safety
- The fact that crews whose foremen were pushers and who exerted excessive pressures to improve productivity were less safe than were those whose foremen were people-oriented
- The fact that crews whose supervisors recognized and praised good work were safer

Persons familiar with modern management approaches may find little new in these results. However, hard-nosed construction managers who advocate strict discipline and adherence to a hierarchy which operates only upward through the foreman may find them both revealing and unsettling.[15]

Individual Attitudes, Aptitudes, and Behavior

How Workers Get into Construction   Without doubt, construction attracts people who like physical exertion, working with their hands, seeing the results of their efforts, and being outdoors. In some instances, choosing construction probably involved following family tradition; in others, it was a result of chance, with a job becoming available when a person was looking for work. Of course, there are instances where individuals planned their careers and entered construction through a variety of educational programs, apprenticeships, and other training. But we really do not know much about either circumstances or motivations which have led people into construction and how these have affected their attitudes about safety.

There are certain almost romantic notions about why workers choose certain crafts—for example, those seeking thrills opt for steel erection. At one time it was proposed that roofing was chosen because it involved high, dangerous tasks, but such a reason proved to be untrue. A study in Southern California found that many roofing workers, mainly Chicanos, entered the trade because roofing was the only work available to them. In general, people do not wind up in construction through carefully thought-out or planned procedures and as a result of specific physical or mental attributes. People seem to come into construction by chance. Those who leave seem to do so because the work runs out, the tasks are too physically demanding, or more attractive alternatives are available. Consequently, we know very little about the how and why people get into construction work. Moreover, in the short run they must be taken as they are—that is, the entire spectrum of people with their strengths and weaknesses tempered by cultural backgrounds and job influences make up the construction work force.

The Macho Problem   Individual attitudes of construction workers about safety most certainly are shaped by their backgrounds, the work situation,

---

[15] For added detail on this study, see N. M. Samelson, "Crew Factors in Safety Performance in Heavy Maintenance Operations," *Technical Report 275,* Department of Civil Engineering, Stanford University, Stanford, Calif., January 1983.

management attitudes and practices, and peer pressures. For example, in the past, so-called manly behavior has sometimes been interpreted by management and worker alike to include working in hazardous situations, handling heavy objects without help, and refusing to use common-sense devices such as safety nets or belts. Overcoming such attitudes is a first step in introducing safe work practices.[16]

The Alcohol and Drug Problem   Alcohol and drugs have created serious problems for all industry and particularly for construction with its less-permanent or casual work force. A recent report (see *Engineering News-Record,* Feb. 6, 1986) stated that estimated losses were $50 billion because of lost productivity, unmet production schedules, absenteeism, tardiness, accidents, security problems, theft, and destruction of property. In behavioral terms, workers who were habitual users of alcohol or drugs:

- Were 3 times as likely to be late to work
- Were 2.2 times as likely to request early dismissal or time off
- Had 2.5 times as many absences of 8 days or more
- Used 3 times the normal level of sick benefits
- Were 5 times as likely to file a workers' compensation claim
- Were 3.6 times as likely to be involved in an accident

Alcohol abuse is of long standing and may affect an estimated 20 to 40 percent of the work force. Alcohol is a depressant, not a stimulant; it reduces individuals' control over their mental and physical reactions. Unfortunately, the nerve centers that cause people to have inhibitions and exercise self-control are affected first. Consequently, judgment is affected before muscular control is lost, so that drinkers who are still able to function physically do so impulsively and irresponsibly. Alcohol cripples the abilities of all individuals, but affects people in varying degrees and in different ways, so that detection of alcohol impairment by observing an individual's behavior is difficult and undependable. To date, more dependable methods, such as breath analyzers, which are widely employed by the police to detect drinking drivers, are seldom used on construction sites.

Recently, drugs have become an increasing problem. In the population as a whole between 1980 and 1985, marijuana use doubled and cocaine sniffing or injecting quadrupled, so that an estimated 25 million people in the United States are involved, or 10 percent of the population.

Experienced supervisors always have a weather eye out, particularly on Mondays or after holidays, for those who may be affected by alcohol or drugs. When the effects on the workers are obvious, supervisors do not let them

---

[16] Such macho attitudes are not restricted to construction workers in the United States. For example, Peace Corps volunteers found that many men in an African tribe had hernias. Attempts were made to teach them to avoid this injury by lifting with bent knees and straight backs. Resistance was strong; the men argued that women lifted that way. Undoubtedly, many examples can be found where American construction workers have stuck to ill-advised practices simply to demonstrate their manliness.

work. Some bosses go so far as to discharge them. When being "under the influence" is merely suspected and these approaches cannot be applied, good supervisors at least put the suspected individuals into jobs where they cannot harm themselves or others. Employers are becoming increasingly aware of the problem and in some cases are adopting measures to counteract it. For example, it was reported that in Dallas programs to train supervisors to identify drug users have been developed, and information and packets are being supplied to AGC members (see *Engineering News Record,* April 28, 1983, p. 178). Some such programs were found to be effective. For example, the Stern-Catalytic Company instituted screening, using urine analysis. Before and during screening, absenteeism was 15 percent; cost effects were measured as 15 percent. After testing was introduced, absenteeism was down to 5 percent and cost effects were 1 percent.

Testing employees for alcohol or drug use is controversial and how it fits into construction management has not been agreed upon. Its general use is being challenged on constitutional grounds as an infringement of privacy, although it may be permitted for those in sensitive positions or where an individual's impairment would endanger others. Whether or not making testing a formal condition of employment is legal or enforceable has yet to be determined. Possibly of greater importance in the short run is that the construction unions have taken a strong stand against it.

Accident Proneness   Within any group of workers, there seem to be some who have more mishaps than others. The notion of accident proneness is hotly debated, with some researchers claiming it does not exist. However, it is clear that there are individual differences with respect to strength, coordination and dexterity, reaction times, or vision, including peripheral vision. The importance of these attributes is dependent on the nature of the work. For example, it is known that more accidents occur when workers are called on to exert heavy manual effort. In this instance, such work assignments should be given to those who have greater strength and better coordination when, for some good reason, job redesign is not feasible.

Assessing Individual Mental Traits That May Affect Safety   Not only may individuals have different susceptibilities to accidents, but each individual may be more susceptible at one time than another. Certainly those under stress resulting from personal problems or pressures on the job will not have a strong focus on safety. Also, boredom may produce inattention, which can lead to accidents.

Today, in some industries, physical and psychological tests are given to determine the suitability of potential employees to carry out the positions for which they are being considered. Also, employment performance records are kept and periodic appraisals of these records along with safety and general health information are conducted to determine their potential for promotion. Little of this is done systematically with workers in construction. In fact, construction unions generally oppose the testing of their members or the keeping or passing on of their performance or accident records on the grounds that employers will use them unfairly. Sometimes it is illegal for insurance companies or governmental agencies to release such information. Thus it is difficult to use tests or past records to determine which workers might be accident-susceptible.

Research has also found that workers who have certain attitudes and behavior tend to have more accidents. These attributes can sometimes be detected by careful observation or from records. The results of two studies, one by G. H. McPherson and the other by L. W. deStwolinski, in the late 1960s, are pertinent and will be described briefly here.

In a detailed study of industrial safety, provided informally to the authors, McPherson of Dow Chemical Company found less-safe performances associated with certain behavioral factors, including the following:

1 Failure to anticipate possible happenings or to pay close attention to the work, or a tendency to become tired or bored

2 A history of home or minor job accidents

3 Working at considerably higher or lower than normal temperatures

4 Poor morale associated with an absence of feedback on performance and low opportunity for advancement

5 Employee indifference as reflected in few and low-quality suggestions

6 Being disliked, an isolate, or lacking the ability to get along with others

7 A background showing instability as measured by poor references on banking records, indebtedness, or arrests

8 A higher than usual record of absenteeism, particularly on Mondays or after paydays

9 Continual complaints about the work situation

10 Frequent medical or dispensary visits

11 Nonconforming behavior such as demands for attention or attempts to attract it by unconventional dress or hair styles

12 Marked changes in reactions during shifts or over different days

Signs such as these should serve as warnings that the individual's potential to have accidents is high.

DeStwolinski's study is based on the answers to questionnaires circulated in the union to over 800 members by Operating Engineers Local No. 3 in Northern California.[17] Workers' perceptions of themselves, their foremen, and job management were probed with a series of questions. They were then also asked to report their accident records over the years. Responses to eight questions in particular showed high correlation with accidents. Answers which placed individuals in the accident-susceptible class were:

---

[17] L. W. deStwolinski, "A Survey of the Safety Environment of the Construction Industry," *Technical Report 114,* Department of Civil Engineering, Stanford University, Stanford, Calif., October 1969.

- My job management does not know its job well.
- I have worked a relatively short time in my present job classification.
- I would like to have an opportunity for a good family life.
- My foreman is stubborn.
- My coworkers are boring.
- My job management does not praise good work.
- My coworkers are not safety-minded.
- Risk-taking is a part of the job.

As with McPherson's work, deStwolinski's findings show clearly that certain perceptions, attitudes, and behavioral patterns are identified with higher accident potential.

It would be wrong to assume that management evaluations of workers, psychological tests, or questionnaires for individuals do more than indicate a greater likelihood that they are susceptible to accidents. It is hoped that by citing such findings, managers and fellow workers alike can be alerted to look for cues which will indicate persons who require particular attention. This might be accomplished through efforts to change attitudes and behavior through training and counseling, or at least watching them carefully at all times.

Foremen can be particularly effective in detecting individuals with adverse attitudes and behavior. One procedure is to use a simple checklist or tabulation which will show telltale signs. Figure 12-4 is such a checklist which combines a few of the items developed by McPherson and deStwolinski. The items chosen and their weightings are arbitrary and can be challenged. The merit of such a scheme is that to develop a rating, the foreman must observe each worker's behavior and attitudes closely, which he otherwise might not do because he is usually very busy and under many pressures.

SITE-SPECIFIC ASPECTS OF CONSTRUCTION SAFETY

To build construction projects safely requires not only that workers and managers alike know that accidents can happen and be alerted to and alert about them, but that commonsense preparations and procedures of a physical nature be followed on-site. These are spelled out in detail in many sources. Among these are handbooks such as those of the AGC, the safety guides and safety orders of the Occupational Safety and Health Administration (OSHA), and those of individual states which administer OSHA-approved programs. An excerpt from the California OSHA standards is given in App. D.

Failure to follow these commonsense or legally established safety and health rules usually means that a contractor is breaking the law and is thereby subject to citation and possible fines. But far more important to the contractor, most of their provisions offer sensible ways of reducing accident or health hazards. A few of these site-specific measures are discussed briefly here under the heading of design, assembly, and use of temporary devices and structures; housekeeping; per-

Name	Frequent Monday absences	Requires supervision	Has personal problems	Has individual abnormalities	New employee	Frequent early quits	Attitude* rating	Total†
Bill	x		x			x	4	7
Mike							2	2
Clark	x	x	x	x	x	x	6	12
Jim	x			x			4	6
Frank		x	x		x	x	4	8

*Estimate attitudes on a scale 1 to 10, where 10 equals low attitude about self, coworkers, foreman, management, and need for risk-taking.

†To compute the worker's rating, add the number of checkmarks in the first six columns to the attitudinal grade in column 7 and enter the total in the last column. Crew members with the highest numbers need closer supervision and should be assigned the least-hazardous tasks.

FIGURE 12-4
Crew safety analysis procedure for foremen.

sonal protection; tools and equipment; trenches; work surfaces, ladders, scaffolds, ramps, and elevators; and monitoring. Far more detail is available from the sources listed above and others, and should be obtained and studied.

Design, Assembly, and Use of Temporary Devices or Structures

Failures in temporary devices or structures sometimes occur on construction projects through errors or omissions in their design, assembly, or use, or a combination of all three. These incidents which may result in injury to people possibly one time in ten are usually corrected by job personnel and go unreported. Examples, among many, are failures in concrete wall or deck forms and breakdowns or collapses of cranes, hoists, scaffolding, or walkways. However, when through a combination of events these incidents result in injuries or death, they attract widespread attention. Often the resulting claims are enormous and, at times, criminal charges may be brought against the contractor's management.

It is not possible in this book nor in safety orders such as those of California OSHA to treat design, assembly, or use of temporary devices or structures in detail. Rather, it is hoped that a brief discussion of one recent occurrence will highlight the complexities of the problem and indicate how important it is to follow instructions and rules and pay attention to other details.

The case chosen is the collapse in April 1982 of false work supporting deck forms for a freeway ramp under construction in east Chicago. This incident killed 13 and injured 15 workers and final costs will run in the millions of dollars.[18]

---

[18] This description is based on a detailed report by N. J. Carino, H. S. Lew, and W. C. Stone of the National Bureau of Standards, published in the *Journal of Construction Engineering and Management,* ASCE, vol. 110, no. 1, March 1984, pp. 1–18.

The completed structure was to be of long-span, cast-in-place, post-tensioned box girders some 50 feet in the air. It was on both a grade and curve. Form supports were traditional and towers were prefabricated welded frames founded on temporary concrete pads. At the top of the frames, U heads cradled steel crossbeams which carried steel stringers. These, in turn, supported timber joists and a plywood deck. From this description, those familiar with bridge construction will affirm that hundreds of such temporary structures of less-grand scale have been built in a similar manner with few failures. In this instance, however, ignoring what appeared to those involved to be insignificant details in design or construction practices brought tragic results.

Two collapses occurred, roughly 5 minutes apart. The detailed study cited contributing factors for the collapse of unit 4 as follows:

1 Wedges between stringers and crossbeams specified in the scaffold design were omitted. This put unbalanced loads on the crossbeams.

2 The concrete pads supporting the tower shoring were underdesigned, which resulted in a factor of safety inadequate to resist anticipated construction loads.

3 The tops of the shoring towers were not stabilized adequately in the longitudinal direction.

4 The quality of the U-head welds was poor.

It was surmised that the first failure was cracking of the concrete pads. This caused the towers to displace downward; the other deficiencies then contributed to the total collapse. The contributing factors to the collapse of unit 5 were:

1 The falsework system was not tied together adequately because of omission of specified bolts between stringers and crossbeams.

2 The falsework system lacked positive longitudinal stability.

With the collapse of unit 4, the west portion of unit 5 was unstable and fell. Many widely applicable lessons can be learned from this example. Possibly the two most important are:

1 In any situation which goes beyond the very usual and conventional, attention must be given to the details of design and to following those details carefully in the field.

2 People involved in either design or construction must not become complacent and neglect important items because of an apparent resemblance to things they had done earlier. They must go back to basics in both design and execution.

Housekeeping

Maintenance of an orderly workplace free of scattered debris, paper and wrappings, small pieces of lumber, and uncoiled hoses, wires, and cords is required by safety regulations (see App. D) because it makes the workplace safer. Unless these precautions are taken, workers are apt to trip, fall, or otherwise hurt themselves.[19] Possibly even more important, insistence that good housekeeping practices be maintained clearly demonstrates and is a constant reminder that management and workers must be concerned about safety. It probably can be generalized that attention to housekeeping actually saves money. For instance, providing barrels or bins for waste and insisting that they be used means that the debris need not be picked up later as a separate operation. Again, neatly stacking stripped form lumber out of traveled paths may save both cost and material that would otherwise be wasted. Sometimes, however, these cleanup efforts can be overzealous without making the site safer. For example, insisting that all nails be pulled immediately from carefully stacked lumber is probably unnecessary. Rather, this is a chore that can be saved for moments when workers would otherwise be idle.

One important housekeeping rule from a safety standpoint is that a crew must never leave a hazard. For example, if it cannot backfill an excavation, it must provide adequate guarding or other protection before moving on.

Housekeeping can be a particularly sticky problem on multishift projects or those involving subcontractors, where the tendency is to leave the cleanup to others. In such instances, higher-level management must take a firm hand and penalize violators and see that all parties abide by the rules.

Personal Protection

Personal protection has two interrelated facets. On the one hand, it involves the workers' desire or willingness to protect themselves and to use devices for such purposes provided by management. On the other hand, management must not only provide the devices and instructions on how to use them (see App. D), but must also encourage and even enforce their use.

Personal protection begins with workers themselves following such simple rules as wearing hard hats, shirts to prevent sunburn, cuffless pants to avoid tripping, and high hard-toed shoes to guard the feet. Moreover, management must follow the rules it sets for workers. For example, managers should always wear hard hats when they go onto a project, even for a short time. The same concerns apply in getting workers to use contractor-supplied equipment such as safety belts, gloves, goggles, unbreakable glasses or face shields, respirators, and ear muffs or plugs. Often these are uncomfortable and workers feel imposed upon if forced to use them. Education regarding the hazards they can face in not using them and management's insistence on their use is essential.

Contractors often complain that OSHA regulations are unfair since even when they supply these devices, they and not workers are cited and penalized if workers do not use them. Counterarguments include (1) enforcing such rules

---

[19] An entire issue of *Ergonomics* (vol. 28, no. 7, July 1985) is devoted to falls, slipping, and tripping in the workplace.

against individuals by OSHA inspectors is not physically possible and (2) insisting that work be done safely is a management responsibility and that workers and supervisors should be disciplined or discharged for safety violations just as they are for being unproductive. Of course, this is more easily said than done.

Guarding Against Back Injuries

In dealing with personal protection, there is a tendency to think primarily of clothing or the devices workers can wear. But personal protection goes far beyond this. To be most effective it requires a careful examination of all aspects of site condition and development of approaches that can be taken to mitigate the effects of a particular hazard. This approach is well-illustrated by the discussion of noise given later in this chapter. Another safety hazard calling for both worker and management attention, used here as to illustrate personal protection, involves injuries to backs.

The human frame and its muscular supporting system evolved from earlier creatures that traveled on four legs, with their backs normally positioned horizontally, but the human body structure has not changed to fit its erect posture and travel on two legs. Lifting or carrying objects was not possible while walking on all four legs with the spine horizontal. This ability came about with erect posture. It should be obvious, then, that construction employers and workers alike should recognize the limitations the human physical structure places on the ability to lift and carry heavy objects. However, this limitation is seldom recognized, as illustrated by the following:

• The weights of many of the objects handled by construction workers far exceed those that research has shown to be safe (see details below).
• As shown in Table 12-3, back and trunk injuries account for a third of all that occur.
• As shown in Table 12-8, back injuries are the most costly, which probably means they are the most severe.
• Back injuries, once they occur, create a permanent weakness, demonstrated by the finding that repeats among those with a previous back injury are two to three times greater than the average.

Because of the frequency and severity of back injuries in all industry, they have been studied intensively by safety and health engineers and scientists.[20] Their findings can be most helpful in construction where to date the problem has often been ignored.

The methods developed for preventing back injuries follow the accident-

---

[20] The importance of back problems in the workplace in made clear by the attention given to them in the literature. For example, an entire issue of *Ergonomics* (vol. 28, no. 1, January 1985) is devoted to them.

chain approach mentioned at the beginning of this chapter. It is attacked by (1) controlling the weight and conformation of objects to be handled, (2) setting up the work situation to eliminate the possibility of overexertion and adverse body reactions, possibly including worker selection, and (3) educating supervisors and workers about the hazards of lifting heavy objects and how to avoid them.

Psychophysical testing offers one way to determine the weight and conformation of objects that can be safely handled. It involves asking individuals about their perceived abilities. Partial results of such a study are shown in Fig. 12-5, which gives findings for the most difficult positioning, one which calls for lifting from floor level. This figure illustrates the wide range of perceived abilities among both men and women, taking into account the distance of the object from the body and how frequently the lift is made. Another variable is the conformation of the object. The values in Fig. 12-5 are for very compact items; for more bulky objects, Fig. 12-5 values should be reduced by roughly 25 percent.[21]

The spread of values for what constitutes maximum acceptable loads to be lifted, as given in Fig. 12-5, indicates the difficulty in determining the upper limits. One report states that "lifting loads greater than 35 pounds when held close to the body, or 20 pounds between 25 and 35 inches in front of the body, would be hazardous for some people."[22] Another finding is less severe. It states that "load handling less than about 44 pounds resulted in very few instances of severe strain."[23] Maximum values proposed by the International Labor Organization in 1962 were the following, for ages 20 to 35: men, 53 pounds; women, 32 pounds.

Determining maximum allowable weights of materials to be handled by construction workers is particularly troublesome, given the many and highly variable conditions under which lifting takes place. Not only do materials when delivered to the site have many different unit weights and conformations, but once on site, they are often combined into larger and more cumbersome pieces. Often handling is from awkward positions. It is understandable why safety regulations in the United States are silent about maximum lifts, although they have been proposed from time to time. To illustrate, the Cal/OSHA Guide for Construction (see App. D) states only that "safe manual lifting practices should be followed."

As of today, materials or equipment delivered to or handled on construction sites far exceed the safe limits shown above. A sampling of common materials as they arrive on site and are handled by individuals would include the following:

[21] Another data base for acceptable weights to be lifted by industrial workers is given by A. Mital in *Ergonomics,* vol. 27, no. 11, November 1984.

[22] See P. B. Chaffin and K. S. Park, *Journal of the American Industrial Hygiene Association,* vol. 34, 1973, pp. 513–525.

[23] See P. B. Chaffin et al., *Journal of the American Industrial Hygiene Association,* vol. 38, 1977, pp. 662–675.

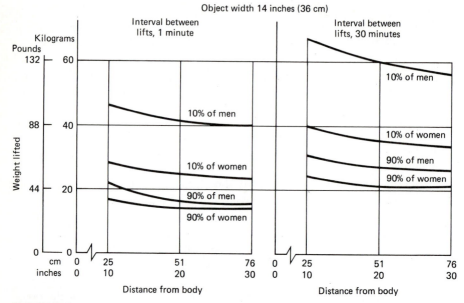

FIGURE 12-5
Range of abilities of men and women to lift weights from floor to knuckle (mid-thigh) height. Based on the psychophysics rating method (asking what is enough). Weights should be reduced roughly 25 percent for objects 30 inches (75 centimeters) wide. (*Source: S. H. Snook, Ergonomics, December 1978, vol. 21, no. 12, pp. 963–985.*)

Cement, in sack	94 pounds
Roofing paper, per roll	60 to 80 pounds
Roofing asphalt, per container	100 pounds
Lumber 6-inch × 8-inch × 10-foot piece	94 pounds (minimum)
Reinforcing steel #8 bar, 12 feet long	32 pounds each
Reinforcing steel #12 bar, 11 feet long	64 pounds each

From data such as these, it should be obvious that to date, contractors and those who design structures or equipment or supply materials for construction are oblivious to or have not taken seriously back-injury problems. Given its cost and possible liability implications for suppliers as well as contractors, this situation may well change.

At the work site, careful planning of the lifting task and positioning of materials can do much to reduce the opportunity for back or trunk injuries. Always the first aim is to avoid any heavy lifting and carrying. Where overly heavy loads cannot be avoided, provision should be made for devices, such as simple conveyors, skids, or handcarts, to lift and move them. Where lifting cannot be avoided, footing should be solid, nonskid, and free of debris. Load position, both before and after lifting, should be as near waist height as possible, rather than on the floor or high up. Handling should not call for awkward body positions or twisting and turning. In sum, lifting procedures should be planned and not left to chance.[24] Both supervisors and workers should know the hazards associated with lifting and be continually alert to avoid them. First of all, it must be made clear that management considers lifting excessive loads to be unacceptable and stupid rather than manly. Workers should be directed to ask for help rather than to handle the heavy load just this once. After all, only one lift of a too-heavy load or from an awkward position can bring a permanent injury, costly to worker as well as contractor.

The view is widely held that proper lifting involves holding the back, which is the weak link, straight and taut, and supplying the lifting force with the legs, which are strong. Some argue that lifting posture does not matter. In any event, focusing attention on safe lifting, however done, offers another way of drawing attention to and keeping it on all jobsite hazards.

Another proposed stratagem for reducing the hazards associated with lifting is to attempt to select suitable people for such chores, in one of three ways. First, assign tasks involving lifting to larger, stronger people. This approach offers only a partial solution because it makes job management more difficult and does not eliminate individuals with back problems or those who claim that their backs have been injured in order to get easier work or compensation. Second, establish and keep medical histories, possibly combining these with preemployment physicals. A plan such as this can bring strong opposition from unions, on the grounds that employers will use them unfairly to discriminate against older or less physically able persons. The counterargument is that it keeps workers out of situations where they may be injured. The third approach is to give preemployment dynamic strength tests.[25] This approach, as well as selection by physical attributes, has run into both political and legal criticism on the basis of sex and race discrimination.

It has also been demonstrated that the problem of back injuries is associated with physical and mental conditioning. There is the well-known medical fact that people's backs go out more frequently when their physical condition is poor, when they are tired, or when they are depressed or mentally upset. Recently a few plans involving programmed physical exercise have been instituted on construction sites. Also, part of a good safety program is recognition that people who are overtired or mentally distracted are likely candidates for injuries, and they should be protected or kept off the job.

It is clear that none of the approaches outlined here, taken alone, constitutes a right way to reduce back or other personal injuries. Awareness of

[24] See *Work Practice Guide for Manual Lifting,* National Institute for Occupational Safety and Health, March 1981. For a summary of the effects of modes of lifting, see J. E. Kholm et al., *Ergonomics,* vol. 25, no. 2, February 1982, pp. 145–161.

[25] See for example, J. L. Pyte and E. Kamon, *Ergonomics,* vol. 24, no. 9, September 1981, pp. 663–672, and *Preemployment Strength Testing,* National Institute for Environmental Safety and Health, reprinted March 1979.

all of them and implementation of those that fit seem to make the best sense.

Tools and Machines

Tools and machines handled or maintained improperly can be dangerous and cause injuries. Each tool or machine has its peculiar uses and hazards (see App. D). Those actuated by electric power, such as a variety of saws, or by air, such as nailers, staplers, or jack hammers, should be employed only by trained personnel who have been fully informed of possible dangers involved in using them. Improperly grounded electric tools or cords serving them can be lethal in wet weather. Again, the moving parts of power tools such as saws, belt or chain drives, exposed shafts, or revolving or reciprocating parts must be positively guarded so that they do not catch fingers, hands, or loose clothing. The list of such potentially hazardous items is almost endless.

Moving equipment such as tractors, trucks, scrapers, backhoes, or cranes offer extreme hazards, particularly to those moving or working around or below them (see App. D). Although warning devices such as backup horns are often prescribed, they are not fail-safe, so that only through extreme care by operators and others can death or severe injuries be avoided.

Trenches and Other Excavations

Soil caving into trenches, shafts, tunnels, or other excavations is a primary hazard in underground work. One reason is that if workers are buried over their heads, they suffocate even if they are not otherwise seriously injured. For this reason, a common rule is that trenches or shafts over 5 feet deep must be supported regardless of their apparent stability. In deeper excavations, although the walls are supported, emergency exits must be provided. In contrast to other possible accident hazards where reliance is placed on workers and supervisors to observe safe practices, in excavation firm rules for protection have been established which do not permit exceptions. (See App. D for added detail.)

Work Surfaces, Ladders, Scaffolds, Ramps, and Elevators

A wide variety of devices are employed to give workers access to the locations where they work. Careful design and safe use is imperative if accidents are to be avoided (see App. D). The rules for job-built ladders provide a simple example. For them, standards are set for lengths and widths, quality of materials, rung spacing, securing, and other details. Users are expected to pitch and support them properly, to be sure they project at least 3 feet above the landing, and to tie them off securely at the top. Ascents or descents are always made facing inward.

Falls from scaffolds and roofs and into openings and elevator shafts and slipping and tripping offer other chances for injuries. Present-day rules and practices require that positive steps be taken to prevent them.[26]

ENVIRONMENTAL HEALTH HAZARDS IN CONSTRUCTION

For some workers, construction involves using or being otherwise exposed to a variety of vapors, liquids, or suspended solids as well as bacteria or viruses. These can have short-run effects such as burned skin or eye irritation and nausea, or long-term consequences such as cancer, lung impairment, and other serious illnesses. Working in excessive heat and exposure to vibration, impacts, and jolting from operating tools or equipment can damage the body over extended periods of time. Furthermore, exposure to high noise levels will diminish one's ability to hear. Although closely akin to injuries, effects such as these are usually considered separately as environmental health problems.

Flammable and Other Dangerous Substances

Flammable liquids and their vapors and certain dusts can ignite or explode, resulting in serious burns or injuries. Places where they are used must be well-ventilated and free from sources of ignition; smoking must be absolutely forbidden. In addition to burning or exploding, contact with flammable liquids or breathing their fumes can also damage skin, eyes, or lungs. Avoiding the vapors and employing skin protection such as gloves or aprons and goggles for eye protection are musts. Spills must be washed from the body and contaminated clothing replaced. Instruments for controlling fires such as well-maintained extinguishers must also be available.

Hazardous Chemicals

Along with flammable liquids and vapors, today's construction uses many other substances that can have short- or long-term health effects, some of them unknown.[27] Among the means for protecting the workers are to require that the manufacturer provide information on the hazards involved and the remedies after exposure and that employers notify their employees of these

---

[26] See the earlier reference on falls, slipping, and tripping in *Ergonomics,* vol. 28, no. 7, July 1985.

[27] Even the more common substances used in construction that may create safety or health hazards are too numerous to list here. Furthermore, many new products are being developed. Manufacturers are required to furnish information about the hazards these products present. Other sources of information include the publications of the National Institute for Environmental Safety and Health, and OSHA and its state counterparts. Also, the Associated General Contractors publishes a *Hazardous Substance Guide for Construction,* and other associations often have publications or issue information to their members.

facts. However, providing this information does not relieve employers of their obligation to maintain safe working conditions.

Often containers of hazardous materials are labeled. Unfortunately, this may not be particularly helpful, given the procedures for supplying and using such products. Even if supplied, labels may be damaged, lost, or ignored.

Fumes produced by chemicals and by hot work such as cutting, heating, and welding can have severe consequences to workers. A cardinal rule is to see that work areas are well-ventilated so that fumes do not accumulate. Entering, much less working in, confined spaces such as manholes or tanks without adequate ventilation is not permissible.

Carcinogens (Cancer-Causing Substances)

Medical research in recent years has demonstrated that exposure over time to a number of materials or supplies employed in construction can cause cancer. Asbestos is much in the news, but other substances including certain dusts have been proved to be carcinogens or are highly suspect. Exposure to some of these is no longer permitted; if allowed at all, workers must be protected either with masks or by limitations placed on the amounts allowed in the air.

Asbestos has been used for many years in wallboard, roofing, and pipe, and for insulating, fireproofing, and soundproofing because of its low cost, strength, stability, and fire-resistant properties. In place and contained, it is not a health hazard. The danger, which was recognized before World War I, comes from inhaling its smaller, airborne fibers, which imbed themselves deep in the lungs. One or more of three serious ailments can result: asbestosis (hardening or thickening of the lung tissue), lung cancer, or mesothelioma (a rare form of cancer), any of which can be fatal. The early symptoms, unlike those of silicosis (see below), may not be recognized as being potentially fatal, but in later stages, which may be delayed from 3 to 35 years, with a mean of 20 years, physical deterioration is rapid.

Today, most of the hazard from asbestos comes when it is on and erodes from exposed surfaces or during its removal and disposal in demolition or remodeling operations. Elaborate procedures, too numerous to outline here, must be followed to protect building occupants or workers involved with it. These are or will be promulgated in the future by the Environmental Protection Agency for schools and by OSHA or its state counterparts for other situations.

Among the strategies that are or will be employed to protect workers and others from the hazards of dealing with asbestos are special licensing of contractors and training of supervisors and workers, and specifications about the procedures and protections to control removal and disposal. Another will be to monitor the operation to ensure that the number of airborne fibers at the work site will not exceed some specified limit. As of 1985, this limit was two fibers for cubic centimeter of air, with the likelihood that this standard will be made more stringent, given the public clamor surrounding the issue. It should be clear that asbestos and its handling presents many issues of unknown cost and health hazards.[28]

Dust

Dust has always been a nuisance on outdoor construction sites and more recently it has been found to be an environmental health hazard as well. In excavating dry ground and moving equipment around construction sites, clouds of dust unless controlled can envelop workers and equipment alike. Also, in dry weather, dust blown intermittently by high winds can hamper and even shut down entire projects. Other operations, such as drilling and blasting, unless controlled can also create serious dust problems.

The most effective method for reducing the effect of dust as an occupational health hazard is that of engineering the dust out of the picture, which can be accomplished in a variety of ways, depending on circumstances. In earthwork, wetting before excavation is sometimes possible. Sprinkling haul roads and open areas with water (possibly with a detergent added), spraying a dust-palliative oil, or applying calcium chloride (which attracts moisture from the air) may be effective. Wet drilling in quarry and tunnel work or installing a continuously running exhaust system and dust collectors are common practices. An interesting bonus was the discovery that wet drilling tends to increase production, since it allows operators to more closely control their rigs while also increasing steel and bit life because of the cooler operating conditions. Personal protective equipment in the form of dust respirators can also be effective in reducing the amount of dust inhaled by the worker. These cause some problems, however, since the respirators are generally not comfortable to wear and are usually discarded after a few hours.

The long-term consequences of exposure to dust, in addition to the possibility of cancer as described above, are usually categorized under the heading of pneumoconiosis, which means literally "dust retained in the lung." The more common disease under this heading is silicosis, caused by silica dust, which results most often from drilling and earth-moving operations.[29]

Silicosis (also called grinders' rot, miners' consumption, miners' phthisis, potters' asthma, and stonemasons' phthisis) is caused by inhaling fine silica-bearing dust. Rocks that can produce this dust (either by drilling and blasting or by erosion) include quartz, granite, feldspars, kaolin, mica, serpentine, shale, slate, and talc. Much research has been done to determine the dangers.

---

[28] For an excellent summary of the asbestos problem and for references see J. and M. M. Hinze, *Journal of Construction Engineering and Management,* ASCE, vol. 112, no. 2, June 1986, pp. 211–219.

[29] This discussion is based on research carried out in California by Fred Ottabani in the late 1960s as reported by L. W. deStwolinski in "Occupational Health in the Construction Industry," *Technical Report 105,* Department of Civil Engineering, Stanford University, Stanford, Calif., 1969.

In general, the results have shown that fatal silicosis can occasionally develop with exposures of a month or less. However, this rapid advancement of the disease is rare; the most common pattern involves an accumulation of dust in the lungs over a period of years. This slow progression can be just as incapacitating and fatal as a rapid buildup.

Open-air work, such as with jackhammers, was once thought to be relatively safe since it was assumed that air currents carried the dust away. However, studies have shown that individual operators must be widely spaced to ensure a dust-safe atmosphere. It has also been shown that pusher-cat operators on grading jobs are consistently exposed to dust levels well above accepted state and federal standards. This is true regardless of soil or weather conditions. Belt and front-end loaders also show high dust levels in the operators' areas. Even jobs that are apparently relatively dust free have produced extremely high dust counts in the operating compartments of earth-moving machinery.

Prevention of disabling disease caused by high dust counts in earth-moving machinery can best be effected by the use of pressurized closed cabs with filtered air. It is important to realize that environmentally controlled cabs are not an unnecessary luxury. These cabs not only offer protection against the long-range effects of silicosis (which raises workers' compensation rates), but also studies of agricultural machinery have shown that such a system increases operator efficiency and productivity by reducing the hazards to health and increasing comfort and safety. Personal respirators for operators are an alternative but a less satisfactory approach. As with so many other personal protective devices, they are uncomfortable, so that workers will not wear them unless educated about the risks of not doing so.

Coccidioidomycosis is an infectious disease prevalent in the arid and semiarid regions of the southwest. It is transmitted by inhalation of the spores of fungus in dust. Although rarely fatal, the disease can be completely incapacitating. Exposure is most common in remote areas where highway, pipeline, or utility construction might take place. In these areas, if the soil has not previously been disturbed, dusty digging operations cause spores to rise and be readily inhaled. The best method for controlling it in areas where it can be expected is through the use of pressurized air-conditioned cabs. For workers digging or laying pipe, respirators are highly recommended during recognized periods of high exposure. There is available a simple skin test that defines immunity to the infection; all workers entering a region where high exposure is likely should be tested. Immune workers or lifetime residents of the endemic areas should be assigned to operations with high-exposure probability. Blacks and Filipinos have been shown to be highly susceptible to the disease and so should be assigned work in areas where spore concentration is likely to be low. A continuing skin-test program is also highly recommended, along with air filtration wherever possible. Thorough cleaning of all equipment before removing it from an endemic area will help to prevent transportation of spores to other locations.

Heat

The physiological aspects of heat in work situations and the dangers associated with exposures to it were discussed in Chap. 9. That presentation makes clear that effects from a combination of internally generated heat and environmental conditions can affect one's ability to work safely and a worker's health as well. Under extreme conditions these heat effects can be life-threatening.

Studies by F. Ottobani and others have shown that in hot weather operators of heavy equipment commonly reach a high level of dehydration because of the combined influence of a hot day, solar radiation reflected from their machinery, and heat generated by the engines.[30] Measurements on the seat platforms of earth-moving equipment to determine these exposures have shown temperatures ranging between 194 and 266 degrees Fahrenheit.

Recommendations to overcome these difficulties, among others, are that operators should drink as much water as is lost in perspiration and urine, which can be as much as a gallon per day. In contrast, the usual daily intake is about one quart. Drinking water should be readily accessible on the equipment so that operators do not have to leave their posts. Since the body, when overheated, will not tolerate liquids that are too hot or too cold, a 55 to 60 degrees Fahrenheit temperature for the water is proposed.

The machine itself should be designed to keep equipment temperatures in the proximity of the driver's compartment as low as possible. For example, hydraulic oil reservoirs and exhaust systems should not be located directly under the operator's seat. Of course, as has been mentioned earlier, an ideal way to compensate for heat, dust, and noise and thereby to aid productivity and worker satisfaction is the installation of air-conditioned cabs on all machines subject to extremes of environment.

Rules such as those for avoiding dehydration just pointed out for equipment operators should apply to all workers. Wherever possible in hot weather or where work is carried out under elevated temperatures and at high relative humidities, it is sensible to provide shade or shielding from direct sunshine or to protect workers from heat from other sources. Providing good ventilation and circulating air which causes perspiration to evaporate also have excellent cooling effects.

It is probably worth repeating that since workers are less productive and more prone to accidents when working at high temperatures, avoiding such conditions is to the advantage of both contractors and workers.

Cold

How cold and wind combined with it affect the body is discussed in Chap. 9. There are also data to indicate that working in cold weather increases accident frequency and severity as well as costs. For example, an analysis of construc-

---

[30] See deStwolinsky, op. cit.

tion accidents in Ohio showed that accident frequencies in January were about 47 percent higher than in August; severity was four times as great.[31] This study also indicated that different factors were involved during the two periods, but did not state what they were. However, it can be presumed that contact with cold tools or objects, frostbite, and the clumsiness of working in gloves and bulky clothing all contributed to the higher accident level in cold weather.

Noise[32]

Excessive noise levels have both short- and long-term impacts on construction safety and health. High noise levels make it difficult for workers to hear warnings or sense danger and may distract their attention away from safety to other matters. In the long term, exposure to high noise levels can drastically accelerate the naturally occurring hearing loss that comes with aging. At one time this hearing damage went unrecognized as an occupational health hazard. This is no longer the case. Today, full hearing loss is compensable as an occupational injury under workers' compensation, and there are a few cases where partial loss has been made compensable. Certainly, those affected lose a valuable asset. It should be clear, then, that diminishing or mitigating excessive site noise demands the attention of the construction industry and its managers. A brief discussion of the health aspects of noise follows.

How People Hear Sound, either desirable or unwanted (which is called noise), results when a vibrating source causes pressure changes in the air. Sound has two dimensions—intensity, or loudness, which denotes the changes in air pressure; and frequency, or pitch, which is the time period over which a cycle of the pressure change occurs. The human ear picks up pressure at the various frequencies through vibrations of the ear drum. It then passes these vibrations on by means of three small bones in the middle ear and they, in turn, cause pressure changes in a fluid in the cochlea, a small, snail-shaped bony structure in the inner ear that is lined with hairlike cells. The cells convert the vibrations at the frequencies at which they were generated into electrical impulses which are carried to the brain. The result of this complex process is hearing. Extremely strong air vibrations can destroy the delicate mechanisms which produce hearing. At somewhat lower intensities, exposure to continuous or long uninterrupted intervals of sound can result in permanent hearing impairment over time. Relief from noise for even short intervals such as 5 minutes will permit almost complete recovery when the intensity is not too great.

It is the higher frequencies which produce not only pure high tones but also

---

[31] See E. Koehn and D. Meilhede, *Journal of the Construction Division,* ASCE, vol. 107, no. C04, December 1981, pp. 585-595.

[32] Construction noise levels and their effects on productivity are discussed briefly in Chap. 9.

overtones which enrich hearing; for example, through them it is possible to distinguish between voices and to recognize the differences among musical instruments. Unfortunately, it is the higher pitches that are most susceptible to damage by excessive noise levels.

Measuring Noise Levels   Sound intensity is commonly measured in decibels, with the common unit being the dBA. This single unit gives the combined intensity from all frequencies that humans can hear, which range from about 15 or 20 to 18,000 or 20,000 hertz (cycles per second; middle C is 400 hertz). The dBA scale gives greater emphasis to frequencies over 1000.

Sound intensities that humans can register vary tremendously. If the pressure difference creating the faintest sound that can be heard by an average young person is given a value of 1, the greatest is almost a trillion. To accommodate this range, the decibel scale of intensity is based on logarithms to the base 10. Thus, this faintest sound is rated at 0 decibels, the highest at about 160. This means, then, that for each increase of 10 in the decibel scale, the intensity increases 10 times over. In the ranges common to construction, 90 dBA is considered by some agencies to be the maximum permissible for 8 hours. On a site where dBA is measured at 100, the intensity is 10 times the 90-dBA level and at dBA 110, the intensity is 100 times the 90-dBA base.[33]

Construction Noise Levels   As already mentioned, the ability to hear deteriorates with age. This is shown in Fig. 12-6. The dotted lines on the figure show the change in hearing thresholds at various frequencies for individuals not exposed to loud noise. For the younger age group a rough averaging would give a dBA threshold value of about 0. However, the older group has a higher threshold, particularly in the higher frequencies. Figure 12-6 also shows substantially higher thresholds for equipment operators. The hearing of the younger group has already been somewhat affected but the hearing of those in the older sample has been seriously impaired. For example, Fig. 12-6 shows that an operator from the older group can barely hear the higher frequencies when people speak to him in a normal voice from as near as 10 feet. It is data such as these that have focused attention on noise on construction sites as an occupational hazard.

Construction sites by their nature are noisy; reported continuous background levels are in the range of 80 to 90 dBA. By comparison, traffic noises are low; a downtown heavily traveled commercial street has levels between 62 and 73 dBA. Power equipment can be far worse. Data gathered in the late 1960s before noise on construction sites became an issue are shown in Fig. 12-7. As noted in the figure, these readings were taken at 6 feet from the noise source so that operators or nearby workers would be fully exposed. Measure-

---

[33] For a discussion of how sound levels are defined, descriptions of several noise-control ordinances, and of pressure levels of certain construction equipment by an ASCE task committee, see *Journal of the Construction Division*, ASCE, vol. 103, no. CO 1, March 1977, pp. 123–137.

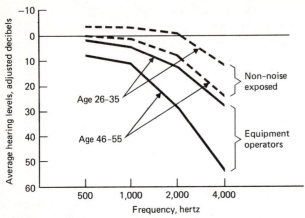

FIGURE 12-6
Average threshold hearing levels, by frequency, as affected by age and by working as an operator of heavy earthmoving equipment. (*Source:* Occupational Health Studies of Heavy Equipment Operators, *California State Department of Public Health, 1976.*)

ments, probably taken in the mid-1970s, reported by the ASCE task force (see reference) include the following:

Scrapers	85–95 dBA
Dozers	87–89 dBA
Dozers—squeaky tracks	90–93 dBA
Trucks—off-highway	81–96 dBA
Concrete saws	88–93 dBA

Even if, as discussed below, noise levels such as shown here are reduced, many older workers may well have suffered irreversible hearing loss.

Issues Involved in Controlling Construction Noise   Noise as a safety and environmental health hazard in construction was largely ignored until the early 1970s, when data such as those cited here became available. For that reason it cannot be assumed that these earlier measurements apply to all construction sites today. In many instances, levels have been lowered through efforts to be discussed in more detail below. Even so, many noise problems are far from resolved. Unanswered questions would include the following:

• At what intensities and frequencies and periods of exposure and in what work situations do various degrees of hearing impairment occur?
• What degree of hearing loss beyond that caused by aging is acceptable?
• If hearing is impaired to an unacceptable level, what are the economic and noneconomic losses to individuals and society?

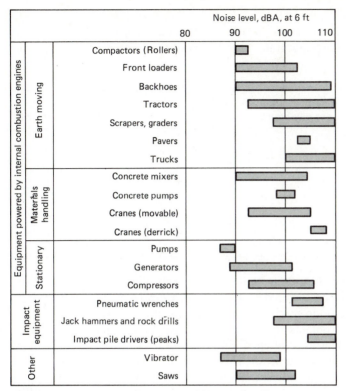

FIGURE 12-7
Construction equipment noise levels. (*Source: U.S. Environmental Protection Agency,* Report to the President and Congress on Noise, *December 31, 1971.*)

• How should economic and noneconomic losses be distributed among worker, employer, and society?

Employment-related total loss of hearing is today compensable in almost all the states just as are other permanent impairments such as the loss of a body part or vision in an eye. Settlements range from about $2000 up to $43,000 for a complete loss of hearing in one ear and from about $11,000 to $165,000 if both are affected. If hearing is only partially affected, claims allowed probably will be treated as a permanent partial disability with compensation based on a wage-loss replacement determination. As of today, however, relatively few claims for partial hearing loss have been paid. If they become prevalent, the total costs will be staggering.

It should be clear that the relationships between hearing damage and noise intensities, durations, frequencies and individual susceptibilities are not known. Also there are concerns about the cost of lowering present-day workplace noise levels. Today's approach is for the Occupational Safety and Health Administration (OSHA) or their state counterparts to adopt arbitrary noise-level ceilings which attempt to control excessive noise and still permit industries to function. These limits are as listed in the following table.

Allowable intermittent or continuous noise total exposure time per day-hours	Sound level, dBA
8.0	90
6.0	92
4.0	95
3.0	97
2.0	100
1.5	102
1.0	105
0.5	110
0.25 or less	115

Allowable Impact of Impulsive Noise   It is likely that present-day acceptable noise levels will be lowered in the future. Certain other countries now have set maximum 8-hour continuous noise levels at 85 dBA.

There are many problems to be faced by contractors attempting to meet these limits or by governmental agencies attempting to enforce them. Among them are that on large construction sites there are many noise sources that should be monitored, and measurement at each location requires special equipment and is time-consuming. Also, determining actual exposures is difficult when conditions vary throughout the workday. In sum, except when there is one principal continuous source or the noise results from impact, determining whether noise standards are exceeded is difficult.

Regardless of the difficulties in controlling construction noise, the industry must anticipate increasing pressures to control it from environmentalists, labor organizations, and government. To illustrate, in California, employers (including constructors), are required to meet the noise standards and can be cited if they fail to do so. In addition, most other industries must develop and maintain noise-control programs. One can ask how much longer contractors can avoid this requirement.

Approaches to Noise Control   As discussed earlier, there are two primary concerns with respect to noise on construction sites. First, noise may contribute to accidents, since it makes hearing difficult and may lead to inattention. The levels at which this can occur are shown in Fig. 12-8. From this figure it can be seen, for example, that at the permissible continuous noise level of 90 dBA, communication between individuals is impossible if they are more than about 12 feet apart. Communication by shouting is possible only at distances of 2 feet or less. It is then highly advantageous for safety's sake to keep noise levels down. Otherwise, workers are deprived of one of their most valuable

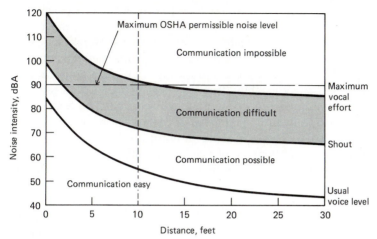

FIGURE 12-8
Relationships among noise intensity, distance, and ability to communicate.

protective senses. Second, a long-range consideration is to prevent or slow hearing loss from exposure to noise over a period of time.

Noise control in construction can be approached in four ways, by (1) reducing noise at the source, (2) interrupting the noise between source and workers, (3) separating noise sources and workers, and (4) providing personal protection.

As shown by the examples in Fig. 12-7, construction equipment and tools have been or can be a major contributor to high noise levels. In recent years, manufacturers have made numerous design changes to lower the noise levels produced by their equipment from those shown here. Examples would include housing which dampens sound and does not vibrate, enclosing engines or compressors, better-designed air intake devices or mufflers, and changes in materials or better lubrication of moving parts or drive mechanisms. Again, there have been substitutions of electricity or hydraulics for air as power sources. Skirts or shielding have been provided for impact tools. Blades on saws have been modified or housing provided to reduce or contain noise. This short list merely suggests the many ways in which source noise levels can be lowered. In some instances, specifications or purchase agreements have stipulated permissible equipment noise levels; these may force product redesigns on otherwise reluctant suppliers. However, the best-designed schemes for noise control can be destroyed by failure to maintain them at the job site, so that proper replacement, repair, and lubrication are essential. Even carelessness, such as failing to close access doors or housings, can destroy the effectiveness of good equipment design.

Interrupting the noise between source and workers has both psychological and physical elements. Research for highway-noise abatement has shown that screening by vegetation or by a thin wooden barrier does little to lower noise levels. It can, however, make the noise seem less intrusive and be psychologically effective. However, to actually lower noise levels, barriers must have mass. It is recommended that for highways, barriers must weigh at least 4 pounds per square foot of surface, which usually calls for materials such as precast or cast-in-place concrete, concrete block, or brick. As a practical matter, merely separating a noisy machine from workers with a ¼-inch plywood barrier will provide little noise reduction, although there may be fewer complaints. On-site offices usually are quieter than the site itself by as much as 10 dBA if windows are closed. This difference in noise levels may at least partially explain the apparent indifference of managers to high noise levels.

For earthmoving and similar operations, pressurized, air-conditioned, sound-insulated cabs provide an effective means for reducing noise levels as well as protecting workers from dust. They may also increase productivity. Contractor resistance to providing them probably stems from the notion that construction workers are he-men and that noise, dust, and other discomforts are all part of the job as they were 20 years ago; but this attitude does not fit today.

In some instances, noise levels can be reduced by physical separation between sources and workers. Measured reductions are roughly 3 dBA with each doubling of the distance. For example, the levels shown in Fig. 12-7 would be 6 dBA lower at a distance of about 25 feet, 9 dBA lower at 50 feet, and 12 dBA lower at 100 feet. This argues strongly for keeping work stations away from earthmoving and other noisy operations and for setting up power sources such as compressors at a distance, unless effective noise barriers are installed.

Personal protection for workers against noise with devices such as earplugs or earmuffs is an expedient but often ineffective solution. A variety of such devices are available and, if worn, can reduce noise levels by as much as 15 dBA.

Because earplugs or earmuffs make hearing less acute, they may be a deterrent to safety by impairing the human alarm system. Stated differently, they may not be fail-safe because they improve one safety device but impair another.

Earplugs are inserted into the ear and block noise entering it. They come in a variety of materials, including foam plastics which can be rolled between the fingers and formed to completely fill the ear channel. Some plugs are attached to a frame for easy holding. These are similar to some that are supplied for the passenger sound systems on commercial airlines. Worn for long periods in the heat, dirt, and dust of construction sites, earplugs can create soreness and, on occasion, cause infection.

Earmuffs cover the entire ear and keep sound from reaching it. They are mounted on frames that go over the head or, alternatively, directly on hard hats. Like earplugs they can be uncomfortable if worn for long periods of time. Also, they can be hot.

Earplugs and earmuffs are particularly useful for short-term noise exposures, as that around jet engines at airports. However, on construction operations where noise levels are continuously high, workers are reluctant to wear them even when they are supplied by contractors whose safety rules require their use. Somehow these workers see short-term discomfort as more important than long-term but nebulous hearing loss. Particularly irritating to contractors who supply hearing protection are the safety-violation citations that the companies (but not the workers) receive if workers fail to use it.

All in all, noise is one of the most vexing problems facing the construction industry. Progress is being made, but much remains to be done and the construction industry will feel increasing pressures to improve its practices.[34]

Vibration and Jolting

Operators of earthmoving equipment, trucks, and certain other construction machinery may be subject to whole-body vibration and jolting. Impacts are generally vertical and fall below the body's natural resonance band of 4 to 8 hertz. Reported health effects include kidney damage; hemorrhoids; reduction in the height of the spinal column; and changes in the dynamics of joints, ligaments, and muscles, which can increase one's susceptibility to injury.[35] Researchers directly concerned with the effects of such vibration on construction workers state that:

> Results suggest the existence of a selection process in which members of the exposed group leave jobs which subject them to whole-body vibration when they become afflicted with certain conditions such as heart diseases, including narrowing of the arteries, obesity, and certain musculoskeletal conditions such as the displacement of the intervertebral discs.[36]

Studies of vibration effects such as those just mentioned indicate that they can result in long-term health problems. However, because these effects are long-range and affected workers often change to other industries or work assignments, little firm data are available to assess the extent or severity of the problem or to support claims for workers' compensation. As with hearing loss, however, the costs of such disabilities which are now borne by workers may become compensable.

WORKABLE APPROACHES TO SAFETY AND ENVIRONMENTAL HEALTH IN CONSTRUCTION

Programs and Sources of Information

The earlier sections of this chapter have outlined the construction safety and environmental health situation as it is today and the known effects of attitudes

[34] See W. Bowlby and L. F. Cohn, *Journal of Construction Engineering and Management,* ASCE, vol. 109, no. 2, June 1983, pp. 169-182, for methods for modeling noise levels in highway construction.

[35] See, for example, H. Mertens, *Aviation and Space Environmental Medicine,* vol. 4, no. 9, 1978, pp. 237–298.

[36] See, for example, T. H. Milby and R. C. Spear, *Relationship Between Whole-Body Vibration and Mobility Patterns Among Heavy Equipment Operators,* National Institute of Occupational Health and Safety, July 1974.

and behavior, certain site-specific conditions, and environmental influences on safety and health. This section discusses what is being done and what can be done to make use of this information.

OSHA or state safety regulations prescribe what might be called a minimum program for safety and environmental health. A portion of the Cal/OSHA regulations are reproduced in App. D. They begin by setting out the requirements for legally acceptable safety programs. Included are provisions that direct every contractor to have an accident-prevention program; to adopt, use, and publicize a code of safe practice; to keep records; and to report serious accidents. The regulations also prescribe many commonsense practices that must be followed. This state-of-the-art information should be known and studied by construction supervisors at all levels.

Effective programs to improve safety and environmental health do not come from paying lip service to governmentally imposed regulations. Rather, they require an organized effort at all levels in a construction company. To get a program going requires a substantial commitment of time and a willingness to spend money, beginning with managers at the top. In turn, their commitment and practices must permeate all levels of a company. Signals on what is expected must be clear and follow-up is essential.

Many research reports, books, and manuals on safety and environmental health have been developed by governmental agencies, universities, trade associations, unions, and companies, so that materials are plentiful. Safety is an integral part of the training programs such as the STP superintendent and foremen series of the AGC (see Chap. 11). It is also integrated into apprenticeship and in-service training efforts. In addition, many large contractors have well-developed safety programs staffed by able professionals. Consulting services are sometimes available through insurance companies and private firms. Their assistance can be useful in getting safety programs going or in improving existing ones.

Some of the elements of programs that have been successful are discussed briefly below. It must be recognized that since construction tasks and the organizations carrying them out vary considerably, no single approach or combination of approaches is necessarily the best. Furthermore, each approach should be fitted to the circumstances peculiar to the project, the craft, and the individual workers.

Safety Activities Aimed at Individuals and Groups

Inasmuch as it is individuals and groups of workers who primarily have accidents, good safety programs must first take this factor into account. Approaches include (1) preemployment screening, (2) orientation at the time of hiring, (3) safety attention-getters, (4) crew (tailgate) safety meetings, and (5) award schemes.

Preemployment Screening   Preemployment screening has to do with procedures that keep workers with poor records or high accident potential off the job or at least in less hazardous work situations. In some industries, records of earlier accidents are available to employers. Again, standardized tests for determining such things as dexterity or ability to lift may be employed. In general, however, these have not been allowed in construction. Rather, workers have been hired as they come to the job site or are sent there from the union hiring hall or other referring agency. Under these conditions and in the face of laws against discrimination, little formal preemployment screening is possible.

Informally, job managers have often been quite successful in screening out workers who are unsafe or unfit. For example, perceptive foremen will carefully observe new hires for a day and then dismiss them if they fail to measure up. In communities where the construction work force is small, malingerers or those who have frequent accidents become identified and are not hired and eventually disappear from the labor pool. Sometimes informal agreements are worked out between contractors and business agents so that such people are not dispatched from the hiring hall. Such screening becomes difficult on large projects where workers come from many locations and are hired quickly and in large numbers.

Orientation at the Time of Hiring   Workers' impressions of company and management attitudes toward them and toward safety are largely set before hiring procedures are completed. They begin when new workers park their automobiles and enter the job office. If the parking area and entry are carefully ordered and free of debris and potholes and the job office is clean and attractive inside and out, workers conclude that management is alert and effective. This conclusion is reinforced if the job site itself is clean and orderly, all workers and supervisors are properly clothed and wearing hard hats, dangerous areas are barricaded or marked, ladders are in good shape and tied off, and scaffold and other appurtenances meet acceptable standards. When new workers join the crew, the foreman introduces them to other members, shows them around the work area, and describes the work assignment, giving particular emphasis to safety.

Ways of introducing workers to their jobs do not just happen; rather, they are carefully planned and executed. Neither are they restricted to large contractors or major projects. They will be done on every job where management is serious about safety (and productivity as well).

A number of large progressive construction companies augment this basic approach to safety for new workers by insisting that all of them, when they first come to a project, view a company-prepared sound movie or narrated slide presentation which clearly outlines company policies and practices. Often these presentations are professionally done. They seem to be most effective if personalized with, among other features, a strong statement about safety by top company personnel.[37] In keeping with the finding that workers

[37] The companies that have developed such presentations are generally pleased to share them with others. Many programs are in existence; safety representatives or construction trade associations usually know how to obtain them.

new to the particular job are more likely to have accidents, it is common to insist that those new to a project be involved in the orientation program, even though they have been employed before by the company.

Safety Attention-Getters Continued activity in accident prevention is imperative to keep accident awareness always in the forefront for workers and managers alike. A number of strategies have been found useful, and a few common ones are described here.

On larger projects that have safety organizations which keep safety statistics, it is common practice to display them prominently on large bulletin boards set where they are visible to those entering the project. It states something like: "This project has worked _____ man-hours without a lost-time accident." Personnel take pride in a favorable record and are very disappointed when the record is marred. In some instances, statistics may be given by areas or by crafts to generate special interest and competition.

Prominently displayed posters calling attention to common safety hazards offer another means of focusing attention on safety. Topics covered are numerous; a few among them would be slips and falls, proper clothing, safe lifting practices, and electrical hazards. Series of such posters are readily available from suppliers, so that they can be changed often. How effective such displays are is an open question. However, if they are left too long or go unmaintained or covered with graffiti, their effect may be negative, since such neglect clearly indicates management's indifference.

A variety of attention-getters aimed at foremen and crews are in use. These provide direct reminders about safe practices and hazards. Figure 12-9 shows the two faces of a plastic foreman safety-reminder card that fits into a shirt pocket. The ideas on it are based on the results of research on how safe foremen operate, as described earlier in this chapter.

Figure 12-10 is a safety checklist suggested for a roofing foreman. It is particularly designed for situations where a foreman supervises several small projects, with the crews for each of these under the direction of a craftsman acting as subforeman. By the time foremen have completed this form they have (1) checked out the job for safety and (2) provided information needed in emergencies. In showing this form, the authors are suggesting that (1) such a check form can be useful and easily developed for each kind of construction, and (2) its ideas can be incorporated in a more comprehensive form that includes other detailed planning information. This could include materials and equipment needed and where to find them and (very important) what the crew is to do when this job is finished.

Figure 12-11 shows faces of plastic cards that can be given to roofing foremen or workers. Their resemblance to playing cards is designed to attract attention. The two card faces at the top of the figure represent the two sides of a single card and cover both accident avoidance and first aid for burns, a common accident among roofers. The card faces at the bottom point out ways of avoiding falls and strains, two other prevalent roofing injuries.

CREW ACTION STEPS

1. Keep your cool. Anger breeds accidents.
2. When the work's not getting done, find out why; don't just push your workers harder.
3. Keep available to your crew. If possible, watch rather than work.
4. Teach the safe methods.
5. Watch for hazards on your job and correct them immediately.

This material was prepared under the grant number E9FOD432 from the Occupational Safety and Health Administration, U.S. Department of Labor. Points of view or opinions stated in this document do not necessarily reflect the views or policies of the U.S. Department of Labor.

developed by
Stanford University, CE,
Construction Management
Stanford, CA 94305 (415) 497-4447

 166

NEW WORKER ACTION STEPS

Everyone new to your crew (no matter how experienced) is a new worker.

1. Ask about last job.
2. Describe the new job.
3. Show worker around site; point out hazards.
4. Introduce worker to the others.
5. Describe your rules.
6. Give worker a test run on tools and equipment.
7. Keep an eye on the new worker during the first few days. Check back to see how the worker is coming along.

From CAL/OSHA Construction Safety Orders:
"When a worker is first employed, he/she must be given instructions regarding job hazards, safety precautions and the employer's Code of Safety Practices."   1510 (a)

 166

FIGURE 12-9
Foreman's safety reminder card. The two faces shown here are printed on a pocket-sized plastic card.

The effects of attention-getters such as those described here can be short-lived if left to stand alone. It is necessary to develop new and different items, as was mentioned above for posters. It is important also that higher-level managers support and encourage them. For example, they might carry cards similar to those shown in their own pockets and display them. Again, they could ask foremen if they have them in their pockets and use them.

As has been mentioned before, the best attention-getter for safety is emphasis by higher management. But supplements such as those described here and many others can extend their efforts even further.

Crew (Tailgate) Safety Meetings

Crew safety meetings to be held at least every 10 days were initially mandated by OSHA and are today required by OSHA or state regulations. Some contractors have largely ignored this requirement or have given it only lip service, at best. They make little effort to see that sessions are held and reported or to

ROOFING
Foreman Safety Checklist

Emergency Phone Numbers

_____

_____

Nearest hospital or clinic

_____

ADDRESS

Roof readiness ☐ clean
   ☐ openings barricaded
   ☐ perimeter protected

☐ Materials handling planned

☐ Proper equipment in working order

☐ Crew clothed in hard hats, safety
   shoes, long sleeves

☐ Workers fit physically and mentally

☐ Special attention given to new workers

☐ First aid kits stocked and accessible

☐ Fire extinguishers available
   and operative

☐ Water available for drinking and burns

☐ Weather conditions deserving
   attention
   ☐ wind
   ☐ heat
   ☐ slippery deck

Special instructions _____

_____

_____

_____

_____

Developed by Construction Safety, Civil Engineering, Stanford University, CA 94305

FIGURE 12-10
Roofing foreman's safety checklist. This is supplied in multiple copies to the foreman who gives
a copy to the subforeman in charge of the crew.

help foremen by providing topics or coaching to make meetings effective. Unfortunately, such negative attitudes diminish the effect of an important tool which other contractors have found to be valuable for developing and implementing safe work practices.

A set of suggestions for carrying out effective crew safety meetings are given in the left-hand portion of Fig. 12-12, which shows one side of a pocket-sized plastic card. The other side of that card, shown on the right-hand side of Fig. 12-12, is a short list of suggested topics for crews doing underground construction. These exhibits are only a small sample from an extensive and readily available literature on crew safety meetings.

There are many ways to conduct effective crew safety meetings. Three suggestions given in Fig. 12-12 deserve mention: (1) pick a relevant single subject related to work in progress or soon to be done, (2) encourage participation, and (3) keep the meeting short. Foremen with directive management styles probably should conduct the meetings themselves. However, if they employ a participative approach, crew members might be designated to select topics and lead the discussions. As mentioned earlier, safety meetings which are non-

FIGURE 12-11
Samples of pocket-sized plastic cards supplied to roofers to remind them of job hazards.

threatening to the foreman can be a useful medium for gaining crew participation in advancing ideas on how to improve productivity as well as safety.

It is common practice to schedule crew safety meetings at the beginning of the workweek to provide an opportunity to bring workers back onto the job and remind them of the hazards in their work. Also, during these meetings the foreman can assess each worker's physical and mental condition and deal with any problems that might ensue. However, sticking to such a rigid schedule or

Suggestions for Tailgate Meetings in Underground Construction

Before the meeting
- Pick a single hazard which you feel is particularly relevant to the upcoming work.
- Consider ways to control this hazard. (Additional ideas might be found in Cal/Osha guide)

At the meeting
- Tell your workers the hazard to be discussed at today's meeting and give an example from your own experience.
- Ask workers by name for their suggestions of what the hazards are and how they might be controlled.
- Summarize what has been said and add your input.
- Encourage your workers to ask questions.
- Keep meetings brief (5–10 minutes).

This material was prepared under grant number EJFOD342 from the Occupational Safety and Health Administration, U.S. Department of Labor. Points of view or opinions stated in this document do not necessarily reflect the views or policies of the U.S. Department of Labor.

SF0880

Suggested Tailgate Meeting Topics in Underground Construction

- Cave-in prevention and protection
- Positive ventilation in confined spaced
- Overhead high voltage lines
- Traffic control
- Pipelaying
- Trenching
- Earthmoving equipment
- Rolling pipe
- Water mains and utilities in trench or area
- First aid and emergency telephone numbers
- Power tools
- Effects of equipment vibration on trenches
- Personal protective equipment
- Poisonous plants: poison oak, etc.
- Pacing work and lifting techniques
- Housekeeping

Developed by Stanford University Civil Engineering Construction Safety and Health Project, Stanford, CA 94305 **(415) 497-4447**

FIGURE 12-12
Suggested procedures and topics for tailgate safety meetings for crews working in underground construction.

restricting the number of meetings may have drawbacks. For example, a brief outlining of dangers should be a part of the instructions for carrying out every new task. Again, if the project is remote and involves driving long distances, a brief session discussing management's concern for the workers' safety and the importance of safe driving practices may be in order just before the weekend, particularly the longer ones that come with holidays.

An occasional alternative to safety meetings for single crews is to assemble the contractor's entire work force. In this instance, job superintendents might conduct the meeting and express their concerns and possibly make safety awards. By inviting subcontractors' crews to meetings, their members could also be initiated into an overall job safety effort.

To encourage participation and interest in safety, some companies may have a rotating crew safety monitor to serve as another pair of eyes for the foreman. As pointed out by one safety consultant, "These individuals would be naturals, if competent and motivated, not only to point out unsafe conditions or acts to crew members and the foreman, but also to conduct safety meetings."

Safety Award Schemes

A wide variety of monetary and nonmonetary schemes for safe performance are employed by contractors. Those using them are usually convinced that their particular schemes offer the best way to encourage safe practices by those striving for first or repeat awards. Companies which give awards involving money or equivalents such as expensive gifts argue that these are essential for the program to be effective because they clearly demonstrate management interest. Those who do not give such awards claim that publicity, recognition, and attention from higher management are the primary motivating forces and other awards are unnecessary. Administered expeditiously and fairly, either approach seems to be among successful contributors to safe work practices.

To have a formal, effective, and fair safety award system, a contractor should be able to determine quickly who deserves recognition. Few can do so, and determinations usually are based on statements from potential awardees. Some checking is possible for small firms where management is in close contact with individual projects. However, the records of large firms almost never include the safety records of individual workers. Even for foremen and superintendents, records of accidents occurring on work for which they were responsible are scarce.[38] That such award schemes work as well as they do is a tribute to the honesty and fairness of construction people.

Reports from small contractors with excellent safety records demonstrate the effectiveness of a variety of informal safety award schemes. Four examples will illustrate this approach.

Contractor A: The owner deliberately does only one, usually high-risk, contract at a time which he directs himself. He adds an hour's pay and a personal commendation to each worker's weekly check if, in the owner's opinion, no unsafe incident which might have led to an accident was detected for the crew during the week.

Contractor B: This firm places a pack of trading (blue chips) stamps in the pay envelope of every worker whose crew had no recordable accidents the previous week. Foremen and superintendents also receive stamps if crews they direct are accident-free. A demonstrated advantage of this scheme is that it develops spouse and family interest in safety.

Contractor C: This firm deliberately has only two or three contracts underway at one time and the owner follows their progress and safety records closely. His policy is to set the regular salaries of job superintendents at a relatively low level but to provide substantial annual bonuses to them if the company has a profitable year. Comparative accident records are a major consideration when the owner sets individual bonuses.

Contractor D: The foreman with the best accident record during the past year is assigned a new deluxe vehicle (pickup or other) for personal and crew use.

---

[38] Exceptions are firms with some form of accident cost accounting system, as discussed below.

In general, large firms attempt to standardize their safety awards through-out the company, as they would do with a suggestion system (see Chap. 11). If given at all, awards to individual workers include items such as decals for hard hats, company belt buckles, or savings bonds or other relatively small mone-tary awards. For foremen, awards may be, for example, a plaque for super-vising, say, 20,000 hours accident-free, and a gold watch or some other valu-able item at 100,000 hours. Somewhat higher targets are set for superintendents. Other appropriate items would be available for repeaters. Presentation of such awards are made, with appropriate publicity, by higher-level company personnel.

Some very large companies periodically publicize the safety records of their various divisions and award those that show exceptional records or decided improvement. These aim to motivate highest-level management. One difficulty with such schemes is to recognize fairly the level of exposure in different seg-ments of construction, for example, building versus tunneling.

It is not possible here to describe the many safety award schemes that are in use in construction. But, from the few noted, it is apparent that successful ones have in common top management's conviction that safety is important and a strong desire to foster it.

Keeping and Using Accident Records and Costs to Promote Accident Prevention

As shown earlier in Table 10-2, the most important satisfier for all levels of construction managers above foreman is making a profit on the current job. More basic still, at the company level making a profit means survival. This is the reason that (as pointed out in Chap. 2), construction managers at all levels place such emphasis on cost records. Yet many in the industry seem to have a blind spot with respect to accident costs; somehow they fail to recognize that accidents have a strong effect on costs and profits. One reason for this blind spot is that in estimating, the contractor puts premiums paid for workers' com-pensation and other insurances in an overhead item, to be added to direct costs at a fixed percentage. When the bills for insurance are paid, this fixed percent-age will annually be adjusted up or down in estimates for upcoming projects. Among other things, this procedure neglects the fact, mentioned earlier, that insurance premiums reflect past accident experience and a bad record will catch up with the firm in later years. In addition, there will be nothing in the estimate to reflect indirect accident costs which, as pointed out earlier, at least equal and are often several times the direct ones. The issue then, is how to overcome this blind spot about accident costs, which are as real as any others, and focus the attention of busy management at all levels on accident costs and their effects on profit.

A strategy that is gaining increasing favor for directing immediate attention to accident costs is to assess them directly and promptly, not only against the individual projects but against the specific operations and their supervisors. More specifically, the strategy is to develop, adopt, and publicize a schedule of the probable cost of a particular kind and severity of accident, and if such an accident occurs, to charge this cost at once against the project and the particular work item and its supervision. In this manner, accidents are brought home forcibly at once to all levels of management on the job and in a clearly recognizable form—costs. This is far more effective than waiting several years when the costs will show up in a rising experience modification factor which will bring higher insurance costs to the entire company.

Suggested values for the cost of common construction accidents, stated in labor-hours, have been given in Table 12-8. By referring to it, one sees that the real costs of construction accidents are high enough to gain immediate attention.

Essential to using an accident cost system is to attribute accidents not only to cost items but directly to the supervisors involved. A number of contractors who have done so have found that when accident costs are assessed directly against certain supervisors, they have been shown to be losing rather than making money for the company. Until the accident costs were evaluated, they had often been given excellent performance and profitability ratings by higher management.

An accident cost system is effective only if put into place and followed through by highest company management. Then the message becomes clear. As with any innovation, instances can be cited where adapting such a system brought adverse reactions from superintendents and foremen. Their argument often was that accidents were primarily the result of chance rather than of poor planning, supervision, and attitudes. Reactions such as these clearly indicate that training in safe work practices and accident prevention should be a part of introducing the cost assessment plan. Craftsmen also have at times been upset when told of the accident cost-accounting scheme by their foremen. They interpret it as a means for increasing company profits rather than as a benefit to them. Here also, higher management must have carefully planned ways for informing the workers of the system rather than having it heard about in garbled form through the job grapevine. It should be made clear to craftsmen that the plan also benefits them, since workers suffer most from job injuries and this plan is a part of an overall effort to protect them.

Provisions for First Aid and Emergencies

Because construction is hazardous, the law and commonsense both require that provisions be made just in case injuries occur. Injuries can, of course, range from a cut finger to situations in which a worker must have emergency medical attention or be rushed to a hospital. Minimums to be provided, regardless of job size, as specified by governmental agencies such as Cal/OSHA, are first-aid kits, readily available medical services, and personnel immediately available holding current Red Cross training certificates. On large projects, registered nurses may be employed. On very large projects, ambulances may be required; and if projects are in remote areas, physicians and even a small hospital may be made available. If these higher-order services seem necessary, it may be advisable that the owner stipulate what is desired in the job specifications or even provide them, in order to put competing contractors on an even basis.

A difficulty with first aid is to convince workers who have what seems to them to be superficial injuries to have them treated. For example, stopping work to treat a cut or skin puncture may not be considered manly, despite the fact that if left untreated it may lead to a serious infection. Workers and probably foremen and superintendents will frown on visits to the first-aid station, particularly if the incident is to be reported. Creating an atmosphere in which such injuries are taken care of without reflecting adversely on workers and supervisors is a dilemma with which higher management must deal carefully.

Monitoring Site-Specific Safety Procedures and Activities

As demonstrated earlier, higher-level management's concern and involvement is an essential element in safety. Among the techniques already mentioned is knowing, talking about, and rewarding good performance. Another is to see that site-specific measures such as those discussed here are carried out promptly and well. This requires the establishment and subsequent use of a formal monitoring and reporting system.

Specific ways of monitoring site-specific safety efforts and lapses will depend on circumstances. On larger projects, a full-time safety engineer would report to higher management, and possibly to a home-office safety official as well. On smaller projects, monitoring would be a part-time assignment but should always require a formal reporting system. These activities are not intended to usurp line-management's safety efforts but to augment and reinforce them.

It is not possible here to give full details of reporting-monitoring systems. Certainly they will involve (1) a detailed investigation and reporting of incidents and accidents, (2) a follow-through to see that arrangements for new worker orientation and toolbox safety meetings are carried out, (3) a check to see that safety and emergency equipment and arrangements are readily available and in good order, and (4) a formal way to periodically assess project conditions. All these should follow a systematic plan which includes reporting to and required reviewing by higher-level management. Many of these items can be covered by a carefully planned inspection report, an example of which is given in Fig. 12-13. This was developed by a large building contractor. Note that the form (1) provides a very useful list of items and hazards to be observed, (2) covers both the general contractor and subcontractors, (3) indicates if corrective action is needed and whether it has been carried out, and (4) calls for review and sign-off by higher-level job and home-office management.

JOB # _____ JOB LOCATION/NAME _____ DATE _____ TIME _____

JOB SUPERINTENDENT _____ JOBSITE SAFETY COORDINATOR _____

R&S SAFETY SECTION _____ REPORT REVIEWED WITH _____

_____

X = Corrective Action Required

This records: _____ Inspection _____ Safety Meeting _____ Other (specify) _____

	R&S	Subcontractor	N/A; Not Insp.	Corrected			R&S	Subcontractor	N/A; Not Insp.	Corrected
1. Personal Protective Equipment						**7. Fire Protection**				
1. Hard Hats						1. Extinguishers				
2. Eye Protection						2. Flammable Materials Storage				
3. Ear Protection						3. Welding/Cutting Equipment				
4. Respirators										
5. Proper Clothing						**8. Tools**				
6. Footwear						1. Condition				
7. Safety Belts						2. Guarded				
						3. Power Cords				
2. Housekeeping						4. Temp. Power Boxes				
1. Exits & Stairs Clear										
2. Piling & Stacking						**9. Site & Public Protection**				
3. Debris Removal						1. Excavations/Trenches				
4. Nails Bent or Removed						2. Earthmoving Equipment				
						3. Forklifts/Cranes				
3. Ladders & Stairs						4. Fences				
1. Ladder Condition						5. Lighting				
2. Ladder Tied Off						6. Barricades				
3. Ladder 3' Above Landing						7. Signs				
4. Stairs						8. Rebar Caps				
4. Railings/Covers						**10. First Aid**				
1. Perimeter						1. Trained Personnel				
2. Floor Openings						2. Kits/Supplies				
3. Stairs/Ramps						3. Sanitation/Water				
4. Walkways										
5. Elevator Door Openings						**11. Programs/Information**				
						1. Twice-Daily Inspections				
5. Scaffolds						2. Orientation:				
1. Railings						New Employee/Haz. Sub.				
2. Tied to Buildings						3. Safety Meetings				
3. Planks & Platforms						4. Posting Requirements/Signs				
6. Electrical						**12. Other (List)**				
1. Lighting										
2. Grounding										
3. Cords, Plugs, and Receptacles										

Comments _____

_____

FIGURE 12-13
A comprehensive jobsite safety inspection report. (*Courtesy, Rudolph and Sletten, Inc.*)

THE UNION'S ROLE IN CONSTRUCTION SAFETY

The building trades unions speak for their members about safety in several ways. Among the most important is in the political arena, where they are strong advocates for safety legislation at the federal and state levels, and lobby to strengthen it. They are always on guard to see that already enacted legislation is not weakened. Furthermore, representatives are actively engaged in advising on and reviewing the regulations that are issued by federal and state safety agencies to implement safety legislation. Of concern are not only matters affecting safety as such, but issues involving various aspects of workers' compensation such as eligibility for and levels and durations of benefits. Since they are so active in the legislative and regulative areas, union safety representatives are among the best informed on their details and on safety in general.

Some union locals have safety staff who visit and monitor the safety situation on individual projects. They are also on call to union business agents or individual members who have safety concerns. In addition, they may work with universities in safety research, conduct seminars on safety, and advise foremen and other union members about safe practices or how to conduct safety meetings. Contractors interested in safety welcome "another pair of eyes" on their jobs. However, at times, some complain that union safety staff and business agents use safety as a form of harassment in order to gain concessions on other issues.

Unions often participate in joint contractor, union training or apprentice programs and are an effective force in teaching workers and foremen safe practices.

Charges that unions are overly aggressive in their safety activities or, on the other hand, that they do too little, are often heard. There may at times be some truth to both charges. However, they have been and will continue to be a positive force in improving safe practices in construction. One difficulty to be faced in improving safety in nonunion construction is that workers have no organized mechanism for dealing with safety problems.

CONTRACTOR TRADE ASSOCIATIONS' ROLE IN SAFETY

National and local chapters of trade associations such as the Associated General Contractors and the numerous subcontractor organizations usually have staff assigned to safety activities. They also devote space to safety topics in their publications and, in some instances, produce safety manuals or handbooks.[39] Some associations hold or promote safety conferences or workshops for their members. And, as mentioned earlier, contractor or contractor-union sponsored programs for apprentices and for supervisors include safety as an important topic.

As do the unions, contractor trade associations follow closely and attempt

---

[39] See, for example, the *Manual of Accident Prevention in Construction* of the Associated General Contractors.

to influence the laws and regulations covering safety, worker compensation matters, and related issues at both federal and state levels.

In addition to employing safety staff directly, construction trade associations may provide an organizing base for committees or ad hoc task groups of safety and environmental health specialists. These can serve a number of purposes, among them evaluating or disseminating information about new approaches to or devices for safety or reporting on possible environmental hazards. Again they may develop and release expert opinions about proposed safety and health regulations, particularly if they seem impractical or unrealistic.

As would be expected, the level of safety activity in construction trade associations varies widely, primarily depending on how much their membership demands and is willing to pay for.

GOVERNMENT'S ROLE IN CONSTRUCTION SAFETY

As indicated earlier in this chapter, the individual states enacted workers' compensations laws, beginning in Maryland in 1902, which provide benefits to workers injured on the job. Basically these laws dealt with accidents that occurred or were expected to occur. Later the individual states enacted other statutes which imposed on employers a responsibility to take steps to prevent accidents or impairments to health. These laws varied greatly among the states in the conditions they imposed; furthermore, enforcement of their provisions was not uniform and was sometimes lax. In early 1970, under pressure to correct what was seen to be an unacceptable situation, Congress enacted the Occupational Safety and Health Act. In doing so it preempted this field of legislation from the individual states and established a uniform law nationwide. It also created the Occupational Safety and Health Administration (OSHA) in the Department of Labor which was empowered to develop and enforce detailed regulations for safety and environmental health for all industry. Included was the authority to assess penalties, both civil and criminal, against employers who violated the laws and regulations.

Since 1970, Congress has amended and modified the original OSHA law several times. A statute of particular concern to construction had to do with mine safety and health. It seemed to transfer to the Mine Enforcement and Safety Administration in the Bureau of Mines of the Department of the Interior jurisdiction over quarrying and certain other construction operations closely related to mining. In 1983, authority on these construction-related activities was definitely assigned to OSHA.

OSHA legislation provided that federal authority could be transferred to individual states when the state (1) had an approved safety organizational plan and staff and (2) agreed to set safety and environmental health regulations equal to or more stringent than federal ones. To aid in administering the state program the federal government agreed to pay 50 percent of program costs.

Federal legislation delegates research activities in safety and health not to OSHA but to the National Institute for Occupational Safety and Health

(NIOSH), an arm of the Department of Health and Human Services. For this reason, OSHA is barred by law from sponsoring and funding research either in house or by outside consultants.

As indicated, OSHA and its state counterparts have developed standards and codes of practice for safety and environmental health. These are typified by the excerpt from the Cal/OSHA requirements and rules given in App. D, which gives some notion of the many hazards faced by construction workers and contractors. Other major efforts of governmental safety agencies include (1) investigating serious accidents, (2) checking complaints of safety violations, and (3) making unannounced site inspections. With tasks 1 and 2 required by law, unannounced site inspections are sporadic at best. It has been estimated that given the present level of financing and staffing, only 3 to 4 percent of the workplaces will have such inspections.

If investigations show that contractors have violated safety or health regulations, citations are issued. These may include a fine, usually in modest amount. If infractions are very serious and deaths or very serious injuries result, criminal charges can be brought against the contractor's supervision. Citations can be appealed to review boards. As a matter of policy, some contractors always appeal; others do so only in serious cases, if at all.

Under the early OSHA regulations, inspectors of the governmental agencies were required to cite contractors if they found infractions. This practice created a strong adversarial climate. In some cases, contractors responded by barring inspectors from their sites unless they came armed with a court order. Charges by contractors that complying with OSHA regulations increased construction costs without compensating benefits have been common. For example, an opinion survey showing the perceptions of representatives of the 400 largest contractors, reported in 1976, attributed, on the average, 2.8 percent of total job costs to OSHA compliance. Results of a repeat survey reported in 1983 lowered this percentage to 1.4, indicating that contractors felt better able to live with OSHA. Replies from contractors in Ohio and Indiana usually gave somewhat higher results. A preponderance of those surveyed felt that OSHA's regulations increased costs and did little to improve safety. They urged that its requirements be modified.[40] More recently, some OSHA or OSHA-state safety agencies have set up separate advisory inspection programs. Under them, a contractor can request an inspection and the observer can point out unsafe conditions but is not empowered to issue citations. Here, again, budgetary and staff limitations confine the size of the program.

A substitute for inspections by safety-agency personnel has been explored on a few large projects by contractors qualified by an already excellent safety record. It involved an employer-union-OSHA agency agreement to have a management-worker project safety committee. Members—possibly two from management and two from the work force—conducted periodic inspections and received and investigated complaints submitted anonymously by workers.

---

[40] For details of these surveys and a bibliography, see E. Koehn and K. Musser, *Journal of Construction Engineering and Management*, vol. 109, no. 2, June 1983, pp. 233–244.

Activities were carried out during working hours as a part of regular work assignments. Membership was rotated. In return for this good-faith effort by the contractor-union team, the safety agency gave up its right to make job inspections. Reports indicate that this nonadversarial approach has worked well.

Statistics on the first 10 years of OSHA showed a disappointingly small decrease in construction and other industry accidents. Also, OSHA's image among many contractors was first seen as obstructionist rather than helpful. It was therefore decided to augment the enforcement or punitive approach with an educational effort called the "New Directions" program. It was intended to demonstrate through workshops and other educational efforts how unnecessary and costly accidents were and what could be done to reduce them. Early results from this effort are promising.

Governmental agencies that buy construction services have been leaders in requiring that their contractors operate safely. Many of these agencies, for example, the Corps of Engineers and Bureau of Reclamation at the federal level, have developed safety codes and manuals and incorporate them in their specifications. These safety provisions are administered by the agency's on-site supervisory staff. On large projects, separate safety personnel may be assigned. Zealousness in forcing contractors to adhere to safety regulations varies, of course, among governmental agencies and with the responsible individuals. Among the difficulties is the degree to which enforcing safety rules appears to interfere with the independent contractor's right to direct the work, even though observance is called for in the contract. It is clear, however, that contract provisions on safety and their enforcement can be a strong force for accident prevention.

THE OWNERS' ROLE IN CONSTRUCTION SAFETY

Owners (buyers) of construction projects, including governmental agencies, ultimately pay the costs of accidents on their projects. As discussed earlier in this chapter, these include not only the charges for workers' compensation and other insurances but the hidden costs of job disruptions and delays as well. At times, also, owners may be held liable for claims which assert that they were negligent in not properly supervising the contractor's activities. They, along with the contractor, receive adverse publicity and suffer a poor public image. In sum, then, owners pay a substantial fraction of the costs of accidents on their projects.

As a group, at least until recently, private owners, although recognizing a moral obligation that workers should not be hurt on their projects, have not focused strong attention on the safety records of the contractors they employ.[41] This attitude is changing rapidly, in part because of a widely publicized Business Roundtable study. It made clear that (1) accidents need not

---

[41] There have been, of course, exceptions. Certain owners, particularly those involved in carrying out hazardous manufacturing or process operations, have imposed strict safety standards and have carefully monitored the practices of contractors working for them.

happen and (2) construction accidents were costly to owners. Also it offered procedures for evaluating contractors' safety records as a part of the selection process.[42] Owners, particularly those awarding cost-reimbursable or profit-sharing contracts, will have tools with which to select contractors with better safety records. In addition, all buyers of construction will see the advantages of engaging contractors with better safety records or of pressuring them to do something about safety. Owner interest in both the private and public sector can have a dramatic effect in reducing accidents. After all, owners control the money and award contracts. To quote an old adage: "He who pays the piper calls the tune."

THE ROLES OF ENGINEERS AND ARCHITECTS IN CONSTRUCTION SAFETY

Even before construction is undertaken, engineers and architects can play several roles in construction safety. Typically they design the facilities to be built. One role for them in construction safety then is to be sure that the design facilitates building without exposing workers to undue hazards. Again, engineers and architects usually prepare the specifications and other bid documents. A second safety role is, then, to impose safe practice requirements on the contractor through these documents. Finally, at least in private work, engineers and architects advise the owner about the relative abilities of potential contractors to complete projects economically and on time. An equally important measure of contractors' abilities should certainly be their safety practices and records.

Engineers who represent the owner or are on the contractor's staff are deeply involved in the details of planning and of designing equipment and facilities for project execution. Many examples of the safety implications of such work can be developed merely by reading through the Cal/OSHA safety guide given in App. D. For instance, engineers are often involved in selecting or approving personal protective equipment and in many procedures involving environmental health. They may design or approve devices needed for access to the work such as stairways, ladders, safety nets, or scaffolds. In fact, Cal/OSHA maintains that scaffolds "must conform to design standards or be designed by a licensed engineer." Finally, engineering judgment is the principal basis for setting the requirements for tools and equipment. In instances such as these, safety should be a primary factor in design or selection.

As is the case with owners, engineers and architects sometimes are held liable when construction accidents cause injury or damage property. This can happen even when the contractor is primarily responsible. The basis of such

---

[42] See *Improving Construction Safety Performance*, The Business Roundtable, 1982. Since that report was issued, a variety of expansions and extensions of the procedures for evaluating contractors' safety performance have been developed, including a book titled *Construction Safety Management* by R. E. Levitt and N. M. Samuelson, op. cit., and a proprietary method employing the expert-systems approach developed by R. E. Levitt.

claims is negligence, defined as a wrongful or negligent act resulting in an injury. One form of wrongful act ascribed to engineers and architects is that of observing an unsafe condition or practice and doing nothing to correct it. Another is failing through lax inspection to detect an unsafe situation. Over the years, design professionals have attempted to limit their liability. In their contracts with the owner for supervising construction, they refuse to assume responsibility for safety. For example, the Standard Form of Agreement between Owner and Architect states that "the architect shall not have control or charge and not be responsible for construction means, methods, techniques, sequences or procedures, or for safety precautions and programs in connection with the work, (and) for the acts or omissions of the contractor, subcontractors, or any other person." Provisions such as these are not necessarily sufficient to remove the liability. And even if the courts rule that engineers or architects have no liability, it is costly to them since they must prepare a defense and then go to court to obtain dismissal from such lawsuits.

All in all, engineers and architects have professional, financial, and humanitarian reasons for being concerned about construction safety. It is encouraging to see their increasing attention.

SUMMARY

This chapter has attempted to summarize the present-day situation regarding construction accidents and the knowledge and practices that can reduce their impact on human victims and on construction costs. It begins by outlining a rational approach to the problem, summarizes the appalling industry-wide statistics on accidents, and emphasizes that construction has been and can be carried out almost accident-free. It then summarizes research that shows how attitudes and behavior of construction people at all levels can affect the number and severity of accidents and explores the site-specific aspects of safety and environmental health. Next it describes several workable approaches that contractors can employ in implementing accident-reduction programs. Finally, it examines the roles that unions, trade associations, governmental agencies, owners, and engineers and architects play in construction safety. From all this it can be seen that sufficient means, proper attitudes, various approaches, and a variety of interests are available to reduce substantially the human and money costs of construction accidents in the years ahead.

CHAPTER 13

COMPUTERS AND OTHER TOOLS FOR IMPROVING PRODUCTIVITY IN CONSTRUCTION

INTRODUCTION

The purpose of this chapter is to survey a few of the many computer, graphical, mathematical, economic, simulation, automation, and expert-systems techniques and tools that are or will be available to on-site construction managers. This presentation cannot and does not attempt to completely cover the field; rather it suggests areas for further exploration for those who feel that particular applications may be useful.

It has been demonstrated repeatedly that analyses made away from the noise and confusion of the work face can improve productivity. The more commonly used techniques, including planning, scheduling, and methods improvement studies, have been discussed earlier in this book. Other less frequently used and often more complex approaches, many of them computer-based, are described here.

The first gain that may result from employing these more sophisticated techniques is that users are forced to think through their problems in advance in order to formalize them into computer programs, graphs, equations, simulations, automated approaches, or expert system analyses. Otherwise, operations may not be analyzed and thought through carefully; commitments that cannot be rescinded may have been made or execution may have begun. A second advantage of such analyses if properly designed and carried out is that they reduce the opportunities for field decisions that are often influenced by preconceived notions, spur-of-the-moment hunches, or personal desires or prejudices.

The approaches described here are not infallible. They are only as good as their factual inputs, their analysis, and the application of their results. Furthermore, users must understand their limitations rather than take the findings blindly. To dismiss them because they defy conventional wisdom or rules of thumb is ill-advised and can be costly. This is illustrated by the common reaction of some dirt foremen or equipment operators when they are shown the results of the simple graphical solution for economical loading times for scrapers described in detail below. This analysis demonstrates that in many situations, hourly production is higher with smaller loads. Yet the conventional wisdom or gut feeling is that productivity is always highest if scrapers are push loaded until dirt spills over the sides. Here, as in many other instances, analysis can pay off if the findings are implemented.

The sections which follow describe several techniques that have proved both valuable and workable or have shown promise for the future. The human problems of getting acceptance were treated earlier, in Chap. 10.

COMPUTERS IN CONSTRUCTION

Computers today are an accepted tool for construction and in many firms will soon be more common in home and job offices and field shacks than typewriters and calculating machines have been in the past. Some among the many ways in which this is happening are summarized in the pages which follow.

In most construction companies, computers were used first for data-processing and record-keeping. In effect, the computer replaced traditional data assembling, computing, summarizing, and records-keeping functions. When first introduced, computers were costly and required programmers with special skills. Larger contracting firms often set up special departments; smaller companies that used computers at all sent raw data to computer centers that batch-processed the information using general or specially developed programs. With either approach, computers took over many number-crunching functions. Today, with the advances in both hardware (computers) and software (programs), there is far less need for either programmers or special outside services.

Acceptance of computers has come gradually as (1) higher-level management in both home office and field have become accustomed to and dependent on computer applications on the business and data-processing sides, (2) mainframe, mini-, and microcomputers have become less costly, and (3) user-friendly software programs have been developed and are more readily available so that line supervisors as well as planners and engineers can use them.

The current computer situation is confusing and is changing rapidly. There are many computer options with varying capacities and costs and a large variety of programs usually but not always matched to the particular hardware.[1] Furthermore, almost every construction firm has chosen its own unique com-

---

[1] *The Constructor,* the magazine of the Associated General Contractors, December 1984, pp. 30–57, lists software packages for construction provided by some 150 vendors. Many of these firms offer orientation and training programs, workshops for individuals or groups, as well as manuals, site visits, and on-line assistance.

bination of machines, programs, and reporting systems. For example, one large company has a mainframe computer in the home office which handles company-wide accounting and other functions. It also communicates with microcomputers on every project. Software to fit the company's needs has been developed in house. The mainframe accepts, processes, and stores data of all sorts that are important to higher management and also monitors and makes payments for materials and payroll for individual projects, using data sent in from on-the-job microcomputers. The mainframe also carries certain details of individual job estimates, schedules, expenses, progress payments, and financial projections. The on-site microcomputers are employed for planning, scheduling, and materials control, among other things, and can call up from the mainframe some but not all of the information about the project that is in the mainframe memory. Other firms may employ a combination of mini- and microcomputers to accomplish about the same purposes.[2] Smaller firms may contract with a service bureau to handle accounting, payroll, and similar financial matters and use a home-office or on-site microcomputer for estimating, scheduling, setting up and processing change orders, and similar functions. A key variable that affects each firm's choice is that the cost of computer hardware of a given capacity has fallen rapidly, some say by as much as 25 percent per year, although prices may be bottoming out. Also, software is proliferating, is more sophisticated, and is becoming interchangeable among makes of hardware. Given all these alternatives and variables, only the most general statements about computers and programs can be made here, since any specifics about this fast-moving field would soon be outdated and no longer applicable. Even so, all this should be viewed merely as a constraint governing computer use; every advance ultimately makes computers a more valuable tool.

With such widespread application of computers there are serious concerns about confidentiality and security. Simple strategies include introducing codes into procedures for gaining access to data which must be satisfied before they will be produced. For example, restrictions can be placed on the levels of management that can see certain information. Another example concerns job payrolls and other access to funds. In the procedures for writing paychecks with jobsite computers, the names, classifications, and payments to individual workers must first be cleared by the home office computer to prevent the addition of fictitious individuals. Similar controls can be exercised over payments for materials, equipment, tools, or supplies.

This brief discussion of the roles computers play or can play in improving the construction-management function does not attempt to separate home-office applications from those aimed at field operations, because they will differ widely among individual firms.

Describing all the ways in which computers can already make managers at

---

[2] See J. L. Rounds and G. Warning, "Impact of Computing on Mid-Sized Construction Companies," *Journal of Construction Engineering and Management,* ASCE, vol. 113, no. 2, June 1987, pp. 183–190.

all levels more effective is impossible. The discussion which follows looks first at a few of the recent applications on a stand-alone basis and then briefly explores management information systems (MIS) by means of which these applications and other isolated activities can ultimately be integrated.

Specific Applications of Computers as a Construction-Management Tool

As already mentioned, computer use in construction is widespread. Hardly a company or large job is without these information-handling tools. In selecting among the many alternatives, smaller organizations often choose microcomputers, while larger companies often combine mainframe or minicomputers in the home office with microcomputers in their other offices and on job sites. Advising on choices in this complex and fast-moving field is beyond the scope of this book; even so, a brief explanation about the alternatives is appropriate.

The trade-offs in selecting among computers are cost, capacity, available software, and anticipated useful life. Rarely can computer hardware be expected to be sufficiently efficient for more than 5 years, given the inevitable expanded use and changing company requirements. Also, new and more efficient and user-friendly software packages come out often. This does not mean that companies should delay using computers but only that any choice should be seen as a short-term trade-off between cost and the new or expanded applications, tailored for construction, that can make any system obsolete.

The sections which follow give an overview of how many organizations are using computers. Footnotes provide references to more extensive articles on particular subjects.[3]

Computers Applied to Payroll and Cost Accounting   Payroll calculations and cost accounting are often an initial jobsite application of computers. In this role they merely offer an easier and more accurate means for carrying out already existing tasks. In general terms, the processes are as follows:

1 Time sheets for individual crews, made out by the foremen or cost engineer, list the hours worked on straight time and overtime, with worker-hours for each individual broken down into cost-accounting classifications. When appropriate, pertinent data are recorded for equipment use and for permanent and consumable materials. These data are entered directly into or transcribed into a jobsite computer, often in tabular form on a spreadsheet.[4]
2 Starting with these raw data and using unit rates or prices that have been established and entered into the computer, it

[3] An excellent general report on small computers in construction, prepared by an ASCE Task Committee, appears in the *Journal of Construction Engineering and Management,* ASCE, vol. 111, no. 3, September 1985, pp. 173–189. It includes a glossary of computer terms and a recommended reading list.

[4] For a discussion of spreadsheet applications in construction, see L. C. Bell and B. C. McCullouch, *Journal of Construction Engineering and Management,* ASCE, vol. 109, no. 2, June 1983, pp. 214–223.

a Assembles and prints out the hours worked straight time and premium time at the various rates from data on the daily-submitted field reports. From this it determines gross pay, deductions for taxes, benefits, and contributions, and net pay. After the computations are made and verified, the on-site computer may actually write the paychecks or, alternatively, transmit the raw data to a central office computer for checking and check writing.

b Computes and tabulates unit costs for individual work items by assigning payroll, equipment, and other charges appropriately to the item and dividing the sum by the reported production.

c Calculates and prepares the required records for appropriate individuals or agencies by summarizing the accumulated data. These include income taxes withheld; workers' compensation and other insurances; and union health, vacation, pension, and similar funds.

These examples show that, once the computer is supplied with the raw data, it can sort, store, accumulate, and print out all this information in any form needed.

Along with information such as that just described, it is common practice to compile and print out tabular or graphical progress reports of all sorts, such as a combination of payroll, progress, and equipment data which can give managers pertinent information about how a project is coming along. For example, programs can be written which extrapolate past costs and in turn project trends into the future. From data such as these it is possible to assess both financing needs and expected profit or loss. Also, items that were overrunning estimated or budgeted unit costs can be flagged in the weekly or monthly summaries to alert busy managers to potential troubles. Correctly used, such data have been found to be most helpful. Unfortunately, instances can be cited where flagging obscured rather than illuminated problems. A common example is that highlighting relatively small items that have costs overrunning the budget may divert attention away from others that appear to be doing well but, studied carefully, could be made even more profitable.

Computers as an Aid in Estimating   Cost data from past projects are a fundamental tool for estimators, and computer records from them can be a most useful tool. These records might be kept in money terms or in labor and equipment hours per unit of output. For simple projects involving the same sort of work, estimates of unit costs often can be taken directly from computer data recorded from past projects. Along with these would go computer tabulations of quantities taken from the drawings or bid schedules for upcoming projects. Combining the past unit-cost data with the new quantities could lead to a preliminary estimate of direct costs for bidding purposes. Procedures for making adjustments in final pricing to recognize differences from earlier work, changes in wage scales, material prices, equipment costs, competition, and other factors can easily be incorporated into the procedure. Advantages from such computer-based estimating include less chance of error and an ability to prepare bids more rapidly, which saves time and facilitates bidding more work.[5]

As projects become more complex and variable, the computer's role in estimating evolves and changes. The programs will be more complex but flexible and adaptable and able to respond to any estimating method. Most certainly, as with traditional methods, they will permit easy comparisons with unit costs from past projects.

Possibly the biggest obstacle to more widespread adoption of computers is that estimators familiar with other approaches and having pride in their knowledge sometimes resist them because they find them to be potential threats to their professional standing or job security.

Computers as an Aid in General Accounting   Computers play a substantial role in general accounting for many contractors. Among the functions can be maintaining records of assets and liabilities, balance sheets, accounts payable and receivable, and other standard accounting needs too numerous to mention here. Not the least among these is that of tracking accounts payable to be sure that payments are made in a manner which permits discounts to be collected.

Computer Access to Personnel Records   Personnel records have many uses and can be employed more effectively if available on a computer. Particularly in larger companies, such summaries can provide a data base from which management can classify employees as a step in establishing equitable and comparable salary scales and pay increases or bonuses. The ability to sort and produce data about employees in a variety of ways can be most helpful; for example, the computer can be used to list personnel by present assignments, abilities, and experience when staffing new projects or reorganizing for phase changes in ongoing projects. Of course, such records or listings cannot fully describe people and are not a substitute for personal ratings and other evaluations. But they do help to put decisions about personnel in a form that is more manageable and less subject to bias.[6]

Computers as an Aid to Safety Monitoring and Improvement   As pointed out in Chap. 12, accidents, which are costly, can be reduced if management works at the problem. An important step is to have computer-based records telling where and how each accident occurred and its consequences, the supervision under which it happened, the worker or workers involved, and its probable cost, defined by measures such as were given in Table 12-8. From such a data base it is easier to develop, summarize, and provide information to

[5] For a recent report and bibliography on microcomputers for estimating see Z. Herbsman and J. D. Mitrani, *Journal of Construction Engineering and Management,* ASCE, vol. 110, no. 1, March 1984, pp. 19–33.

[6] An unusual use of personnel data on an atomic project was to determine the hours of exposure of workers to radioactivity so that they did not exceed the allowable maximum.

management in a variety of forms so that it can see ways to improve performance in safety. Among the uses are: (1) information for publicity and (2) a means of introducing accident records into individual personnel files so that they become an element in performance ratings when considering performance, promotions, or bonuses.

Computers for Activity Planning and Scheduling[7]   Planning and scheduling entire construction projects and the operations within them require thinking through how they will be built step by step. Concerns include the resources needed (e.g., drawings, workers, managers, materials, equipment, tools, and access), the time available, the sequence and interdependencies among operations that must be observed, and environmental and external constraints. With these variables taken into account, schedules can be developed which set the timing for carrying out individual actions and activities.

The complexity in planning and scheduling entire projects varies enormously depending on factors such as the nature, size, and interdependencies to be expected. As pointed out in Chap. 5, analytical approaches (except for simple projects where verbal orders will suffice) range from bar (Gantt) charts to very complex computer-based CPM (arrow diagram), PERT (Program Evaluation and Review Technique), or PDM (precedence, node, and connecting line) diagrams involving literally thousands of individual activities. The details of these project-level techniques, including computer applications, are spelled out in numerous books, technical articles, and reports and are outside the scope of this book. But their completeness, accuracy, and timeliness can noticeably enhance on-site productivity and therefore cannot be ignored. In the first place, on-site operations will be severely affected if project planning and scheduling is not adhered to or is incomplete or out of phase in any respect. Short-run scheduling, an important subject in this book, must follow the overall job plan if it is to be effective. In technical terms, short-run schedules are subnetworks of the overall CPM, PERT, or PDM networks for the entire project. Often these are presented in bar chart rather than network form; but either can be displayed on or plotted by computer.

With the coming of microcomputers on job sites, planning and scheduling can be better integrated. Furthermore, in the future the actual short-run job schedules in the microcomputer at the job site will be fed back into the overall project schedule so that the latter reflects what is actually happening. In this way, the common complaint that most overall project schedules are out of date and useless may be silenced.

Computer Applications on Job Sites

As already indicated, microcomputers suitable for and cheap enough to provide for construction job sites are readily available. The degree to which they

---

[7] For an early but thought-provoking discussion of computers in planning and scheduling and the directions they may take, see B. C. Paulson, Jr., *Journal of the Construction Division,* ASCE, vol. 98, no. CO2, September 1972, pp. 275–286.

will be applied depends, therefore, on a willingness of field supervisors to develop knowledge of what computers can do and then with encouragement by management make an effort to use them. Questions will naturally arise about the kind and extent of the system (one computer or separate ones for different functions), who will be responsible for the completeness and accuracy of the inputs to the system, and who will have access to the information that is stored or produced. A typical question might be, "Who should account for materials that have been or are to be delivered to the site?" Should this be an on-site engineering or purchasing function, or should project line management have its own computer and keep its records so that each supervisor has ready access? These and many similar questions must be answered separately by each firm for each job, but the answers will be increasingly important as the use of management information systems (MIS) advances.

The paragraphs which follow briefly describe a few among many actual or possible applications of computers at job sites.

Gathering Job Payroll and Other Pertinent Data   Job payroll is commonly reported daily by the foreman of each crew, although timekeepers or job engineers may sometimes have this duty. The information is usually first recorded on a prepared form that shows the name of each worker, the hours of straight and premium time worked, and a breakdown into cost accounts to which the hours are allocated. Equipment or materials usage is reported in a similar manner. This information later is summarized in separate reports serving payroll, cost-accounting, and other functions. In many companies, foremen are saddled with the responsibility of reporting this information as an after-work task. Sometimes they do not take the cost breakdown portions seriously, or they may distort the information in an effort to protect themselves.

Payroll, equipment, materials, and production reporting tasks can be simplified and probably made more accurate with a microcomputer. Foremen at job terminals could call up software programs that would project cost-reporting forms on the viewing screen. For payroll, these displays would include a column showing appropriate cost codes. Also, they could prerecord the names of the usual members of the crews. Foremen could then quickly record hours worked in the appropriate spaces. The computer would then check the data and extend it as desired. Equipment, materials, and production reporting could follow a similar pattern. Finally, this information could be transferred into permanent records and summarized. Automation of record keeping is easily within the capability of present-day microcomputer hardware and software; implementing a suitable system is another matter.

Recording Job Happenings   On well-run projects, managers at all levels keep progress reports and diaries or their equivalent. In them, they record events of the day, including descriptions of work accomplished, changes, accidents, or other important happenings. Here the computer could serve to record and store the data and, if appropriate, transmit it to other people. Once an individual has learned to record messages in the computer by acquiring rudimentary typing skills, keeping and storing such records will be easier than writing them out longhand. The hardware and software are there; what is required is for managers at all levels to develop new habits which incorporate computers.

Intrajob Communications   Projects of any size today use portable, hand-carried, or vehicle radios which provide verbal communication around projects. Computers with terminals at strategic locations can both support and formalize these information exchanges. For example, messages to appropriate people can be introduced into the system (commonly referred to as electronic mail) and intended receivers can have access to them at their convenience. The list of potential uses for such communication is almost endless. If desired, these interchanges on radio, telephone and computer can be recorded.

Dissemination of Job Schedules and Similar Information   Details of schedules such as when certain crews are to do what can readily be made available on call at conveniently located terminals. In addition, changes can be made known to all and recorded for future reference. This is far better than relying on word of mouth or written plans, some of which may relay out-of-date information.

Tracking Resources: Drawings, Ordered or Stored Materials, Tools, and Supplies   As discussed earlier, crews can be far more productive if they have the necessary drawings, materials, tools, equipment, and access to the work area. On complex projects, this need can be satisfied only by tracking the availability and location of each resource needed for each task. This has traditionally been done by the superintendent or foreman, who locates items either by physically looking for them or by hunting through written records. Often the records are incomplete, not in usable order, or not readily accessible. Computers can make the process of finding things quicker and easier.

That the status of materials management has been far from ideal is supported by a study conducted for the Business Roundtable.[8] This study found that materials management is generally neglected. It states that "Materials management lacks definition, boundaries, credibility and acceptance much as 'Scheduling' and Cost Engineering did 15 to 20 years ago." This statement is supported by information from the following table in the report which illustrates how responsibility is often dispersed among members of construction organizations.

[8] See *Business Roundtable Report A-6,* app. 5, February 1983.

	Performing member of management team			
Materials activity	Management (purchasing)	Project management	Construction	Engineering
Materials planning		X		
Requisition schedule	X			
Specifying goods and services				X
Materials takeoff				X
Order placement	X			
Lay down and warehouse space planning			X	
Materials liaison		X		

The Roundtable report proposes establishing the separate function of materials management, which would administer or at least coordinate and monitor all materials activities so that all aspects of the problem would be covered. Whether this or some other strategy is adopted and should be expanded to cover other resource tracking is debatable. It is clear that computers can provide a solution.

As a result of the Business Roundtable report, follow-up studies of the benefits of computer-based materials management systems were made under the auspices of the Construction Industry Institute. Among their findings were that:

• Such a system can bring a 6 percent reduction in craft labor costs.
• Such a system will reduce the bulk material surplus on a project from 5 to 2 percent; a savings of 3 percent. This is significant, particularly when such bulk materials are usually sold at the end of the project as scrap.
• Other cited benefits included reductions in project cash flow and reductions in the size of the management work force.[9]

A specific example of how computer monitoring can improve job and crew productivity involves the handling of prefabricated components for a large industrial project on a crowded site. These are ordered several months in advance and often a single order includes items destined for several places in the project. Under traditional record keeping, tracking individual items had been very difficult and time-consuming. First of all, the contractors' records were set up and filed by purchase order number, done so to facilitate the payment of vendors in time to collect discounts. Next the records were filed by the pur-

---

[9] For further detail on these studies see "Costs and Benefits of Materials Management Systems," *CII Publication 7-1,* November 1986; "Attributes of Materials Management Systems," *CII Source Document 1,* April 1985; "Cost and Benefits of Materials Management Systems," *CII Source Document 17,* July 1986; and "Project Materials Planning Guide," *CII Source Document 27,* June 1987. L. C. Bell and G. Stukhart, *Journal of Construction Engineering and Management,* ASCE, vol. 112, no. 1, March 1986, pp. 14–21, and ibid., vol. 113, no. 2, June 1987, pp. 222–234 summarize these documents.

chasing department. Field personnel could determine if the item was on hand and locate it only by laboriously going through receiving records or by undertaking a hunt through the storage area.

With an integrated computer system, each component can be cataloged in several ways, such as by: (1) purchase order number; (2) its final position in the project, identified by coordinates on an isometric drawing of the site; (3) scheduled arrival and installation dates; and (4) the craft designated to perform the work. At a stated interval before an item is scheduled for arrival or installation, the computer issues a check-availability warning on items that have not reached the job site so that they can be expedited or other actions can be taken. When deliveries of fabricated units from each supplier arrive on site, purchasing is not only alerted, but storage locations are noted in the computer records so that they can be found later. All this information is available for foremen at an on-site computer terminal. With this system, the master-computer record (data base) might be based on, for example, purchase order numbers. However, it can be quickly sorted and reproduced in other arrays such as item number, location in the finished plant, location in storage, proposed arrival or installation date, or designated craft. Through displays or printouts arranged in appropriate form, field personnel can count on knowing that all needed items are on hand and can be easily located.

In addition to permanent materials, many consumable items are required on construction sites. Examples would be rod and other welding supplies, pipe and fittings for plumbers, conduit and other electric supplies, and lumber and other standard items for carpenters. Through some simple sign-out arrangement, recorded in the computer, a record of stock on hand can be maintained along with an automatic reminder to reorder at appropriate times.

Tools could also be assigned and tracked by a simple computer program. As discussed elsewhere in this book, unavailability of tools is among the most common causes of worker delays, with accompanying problems far deeper than a simple lack of records to show which workers, foremen, or superintendent have them. But readily accessible tool information can be an important adjunct to a tool supply and control program.

The examples cited illustrate how field management can use computers to simplify one of its tasks, thereby saving time and energy and avoiding false starts and work delays. Parallel systems can be and have been developed to keep track of and control other work elements such as drawings, special equipment, scaffolding, special tools, and commonly used items of all sorts. To what degree this is done depends primarily on the willingness of managers to be involved.

Other Possibilities for Job-site Computer Applications   Opportunities for employing computers on job sites can develop whenever there are needs to share, exchange, process, or store information. Some of these involve doing present assignments differently, as illustrated by the examples given above. Other possibilities include using the computer to gain new insights from data now being gathered. Here, the ability to program the computer to develop and plot graphs and charts of items as diverse as cash flow or crew time losses can be valuable. Finally, the computer makes possible the ability to explore and evaluate new or different approaches to job problems. This is typified by simulation and expert systems, which are discussed later in this chapter.

Management Information Systems (MIS)[10]

A management information system is a coordinated computer-based information-gathering and record-keeping arrangement which can provide information from a common data base to all units of an organization. Such a system has the advantage that everyone has the same information; furthermore, if corrections are made to it, the change becomes instantly available to all. This obviates the need to separately inform top management, personnel, engineering, purchasing, warehousing, accounting, construction, and others.

Many construction organizations are developing and using some elements of MIS data bases. The more advanced systems incorporate information on labor; purchasing; material and equipment inventory and availability; major project quantities; schedules; cash-flow projections; and certain other important items. Access to the data is controlled and made available on a selective basis. Likewise, changes can be made only by authorized personnel. Top management would have full access. However, data to a project manager might be limited to information concerning only his project, and field personnel could call up only the information peculiar to its operations. Similar restrictions would cover the various departments of the home office.

An ideal and complete management information system requires inputs of vast amounts of data pertaining to all aspects of a company's planning and operation into a large computer. These are then processed and released in readily usable form for decision making. Much of this reporting is on a scheduled basis, fitted to the particular use; however, prompt responses to special needs is possible. Proponents of such a system refer to it as a glue that will fuse a construction organization and its operations into an integrated whole.

As of today, no such integrated management information systems exist as ready-to-use packages; they remain a target for the future. As described above, bits and pieces, which in time will make up such a system, are now in everyday use; but they have not been linked into an overall scheme.

To design and implement a complete MIS system for a large construction company can be a formidable task. Working toward it can be expected to take

---

[10] For a brief review of the status of the management information systems of design, project management, and construction firms see *Engineering News-Record,* May 30, 1984, pp. 45–58 and Sept. 25, 1986, pp. 22–32, the latter of which summarizes a roundtable discussion among the MIS managers for several large construction companies. It indicates that there are many approaches based on many combinations of hardware and software. Progress is being made, but full realization of the potential of such systems is yet to come.

large amounts of time, money, thinking, and cooperation, and probably calls for outside expertise. The first step is for each class of user clearly to define its information needs, not only by subject but in terms of amount of detail. From these needs will come the specifics of the data base to be assembled and stored in the computer. Only then will it be possible to call up the appropriate data, assemble and analyze it, and provide the findings in suitable form at a level of detail necessary to permit action by the parties to which it is provided.

To put the demands that would be made on a complete MIS system into perspective, a listing of a few of the activities for which it would provide information in suitable detail is helpful. These must be arranged in ways that make comprehension and action possible. For many of these items, the same sort of information is needed but the level of detail provided and the presentation format will vary with the user's position in the organizational hierarchy. Furthermore, the various ways the information is to be used, among them control, appraisal, approval, or routine communication, may affect both content and schedule for reporting the results.

A partial listing of activities that would provide inputs to or draw information from a comprehensive MIS system could include:[11]

For bidding or otherwise obtaining work

- Quantity survey
- Unit costs on similar work from the past
- Unit and other costs established for this project
- Anticipated schedule
- Resources required—people, equipment, materials, etc.
- Anticipated cash flow
- Overhead and other add-ons

For setting up, administrating, and monitoring a going project

- Data (see above) from earlier projections of project activity
- Detailed budget and variations from it
- Detailed schedule (critical path or precedence diagram) with changes as work progresses
- Progress and final payment records, including revised quantities
- Cash-flow records
- Purchasing: ordering, paying for, and tracking permanent and other materials and collecting discounts
- Payroll, including withholdings and contributions to various accounts
- Unit costs compared with budget
- Equipment use and costs
- Safety performance by tasks, supervisors, and individual workers

---

[11] For a more comprehensive listing of activities to be covered by an MIS system and of the parties that should receive information about each, see K. A. Tenah, *Journal of Construction Engineering and Management,* ASCE , vol. 110, no. 1, March 1984, pp. 101–118.

Home-office records

* General financial data—income, disbursements, etc.
* Personnel information including talents, availability, service records, vested rights
* Safety performance records of permanent supervisors at all levels including foremen

This partial listing of activities supplying information to and needing feedback from an MIS system demonstrates why implementing a complete program is possible only with very substantial computer capacity for data storage and processing. Furthermore, retrieval as needed from such a system will call for a number of stations or terminals. One possibility could involve a number of terminals and printers for a mainframe computer in the home office, coupled with microcomputers for both inputs to and outputs from the system at job sites. All in all, the goal of having complete MIS systems is a challenging one; achieving it will not be easy.

CONSTRUCTION APPLICATIONS OF GRAPHICAL MODELS

Graphs give data in pictorial form. They are particularly helpful because they can include significant amounts of data in charts or graphs which the mind perceives as single chunks; in contrast, data in tables requires a chunk of memory for each number. Where graphs have heretofore been plotted by humans, today, computers develop them directly from data stored in their memories.

Graphical models go a step beyond graphs, which merely present data; they permit users to solve problems without resorting to computations at all. They also permit direct comparisons among solutions and sensitivity analyses by which the relative effects of changing one or more variables can be seen.

Many examples of graphical models for solving construction problems can be cited. One of the most interesting, developed some years ago by the Caterpillar Tractor Company but pertinent today, is used to determine the optimum time to push-load scrapers. As mentioned earlier, there is the prevalent notion that pushing scrapers until they spill must be the most economical way because each then carries its maximum load. What the graphical model shows is, first, that there is no right push-loading time, and second, that it provides a quick way for determining the optimun push times under different operating situations.

The graphical analysis for the scraper-loading problem begins with the development of a load-growth curve. Such a curve for a particular soil and a large scraper and suitable pusher is shown in Fig. 13-1. This curve demonstrates that as pushing begins, the load increases rapidly, but as scraper capacity is approached, the loading rate decreases. For example, Fig. 13-1 indicates that the first 5 cubic yards are loaded in about 0.07 minutes (4 seconds) but that 0.24 minute (14 seconds) are required to increase the load from 35 to

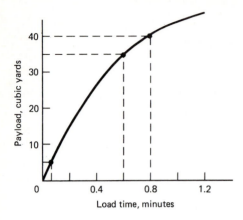

FIGURE 13-1
A typical load-growth curve for a bottom-loading scraper pushed by a track-laying tractor.

40 cubic yards. Stated differently, early loading per unit is three times faster than late loading. The question becomes then, "Is overall production per unit time for this tractor-scraper spread improved by continuing to push at the slower rate of loading and if so, for how long?"

Figure 13-2 offers a graphical solution to the scraper-loading problem, recognizing two different situations. On the one hand, if there are too few scrapers and the pusher is idle between loadings, scraper output will determine pro-

FIGURE 13-2
Use of the load-growth curve to maximize production of a scraper system. AO is the cycle time of the scraper less loading time. CO is the cycle time of the pusher less its loading time.

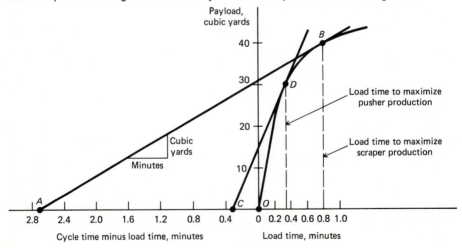

duction. On the other hand, if scrapers are waiting to be loaded, then pusher output will set the upper limit.

Referring to Fig. 13-2, we can see that the ordinate is payload per scraper; the abscissa is time. Thus, the slope of any line drawn on the figure to connect a point on the time scale baseline to a point on the load-growth curve gives output per unit time. Furthermore, it can be seen that such lines have their steepest slopes if they are drawn tangent to the load-growth curve. Using the plotted data on Fig. 13-2, we find the line *AB* represents a situation where the pusher waits for scrapers. In this instance, cycle time less load time is 2.7 minutes. The plot shows that the optimum load is about 40 cubic yards and push time after contact is 0.8 minute (48 seconds). Note that even though the pusher is waiting for scrapers, production per unit time for the spread is lower if pushing is continued beyond 0.8 minute.

Line *CD* on Fig. 13-2 is a solution to the situation where scrapers are waiting for the pusher. Note, as shown by the position of point *C,* that about 0.3 minute (18 seconds) is consumed as the pusher manuevers from loaded to empty scrapers. Then drawing the tangent line *CD* (the steepest), the optimum load becomes about 30 cubic yards and push time after contact about 0.35 minute (21 seconds). Here, optimum output results if scraper loads are less than three-fourths capacity.

To use graphical methods of analysis as illustrated by Fig. 13-2 requires realistic inputs. In the case of the pusher-scraper combination, this involves obtaining data for the scraper on (1) time to maneuver out of the pit, (2) time to travel loaded, (3) time to maneuver and dump, (4) time to return empty, and (5) waiting-to-load time, if any. For the pusher, the individual elements are (1) time for the pusher to accelerate the loaded scraper, (2) time to maneuver to the next scraper, (3) waiting time if any, and (4) position and contact time. These time measurements may be obtained using a stopwatch or second hand on a regular watch or even by counting one little second, two little seconds, etc., which is suprisingly accurate. It is more difficult to obtain a load-growth curve because it varies with the design and power of the machines, the loadability of the material, the degree to which it bulks on loading, and whether and how well it has been loosened with a ripper. On large projects, load-growth curves are physically measured by weighing scrapers after pushing for various intervals. In other instances, estimates based on observation or volume measurements may be employed. Even with these limitations, the graphical solution is helpful in that guessing is replaced by an orderly thinking process.

A first use of a graphical method often leads to other applications. Extensions of the ideas in the pusher-scraper analysis are typical, including the following.

1 The graphical analysis provides a convenient method for estimating output with a given number of scrapers. This combined with data on equipment costs will help management decide how many scrapers to have in a spread.

Often it is found that unit cost is reduced if fewer are used. This conflicts with the traditional wisdom which says that the pusher must always be busy. Those following this rule of thumb have often been surprised to find unit costs were reduced when a scraper broke down or was out of the spread for some other reason. (For added discussion, see the section on simulation later in this chapter.)

2 It can be seen that exactness in pusher time is not critical. In either of the two instances, pusher times can change substantially without significantly altering the slopes of lines *AB* or *CD,* which means that productivity per unit time is not affected greatly with less careful attention to pusher times.

3 It becomes clear that adjusting loading times to fit job situations can increase production. For example, if the scrapers are bunched up and some are waiting, as may happen at the beginning of a work period, loading time would be shortened until the queue of scrapers disappeared. It would not, however, be shortened below the level of maximum pusher production (point *D* in Fig. 13-2). Again, in the absence of a queue or with reported scraper delays, longer pushing would be instituted.

4 By developing load-growth curves for single vs. twin pusher operations, the less costly approach can be determined. Twin pushing steepens the load-growth curve and this increases the slope of the lines showing productivity per unit of time. With the diagram and cost computations, questions that can be answered include:

a Is twin pushing cost-effective, given the cost of another pusher?

b Are there situations involving long hauls or a queuing up of scrapers when two separate single-pusher loading operations are less costly than double pushing?

5 In a manner similar to 4 above, it is possible to evaluate whether ripping the borrow area before loading scrapers is economical.

Graphical analysis does not necessarily give absolute answers, and other factors may come into play. For example, in the pusher-scraper case, the effects of pushing time on equipment operating and maintenance costs must be considered. Loading scrapers less heavily will reduce hourly fuel and tire costs and this argues for even lighter loads. Likewise, loading scrapers with double pushers or in unripped materials may increase their operating and maintenance costs, which may make such procedures less attractive. Even so, the graphical analysis provides a means for quantifying otherwise uncertain costs.

Given that a more complex approach to pushing times as described above is appropriate, arrangements must be made to see that recommended loading patterns are followed by operators. One strategy in use for many years is to have a timing mechanism which sets off a signal or horn when the stipulated pushing period has elapsed. With automation (see the discussion later in this chapter), this could be done without the operator's being involved except to set a control.

The load-growth curve concept often does not directly apply to the loader-truck problem. Such is the case where a given number of buckets almost fills the trucks. In this case, the load-time curve progresses upward in equal discrete steps, once loading begins. This means that as another bucket is loaded into the truck, the slope of the unit-cost lines will always increase. If for some reason, however, partial buckets are needed to completely fill trucks, a graphical analysis paralleling Fig. 13-2 may be employed to determine if and when it is economical to top out the load with a partial bucket.

Another group of graphical displays or models center on planning and scheduling (see the earlier discussion in Chap. 5). They may be either for the project as a whole or for detailed planning levels. Plots showing the scheduling networks are very helpful in establishing interrelationships of all sorts, particularly if given on a time scale. Related to and stemming from networks or similar planning tools are plots showing resource use; for example, plots of how many workers are required for each craft at given stages of the project.

Computers will play an increasing role in deriving data for and producing and displaying the results of basic graphical analyses. How this is done depends on the needs and ingenuity of the user. Examples are given in the discussion of simulation later in this chapter.

CONSTRUCTION APPLICATIONS OF MATHEMATICAL MODELS

The use of mathematical models to evaluate solutions to technical and management problems is today commonly called operations research. Among the techniques are linear and dynamic programming and allocation and probability theory. Often network-analysis techniques are considered a part of operations research. Computing applications abound. The paragraphs which follow examine very briefly three examples where a relatively simple mathematical technique from operations research, referred to as allocation theory, can be applied to optimize the use of construction resources.

Mathematical Model for Determining Equipment Assignments

The first example deals with assigning three available trenching machines of different capabilities to three projects which must be done concurrently. Assignments are to be made on the basis of minimum cost to the company. Total estimated cost for having each machine do each project, including mobilization, on-site operations, and demobilization are shown in Table 13-1.

There are six combinations for doing the three jobs with the three machines. This simple matrix could be solved by calculating the total cost of each alternative or possibly by intuition or inspection. However, with more projects and more machines, this approach becomes very unwieldly; for example, with 6 projects and 6 machines there would be 720 possible combinations. With the allocation procedure, only 6 repetitions are required.

The allocation method for solving such least-cost problems is to find the dif-

TABLE 13-1
TRENCHING PROBLEM: COST OF OPERATING BY PROJECT AND MACHINE

	Project C, $	Project D, $	Project E, $
Trencher X	12,000	20,600	6,400
Trencher Y	12,600	25,400	6,800
Trencher Z	15,800	30,600	9,200

ference between the two lower costs in each row and each column and to start allocating equipment to the jobs in order, starting with the largest incremental differences. The matrix in Table 13-1 is shown in Table 13-2 with six differences calculated, one each for the two lower costs in each row and each column. Since the largest difference is obtained in a row, that for trencher Z, the first assignment would be made to the least project cost for trencher Z, which is project E. Then with trencher Z and project E assigned, a new matrix is made with the remaining elements and the process is then repeated. Table 13-3 shows this matrix and the differences in the rows and columns. The maximum difference is still in a row and is for trencher Y. Thus the assignment is made for its least project cost, project C. The remaining assignment is made for trencher X and project D. The total of all project costs—the least possible—is $42,400.

The process described above uses a simple mathematical technique where assignment of single items (e.g., machines, superintendents) are made to an equal number of positions or locations in order to maximize a particular measure of value (e.g., dollars, production, time). Where more variables are involved, the steps would be repeated until all assignments were made.

Mathematical Model for Choosing Among Sources of Material

Another class of problems which can be solved by simple mathematical approaches involves cost comparisons among multiple sources that can serve a project when the full capacity from all sources is not needed. This can be illustrated by the case where four concrete placements must be scheduled in a

TABLE 13-2
TRENCHING PROBLEM: FIRST DIFFERENCE CALCULATION

	Project C, $	Project D, $	Project E, $	Cost difference, $
Trencher X	12,000	20,600	$6,400	5,600
Trencher Y	12,600	25,400	6,800	5,800
Trencher Z	15,800	30,600	9,200	6,600
Cost difference	600	4,800	400	

TABLE 13-3
TRENCHING PROBLEM: SECOND DIFFERENCE CALCULATION

	Project C, $	Project D, $	Project E, $	Cost difference, $
Trencher X	12,000	20,600		8,600
Trencher Y	12,600	25,400		12,800
Trencher Z				
Cost difference	600	4,800		

single day, with concrete available from three ready-mix suppliers. Concrete prices, delivered, vary depending on such factors as the distances from plants to projects. Data on suppliers' capacities, project requirements, and unit prices for delivered concrete are given in Table 13-4.

The solution aims to find the least total cost of concrete delivered to all four projects. The approach is the same as in the previous example. As shown in Table 13-4, the difference between the lowest cost and next lowest cost is determined for each row and column. The greatest difference, $21, is in row 1. The first allocation is then in row 1 and to the lowest-cost item, project B, at $42. Since project B requires 540 cubic yards and plant 1 has a capacity of 600 cubic yards, 60 cubic yards of unused capacity remain.

With project B satisfied, it is eliminated from the matrix, as shown in Table 13-5. For the second allocation, the greatest difference is $24 in row 3; 450 cubic yards from plant 3 is given to project C, leaving 150 cubic yards still available.

TABLE 13-4
CONCRETE DELIVERY PROBLEM: PLANT CAPACITY, PROJECT REQUIREMENTS, AND UNIT PRICES

		Requirements and unit prices				
		Project A	Project B	Project C	Project D	Cost difference
	Available capacity, cubic yds.	Requirements, cubic yards				
		300	540	450	760	
Plant 1	600	$69	$42	$66	$63	$21
Plant 2	900	60	54	36	48	12
Plant 3	600	75	48	42	66	6
Cost difference		9	6	6	15	

TABLE 13-5
CONCRETE DELIVERY PROBLEM, SECOND STEP

		Requirements and unit prices				
		Project A	Project B	Project C	Project D	Cost difference
	Available capacity, cubic yds.	Requirements, cubic yards				
		300		450	760	
Plant 1	60	$69		$66	$63	$ 3
Plant 2	900	60		36	48	12
Plant 3	600	75		42	66	24
Cost difference		9		6	15	

TABLE 13-6
CONCRETE DELIVERY PROBLEM, THIRD STEP

		Requirements and unit prices				
		Project A	Project B	Project C	Project D	Cost difference
	Available capacity, cubic yds.	Requirements, cubic yards				
		300			760	
Plant 1	60	$69			$63	$ 6
Plant 2	900	60			48	12
Plant 3	150	75			66	9
Cost difference		9			15	

The next step is to provide another matrix (Table 13-6) with project C eliminated and showing revised available capacities and cost differences. From this it can be seen that with the greatest difference of $15 in the column for project D, that project should be served 760 cubic yards at $48 by plant 2, leaving 140 cubic yards of untaken capacity.

The fourth assignment, as shown by Table 13-7, indicates that the 300 cubic yards required for project A would be served by 140 cubic yards from plant 2, 60 cubic yards from plant 1, and 100 cubic yards from plant 3. Unused capacity would be 50 cubic yards in plant 3.

TABLE 13-7
CONCRETE DELIVERY PROBLEM, FINAL STEP

		Requirements and unit prices				
		Project A	Project B	Project C	Project D	Cost difference
	Available capacity cubic yds.	Requirements, cubic yards				
		300	540	450	760	
Plant 1	60	$69	$42			
Plant 2	140	60			$48	
Plant 3	150	75		$42		

Summing up the results of the analysis, it can be seen that the minimum cost combination allocates the following

540 cubic yards from plant 1 to project B at $42	$22,680
450 cubic yards from plant 3 to project C at $42	18,900
760 cubic yards from plant 2 to project D at $48	36,480
140 cubic yards to project A from plant 2 at $60	8,400
60 cubic yards to project A plant 1 at $69	4,140
100 cubic yards to project A from plant 3 at $75	7,500
Total cost	$98,100

No other combination of deliveries would cost less. Whether or not this arrangement meets other job constraints cannot be answered by the analysis. These could be recognized by changing the original data and making another analysis.

Mathematical Model for Optimizing Choices Among Multiple Origins and Destinations

The model, often called the transportation problem, is another operations research technique that has been fitted to construction situations. This form of analysis was originally applied in planning travel for sales representatives who had to visit a number of customers in different places. The aim of the analysis was to determine a route which would minimize travel cost, distance, or time.

One among many construction applications of the transportation problem involved moving material from a number of borrow pits at different locations and elevations into an earth-fill dam. Procedures would have been straightforward with no special analysis required if the movement were directly from borrow pit to dam along a single path. However, when complexities arose such as constraints on which materials could be used when or if haul road layouts had to be manipulated, the optimum solution was no longer obvious. The approach that was employed to find a minimum cost solution was to develop a matrix which showed the unit cost of moving material from every source to every destination. By longhand in a preliminary attempt and later by computer, a minimum cost solution was then found using standard transportation-problem procedures. The optimum solution, found after the dam was completed, indicated that costs could have been reduced by 10 percent had the model been employed.

In solving this problem, several strategies were developed to overcome complications. As one instance, the construction plan called for flooding an upstream borrow pit on a given date. This difficulty was recognized in the analysis by considering this pit as two separate origins. Unit costs to the various destinations were calculated as usual for the time when the pit would be available. For later times, these costs were set so outrageously high that these alternatives would never appear in a minimum-cost solution.

Mathematical Modeling to Optimize On-site Positioning of Equipment

Another construction problem that has been treated mathematically is that of determining how to optimize the positions of tower cranes for unloading materials. For each feasible location of crane base and of pickup and delivery points, the crane boom must swing through a given angle and the trolley must travel along the boom for a given distance. Knowing the hookup and release times, the angle and angular velocity of swing, and the distance and rate of travel of the trolley for each position of the crane base, one can determine cycle times. This simple problem can be solved graphically.[12] It is worth noting that this is only one small element in the larger problem of site layout, which is amenable to analysis by dynamic programming or expert systems.

This discussion of mathematical models and operations-research techniques has been intended primarily to point out that such tools are available. Unfortunately, they are underutilized in construction today because few construction people, including construction engineers, know that they exist or, if they do know, they lack the educational background or time necessary to apply them.

CONSTRUCTION APPLICATIONS OF ECONOMIC MODELS

Economic and financial analyses define the consequences of alternative uses of money over time, recognizing the time value of money through appropriate

---

[12] See W. E. Rodriquez-Ramos and R. L. Francis, *Journal of Construction Engineering and Management*, ASCE, vol. 109, no. 4, December 1983, pp. 387–397.

interest rates. Such analyses are essential in any well-managed business. Most of them are outside the scope of this book, but it seems appropriate to discuss briefly the economic aspects of investments or other expenditures in construction equipment and plant, which make up a major element of the cost in many important phases of construction. Such equipment is expensive to provide and to operate; for example, it is not uncommon to find earth-moving equipment costing more than $500,000 per unit, with a cost, when fully operated, in the range of $150 to $200 per hour. Thus a systematic and detailed analysis of equipment costs and an understanding of the factors that are involved can be very important.

The techniques used by most contractors for determining equipment costs have greatly oversimplified an exceedingly complex problem. In general, they have relied on averages for variables such as output, useful lives, fuel consumption, depreciation, repair and maintenance costs, interest rates, and taxes. These approaches may be useful as guides but with improper assumptions or erroneous analysis can give wrong answers.

To a degree, this lack of attention by contractors to the issues underlying equipment costs is understandable, given their complexity, the difficulty of maintaining adequate records, and the other claims on their time. Also, it is possible to fall back on data developed by specialists in this field who gather information and from it publish schedules of equipment ownership, rental, and operating costs.[13] These are widely accepted by private owners and public agencies as well as contractors as a basis for setting equipment charges or settling claims. But serious errors can result if they are used blindly rather than as a guide in making decisions in specific cases.

Two fundamental and separable analyses are basic to decisions about construction equipment or plant. The first rests on engineering economy, the second on finance—or who pays what and when, a topic not discussed here.

The discipline of engineering economy offers several methods for evaluating the economic consequences of procuring the services of equipment[14]. One among the many decisions to be made involves replacing currently available units or fleets with new, different, and possibly more-productive machines. Another might be to evaluate the relative costs of ownership versus rental or of subcontracting the work to others. The second and separable analysis involves finance, that is, the sources of money to implement each solution. For a variety of reasons, these two analyses, although both are stated in money terms, have different inputs and may lead to different conclusions. For exam-

---

[13] As an example, the Associated General Contractors makes available an annual Contractor's Equipment Cost Guide, developed in conjunction with Dataquest Inc., the publishers of the Rental Rate Bluebook. In this guide are data on ownership and operating costs for a wide range of construction equipment.

[14] There are a score or more textbooks on engineering economy. Among the most widely used is 8th Ed. by E. L. Grant, W. G. Ireson, and R. S. Leavenworth, *Principles of Engineering Economy*, Wiley, New York, 1989.

ple, the least-cost solution from an economy-study point of view may demand financing that is out of reach.

The methods for making economic comparisons among the feasible alternatives are well known; the difficulty is that analyses must rest on a host of underlying predictions about the future. In earlier years, detailed analyses reflecting these uncertainties were very time-consuming. Today, the computer makes them easier to do. For example, it is now common practice to recognize uncertainty directly in the analysis. Comparisons are based on estimates of the probabilities associated with the underlying assumptions and outcomes as a way of quantifying favorable and unfavorable viewpoints.

Typical of the economic decisions about major pieces of equipment, equipment fleets, or construction plant is to determine when replacement is justified, if at all. This can be an extremely complex problem. Questions that must be answered can include the following:

1 Will there be a continuing need for the services of this machine or plant or a replacement to perform similar services?
2 What alternatives to replacement, such as rental, subcontracting, or purchased services should be evaluated?
3 What interest rate is appropriate in considering the available alternatives, in light of the risk involved and other uses that can be made of the company's financial resources or borrowing capacity?
4 In considering replacement, what are the future consequences if the old machine or plant (called the defender) is retained compared with acquiring a new machine (called the challenger). Among the variables to be considered are:
  a The investment required to acquire each possible challenger
  b Technical advances which lead to improved productivity or reduced operating costs of the challenger or challengers as compared with the defender
  c Salvage value of defender and challenger or challengers
  d Costs of maintenance and periodic overhaul for defender and challenger or challengers
  e Costs related to the inadequacy or obsolescence of the defender
  f Costs associated with the reliability or frequency and duration of failure of the defender and challenger or challengers, which involve:
    (1) Cost of lost productivity if the machine is completely out of operation
    (2) Costs involved when an operation in progress must be shut down or its efficiency is impaired because of breakdown
  g The impact of overall or differential inflation or price changes
  h The effects of governmental tax policies or changes in them in areas such as depreciation allowances, investment tax credit, or gain or loss on sale or disposal

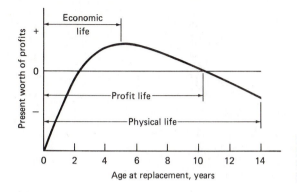

FIGURE 13-3a
Diagram illustrating an economic model for equipment replacement, profits vs. age of replacement.

Analyses to determine the economic consequences of choosing different construction equipment or plant alternatives often compare uniform equivalent annual costs or the present worth of costs (or benefits less costs). Another approach determines the interest rate at which the alternatives are equally attractive.

James Douglas approaches the economics of construction equipment replacement employing a computer model which makes it possible to evaluate the effects of the factors listed above, plus several others.[15] The analysis assumes an infinite number of successive replacements of a machine, assessing the effects of replacement in each year separately. Results are stated in terms of the present worth of accumulated profit or loss to the owner if replacement is made over and over again in each of these years. A plot illustrating the results of such an analysis is given in Fig. 13-3a. Beginning at the left, it shows substantial losses with early replacement. These decrease to zero with time. Then there is a period during which the machine shows a profit, which peaks at economic life, the time when the investment shows the greatest overall profit. After this point, profits decrease. They may, as suggested by Fig. 13-3a, turn into losses before physical life is reached. In actual situations, the curve would not be smooth and continuous as shown in the figure; rather there would be abrupt breaks at specific times, reflecting major costs such as overhauls or favorable tax allowances tied to the age of the equipment.

M. C. Vorster has proposed another method for evaluating and presenting the economics of construction equipment replacement. He calls it the cumulative expenditure model. It is illustrated in Fig. 13-3b. The basis for the ordinates on the figure is, in each case, the sum of present values for each year for

---

[15] See *Construction Equipment Policy,* McGraw-Hill, New York, 1975. In addition to equipment economics, this excellent book also treats subjects such as equipment records, financing, standardization, inventory, maintenance management, and safety. See also, J. Douglas, *Journal of the Construction Division,* ASCE, vol. 104, no. CO 2, June 1978, pp. 191–205.

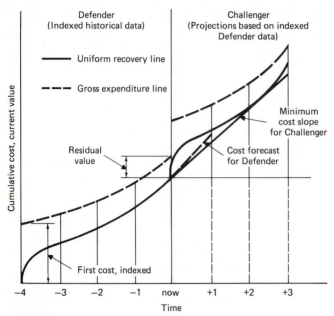

FIGURE 13-3b
Diagram illustrating an economic model for equipment replacement,
cumulative cost model, defender and challenger. (*After M. C. Vorster.*)

all expenditures, adjusted to year 0 (now) at a chosen interest rate. Expenditures for the defender from year −4 to year 0 (now) are from field records. Those from year 0 (now) into the future are projections of anticipated costs for the defender or challenger. The slopes of lines drawn from the base points to the uniform recovery line represent equivalent uniform annual costs in year 0 (now) dollars. For the example given on Fig. 13-3b it can be seen that from an economic viewpoint, replacement now will be better than retaining the defender, given that the defender's line showing the projected cost forecast for the year 0 to +1, lies above the minimum cost line for the challenger. Vorster's method for presenting the results of equipment economy studies, being in graphical form, has the same advantage as the Douglas method; each is easy to explain to busy decision makers.

As a part of his research, Vorster examined field records of construction equipment performance and cost over time to determine the relationship between equipment age, maintenance effort, and performance. He found that the common assumption that managers increased maintenance effort more or less uniformly with the age of the equipment (between major overhauls) did not apply. Rather, when downtime increased, jobsite management intensified its maintenance efforts. After this had decreased downtime, managers relaxed the maintenance efforts and performance worsened again. Often this cycle of neglected maintenance followed by strong attention to it was repeated several times with similar results. Finally, as the equipment became badly worn, the pattern broke. Then field managers insisted on replacement. This example illustrates one among many influences that management's actions can have on equipment costs. In this instance, because field managers were not constrained by a uniform company equipment maintenance policy, they devised their own strategies to meet situations that affected their operations and personal well-being. In terms of equipment-replacement economics, this example calls into question the common assumption that equipment maintenance costs increase more or less uniformly with the equipment's age.

Vorster and G. A. Sears have addressed the issue of failure costs, which are not taken into account in the usual replacement-economy studies.[16] The point is that the consequences of a failure in terms of overall job costs differ with the function which the particular machine fulfills. For example, if the loader in a loader-truck spread breaks down, all activity ceases, whereas if a truck goes out of service, output is diminished but only by the output of one truck. The Vorster-Sears paper offers a model for dealing with these sorts of situations.

Another study of the economics of construction equipment replacement made in Israel involves tower cranes, concrete mixers, and hoists in on-site building construction. It also examines cranes and concrete mixers in centralized plants for industrialized construction. For these kinds of equipment, economic life is in the range of 9 to 12 years. Equivalent uniform annual equipment costs, as a percentage of purchase price of the equipment, range from 14 to 16 percent.[17]

This brief discussion of economic models suggests strongly that, as in many other areas, policies regarding construction equipment and plants can have important implications for overall construction costs and productivity. It is beyond the scope of this book to treat the topic in detail, but the authors would be remiss if they failed to indicate its importance.

SIMULATION AS A CONSTRUCTION-MANAGEMENT TOOL

Simulation in its broadest sense means imitating or representing reality. As used here it can be defined as a way to generate happenings before they occur. This concept is not new in construction. For example, estimators think through (simulate) each step in a proposed project by building it in their heads. This involves setting the number and makeup of the crews and selecting equipment and other elements. Following this, cost of completion per unit of time is computed, combined with an estimate of the time needed to carry out the task to give its total cost.

[16] See *Journal of Construction Engineering and Management,* ASCE, vol. 113, no. 1, March 1987, pp. 125–137.
[17] See S. Selinger, *Journal of Construction Engineering and Management,* ASCE, vol. 109, no. 4, December 1983, pp. 398–405.

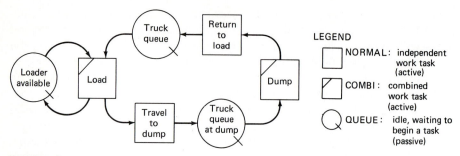

FIGURE 13-4
Flow diagram for a typical loader-truck spread (Halpin-Woodhead CYCLONE notation).

An example of how this estimating approach can be diagrammed as a first step in simulation is given in Fig. 13-4. It involves loading material into trucks and hauling it to a disposal area. For each step in each truck cycle, including waiting for the loader or to unload at the dump, times are assigned or developed. Given these cycle times and hourly costs for the equipment, one can compute output and costs.

The Deterministic (Conventional) Approach to Planning and Analysis

With the deterministic approach to planning and analysis, single values of time are assigned to each operation and delay in each cycle. Often these are diagrammed, as shown for a loader-truck spread in Fig. 13-4.[18] In this instance, if too few trucks are provided, it will follow that the loader will have idle time after each loading. The productive capacity of the trucks will determine overall output. This output will increase linearly as trucks are added until loader productive capacity is reached. This result is shown graphically by the solid lines in Fig. 13-5. For the situation diagrammed, which is based on a specific set of assumptions about loader capacity and cycle time and truck capacity and haul time, this balance point is reached with slightly under 12 trucks. Up to that number, unit costs will decrease as the loader is occupied a greater percentage of the time and loader costs per unit of output will decrease. Beyond this number, trucks will begin to queue up, with waiting time increasing as more trucks are added so that unit costs will increase. Usually this deterministic analysis includes allowances to recognize (1) loader-truck interdependencies and (2) breakdown and delays external to the operation, both of which reduce output. These effects are customarily lumped by assuming that the system operates at

---

[18] This system of designation is called CYCLONE for Cyclic Operations System Network. It was developed by D. W. Halpin and R. W. Woodhead. (See their *Design of Construction Process and Operations*, Wiley, New York, 1976, or D. W. Halpin, *Journal of the Construction Division*, ASCE, vol. 103, no. 3, September 1977, pp. 489–499, for details.) Others have employed a link-node system with links corresponding to NORMAL and nodes to COMBI.

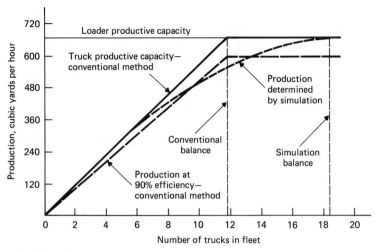

FIGURE 13-5
A plot of loader-truck output for a typical situation, as determined by conventional estimating methods and by simulation.

a rate such as 90 percent efficiency or 54 minutes per hour. Output under this assumption is shown by the dashed lines in Fig 13-5.

The Simulation Approach to Planning and Analysis

Contrasted to deterministic fixed-time approaches, simulation will model the system by directly recognizing the inherent randomness in the times required to carry out each individual activity. The results of simulation for the loader-truck example are shown by the dotted curve in Fig. 13-5. It demonstrates that (1) as the separate capacities of the trucks and loader approach a balance, actual output falls below that set by the deterministic model, and (2) to fully use the capacity of the loader will require several more trucks than indicated by the deterministic approach. It can be seen in Fig.13-5 that the assumption of a 90 percent output for the deterministic model does not truly reflect the effect of randomness in the operation. But it also must be recognized that maximum output as determined by simulation (more than 18 trucks in the example of Fig. 13-5) would seldom if ever result in lowest unit cost. Instead, this must be determined by computing costs with different numbers of trucks.

In simple terms, analyses by simulation involve the following steps:

1 Develop a flow diagram such as Fig. 13-4 for the operation to be simulated. Note in the diagram that a queue always precedes a COMBI operation.

2 From past experience, observation, or other sources of data, set the range and distribution of times required to carry out each of the NORMAL or COMBI operations (squares in Fig. 13-4). This step is discussed further below.

3 Input the data from steps 1 and 2 into an appropriate computer program (see below) which generates the time sequences for the operation for a stated number of cycles or for a given period of time such as a shift or half-shift. Results can be stated in appropriate units such as number of loads, output per time interval, or unit or total cost. Because of the randomness of the inputs, results from individual computer runs will differ, so that a number of iterations are required to obtain average values and the spread among results.

4 Where comparisons to measure the effects of possible changes in the operation or other assumptions are desired, new runs using revised approaches, relationships and values are made. Examples from the simple loader-truck operation might include changing bucket size or the cycle-time distribution of the loader or the number, capacity, or travel time distributions of the trucks.

As indicated, simulation requires development of a range and distribution of times to carry out each step in an operation. This is commonly done as follows:

1 Data are developed on the observed (or estimated) times required to carry out each repetition for each NORMAL or COMBI work task. These must, in some manner, show the distribution among the individual times.

2 To facilitate analysis by computer, these data are fitted to the most appropriate mathematical probability-distribution curve either by using overlays on the computer screen or by curve fitting by the computer directly. The distributions and cumulative distributions of values given by four of the most used curve types are plotted in Fig 13-6. Other distributions that may be appropriate include the Erlang, Poisson, BETA, GAMMA, and PERT.

The probability distribution curves shown in Fig.13-6 have been plotted in two ways. Figure 13-6a gives the probability that a particular event will occur at a given time T. (For example, the next truck will be loaded in 60 seconds.) For particular-event curves, the area under the curve equals 100 percent. The second form of curve, (Fig.13-6b), represents the accumulated area under the probability curves shown in Fig.13-6a. Each curve can be described by an equation. Mathematical definitions of the distribution curves are not given here since they are available as needed in simulation software programs. Any among the many distribution curves can be described by a set of numerical parameters. For example, the parameters for the normal curve are its mean and standard deviation. Those for the log-normal curve (which often fits the data for equipment combinations), are mode, mean, and standard deviation.

In the computer-simulation procedure, specific values to apply to each step in each cycle are generated randomly from the selected and defined time-distribution curve. This approach is based on the assumption that chance alone determines what the next time will be but that when all values are set, their summation will match the stipulated curve.

Several software packages for computer simulation with the common com-

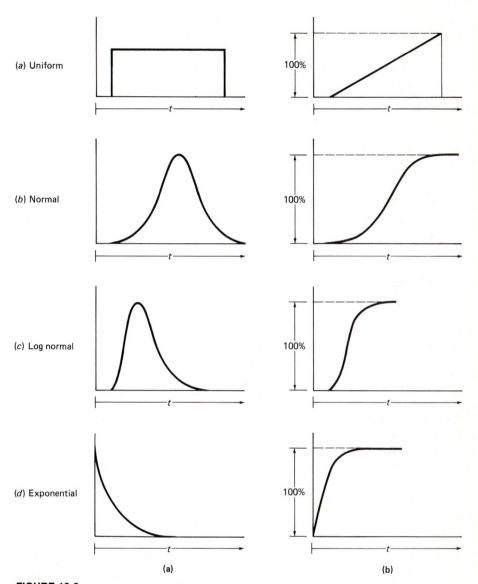

(a) Uniform

(b) Normal

(c) Log normal

(d) Exponential

100%

100%

100%

100%

(a)

(b)

FIGURE 13-6
Examples of curves used in random distributions of events for simulation programs. Ordinates are (a) probability and (b) cumulative probability, plotted versus time, and abscissa.

puter hardware are available. To use them requires prescription of operating conditions and constraints. For example, for the simple loader-truck simulation outlined in Fig. 13-4, program restrictions might be categorized as follows:

1 Transfer restrictions, such as (a) A truck cannot finish loading and start a haul trip until it has been fully loaded; (b) if the trucks form a queue waiting to load, they are serviced on a first-come, first-served basis; and (c) a truck cannot take on a load without the loader dumping into it.

2 Travel restrictions, such as the possibility that faster carriers can pass slower ones; the programs accept either alternative.

In addition, the number of repetitions in a given computer run is stipulated or, alternatively, runs may be stopped when a given period of time such as half a shift has been replicated.

The level of detail in the inputs to and outputs from simulation programs can usually be stipulated in the computer program. In the simplest form for the loader-truck example, it might show the average number of loads hauled in the prescribed time and the variation in this number among the different individual runs, possibly stated as the standard deviation. By incorporating the unit costs of equipment into the computer program, it can develop total and unit costs, as is done in conventional deterministic estimating.

Simulation of Earthmoving Operations

Several applications of simulation to earthmoving operations were made in the 1960s.[19] They involved both shovel-truck and pusher-scraper combinations. Tiecholz began by analyzing data collected by the (then) Bureau of Public Roads covering approximately 50,000 complete equipment cycles. Among his findings were:

1 The log-normal curve best approximated actual field observations for these equipment combinations.

2 As the coefficient of variation (standard deviation divided by the mean) increased among the results of individual simulation runs, average productivity fell substantially. This reflected greater imbalance in the coupled operations.

3 The form of distribution curve assumed for the analysis, whether normal, long-normal, exponential, or uniform (see Fig. 13-6) had a much less drastic effect on productivity than did spreads in the distributions in the basic performance curves.

Tiecholz also examined the Bureau of Public Roads data to find the time distribution of external delays not associated with queuing. For power shovels and trucks the percentage of cycles suffering delays were 27 and 49, respectively. The percentage of delays lasting 5 minutes or less were 97 and 88 percent, respectively. A very small percentage of the delays were longer than 15 minutes; these averaged about 1.5 hours. In general, external delays were best portrayed by an extremely skewed log-normal distribution. Data on external delays such as these are needed if simulation studies are to be extended to in-

---

[19] See P. Tiecholz, *Technical Report 26,* June 1963; P. Tiecholz and J. Douglas, *Technical Report 29,* August 1963; J. Douglas, *Technical Report 37,* June 1964; and A. Gaarshev, *Technical Report 111,* August 1969, the Construction Institute, Department of Civil Engineering, Stanford University, Stanford, Calif.

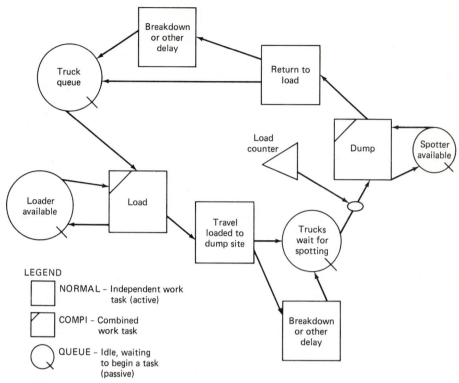

FIGURE 13-7
Flow diagram for a typical loader-truck spread, expanded to show possible delays in the truck cycle (Halpin-Woodhead CYCLONE notation).

corporate the possibility of delays in the truck-travel or truck-dump steps in the truck cycle. A flow diagram to incorporate delays in the truck cycle is given in Fig. 13-7. Factors that might be introduced include breakdown, traffic congestion on public highways, or pit stops for scheduled or unscheduled maintenance.[20] To incorporate delays, probability functions (to show the possibility of delay and its duration) would be stipulated. Similarly, this approach could be applied to the spotter-available function to account for access or other difficulties at the disposal site.

Another study on earthmoving by simulation provided a means for directly adjusting conventional deterministic estimates for loader-truck outputs in order to recognize production losses resulting from variability inherent in the system. The effect is captured in the D factor shown in Fig. 13-8 (see Douglas, 1964, op. cit.). To measure this effect, the production of the shovel working alone without delays and that of an increasing numbers of trucks operating free

---

[20] Note that Fig. 13-7 also incorporates a load-counting routine. It records output and stops the simulation after a given number of loads or a stated time interval.

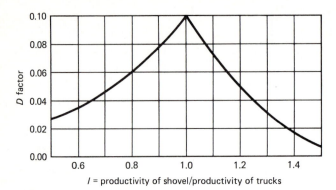

FIGURE 13-8
"D" factor for adjusting shovel-truck output to recognize the effect
of the variability of shovel and truck cycle times.

of waits for the shovel are computed as is done for conventional estimates.
The ratio of these idealized shovel outputs to truck outputs is labeled I. D
factors for usual values of I are shown in the figure. Note that when
$I = 1.0$, D is 0.10 or 10 percent, but that its effect falls off as I values move
away in either direction from 1.0. The D factor generalizes for all shovel-
truck operations the principle illustrated by the simulation results shown in
Fig. 13-5.

The value of the concept underlying the D factor is that it permits separation of the effects of randomness in shovel-truck interactions from other
factors affecting output. Examples of these other factors include complete
shutdowns of the operation for equipment maintenance or for major moves
among loading sites. Separating the randomness inherent in all such operations from those for specific situations gives a more systematic basis for
estimating shovel output than, for example, assuming 10 minutes delay per
hour from all causes.

As has been indicated, the interdependence of loader and haulers reduces
output. Among the ways for partially breaking this linkage is to introduce a
hopper into the system. In this situation, the loader works continually unless
the hopper is full. Trucks load from it immediately and quickly on arrival without waiting to be spotted and for the several loader cycles needed to supply a
full load. Offsetting this potential gain in output is the cost of providing the
hopper and, if necessary, a person to operate it, possibly longer or shorter
times for the loader, and costs and delays associated with moving the hopper
around the loading area.

Figure 13-9 gives the effects on productivity of introducing a hopper into the
loader-truck material-transfer process. Definitions needed to use the figure are
given on it. An example demonstrating its use is as follows:

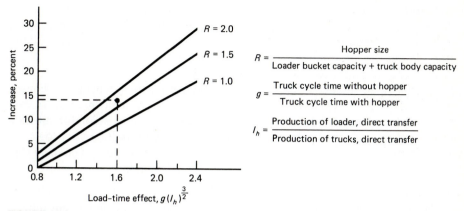

FIGURE 13-9
Percentage increase in production rate when a hopper is introduced between a loader and the trucks it serves. (*From Tiecholz and Douglas, op. cit.*)

Loader data: Bucket capacity, loose   = 6 cubic yards
                    Capacity per hour at 30-seconds
                    cycle time   = 720 cubic yards

Truck data: Body capacity, loose   = 24 cubic yards
                    Cycle time without
                    hopper = 5.0 + 2.0   = 7.0 minutes
                    Cycle time with hopper = 5.0 + 0.5   = 5.5 minutes
                    Number of trucks   = 3

Capacity of system with 3 trucks, direct transfer

$$\frac{3 \text{ trucks} \times 24 \text{ cubic yards} \times 60 \text{ minutes}}{7 \text{ minutes/trip}} \qquad = 617 \text{ cubic yards/hour}$$

Hopper capacity, loose   = 50 cubic yards

$$R = \frac{50 \text{ cubic yards}}{(6+24) \text{ cubic yards}} \qquad = 1.67$$

From these data, given that

$$R = 1.67 \quad \text{and} \quad g = \frac{7.0 \text{ minutes}}{5.5 \text{ minutes}} \qquad = 1.27$$

$$I_h = \frac{\text{Loader capacity/hour}}{\text{Truck capacity/hour}} = \frac{720}{617} \qquad = 1.16$$

$$\text{Load time effect} = g\,(I_h)^{3/2} = (1.27)\,(1.16^{3/2}) = 1.59$$

Referring to Fig. 13-9, we see that the dotted lines show that output would be increased by 14 percent by introducing the hopper. Factors such as those listed above would then be considered in deciding whether or not to include it. But without the information developed by the simulation study, a rational decision about hopper use would not be possible.

Results of the hopper study should be most useful when considering longer-term decisions; for example, when a dragline is to reclaim material from a below-water aggregate source, a hopper can be left in one position for substantial periods so that the costs of providing and moving it will be very low.

A later application of simulation to earthmoving was done by Garslaav (op. cit.). He extended the models described above to include:

1 Two link systems with multiple servers; for example, more than one pusher tractor loading scrapers in either an independent or dependent mode
2 Multilink systems involving more than one queuing activity, as illustrated by Fig. 13-4

One of the important results of these early studies of earthmoving is that they demonstrate that field operations of many varieties can be realistically simulated by computer. This then makes it possible to examine and consider alternatives at the planning stage. During the 1960s, when these early applications of simulation were being carried out, the repetitive data processing could be done only at high cost on large mainframe computers. Furthermore, they demanded a high level of programming skill. At that time, neither was readily available to the construction industry. Some of the concepts were adopted by the research arms of equipment manufacturers and made available on a limited basis. Even so, as will be pointed out subsequently, these methods coupled with developments in computer hardware and user-friendly software have laid the groundwork for a widespread application of simulation to construction.

Later Simulation Studies

A variety of ways to apply either deterministic or probability-based simulation to planning or modifying construction operations has appeared in the literature in recent years. For example, articles, among others, in the Journal of the Construction Division, ASCE, each with extensive bibliographies, include the following:

- N. B. H. Benjamin and T. W. Greewald, "Simulating Weather Effects," vol. 99, no. CO 1, July 1973, pp. 175–190.
- F. Moavenzadeh and J. Markow, "Simulating Tunnel Construction Costs," vol. 102, no. CO 1, March 1976, pp. 51–66.
- D. G. Woods and F. C. Harris, "Truck Allocation Model for Ready-Mix Concrete Distribution," vol. 106, no. CO2, June 1980, pp. 131–139.
- D. B. Ashley, "Repetitive Unit Housing Construction," vol. 106, no. CO

2, June 1980, pp. 185–194. This example involved converting CPM or PERT and learning curves into a simulation program.

• M. A. A. Dabbas and D. W. Halpin, "Integrating Project and Process Management in High-Rise Construction," vol. 108, no. CO 3, September 1982, pp. 361–374.

Without question, many other instances of simulation applied to construction have been carried out but were proprietary or were never written up. Brief descriptions of a few are:

1 Selection of construction equipment spreads for earthmoving. For example, by simulation, travel times and outputs have been determined for scrapers or trucks traveling over roads which have combinations of level and up- and downgrades. These analyses take into account loads, vehicle characteristics such as engine horsepower and transmission gear ratios, and the rolling resistance of the haul road.

2 Comparisons of the relative costs of earthmoving with loader-truck combinations of various sizes and conformations.

3 Comparisons of one-lane vs. two-lane bridges on haul roads. By simulating the one-lane operation, the delays and costs introduced by stopping and waiting were determined for comparison with the outlay for building a second lane which permitted uninterrupted travel in both directions on the bridge. In a similar approach the U.S. Forest Service, in planning its logging roads, simulates and prices operating costs for one-lane and two-lane road conditions.

4 Study of loading, transporting by tug and barge, and unloading rock for two breakwaters.[21] Design or operational controls which affected the project included:

   a Three types of materials, *A, B,* and *C,* destined for different zones in the breakwaters

   b Types *A* and *B* materials loaded to barges at the quarry by bulldozer and conveyer; type *C* loaded by crane; type *A* material unloaded by dozer on barge; type *B* and *C* unloaded by crane

   c Construction of the second breakwater deferred until the first was finished

   d Barge movement independent of work shifts; barge loading and unloading on a shift basis

The alternatives that were analyzed to determine times, delays, and costs were:

1 A basic condition employing two shifts at the quarry and three barges

2 Two rather than three-barges–two-shift quarry operation

---

[21] This simulation was carried out through an IBM computer program called GPSS (General Purpose System Simulator). This program, as is the case with many others, can be applied to a variety of problems. The criteria for them is that the application must involve a series of interdependent happenings which vary in a manner that can be defined mathematically.

3 Single-shift quarry operation with three barges
4 A second unloading crane, three barges, and a two-shift quarry operation

The report on the barge study showed substantial cost differences among the possible alternatives, but did not state which, if any, was adopted. This summary has been included to illustrate by example the usefulness of simulation as a decision-making tool in a variety of construction applications.

In all the examples cited here, simulation brought information on alternatives that decision makers could use in planning specific construction operations. With it, they were able to quantify variables under different assumptions of what reality would be and predict their outcomes, rather than having to rely on a deterministic single answer which would give far less insight into the possible choices.

New Directions for Simulation as a Planning Tool

The brief discussion just given indicates that simulation has been a valuable tool for evaluating alternative ways for carrying out a variety of construction operations. These approaches have been made possible by a combination of the speed with which large mainframe computers can process data and the imaginative applications and approaches developed by persons who understand both construction and computers. Constraints on broader use of the techniques have been a lack of access to or knowledge about computer hardware and software by people with knowledge of and desire to improve construction operations at the field level.

Today and in the near future the use of simulation as a field management tool for planning and operational analysis should increase dramatically, because among other reasons there is:

• An increased knowledge of simulation and its values through publications and examples such as those cited above and through teaching them both in educational institutions and at in-service programs
• Decreased cost and increased power of microcomputers which will make them readily available to both office and field personnel
• User-friendly and readily available software which reduces the need both for a detailed knowledge of computer programming or how to develop the programs

Any writing on new directions that simulation of field construction projects will take is at best dated and merely a progress report. Two areas in which research is underway to be described here are: (1) integration of early planning and scheduling with detailed jobsite planning and scheduling and (2) simulation as a mechanism for planning or modifying specific construction operations.

Area 1. Integration of Overall Early Planning and Scheduling with Detailed Job-site Planning and Scheduling   For all but the most complex projects, prebid and preconstruction planning and scheduling have primarily been done by higher-level executives or estimators. These individuals usually are under many other pressures which develop when jobs are starting up and often lack the time to develop detailed programs. Furthermore, their thinking for this preliminary planning often follows the typical estimating and cost-control patterns and fails to consider the details of scheduling, which integrates and sequences the various work activities; so the details are left to on-site management. As a result, certain crucial decisions are made at the preconstruction planning stage which overlook important details that may later haunt field personnel. Two among the many examples that can be cited are (1) the difficulties involved in allocating and arranging for on-site access to work areas or storage space and (2) the complexity of providing joint facilities such as hoists, cranes, elevators or scaffolding which must be shared by the various crafts and subcontractors. Procedures for assigning responsibilities for these detailed decisions to individuals are discussed in Chap. 5. But task assignment alone is not sufficient. The people responsible for implementing decisions should not rely solely on past experience or hunches to provide the details but should have tools for analysis which will permit sensible comparisons among the alternatives.

Simple and fast computer programs for simulating the major coordinating and sequencing problems early in project planning have been proposed and a few have been described in the literature. In effect, they demonstrate ways by which planners can ask such questions as, "What if we adopted this or that plan, sequence, or equipment?" Because of the lack of time, the approach common today is to disregard these "what-if" questions, and follow hunches or precedents in making such decisions.[22]

M. A. A. Dabbas and D. W. Halpin give an example of how computer-based planning and sequencing approaches can be used during initial planning to select the number and capacity of cranes or hoists to be employed in constructing the frames for high-rise reinforced-concrete buildings.[23] Operations included in their analysis are placing and stripping slab and wall forms and placing reinforcing steel and concrete. One among the effects considered is the increase in hoist times as construction moves successively to higher floors. From such an analysis it is possible to consider the number and size of cranes, the number and size of crews needed, by task, and the crane or hoist time available for other than the specifically stated purposes.

One has only to think through the start-up of any large project to

---

[22] For added discussion and a detailed bibliography on this topic see D. P. Kavanagh, *Journal of Construction Engineering and Management,* ASCE, vol. 111, no. 3, September 1985, pp. 308–323.

[23] See *Journal of the Construction Division,* ASCE, vol. 108, no. CO 3, September 1982, pp. 361–374.

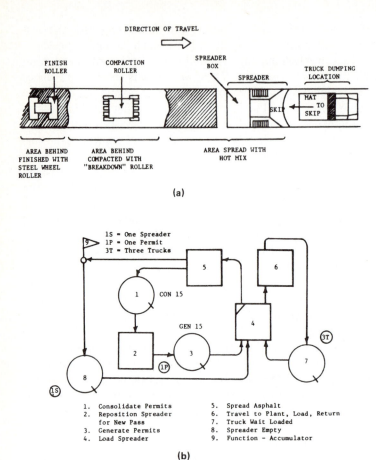

FIGURE 13-10
Diagram of an asphalt paving operation: (*a*) plan view; (*b*) CYCLONE
network.

develop many other examples of decisions made early that needed more careful consideration, which would have been facilitated by the use of simulation.

   **Area 2. Simulation as a Mechanism for Planning or Modifying Specific Con-
struction Operations**   J. Lluch and D. W. Halpin, using CYCLONE diagramming, have simulated a paving operation (see Fig. 13-10) and a precasting
scheme (see Fig. 13-11).[24] Their examples demonstrate how field personnel
can evaluate, at the operation-planning stage, several ways of equipping and
conducting these operations without actually carrying them out. The value of
this approach is that at the early planning stage it can provide comparisons

[24] See *Journal of the Construction Division,* ASCE, vol. 108, no. CO 1, March 1982,
pp. 129–145.

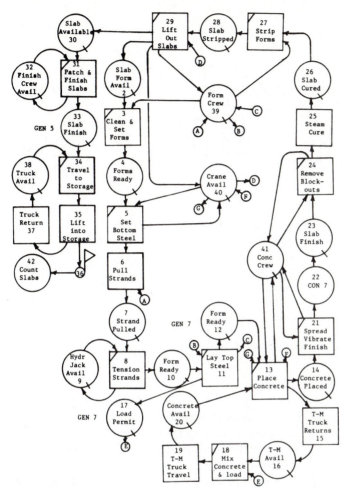

FIGURE 13-11
CYCLONE network for a precasting operation.

among several alternatives easily and quickly so that little added time is required over and above that required to set up a single operating plan.

Paulson and others, doing research sponsored by the National Science Foundation, have developed a workable interactive simulation program linking field observations to computers.[25] It is called INSIGHT (Interactive Simulation of Construction Operations Using Graphical Techniques) and it is now available for both classroom and professional use.

---

[25] See B. C. Paulson, W. T. Chan, and C. C. Koo, *Journal of Construction Engineering and Management,* ASCE, vol. 113, no. 2, June 1987, pp. 302–314. This reference spells out step by step the procedures to be followed in implementing the program.

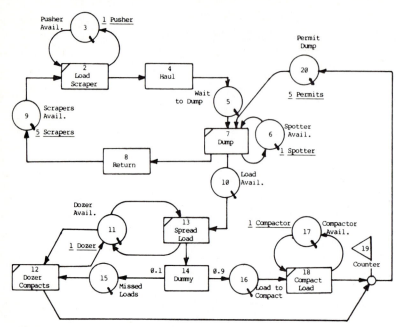

FIGURE 13-12
INSITE model of an earthmoving operation (CYCLONE notation).

Originally run on a PDP ll minicomputer, INSIGHT has been adapted to the new generation of microcomputers. With it, repetitive field operations are recorded on videotape. From this tape, durations for each individual step each time it is carried out can be observed and recorded in the computer, which summarizes the data in the most usable form for analysis, as described below. Given this knowledge of the present operation, it is then possible to develop data and to simulate the effects of changes in the process.

In more detail, INSIGHT operates as follows:

1 The operation to be studied, which can be either for an operation under review or from a similar operation recorded at an earlier time, is recorded on videotape.[26] If desired, the operation can be observed in real time directly on the cathode ray (television style) screen.

2 The recording is viewed rapidly on the CRT screen to clarify the individual steps in the operation selected for analysis.

3 A flow diagram for the recorded operation using CYCLONE notation is developed. Figures 13-4, 13-7, 13-10, and 13-11 would be typical of such diagrams. Figure 13-12 represents an INSIGHT application to earthmoving. It in-

---

[26] With earlier versions of INSIGHT, the operations were filmed using time-lapse techniques (see Chap. 7). With videotape, the delay while waiting for film development and the difficulties in getting data into the computer are avoided so that an operation can be examined more quickly and easily.

volves push loading five scrapers, hauling to a dump site, and spreading the loads, followed by compaction, either with a dozer or a compactor. Recording this particular operation probably would require two or possibly three video-camera setups, probably made in sequence, to cover all phases of the operation.

4 By means of a linkage between the videotape player and the computer, an analyst is able to observe the operation on the CRT screen and identify to the computer the points in time at which each specific event in the flow diagram begins and ends each time it occurs. It is possible to program the computer to quickly position the videotape close to important break or reference points which the analyst has previously identified.

5 Time data stored in the computer is then analyzed using a special computer program which determines the mean time value for each step in the operation and the coefficients for a preselected statistical distribution (either constant, normal, or log-normal). If desired, the data can be displayed on the CRT screen and a best-fit curve can be selected. Furthermore, it is possible, by appropriate programming, to eliminate individual data points which represent aberrations such as an extremely long elapsed time in a given step in one cycle of the operation. After these manipulations, the computer displays the characteristics of each step in the operation.

6 Results of the analysis of each individual step in the operation can be run through the simulation program and if desired, a computer program which develops costs for doing the operation as recorded or modified can be developed. As with any other study involving selections of times for the individual steps from statistical distributions, the summations found in the individual simulation runs will not be identical. Rather, several runs must be made and composite values for the mean and other significant factors must be computed.

One of the many valuable uses for the simulation of current operations is to locate and quantify choke points, long waits, or idle periods, as a first step in finding ways to make improvements. The findings, in either numerical or graphical form, can be given on the CRT screen or in a printout.

7 Using the data developed for the existing operating plan, supplemented as required by estimates of the mean times and time distributions for any new or altered steps, the effects of promising changes can be explored and their results predicted. The first step is to modify the INSIGHT model to fit the new scheme. As an example, in the earthmoving example portrayed in Fig. 13-12, it might be desired to determine the effects of increasing the number of scrapers, eliminating the compactor from the spread, or, alternatively, doing all the compaction with it. Results in time and costs for new simulations can then be compared with the observed operation to determine possible changes in the current operation.

The decreasing costs of computer hardware and software will offer opportunities for many contractors to use simulation to explore job-level innovations based on programs such as INSIGHT. Possibly the most serious obstacle is the confusion brought on by the proliferation of computers, their appurtenances, and their software, which make doing anything seem difficult.

Among the anticipated improvements that can make techniques such as INSIGHT even more effective will be the ability to record data on read-write videodiscs, where they are more accessible than on videotape, which requires data recording and review in sequence. Another is computer graphics, which can offer displays with which users can interact. For example, it may become possible to change the flowchart or modify specific data inputs and explore new schemes without completing the simulation of less-promising approaches. Here, the difficulties lie primarily in developing less-cumbersome ways of interaction between users and computers. One can foresee employing light pens or other mechanisms for transferring user ideas into computer action. It also can be envisioned that for some studies there will be no need to record operations on videotape or for human analysts to view it. Rather, electronic or other devices will register the points in time when given events occur and transmit this information directly to the computer. An example could be that of having instruments which identify the times at which each particular vehicle passes a checkpoint or a crane completes a step in its cycle. Again, certain measurements could be taken and the data passed along, for instance, the positioning of equipment in a work area. Again, transmitters activated by workers or their body movements might provide data on their activities. Exploration and adaptation of procedures and devices already developed in such fields as robotics (see below), electronics, space, and even medicine may provide a way to make quick advances in this area.

This forward look at simulation and computers and their possible applications is oversimplified, but it demonstrates another direction in which improvements in construction planning and implementatiuon will be moving in the future.

CONSTRUCTION APPLICATIONS OF AUTOMATION AND ROBOTICS[27]

Automation describes any process or operation in which a machine or device operates without human intervention. Robotics goes the next step: it involves the use of robots (machines that function as replacements for human beings) which carry a process through from beginning to end, all without human involvement.

Controls which automatically direct machines to perform certain activities and preserve a record of them have been in use for many years in fixed installations that serve construction. Examples include machines which dispense and record the individual weights of cement or asphalt and aggregates and sub-

---

[27] Any discussion of this topic is at best a progress report of a fast-moving field. This particular discussion draws on articles by Boyd C. Paulson, Jr. and A. Wiszawski and D. A. Sangrey, *Journal of Construction Engineering and Management*, ASCE, vol. 111 no. 3, September 1985, pp. 190–207 and 260–280, respectively, and more recent unpublished material.

sequently mix and discharge the mixture in batches as Portland cement or bituminous concrete for pavements. In this instance, production of the plant is usually higher and the product more uniform than if all steps are carried through by humans. But this process is not fully automated; humans must be involved at least as monitors to see that all goes well. Far less automation will be found in the preceding and succeeding steps required to produce a finished pavement. Materials are transported to and unloaded at the plant by or under the direction of humans, and, after mixing, to the laydown site by truck. Laydown and finishing is done by traveling machines, with all the steps also involving humans. Almost every construction sequence follows something of this same process: some automation where a step or a few steps can be carried out at a fixed location but preceded and followed by labor-intensive or human-directed operations. Instances in construction where an entire process is carried out without human intervention (robotics) are rare.

As with the example of the fixed proportioning and mixing plant, automation is often more fully developed in the quality-control functions in construction than in the actual construction itself. Also, there is a growing use of automatic devices which employ strain gauges or deflection-measuring devices to monitor the behavior of structures or the load-carrying elements in hazardous installations such as deep excavations or tunnels. Another application is for cranes, where overloading or improper operating procedures can have disastrous results. With them, because failure can result from many causes, there is no simple guide for a human operator to follow. Rather, instrumentation to measure critical loads or stresses under the several adverse conditions can sound a warning or even override the operator or shut down the machine. Overcoming prejudices against the use of such devices has been difficult, but the high costs of liability settlements and more aggressive enforcement of safety regulations are changing older attitudes.

An application of partial automation to data gathering was cited earlier in this chapter in the discussion of INSIGHT. As was described, an individual viewed time-lapse pictures or videotape and recorded in a computer the points in time when various breaks in the sequence of equipment operation occurred. The computer then analyzed the data so that an analyst, working from it, could determine more efficient modes of equipment selection and operation. Research is underway to find means to automate this data-collection effort directly onto magnetic tape or a recording or transmitting device so that it might be transferred directly into a computer. In turn, an analyst could examine the operation and direct changes or, in some instances, if suitably programmed, the computer could carry out this function. As an example, it should be possible to install devices on scrapers which would develop load-growth curves such as the one shown in Fig. 13-1 under the differing soil conditions and operating procedures of each job site. Comparisons among these curves could lead to ways to optimize performance. Going further, actual control of the scraper might be taken over by an on-board computer which would receive and analyze the data and even operate the scraper controls. It can be foreseen that computers could some day be programmed to continuously optimize, control, and direct the operation of entire equipment spreads.

Lasers and other forms of guidance devices represent an excellent example of partial automation of construction operations. A simple example is a laser beam which sweeps around at constant elevation on a building site. An instrument man is not needed to set elevations; rather they are found by measuring up or down from the rotating light beam. Similar devices are employed to set grades in trenches or give precise line and grade in tunnels.

Gains from the use of lasers to control grading operations can be substantial. For example, electrohydraulic feedback systems mounted on motor graders can achieve base thicknesses to 2 percent tolerances vs. 10 to 20 percent with regular operators. In addition, the work has been accomplished in 20 percent of the usual time. Another cost-saving alternative is to substitute less-costly small bulldozers for blade graders since the dozers can do sufficiently accurate work with these controls.

It can be expected that in time, machines operating in a three-dimensional space environment under remote or computer control will take over certain construction operations, particularly hazardous ones. For example, remotely controlled tractors have been developed that will work under water. Another valuable application will be for handling hazardous wastes, to which humans should not be exposed. Actually, the possible applications of remote control seem endless.

More fully programmed automation of construction operations has been slow in coming. An outstanding example, which can almost be classified as a robot, has been developed by the Shimizu Company in Japan to provide, blend the ingredients, and spray fireproofing materials on steel beams and girders in buildings under construction. It is a computer-controlled, wheeled, self-guiding platform on which is mounted an articulated arm supporting the spray nozzle. Sensors near the nozzle relay signals to the computer which, in turn, actuates mechanisms to properly position the nozzle.

Manufacturers of highly automated machines, robots, or androids (automatic machines in human form) have in general shied away from construction as a market, as evidenced by the fact that at a 1983 robotics exhibition, not one of the 260 robots on display had construction applications. The principal explanations given by the manufacturers for this neglect were (1) the lack of repeatability in construction operations at one location, (2) the influence of on-site variables, (3) the lack of concentration of the market in a relatively few firms, and (4) the individualism of construction managers which causes them to resist this kind of thing.

With a few exceptions, construction applications of robots are largely in the research stage. Actual applications will probably come first in situations where portions of operations can be prefabricated, preassembled, or modularized (see the discussion in Chap. 5) and will be carried on off site or at least away from the work face so that external variables are few, as in manufacturing. But in a field that is moving so fast in so many directions, predictions are risky. Many unforeseen applications of automation and robotics may be on the horizon.

CONSTRUCTION APPLICATIONS OF KNOWLEDGE-BASED EXPERT SYSTEMS

Knowledge-based expert systems are problem-solving computer programs that have developed from research in artificial intelligence (AI), a subfield of computer science that is concerned with symbolic reasoning about and representation of knowledge. Knowledge-based expert systems focus more specifically on solutions to complex problems which require the competence of an expert or experts or where the usual analytical procedures or rules of thumb are not sufficiently reliable because there are too many variables or uncertainties. The systems are designed to capture and store a large body of knowledge known by experts and to organize it in such a form that it can be used by those with a lower level of skill in a particular area. In addition, the programs can explain to users the reasoning by means of which the experts make decisions, given a particular set of circumstances. This makes it possible to offer flexibility so that changes in anticipated conditions can be accommodated by drawing further on the knowledge base provided by the experts and stored in the computer.

Some have explained the expert-systems approach in terms of a team. One party in the team is a "domain expert" who has detailed knowledge in the specific area of interest, and can therefore define the applicable conditions and corresponding actions to be taken in a given situation or when there are variations in the situation. The other party has sometimes been designated as a "knowledge engineer" whose skills lie in the ability to extract data about every detail of the problem from the expert, classify it, and fit it into a computer software package which is then available to others facing the same or similar problems. In situations where the problem being addressed is very complex, the participation of a specialist in knowledge engineering is essential, since the domain expert would not be skilled in the probing techniques required to explore all aspects of the problem and then to organize this information in a form suitable for presentation in a computer program. However, with their increasing familiarity with the microcomputer, it has been possible for engineers who are domain experts to bridge the knowledge-engineering gap and produce expert systems programs useful in the construction industry.

There are two general classes of expert-systems problems. One starts with inputs of known facts in the domain covered by the knowledge base and from these facts draws a conclusion or possibly several conclusions. Explanations of the evidence and reasoning supporting the conclusions would be readily accessible in the program. An example would be a medical diagnosis, where data about a patient's vital functions are the input and identification of the possible ailments the output. In simple terms, it starts with facts and ends with a rule.

The second general class of expert-systems techniques starts not with specific data but with a problem to be solved. It examines the range of possible solutions and after developing the facts pertinent to the particular situation recommends the appropriate choice or choices. Simply put, it starts with a possible rule, develops facts for a particular situation, and supplies arguments to support the recommended solution. A medical example might be as follows: Given a patient with an unknown ailment, prescribe a series of laboratory tests that seem to be the most likely for making a correct diagnosis and by analyzing their results, offer a diagnosis. In some instances, the two approaches might be used in sequence. First, plausible hypotheses (the possible diseases) could be established from the data and given these, the added data needed (additional laboratory tests) to reach a specific diagnosis would be proposed. Finally, a specific program of medication or treatment would be offered.

Program languages for solving expert-systems problems have been developed in a variety of ways. Some are based on general-purpose programming languages, and others, on general-purpose representative languages developed specifically to solve expert-systems problems. Still others are domain-independent and provide a base from which domain-specific programs can be written. No attempt will be made here either to list or to explain the working of these programs or languages, since for must construction applications, specific programs are available.

Only with the coming of microcomputers, beginning in the early 1980s, has attention to knowledge-based expert systems been directed to off-site or on-site construction problems. In time, however, such programs will become as common as some of the other computer applications described earlier in this chapter. A few of the many possible applications already in use or proposed are:

* Monitoring areas such as cost, time, purchasing, and inventory control.[28]
* Making more accurate predictions of the time required to complete tasks by adjusting traditional estimates to recognize such factors as learning, labor availability and competence, quantity of work, weather, and downtime.[29]
* Developing enhanced scheduling techniques. Currently computer programs for developing precedence diagrams or other critical-path methods (and bar charts based on them) are available on all microcomputers. The assumption underlying them is that each event has a fixed time. This does not recognize and explain the uncertainties considered by the expert in setting the most probable time interval or the possible variations from it. An enhanced program could recognize the reality of these uncertainties and their consequences in

---

[28] See M. R. McGartland and C. T. Hendrickson, *Journal of Construction Engineering and Management,* ASCE, vol. 111, no. 3, September 1985, pp. 293–307, for added discussion. This article also offers an excellent introduction to expert systems as a working tool.

[29] See C. Hendrickson et al., *Journal of Construction Engineering and Management,* ASCE, vol. 113, no. 2, June 1987, pp. 288–301, for an example focused in part on bricklaying.

time and money, as described by the expert when the original plan was made. From these inputs, probabilities and possibly the costs of meeting various deadlines or targets could be predicted. Also, as the project progressed, the original schedule and the cost estimate could be updated, making it a usable rather than a neglected tool, as is often the case at present.

• Analyzing contract disputes, in this case specifically aimed at disputes arising from differing site conditions. This program follows step by step the phases of such disputes. It presents the range of factual and technical situations covered by the individual contract clauses and offers the legal precedents that might govern the settlement of the claim.[30]

• Planning complex construction operations where several choices of methods should be considered but where the usual decision rules do not probe deeply enough. As an example, in high-rise construction, there might be several possible erection schemes or sequences to be chosen among.

• Planning for multiple uses of cranes. This operation is too complex to be completely analyzed using the typical planning methods. But expert-systems tools may make this task far easier.

• Handling the layout of temporary facilities on a construction site. SIGHTPLAN makes it possible to plan the layout of temporary facilities in terms of the available space as the needs change over time.

• Evaluating the safety of a construction firm. HOWSAFE based on the findings reported in *Business Roundtable Report A-3,* offers a questionnaire, accompanied by explanations, which deeply probes the records of and attitudes about safety of a construction company and its officials and from this develops a rating scheme.[31]

This very elementary discussion of knowledge-based expert systems is intended merely to suggest the value of a relatively new discipline in which exciting research and applications to construction are underway. The field is moving very fast, and the list of already developed applications is too long to give here. Furthermore, because the field is so dynamic, references in the traditional sources are scarce, and only a few illustrative and readily available ones are offered. It follows that interested readers should dig deeply on their own.

SUMMARY

This chapter looks at how computers, graphical and mathematical models, simulation, automation and robotics, and knowledge-based expert systems can offer ways, separately or in combination, to improve productivity in construc-

---

[30] See J. E. Diekmann and T. A. Kruppenbacher, *Journal of Construction Engineering and Management,* ASCE, vol. 110, no. 4, December 1984, pp. 391–408.

[31] See R. E. Levitt in *Expert Systems in Civil Engineering,* ASCE, April 1986, pp. 55–66. A computer-based program covering this topic, labeled Howsafe, is available for purchase from Building Knowledge Systems, Inc., Stanford, Calif. 94305.

tion. Its purpose has been to give an overview, excite curiosity, and, where possible, offer references so that those who are interested can probe more deeply. In closing, the authors warn that these approaches and many other proposed in this book can be a threat to those who are comfortable with the status quo. Only as this threat is removed, can these techniques gain acceptance in a hardnosed, conservative industry.

METHODS-TIME MEASUREMENT (MTM)[1]

Exhibits A-1 and A-2 are reprints of a sampling of the data-cards prepared by the MTM Association for Standards and Research. They are included here to illustrate the MTM system, which is widely employed in industry for estimating worker performance. The data are entirely factual and represent the results of observations over a period of many years of the times required for the common body, limb, hand, and eye movements to be executed.

Exhibit A-1 shows the basic application data for such movements in a highly organized and repetitive work situation. These assume a suitable layout and the proper tools. A planner may, by informed and careful use of such tables, develop and accumulate the periods of times required for an individual to carry out an activity through a series of successive body movements. Furthermore, it is possible to compare the times required to carry out a given operation following different procedures or sequences, and thus simplify a present method or synthesize a new and more efficient one.

Exhibit A-2 is for the MTM-MEK system which was developed specifically for the measurement of activities that are "one of a kind" or that involve small lots. It is universally applicable. It was developed specifically for tasks that are not highly organized nor highly repetitive, and that do not have a set and precise work method. It is particularly useful as a method for establishing preproduction labor standards for customized products and for activities with ongoing methods variation. It is reported that a trained practitioner can apply this simplified system in one-fiftieth of the time required for MTM-1.

---

[1] The material in this Appendix is copyrighted by the MTM Association for Standards and Research. No reprints can be made without written consent from the MTM Association, 16-01 Broadway, Fair Lawn, NJ 07410.

As with all the MTM systems, one should not attempt to use the charts and tables unless trained in the proper application of the data. Without the proper training, the misapplication of the tables can lead to improper results.

Table 5-1, page 98, in the text is a simple example showing how the US Navy has applied MTM to its standard estimating procedure for in-house construction and maintenance operations.

METHODS-TIME MEASUREMENT
MTM-I APPLICATION DATA

1 TMU	=	.00001	hour	1 hour	=	100,000.0 TMU
	=	.0006	minute	1 minute	=	1,666.7 TMU
	=	.036	seconds	1 second	=	27.8 TMU

Do not attempt to use this chart or apply Methods-Time Measurement in any way unless you understand the proper application of the data. This statement is included as a word of caution to prevent difficulties resulting from mis-application of the data.

MTM ASSOCIATION FOR STANDARDS AND RESEARCH

16-01 Broadway
Fair Lawn, N.J. 07410

ILLINOIS OFFICE
1411 Peterson Avenue
Park Ridge, Illinois 60068
Telephone: 312/823-7120
Telex No. 883094

CALIFORNIA OFFICE
2043 Westcliff Drive
Newport Beach, California 92660
Telephone: 714/631-3113

MTMA 101
PRINTED IN U S A.

© Copyright 1973

EXHIBIT A-1
MTM application data.

TABLE I – REACH – R

Distance Moved Inches	Time TMU A	B	C or D	E	Hand In Motion A	B	CASE AND DESCRIPTION
3/4 or less	2.0	2.0	2.0	2.0	1.6	1.6	**A** Reach to object in fixed location, or to object in other hand or on which other hand rests.
1	2.5	2.5	3.6	2.4	2.3	2.3	
2	4.0	4.0	5.9	3.8	3.5	2.7	
3	5.3	5.3	7.3	5.3	4.5	3.6	**B** Reach to single object in location which may vary slightly from cycle to cycle.
4	6.1	6.4	8.4	6.8	4.9	4.3	
5	6.5	7.8	9.4	7.4	5.3	5.0	
6	7.0	8.6	10.1	8.0	5.7	5.7	
7	7.4	9.3	10.8	8.7	6.1	6.5	**C** Reach to object jumbled with other objects in a group so that search and select occur.
8	7.9	10.1	11.5	9.3	6.5	7.2	
9	8.3	10.8	12.2	9.9	6.9	7.9	
10	8.7	11.5	12.9	10.5	7.3	8.6	
12	9.6	12.9	14.2	11.8	8.1	10.1	
14	10.5	14.4	15.6	13.0	8.9	11.5	**D** Reach to a very small object or where accurate grasp is required.
16	11.4	15.8	17.0	14.2	9.7	12.9	
18	12.3	17.2	18.4	15.5	10.5	14.4	
20	13.1	18.6	19.8	16.7	11.3	15.8	
22	14.0	20.1	21.2	18.0	12.1	17.3	**E** Reach to indefinite location to get hand in position for body balance or next motion or out of way.
24	14.9	21.5	22.5	19.2	12.9	18.8	
26	15.8	22.9	23.9	20.4	13.7	20.2	
28	16.7	24.4	25.3	21.7	14.5	21.7	
30	17.5	25.8	26.7	22.9	15.3	23.2	
Additional	0.4	0.7	0.7	0.6			TMU per inch over 30 inches

TABLE II – MOVE – M

Distance Moved Inches	Time TMU A	B	C	Hand In Motion B	Wt. (lb.) Up to	Dynamic Factor	Static Constant TMU	CASE AND DESCRIPTION
3/4 or less	2.0	2.0	2.0	1.7				
1	2.5	2.9	3.4	2.3	2.5	1.00	0	
2	3.6	4.6	5.2	2.9				**A** Move object to other hand or against stop.
3	4.9	5.7	6.7	3.6	7.5	1.06	2.2	
4	6.1	6.9	8.0	4.3				
5	7.3	8.0	9.2	5.0	12.5	1.11	3.9	
6	8.1	8.9	10.3	5.7				
7	8.9	9.7	11.1	6.5	17.5	1.17	5.6	
8	9.7	10.6	11.8	7.2				
9	10.5	11.5	12.7	7.9	22.5	1.22	7.4	**B** Move object to approximate or indefinite location.
10	11.3	12.2	13.5	8.6				
12	12.9	13.4	15.2	10.0	27.5	1.28	9.1	
14	14.4	14.6	16.9	11.4				
16	16.0	15.8	18.7	12.8	32.5	1.33	10.8	
18	17.6	17.0	20.4	14.2				
20	19.2	18.2	22.1	15.6	37.5	1.39	12.5	
22	20.8	19.4	23.8	17.0				
24	22.4	20.6	25.5	18.4	42.5	1.44	14.3	**C** Move object to exact location.
26	24.0	21.8	27.3	19.8				
28	25.5	23.1	29.0	21.2	47.5	1.50	16.0	
30	27.1	24.3	30.7	22.7				
Additional	0.8	0.6	0.85				TMU per inch over 30 inches	

EXHIBIT A-1
MTM application data (*continued*).

490

TABLE III A — TURN — T

Weight	Time TMU for Degrees Turned										
	30°	45°	60°	75°	90°	105°	120°	135°	150°	165°	180°
Small — 0 to 2 Pounds	2.8	3.5	4.1	4.8	5.4	6.1	6.8	7.4	8.1	8.7	9.4
Medium — 2.1 to 10 Pounds	4.4	5.5	6.5	7.5	8.5	9.6	10.6	11.6	12.7	13.7	14.8
Large — 10.1 to 35 Pounds	8.4	10.5	12.3	14.4	16.2	18.3	20.4	22.2	24.3	26.1	28.2

TABLE III B — APPLY PRESSURE — AP

FULL CYCLE			COMPONENTS		
SYMBOL	TMU	DESCRIPTION	SYMBOL	TMU	DESCRIPTION
APA	10.6	AF + DM + RLF	AF	3.4	Apply Force
			DM	4.2	Dwell, Minimum
APB	16.2	APA + G2	RLF	3.0	Release Force

TABLE IV — GRASP — G

TYPE OF GRASP	Case	Time TMU	DESCRIPTION	
PICK-UP	1A	2.0	Any size object by itself, easily grasped	
	1B	3.5	Object very small or lying close against a flat surface	
	1C1	7.3	Diameter larger than 1/2"	Interference with Grasp
	1C2	8.7	Diameter 1/4" to 1/2"	on bottom and one side of
	1C3	10.8	Diameter less than 1/4"	nearly cylindrical object.
REGRASP	2	5.6	Change grasp without relinquishing control	
TRANSFER	3	5.6	Control transferred from one hand to the other.	
SELECT	4A	7.3	Larger than 1" x 1" x 1"	Object jumbled with other
	4B	9.1	1/4" x 1/4" x 1/8" to 1" x 1" x 1"	objects so that search
	4C	12.9	Smaller than 1/4" x 1/4" x 1/8"	and select occur.
CONTACT	5	0	Contact, Sliding, or Hook Grasp.	

EFFECTIVE NET WEIGHT			
Effective Net Weight (ENW)	No. of Hands	Spatial	Sliding
	1	W	W x F_c
	2	W/2	W/2 x F_c

W = Weight in pounds
F_c = Coefficient of Friction

EXHIBIT A-1
MTM application data (*continued*).

TABLE V – POSITION* – P

CLASS OF FIT		Symmetry	Easy To Handle	Difficult To Handle
1—Loose	No pressure required	S	5.6	11.2
		SS	9.1	14.7
		NS	10.4	16.0
2—Close	Light pressure required	S	16.2	21.8
		SS	19.7	25.3
		NS	21.0	26.6
3—Exact	Heavy pressure required.	S	43.0	48.6
		SS	46.5	52.1
		NS	47.8	53.4
SUPPLEMENTARY RULE FOR SURFACE ALIGNMENT				
P1SE per alignment: $>1/16 \leqslant 1/4''$		P2SE per alignment: $\leqslant 1/16''$		

*Distance moved to engage—1" or less.

TABLE VI – RELEASE – RL

Case	Time TMU	DESCRIPTION
1	2.0	Normal release performed by opening fingers as independent motion.
2	0	Contact Release

TABLE VII – DISENGAGE – D

CLASS OF FIT	HEIGHT OF RECOIL	EASY TO HANDLE	DIFFICULT TO HANDLE
1—LOOSE—Very slight effort, blends with subsequent move.	Up to 1"	4.0	5.7
2—CLOSE—Normal effort, slight recoil.	Over 1" to 5"	7.5	11.8
3—TIGHT—Considerable effort, hand recoils markedly.	Over 5" to 12"	22.9	34.7
SUPPLEMENTARY			

CLASS OF FIT	CARE IN HANDLING	BINDING
1— LOOSE	Allow Class 2	———
2— CLOSE	Allow Class 3	One G2 per Bind
3— TIGHT	Change Method	One APB per Bind

TABLE VIII – EYE TRAVEL AND EYE FOCUS – ET AND EF

Eye Travel Time = $15.2 \times \frac{T}{D}$ TMU, with a maximum value of 20 TMU.

where T = the distance between points from and to which the eye travels.
D = the perpendicular distance from the eye to the line of travel T.

Eye Focus Time = 7.3 TMU.

SUPPLEMENTARY INFORMATION

— Area of Normal Vision = Circle 4" in Diameter 16" from Eyes

— Reading Formula = 5.05 N Where N = The Number of Words.

EXHIBIT A-1
MTM application data (*continued*).

TABLE IX — BODY, LEG, AND FOOT MOTIONS

TYPE		SYMBOL	TMU	DISTANCE	DESCRIPTION
LEG–FOOT MOTION		FM	8.5	To 4″	Hinged at ankle.
		FMP	19.1	To 4″	With heavy pressure.
		LM__	7.1	To 6″	Hinged at knee or hip in any direction.
			1.2	Ea. add'l inch	
HORIZONTAL MOTION	SIDE STEP	SS__C1	*	<12″	Use Reach or Move time when less than 12″. Complete when leading leg contacts floor.
			17.0	12″	
			0.6	Ea. add'l inch	
		SS__C2	34.1	12″	Lagging leg must contact floor before next motion can be made.
			1.1	Ea. add'l inch	
	TURN BODY	TBC1	18.6	——	Complete when leading leg contacts floor.
		TBC2	37.2	——	Lagging leg must contact floor before next motion can be made
	WALK	W__FT	5.3	Per Foot	Unobstructed.
		W__P	15.0	Per Pace	Unobstructed.
		W__PO	17.0	Per Pace	When obstructed or with weight.
VERTICAL MOTION		SIT	34.7	——	From standing position.
		STD	43.4	——	From sitting position.
		B,S,KOK	29.0	——	Bend, Stoop, Kneel on One Knee.
		AB,AS,AKOK	31.9	——	Arise from Bend, Stoop, Kneel on One Knee
		KBK	69.4	——	Kneel on Both Knees.
		AKBK	76.7	——	Arise from Kneel on Both Knees.

TABLE X — SIMULTANEOUS MOTIONS

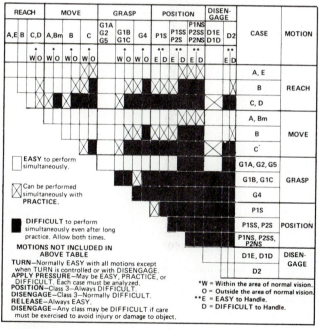

EASY to perform simultaneously.

Can be performed simultaneously with PRACTICE.

DIFFICULT to perform simultaneously even after long practice. Allow both times.

MOTIONS NOT INCLUDED IN ABOVE TABLE
TURN—Normally EASY with all motions except when TURN is controlled or with DISENGAGE.
APPLY PRESSURE—May be EASY, PRACTICE, or DIFFICULT. Each case must be analyzed.
POSITION—Class 3—Always DIFFICULT.
DISENGAGE—Class 3—Normally DIFFICULT.
RELEASE—Always EASY.
DISENGAGE—Any class may be DIFFICULT if care must be exercised to avoid injury or damage to object.

*W = Within the area of normal vision.
O = Outside the area of normal vision.
**E = EASY to Handle.
D = DIFFICULT to Handle.

EXHIBIT A-1
MTM application data (*continued*).

SUPPLEMENTARY MTM DATA

TABLE 1 — POSITION — P

Class of Fit and Clearance	Case of † Symmetry	Align Only	Depth of Insertion (per ¼")			
			0 >0≤1/8"	2 >1/8≤¾	4 >¾≤1¼	6 >1¼≤1¾
21 .150" − .350"	S	3.0	3.4	6.6	7.7	8.8
	SS	3.0	10.3	13.5	14.6	15.7
	NS	4.8	15.5	18.7	19.8	20.9
22 .025" − .149"	S	7.2	7.2	11.9	13.0	14.2
	SS	8.0	14.9	19.6	20.7	21.9
	NS	9.5	20.2	24.9	26.0	27.2
23* .005" − .024"	S	9.5	9.5	16.3	18.7	21.0
	SS	10.4	17.3	24.1	26.5	28.8
	NS	12.2	22.9	29.7	32.1	34.4

*BINDING—Add observed number of Apply Pressures.
DIFFICULT HANDLING—Add observed number of G2's.

†Determine symmetry by geometric properties, except use S case when object is oriented prior to preceding Move.

TABLE 1A — SECONDARY ENGAGE — E2

CLASS OF FIT	DEPTH OF INSERTION (PER 1/4")		
	2	4	6
21	3.2	4.3	5.4
22	4.7	5.8	7.0
23	6.8	9.2	11.5

TABLE 2 — CRANK (LIGHT RESISTANCE) — C

DIAMETER OF CRANKING (INCHES)	TMU (T) PER REVOLUTION	DIAMETER OF CRANKING (INCHES)	TMU (T) PER REVOLUTION
1	8.5	9	14.0
2	9.7	10	14.4
3	10.6	11	14.7
4	11.4	12	15.0
5	12.1	14	15.5
6	12.7	16	16.0
7	13.2	18	16.4
8	13.6	20	16.7

FORMULAS:

A. CONTINUOUS CRANKING (Start at beginning and stop at end of cycle only)

$$TMU = [(N \times T) + 5.2] \cdot F + C$$

B. INTERMITTENT CRANKING (Start at beginning and stop at end of each revolution

$$TMU = [(T + 5.2) F + C] \cdot N$$

C	=	Static component TMU weight allowance constant from move table
F	=	Dynamic component weight allowance factor from move table
N	=	Number of revolutions
T	=	TMU per revolution (Type III Motion)
5.2	=	TMU for start and stop

EXHIBIT A-1
MTM application data (*continued*).

			TIME UNITS			

MTM FOR ONE OF A KIND AND SMALL LOT PRODUCTION ANALYZING SYSTEM AND STANDARD DATA BLOCKS

JANUARY 1986	IF TRAINING IN MTM AND MEK IS LACKING, USAGE OF THIS TABLE LEADS TO WRONG RESULTS	TIME UNITS			
		TMU	SEC	MIN	HR
		1	0.036	0.0006	0.00001

TIME VALUES IN TMU

ANALYZING SYSTEM			DISTANCE RANGES IN.	≤ 8 IN.	>8 TO 32 IN.	> 32 TO 80 IN.	
						WITHOUT	WITH
						BEND	
MEK ELEMENT	OBJECT CHAR-ACTERISTICS	PLACE ACCURACY	CODE	1	3	4	5
GET AND PLACE	≤ 12 / 32 IN. ≤ 18 LBS.	APPROXIMATE	AA	30	50	120	150
		EXACT	AB	50	90	160	190
	> 12 / 32 IN. > 18 - 48 LBS	APPROXIMATE	AC	✕	150	220	250
		EXACT	AD	✕	190	260	290
HANDLE TOOL		APPROXIMATE	HA	✕	70	140	170
GET, PLACE AND PLACE ASIDE		EXACT	HB	✕	100	160	190
PLACE		APPROXIMATE	PA	20	30	40	70
		EXACT	PB	40	50	60	90
OPERATE		SIMPLE	BA	20	30	50	110
		COMPOUND	BB	40	50	70	130

MOTION CYCLES		CODE	TMU
WITHOUT REPOSITION (MOTION LENGTH)	≤ 4 IN.	ZA	10
	> 4 TO 12 IN.	ZB	20
	> 12 TO 32 IN.	ZC	40
WITH REPOSITION (LEVER LENGTH)	≤ 8 IN.	ZD	40
	> 8 TO 18 IN.	ZE	60
	> 18 TO 40 IN.	ZF	120

FASTEN OR LOOSEN	ZZ	30

BODY MOTIONS	CODE	TMU
WALK/40 INCHES	KA	25
BEND, STOOP	KB	60
SIT AND STAND	KC	110

VISUAL CONTROL	VA	40

MTM ASSOCIATION COPYRIGHT 1984

EXHIBIT A-2
MTM-MEK data.

MEK - STANDARD DATA BLOCKS

FASTEN (TIGHTENING OR LOOSENING) BOLTS/NUTS DIAMETER	CODE	BOLT OR NUT WITHOUT LOCKING WRENCH									ADDITION FOR DISTANCE RANGE				
		EASY OF MOTION/OF ACCESS				HARD OF MOTION/OF ACCESS			ADDITION FOR		TOOL			PART	
		A WITHOUT TOOL WING NUT KNURLED-HEADBOLT	B SCREW-DRIVER PANORAMIC LEVER-TOOL, RATCHET	C OPEN-, BOX-END-, ALLEN-WRENCH	D MACHINE-, DRILL-SCREW-DRIVER	E SCREW-DRIVER, PANORAMIC LEVER-TOOL, RATCHET	F OPEN-, BOX-END-, ALLEN-WRENCH	G MACHINE-, DRILL-SCREW-DRIVER	H ADDI-TIONAL TIGHTEN	I DAMAGED OR DIRTY THREADS BECAUSE OF REPAIRS	EH 3	EH 4	EH 5	ET 4	ET 5
— ≤ 3/8 IN.	A-SA	260	300	400	190	370	890	220	80	110					
>3/8 — ≤ 3/4 IN.	A-SB	280	370	590	220	530	1150	250	90	110					
>3/4 — ≤ 1-3/8 IN.	A-SC		640	1050	310	1000	2550	370	110	600	60	120	150	70	100
>1-3/8 — ≤ 2 IN.	A-SD		740	1350	360	1200	3150	440	150	1250					
>2 — ≤ 2-3/8 IN.	A-SE		800	1500	490	1300	3400	590	190	3000					

	CODE	E	C	H	EH 3	EH 4	EH 5	ET 4	ET 5
SECOND SCREW PART (BOLT OR NUT) WITH ADDITIONAL TOOL	A-SGA	130							
STUD SETTER	A-SHA		1000	40	60	120	150	70	100
APPLY SECOND TOOL TO 2ND SCREW PART	PB1	90							
ADDITIONAL PARTS (WASHER, SPRING)	AB3								

	LENGTH	CODE	E	F	G	EH 3	EH 4	EH 5	ET 4	ET 5
WOOD- OR SELF-TAPPING-SCREWS, (ROUGH-DRILLED, 1 TOOL)	≤ 1 IN.	A-SN	400	1000	230					
	≤ 2 IN.	A-SP	1000	1800	290	60	120	150	70	100
	> 2 IN.	A-SQ	1400	2600	410					
WOOD SCREWS WITH PIERCING (2 TOOLS)	≤ 1 IN.	A-SR	730	1650	460					
	≤ 2 IN.	A-SS	1650	2850	520					
	> 2 IN.	A-ST	2250	4000	640					

EXHIBIT A-2
MTM-MEK data (*continued*).

CLAMP AND UNCLAMP

			CODE	TMU
PORTABLE CLAMPING DEVICES	GRIPPING PLIERS		A-FAA	290
	HAND GRIP		A-FBA	340
	SCREW CLAMP		A-FCA	390
	C-CLAMP		A-FDA	630
	PARALLEL SCREW CLAMP		A-FEA	1100
FIXED CLAMPING DEVICES	CLAW (T-CLAMP)		A-FLA	860
	GRIPPING LEVER CLAMP	MANUAL	A-FMA	100
		PNEUMATIC	A-FNA	140
	VISE	APPROXIMATE	A-FPA	400
		EXACT	A-FQA	670
	JAW CHUCK		A-FRA	520
	MAGNETIC PLATE		A-FSA	40

DISTANCE RANGE		
TOOL		
EH		
3	4	5
60	120	150

	CODE	TMU
CLAMP WITH ADDITIONAL TOOL	A-FZA	250

60	120	150

CLEAN AND/OR APPLY LUBRICANT

			ORDER OF MAGNITUDE				ADDITION FOR DIST RANGE		
			POINT	AREA IN INCH			TOOL		
				8X8	20X20	40X40	EH		
		CODE	A	B	C	D	3	4	5
CLEAN	COMPRESSED AIR, BRUSH, BROOM, RAG, CLOTH	A-BA	—	80	180	340	60	120	150
	WIRE-BRUSH	A-BB	—	220	820	4000			
	SCRAPER, SMOOTHER	A-BC	—	340	1400	6000			
APPLY OIL, PAINT, CLEANSING AGENT	PAINTBRUSH, OIL-CAN, RAG, CLOTH	A-BH	80	150	260	—			
	SPRAYING-CAN	A-BI	—	170	470	1800			
	GREASE-GUN	A-BK	80	—	—	—			
APPLY ADHESIVE, SEALING-MATERIAL	PAINTBRUSH OR SMOOTHER	A-BN	100	400	830	—			
	TUBE OR PIN	A-BQ	70	180	420	—			

		WEIGHT IN OZ				
		$\leqslant$ 0.2	$>$0.2 $\leqslant$ 1			
ADDITION	CODE	A	B			
MIX TWO-COMPONENT ADHESIVE	A-BZ	990	1750			
OPEN CONTAINER AND CLOSE IT	A-BZG	310				
HANDS	WIPE OFF	A-BZH	260		60 120 150	
	CLEAN/WASH	A-BZI	890			

EXHIBIT A-2
MTM-MEK data (*continued*).

ASSEMBLE STANDARD PARTS		CODE	TMU	ADDITION FOR DISTANCE RANGE				
				TOOL			PARTS	
				EH			ET	
				3	4	5	4	5
WITH ONE TOOL	SPRING PIN, PIN, NAIL, COTTER, CLIP, PLASTIC BINDING, SPRING	A-NAA	210					
WITH TWO TOOLS	COTTER, PIN, RIVET, DOUBLE COTTER	A-NBA	260	60	120	150	70	100
SPECIAL TOOL	E-RING, SNAP RING, POP RIVET	A-NCA	200					
ADDITION	USE ADDITIONAL TOOL; SNAP TOOL	A-NZA	100					
	DIFFICULT OR CAREFUL HANDLING	A-NZB	80	60	120	150		

MARK		CODE	TMU	ADDITION FOR DISTANCE RANGE				
				TOOL			PARTS	
				EH			ET	
				3	4	5	4	5
CENTER PUNCH OR	MARK	A-MAA	100					
FIGURE STAMP	ON CROSS MARK	A-MAB	230					
COMPASS ≤ 12 IN. DIAMETER	WITHOUT ADJUST	A-MBA	140					
	ADDITION: ADJUST	A-MBB	160					
RULER, TEMPLATE	SCRIBING- LENGTH ≤ 40 IN.	A-MCA	250					
PARALLEL SCRIBE	SCRIBING- LENGTH ≤ 20 IN.	A-MDA	80					
SLIDING GAUGE FOR SCRIBING	ADDITION FOR ADJUST	A-MDB	470					
WRITING IMPLEMENT	1 CHARACTER	A-MEA	60					
	GROUP OF CHARACTERS	A-MEB	180					
BRUSH, OUTLINE TEMPLATE	10 CHARACTERS	A-MFA	380	60	120	150		
LABEL MARKER	10 CHARACTERS	A-MGA	480					
FRONT LAY GAGE PROTRACTOR	SCRIBING LENGTH ≤ 8 IN.	A-MHA	160					
ELECTRIC WRITER	PER CHARACTER	A-MIA	140					
METAL MARKING DEVICE	PER CHARACTER	A-MKA	80					
	ADD.: ADJUST	A-MKB	1200					
STICKER LABEL TAG	APPLY	A-MLA	230				70	100
	ADDITION: REMOVE SUP-PORTING FOIL	A-MLB	110					
CARPENTER'S LINE CAN ONLY BE PERFORMED BY 2 WORKERS *VALUE PER WORKER	UP TO 13 FT	A-MMA	380*					
	ADDITION: HANDLE ROPE	A-MMB	720*					

EXHIBIT A-2
MTM-MEK data (*continued*).

498

INSPECT OR MEASURE			CODE	ORDER OF MAGNITUDE			ADDITION FOR DISTANCE RANGE		
				SMALL < 6 IN.	MEDIUM > 6 IN. ≤ 20 IN.	LARGE > 20 IN. ≤ 80 IN.	TOOL		
							EH		
				A	B	C	3	4	5
GAUGES	DIMENSION OR SHAPE		A-PA	100	160	200	60	120	150
	THREAD		A-PB	280	520	—			
DIAL TYPE MEASURING INSTRUMENTS	INSPECT DIMENSION		A-PC	100	160	—			
	DETERMINE DIMENSION		A-PD	140	200	—			
DIRECT-READING MEASURING INSTRUMENTS	WITHOUT MOVABLE PARTS		A-PE	120	120	230			
	WITH MOVABLE PARTS		A-PF	270	300	—			
	CRANK ROLL TAPE MEASURE ≤ 66 FT*		A-PGA	1600					
	ROLL TAPE MEASURE/FOLDING RULE ≤ 80 IN.		A-PGB	200					
PLUMB BOB WITH 2 WORKERS DOUBLE VALUES ≤ 8 FT*			A-PHA	1900					
VISUAL MEASURING INSTRUMENTS	PROJECTORS (MAGNIFIER)		A-PIA	750					
FEEL OR NOISE INSPECTION			A-PKA	80					

*IF 2 WORKERS ARE ASSIGNED
DOUBLE THE VALUE
NOTE FOR MEASURES > 66 FT: EVALUATE "WALK" SEPARATELY

		CODE	VALUE	ADDITION FOR DISTANCE RANGE		
ADDITION FOR PREPARING AND TERMINATING MEASURING INSTRUMENT AFTER ADJUSTING	STANDARD END MEASURING (UP TO 3 BLOCKS)	A-PZA	2500	60	120	150
	PROTRACTOR	A-PZB	560			
	DIAL TYPE MEASURING INSTRUMENTS	A-PZC	2500			
	MICROMETERS	A-PZD	4000			
	VISUAL MEASURING INSTRUMENT	A-PZE	1600			
ADJUST AREA OR SET TO ZERO	MEASURING RANGE ≤ 6 IN.	A-PZF	180			
	MEASURING RANGE > 6 IN. ≤ 20 IN.	A-PZG	210			

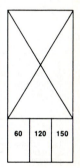

EXHIBIT A-2
MTM-MEK data (*continued*).

499

TRANSPORT				CODE	PRE-PARE	DISTANCE RANGE (FT) -7	7—33	33—165
					A	B	C	D
WITHOUT MEANS OF TRANSPORT	SHIFT			A-TAA		190	440	1650
	EXCHANGE			A-TAB		320	690	2500
HAND TRUCKS	WITHOUT PICKING UP AND SETTING DOWN			A-TBA			450	1650
	WITH PICKING UP AND SETTING DOWN	MANUAL	FLOOR-FLOOR	A-TBB			1350	3150
			FLOOR-TABLE	A-TBC			3600	5400
			TABLE-TABLE	A-TBD			5800	7600
		MOTOR-DRIVEN	FLOOR-FLOOR	A-TBE			1200	3500
			FLOOR-TABLE	A-TBF			2050	4300
			TABLE-TABLE	A-TBG			2650	5000
WITH LIFTING APPARATUS OPERATED FROM THE FLOOR	EQUAL WORKING HEIGHT	HOOK		A-TCA		720	1550	5000
		SLING		A-TCB		1100	1950	5400
	CHANGING WORKING HEIGHT	HOOK		A-TCC		1100	1850	5300
		SLING		A-TCD		1500	2250	5700
TRAVELING CRANE ≤ 80 T	APPROXIMATE	HOOK		A-TDA	2500	3000	3200	4400
		SLING		A-TDB		3350	3600	4800
	EXACT	HOOK		A-TDC		3950	4200	5400
		SLING		A-TDD		4300	4600	5800
PORTABLE CRANE ≤ 16 T	APPROXIMATE	HOOK		A-TEA	3650	770	960	1950
		SLING		A-TEB		1150	1350	2300
	EXACT	HOOK		A-TEC		1750	1950	2900
		SLING		A-TED		2100	2300	3300
LEVERJACK (LIFTING GEAR)	≤ 1.5 T			A-TFA	5500	12900	—	—
	> 1.5 T ≤ 6 T			A-TFB		33700	—	—
CABLE LINE (GRIPPING GEAR)	≤ 1.5 T			A-TGA	14600	7200	28800	144000
	> 1.5 T ≤ 3 T			A-TGB		22400	89600	448000
SET OF PULLEYS	PULLEY CHAIN HOST ≤ 2 T			A-THA	15200	20000	80000	—
	ROLLER PULLEY ≤ 0.5 T			A-THB		1000	4000	—
	SIMPLE PULLEY ≤ 0.25 T			A-THC	3250	400	1600	—
LIFTING PLATFORM	≤ 0.3 T			A-TIA	28600	—	5200	—
ADDITIONAL FASTENING	SHACKLE			A-TKA	1700	—	—	—
	EYEBOLT			A-TKB	940	—	—	—

ADDITIONS	CODE	TMU
DECANT PARTS	A-TZAA	370
ACCESSIBILITY	A-TZBA	130
DIST RANGE >165 FT WALK EVERY EXTRA 33 FT	A-TZCA	250
DIST RANGE >165 FT EVERY EXTRA 33 FT PROCESS TIME	A-TZDA	480
ADDITIONAL HOOKING AND UNHOOKING OF HOOK	A-TZGA	150
ADDITIONAL FASTENING AND UNFASTENING OF SLING	A-TZHA	530
IDLE STROKE PER 3 FT		
CABLE LINE	A-TZKA	700
LEVERJACK	A-TZLA	690

EXHIBIT A-2
MTM-MEK data (*continued*).

PREPLANS

The construction preplans reproduced on the following pages are representative of many that might be supplied to field supervisors. Their application can lead to better utilization of labor and other resources and thereby improve productivity. The examples have been updated from a set provided by a large petrochemical firm which for many years has participated actively in the construction activities required to carry out its operations.

Preplans such as these are not intended to take the initiative toward productivity improvement away from field supervisors or foremen but to make certain that their upcoming tasks have been thought through before they are undertaken. Managers should use them as a base from which to start their planning and should make improvements to them at the base level when the operation is first undertaken and with every subsequent repetition.

The titles of the preplans given here are as follows:

Exhibit B-1. Installing service piping headers and service drops
Exhibit B-2. Erecting floor-slab forms
Exhibit B-3. Shop fabricating material for single-thickness plywood wall forms
Exhibit B-4. Fabricating and assembling integrated wall forms (including re-steel)
Exhibit B-5. Installing prefabricated wood stud and gypsum board partitions
Exhibit B-6. Fabricating and installing preassembled wall forms
Exhibit B-7. Shop fabricating, transporting, and installing wood stud and gypsum board partitions
Exhibit B-8. Installing mechanical-joint ductile-iron pipe
Exhibit B-9. Erecting and aligning structural steel
Exhibit B-10. Handling and protecting preinsulated pipe and equipment

The first five of these preplans are relatively simple, and are typical of checklists prepared by a preplanner on the job site. The other five lay out every step of each operation in detail and are representative of the instructions that the home office might provide as guides to follow on every job site. The difficulty with these long and detailed preplans has been that unless higher management is insistent, field supervisors who are anxious to get on with the work and pressed for time tend to ignore them, regardless of their value.

There is some duplication in the Exhibits: B-6 is an expansion of B-4 and B-7, of B-5. This overlap was included on purpose—to show the different levels of treatment that can be appropriate under different circumstances.

In actual situations, a company might have preplans such as these that could be furnished to individual projects, including such things as still photographs, time-lapse film, or videotape to make them more understandable or to clarify certain details. Alternatively, if microcomputers and printers are available at the project, specific preplans that are stored in a home-office computer could be transmitted to the field when requested.

PROJECT __5268__

TITLE __Install service piping headers & service drops__

PREPLAN NO. __2461__

| STEP NO. | JOB AND METHOD | CRAFT | CREW SIZE | SAFETY AND GENERAL REMARKS | BLOCK CODE | | | ESTIMATED TIME & QUANTITY |
					BLDG.	COST	BLOCK	
1.	Fabricate table for pipe fabrication Table to be 60 ft long by 42 in. wide. Top must be plumb & level. Place along East side of 611 Bldg. running North & South. Use 4x4 posts & 2x6 planking.	Carp	2		620	450	2461	
2.	Lay out column lines across table. Mark for easy reference. Column #6 at North end of table.	Layout	3					
3.	Lay out trapeze hangers in building. In-stall 2x2x4 angle support for unistrut. Fasten to joists with weld. Accurately place and clamp unistrut hanger on table. Assemble & bolt all piping in main header to unistrut hangers. Install complete with valves, tees & El's. When header is completed, wire 2x4 lumber 6 ft long at each unistrut. Pick up with 12 men, two at each 2x4, carry to position in the building. Position on horses under hang-ers, lift to position with blocks rigged from supports at roof. Bolt hangers to previously installed 2x2x1/4 angle sup-ports. Remove all bracing & blocks to storage. Install valved service drops to below ceiling at all locations. Test lines to valves to release for insulation & ceiling installation. For piping at Lab. furniture see later plan which will cover assembly & piping of furniture. For material specifications & installa-tion see Spec. 3717, W-185336 & W-185337.	Plumb	Varies	Foreman to give all signals on lifting & carrying of pipe.				

EXHIBIT B-1
Installing service piping headers and service drops.

PROJECT 5268

TITLE Install service piping headers & service drops

STEP NO.	JOB AND METHOD	CRAFT	CREW SIZE	SAFETY AND GENERAL REMARKS	BLOCK CODE			ESTIMATED TIME & QUANTITY
					BLDG.	COST	BLOCK	
3. (Cont.)	Refer to Eng. Stds. as listed on W-185 337 for plumbing & sanitary installation. Store no material in building beyond that used in immediate installation. Store all material in area at craft box East of building.	Plumb	Varies		620	450	2461	Units 2650 LF 1/2-3" pipe 6 sanitary units
4.	Insulate steam & cold water lines on table before erection of headers. Refer to insulation Spec. 3682 for material specifications. Insulate all drops from ladders. Insulation of pipe under furniture will be covered in a later plan.	AW (Asbestos Workers)	2					300 LF 1" CW 300 LF 1-1/2" steam
5.	Erect scaffold as required by other crafts.	Carp	2					

EXHIBIT B-1

Installing service piping headers and service drops (continued)

PREPLAN AND SUMMARY, SUMMARY SHEET

LOCATION OF WORK   Job 3656   AREA   6   PREPLAN NUMBER   6-B-15

BUILDING Finishing   FLOOR   First   ACCOUNTING CODE   I D

DESCRIPTION AND QUANTITY OF WORK   PIPE LINE NUMBERS   EQUIPMENT NUMBERS

Erecting 13,104 S.F. of Floor Slab Forms   _____   _____
                                             _____   _____
                                                            _____

OUTLINE OF CONDITIONS

Floor slabs forms are to be 3/4" plywood panels set on 2"x8" joists.

The joists will be hung with wire beam saddles from the structural steel 18" W F beams

This preplan includes the area bounded by column lines 3 & 11 and the first inside beam from column lines D & J.

ENGINEERING STATUS: MANPOWER REQUIREMENT SUMMARY   DATE OF PREPLAN 6-5-88

Dwgs.	Specs.	F.I.'s
E-3656-123	_____	_____
_____	_____	_____
_____	_____	_____
_____	_____	_____
_____	_____	_____
_____	_____	_____
_____	_____	_____
_____	_____	_____

DATE START WORK 6-15-88

DATE WORK COMPLETE 7-7-88

CRAFT	M-HR.	CRAFT	M-HR.	CRAFT	M-HR.
CARP	534	*			
LAB.	209	**			
T.D. (truck driver)	$\frac{17}{760}$				

Original est.   Rev. est.   Check after completion

PLANNED AND ESTIMATED BY:

Approved by
Job Supt. _____ 6-10
Area Supt. _____
Chief C.P. _____ 6-8
                Initials   Date

W.O. No. _____
PAGE 1 of 3

EXHIBIT B-2
Erecting floor-slab forms.

505

WORK SHEET

PREPLAN NUMBER   6-B-15

STEP NO.	SEQUENCE, DESCRIPTION AND REMARKS ON WORK OPERATIONS	CRAFT	ESTIMATED MAN-HOURS	ESTIMATED QUANTITIES
1.	Lay 2"x10" planks on first floor steel between column lines 11 & 12, and E.F.G. to form temporary storage platform.	Labor	32	66 - 2"x10"x16'

2.	Cut on the Dewalt saw 150 - 4"x8' and 75 - 6'4"x11-3/4" 3/4" plywood strips—use the strips cut for the wall forms before cutting up the full-size 4'x8' sheets.
	  Move plywood to first floor of bldg.
	  Cut 1500 wedges—approx. 1"x2"x4" (sketch #1A).
	  Seal all edges of plywood with plastic paint.
		Carp
		Labor
		T.D.
			16
			16
			2	

3.	Erect floor slab forms—start in bay 10-11 and work north as far as column line 3—do not form the outside 7' panels. #4 wire beam saddles are to be set over the 18" WF beam at 16" O.C. and 12" at the 11-3/4" wide strip (sketch #1). Cut the 2"x8" joists (ordered in 14' lengths) to 6' - 9-1/2" long and hang in beam saddles as per sketch #1A. Use ledgers to support joists in outside panels. Lay 3/4" ply panels over joists as per sketch #1 and tack in place.—Use 6 J common nails.	Carp Labor T.D.	458* 131* 13*	13,104 S.F.
4.	The pour areas will be about 5000 S.F. and will be bulk-headed along the centerline of beams. The pours will be determined later depending upon the progress of steel erection & weather.	Carp Labor T.D.	60* 30* 2*	600 L.F.

EXHIBIT B-2
Erecting floor-slab forms (*continued*).

506

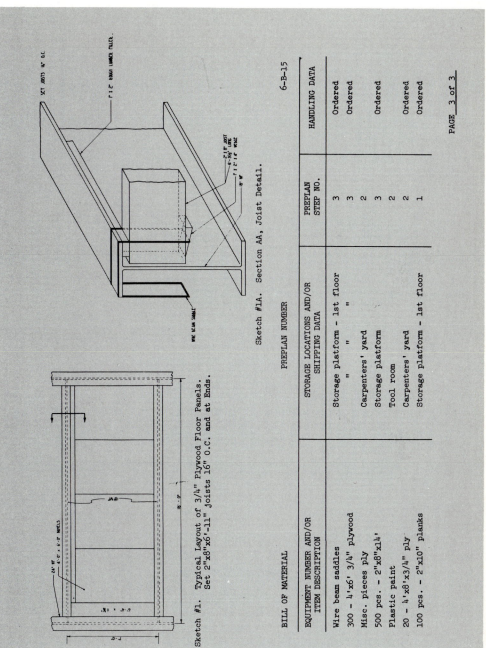

Sketch #1. Typical Layout of 3/4" Plywood Floor Panels.
Set 2"x8"x6'-11" Joists 16" O.C. and at Ends.

Sketch #1A. Section AA, Joist Detail.

BILL OF MATERIAL

PREPLAN NUMBER   6-B-15

EQUIPMENT NUMBER AND/OR ITEM DESCRIPTION	STORAGE LOCATIONS AND/OR SHIPPING DATA	PREPLAN STEP NO.	HANDLING DATA
Wire beam saddles	Storage platform – 1st floor	3	Ordered
300 – 4'x6' 3/4" plywood	" " "	3	Ordered
Misc. pieces ply	Carpenters' yard	2	
500 pcs. – 2"x8"x14'	Storage platform	3	Ordered
Plastic paint	Tool room	2	
20 – 4'x8'x3/4" ply	Carpenters' yard	2	Ordered
100 pcs. – 2"x10" planks	Storage platform – 1st floor	1	Ordered

EXHIBIT B-2
Erecting floor-slab forms (*continued*).

Begin--Plywood, lumber, snap ties and miscellaneous materials available in shop storage areas.

End--All wall form components fabricated, assembled as practicable, ready to go to field on farm wagon.

Steps:

1. <u>Layout</u> accurately and make two plywood templates, one for the start or bottom panels and one for the upper panels.

2. Stack up to 20 pieces of plywood on work bench. Align pieces over each other vertically, and layout holes on top piece of plywood using respective templates.

3. Drill holes through stacked plywood using an 11/16" drill bit in drill motor held plumb by a suitable drill holding jig.

4. Move drilled plywood sheets from work area onto dunnage or pallets, band as required for safety.

5. Cut spacers for double walers using <u>cut-off saw</u> equipped with a remote measuring device.

6. Cut walers to length using cut-off saw above.

7. Assemble double walers with spacers by nailing securely.

8. Cut 2" and 4" stiff backs to lengths required as determined from sketches. <u>Cut</u> miscellaneous stakes as required for soil conditions.

9. Stack walers, stiff backs, bracing materials, stakes and snap ties on farm wagon ready for delivery to field.

EXHIBIT B-3
Shop fabricating material for single-thickness plywood wall forms.

<u>Begin</u>--Material on hand in shop storage.

<u>End</u>--Fabricated wall form units complete with re-steel ready for transport to field.

<u>Steps</u>:

1. <u>Using</u> standard sketches in CM 3.0-1 [not reproduced here], <u>prepare</u> sketches for wall form components to show extent of fabrication and assembly. Units should be made in largest practical sections considering handling and placement. <u>Prepare</u> cut lists for walers, strongbacks, plywood, etc. arranged to permit minimum changes to cut-off saw set-ups. Coordinate sketches with re-steel shop.

2. <u>Prepare</u> cut lists for re-steel arranged to permit maximum cuts for similar lengths.

3. <u>Cut</u> forming lumber using a <u>radial arm cut-off saw</u> equipped with a <u>remote measuring device</u>. <u>Cut</u> plywood sections using a <u>two-way panel saw</u>.

4. <u>Cut</u> re-steel bars on <u>power shear</u> equipped with a <u>remote measuring device</u>. If mesh is required, cut mesh sections with bolt cutters.

5. <u>Assemble</u> one side of wall forms with walers, strongbacks, and snap ties on work table or benches, using <u>automatic nailer</u>. Place form in vertical position and brace.

6. <u>Assemble</u> re-steel mats, place on form and tie off to form snap ties.

7. <u>Assemble</u> remaining side of wall form and secure to form above using snap ties.

8. <u>Load</u> completed wall form units on farm wagon using power equipment and <u>identify</u> for field placement.

EXHIBIT B-4
Fabricating and assembling integrated wall forms (including re-steel).

<u>Begin</u>--Sections of partitions on transport rack ready for installation. Epoxy adhesives, temporary bracing, handling equipment and miscellaneous materials available at work site.

<u>End</u>--Walls installed ready for finish painting.

<u>Steps</u>:

1. <u>Layout</u> partition location on floor with chalk line, laying out cross partitions and openings.

2. <u>Secure</u> the lifting bar to top plate of partition section, <u>bolting</u> through shop drilled holes.

3. <u>Lift</u> partition section with hook of Ruger hydraulic hoist <u>secured</u> to·lifting bar and <u>move</u> section to designated wall location.

4. <u>Apply</u> Epoxy adhesive to floor and adjoining wall and to contact surfaces of wall section.

5. <u>Set</u> the partition section in place and brace temporarily to overhead trusses or beams to allow adhesive to set.

6. <u>Move</u> in adjoining section, <u>apply</u> adhesive to all contact surfaces, <u>set</u> section and again <u>brace</u> to allow adhesive to become hard. Repeat until all wall sections have been moved in and set.

7. <u>Align</u> wall sections as work progresses and before adhesive has set.

8. <u>Tape</u> and <u>spackle</u> joints in preparation for painting.

9. <u>Procure</u> preassembled hinged door and frame assemblies and install in respective openings. <u>See</u> applicable method outline.

10. Partitions are ready for painting. See applicable method outline.

EXHIBIT B-5
Installing prefabricated wood stud and gypsum board partitions.

SCOPE

Prefabricate plywood wall form sections complete with reinforcing steel, assemble into largest units practical and field install units on footings.

ENGINEERED PLANNING

Information:

1. The practical extent of shop work will depend on the site facilities available and the size of the form units to be prefabricated. The twenty and thirty-foot units shown in this method would be very difficult to transport completely assembled from the shop's area to the point of installation. All cutting of form material was performed in the shop and the walers and strongbacks were shop assembled. All reinforcing material was pre-cut, bent and assembled in the shop. Final assembly of units was performed adjacent to the point of installation. Size of form assemblies is frequently determined by pier spacing.

2. Evaluate the use of reinforcing mesh where only temperature reinforcing is required.

3. Supervision must coordinate shop activities so that shop prefabricated form components and reinforcing steel arrive at the unit assembly area as required to provide a continuous assembly operation.

4. See CM 3.3-1 [not reproduced here] for fabrication and assembly details for single plywood forms.

Tools:

1. Depending on the site terrain, a yard crane or mobile truck crane will be required to handle the large preassembled form units to the final installation point.

2. The normal carpenter and reinforcing steel hand and shop tools will be required.

3. The De Walt Saw used in the carpenter shop should be equipped with a dynamic brake. A combination 44-tooth, 16" blade should be used instead of the "standard" 74-tooth blade. Both blades cost the same but the site cost of filing and setting the teeth on the combination blade is considerably reduced. Speed and smoothness of cut are comparable.

Materials:

1. Plywood should be ordered in the standard size sheets best suited for the job rather than purchasing 4'x8' sheets out of habit.

EXHIBIT B-6
Fabricating and installing preassembled wall forms.

Exterior grade form plywood (EXT-DFPA, Concrete Form, B-B) should be purchased in 3/4" or 5/8" thickness. This plywood is easily identified by its red edge seal material.

2. Lumber is purchased in unit banded loads arranged for fork truck handling.

3. Purchase snap-ties and fasteners to accommodate dimensions of wall forms.

<u>Work Area Arrangement</u>:

1. See Figure 1 for a typical carpenter-ironworker shop layout for fabricating forms and reinforcing steel. This layout also permits complete shop assembly of form units where size of units is such that practical transportation can be arranged to field installation locations.

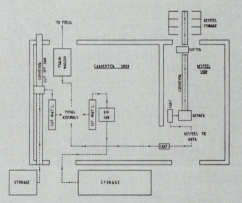

Figure 1.   Typical Carpenter-Ironworker Shop
Arrangement for Fabricating Forms
Complete with Reinforcing Steel.

2. This same shop layout can also be used to fabricate form unit components which are assembled at a field location adjacent to the installation area...

HOW:

1. Prepare detail sketches for all wall form work. The sketches clearly show the extent of shop fabrication and assembly of form units and designate the final field assembly components for the complete form unit. A copy of all sketches is sent to the reinforcing steel fabrication shop so steel can be fabricated in units to match form assemblies.

EXHIBIT B-6
Fabricating and installing preassembled wall forms (*continued*).

2. Shop material lists are prepared from detail form work sketches. See Figures 2, 3, and 4 for typical examples. "Cut Lists" are grouped by size of material for production shop cutting of lumber and plywood.

PRODUCTION NO.	TYPE	SIZE	DWG. NO.	QUANITY	DATE REQ'D.	REMARKS

Figure 2. Typical Shop Form Fabrication Schedule.

LUMBER				PLYWOOD	
UNCUT LUMBER SIZE	CUT LENGTH	QUANITY	USE FOR PROD. NO.	DIMENSION	QUANITY

Figure 3. Lumber "Cut List" for Shop Use.

CUTTING INFORMATION				BENDING INFORMATION		
BAR SIZE	CUT LENGTH	QUANITY	USE FOR PROD. NO.	SHORT SIDE DIM.	LONG SIDE DIM.	REMARKS

Figure 4. Reinforcing Steel "Cut List" for Shop Use.

3. Using CM 3.3-1 [not reproduced here] as a guide, fabricate components for a single plywood form system. Load prefabricated form unit components on wagons and move to field assembly area... Note: Form units less than 18' in length and 6' in height should be completely assembled in the shop.

4. Place horses, in final field assembly area, to accommodate length of form unit to be assembled... This assembly location should be as close as possible to the final form location. Assemble the form units for an entire wall at one time, assigning one crew to place prefabricated walers and tack to plywood, one crew to nail and button up forms and another crew to place reinforcing steel.

5. Reinforcing steel or mesh is held in place with nails. The reinforcing steel should be wired to the snap-ties if the finished wall is exposed. This technique will minimize exposure of steel and patching...

EXHIBIT B-6
Fabricating and installing preassembled wall forms (*continued*).

6. Install top plywood walers, strongbacks, and snap-ties. A crew for each of these operations is desirable on a quantity job. Leave completed formwork units on horses until installed...

7. Raise form unit to vertical position and replace one leg of sling with a chain hoist...

8. Lower wall form unit to foundation and shim as required. The chain hoist provides an easy, rapid means of leveling and positioning the form... Tie in wall units at piers.

9. Install bracing, as determined by height of wall and soil conditions, to provide form stability. Re-check final elevation and position of forms.

10. After concrete cures, strip forms but do not dismantle walers and strongbacks as they can usually be re-used. Clean and re-oil plywood if it has been used more than twice since last oiling.

Inspection and Quality Control:

1. Check final location and elevation of forms prior to pouring concrete and at end of pour.

2. Check location and type of reinforcing steel against specifications prior to pouring concrete.

3. Is form bracing adequate?

EXHIBIT B-6
Fabricating and installing preassembled wall forms (*continued*).

SCOPE

Covers the shop fabrication and installation of gypsum wall board and wood
stud building partitions for a typical office building. This method is
intended to be a guide in setting up fabricating production lines for walls,
partitions, framework, etc. which are repetitive in nature.

ENGINEERED PLANNING

Information:

1. Working from the architectural floor plan drawings, a sketch is prepared showing the installation sequence for all prefabricated partitions
   in the building. Figure #1 is a typical Installation Sequence Sketch
   The sequence for installing the prefabricated partitions must be carefully thought out and the sequence correctly indicated on the Sketch.
   The sequence should begin the opposite side of the building from the
   entrance through which the partitions are brought into the building.
   The corridor partitions are fabricated in panels 16"-0" in length.
   The office partitions are fabricated in one piece and generally range...
   8'-16' in length. The office partitions and corridor partition panels
   should be alternated in the installation sequence to provide an obstacle
   free installation. The Installation Sequence Sketch is used by carpenters installing the partitions.

2. Fabrication sketches are prepared for each partition to be prefabricated.
   Figure #2 is a typical Fabrication Sketch showing an office partition
   and a corridor partition panel. These sketches are prepared from the
   dimensions shown on the architectural floor plan and by sequence number
   as given on the Installation Sequence Sketch (Figure #1). All office
   partitions, if less than 16' in length, are sketched to be fabricated
   and installed in one piece. The corridor partitions are sketched and
   installed in panels 16' in length where possible. The fabrication
   sketches are prepared in reproducible form. Four or five typical partition sketches are selected, prepared and copies made. The proper dimensions and identification marks are then added to complete each sketch.
   This is a time-saving method and eliminates the work of preparing individual sketches for each partition or partition panel.

3. Prepare fabrication sketches for installing lighting and telephone
   facilities in the prefabricated partitions. Figure #3 is a typical
   sketch of this type. A tracing is made from the Installation Sequence
   Sketch and the lighting and telephone facilities are added to it. The
   location of telephone and electrical outlets should be standardized
   throughout the building as much as possible. This greatly facilitates
   the productionizing of the fabrication process. A Fabrication Schedule
   for Switches, Receptacles, and Telephone jacks is also prepared and is
   used in conjunction with the Lighting and Telephone Facilities Sketch.
   Figure #4 is a typical example of this schedule. The schedule should

EXHIBIT B-7
Shop fabricating, transporting, and installing wood stud and gypsum board partitions.

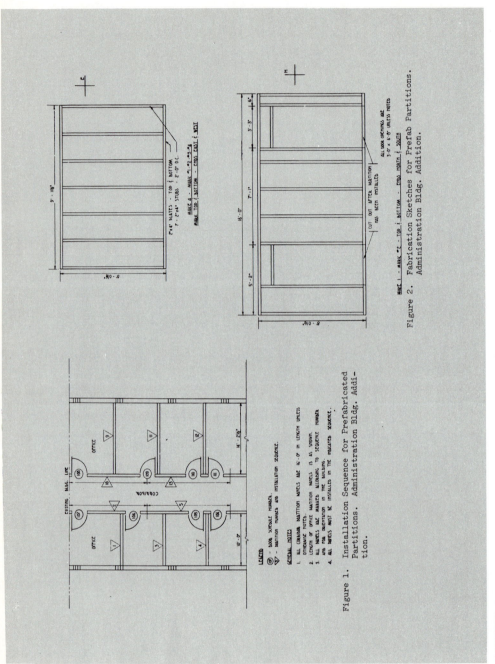

Figure 1. Installation Sequence for Prefabricated Partitions. Administration Bldg. Addition.

Figure 2. Fabrication Sketches for Prefab Partitions. Administration Bldg. Addition.

EXHIBIT B-7

Shop fabrication, transportation, and installation; and stud and gypsum board partitions (continued)

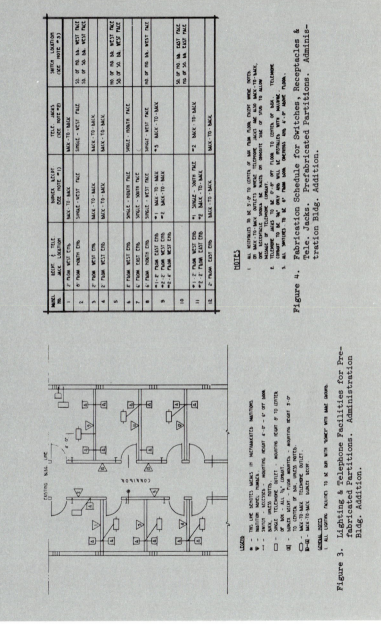

Figure 3. Lighting & Telephone Facilities for Prefabricated Partitions. Administration
Bldg. Addition.

Figure 4. Fabrication Schedule for Switches, Receptacles &
Tele. Jacks. Prefabricated Partitions. Administration Bldg. Addition.

EXHIBIT B-7
Shop fabricating, transporting, and installing wood stud and gypsum board partitions (*continued*).

517

include the partition panel number with the type and location of each outlet, switch, or phone jack to be installed. Both the Fabrication Sketch (Figure #3) and the Fabrication Schedule (Figure #4) are provided for electricians working on the fabrication assembly line.

Safety:

1. If the partitions are "man-handled" along the assembly line, correct lifting techniques should be employed to avoid undue strain.

2. In using the epoxy adhesive as a fastening medium the following safety considerations apply:

   (a) To minimize the danger of skin irritation rubber gloves and a protective bland ointment should be used at all times when mixing and dispensing epoxy adhesive.

   (b) If the skin comes in contact with the adhesive, wash immediately with warm water and soap.

   (c) Provide good ventilation when mixing and dispensing the adhesive.

3. An acid face shield and rubber gloves are worn when etching concrete floor with muriatic acid.

4. IMPORTANT: All staplers must be equipped with a positive safety device to prevent firing except when the stapler head is pressed against the surface to be stapled. All sites should survey their staplers to make sure they meet the above requirements.

Tools:

1. Fastening gypsum board to wood studs: A Mark II Cyclamatic Nailer Model 812-989 equipped with contact trip safety device manufactured by the Bostitch Co. or equal is used. 1-1/2" long steel Tee-nails, .097 dia. are used with the nailer... The nailer operates at approx. 85 psi air pressures.

2. Transporting the prefabricated partitions from the shop to the field for installations: A transport rack handled by a fork-lift truck is used. The transport rack is built according to the details shown in Figure #9... The rack will hold six partitions up to 16' in length.

3. Handling the partitions in the building during Installation: A 2-ton Ruger Hydraulic Hoist, fitted with a lifting bar...

4. Locating electrical receptacles and phone jacks: Templates are site fabricated from 1" thick wood material. Figure #8 [not reproduced here] shows the templates in use. This eliminates measuring with a tape or rule.

Materials:

1. All studs and plates for the partition framework are pre-cut in the carpenter shop, identified, and stored on a farm wagon to be delivered to the fabrication assembly line...

EXHIBIT B-7
Shop fabricating, transporting, and installing wood stud and gypsum board partitions (*continued*).

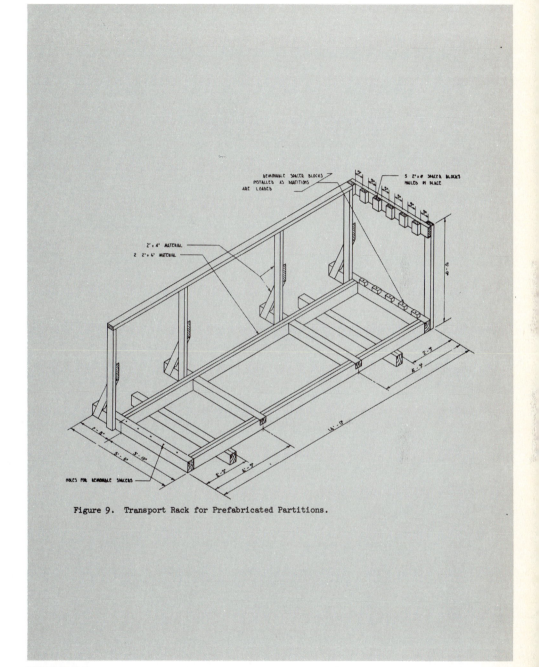

Figure 9. Transport Rack for Prefabricated Partitions.

EXHIBIT B-7
Shop fabricating, transporting, and installing wood stud and gypsum board partitions (*continued*).

2. All telephone jacks are prefabricated in the electrical shop complete with conduit and delivered to the assembly line. Metallic raceways for telephone wires should be eliminated wherever possible. Non-metallic sheathed wires for receptacles is also pre-cut to desired lengths in the shop and delivered to the assembly line.

3. Gypsum board is purchased in standard 4'x8' sheets and stored adjacent to the assembly line.

4. An epoxy adhesive is used for fastening the partitions to the concrete floor, the exterior concrete walls, and to abutting interior partitions. A recommended adhesive is "Clinco Tie-Bond." This is an all-purpose two-part epoxy and adhesive manufactured by the Clinton Company of Chicago, Illinois. It consists of two parts of a Thixotropic Epoxy Adhesive and one part of a Thixotropic Hardener. It has a pot life of 2-1/2 hours at 77°F and obtains full strength in two days at this temperature.

<u>Work Area and Arrangement</u>:

1. A typical assembly line layout for prefabricating partitions is shown in Figure 13...

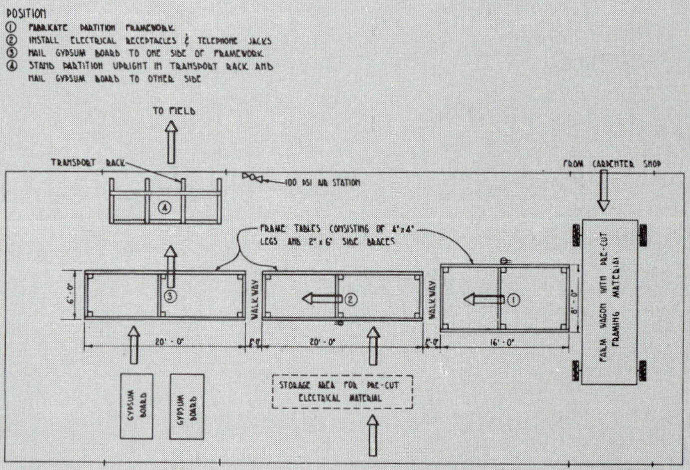

Figure 13. Assembly Line Layout for Prefabricating Partitions.

2. A closed type building is preferred for establishing the assembly line. The gypsum board must be protected from the weather.

EXHIBIT B-7
Shop fabricating, transporting, and installing wood stud and gypsum board partitions (*continued*).

3. The basic assembly line consists of one 16' frame type table for fabricating the partition framework and two 20' frame type tables, one for installing the electrical facilities and the other for applying the gypsum board.

4. During fabrication, the partitions are manually slid along the frame table to the next position. If the quantity of partitions to be fabricated warrants it, a conveyor system should be used for handling the partitions.

5. The pre-cut framing material, prefabricated electrical material, and gypsum board should be stored as close to the assembly line as possible to be within easy reach of the fabricators.

6. Electrical outlets are located on the assembly table for power saws and drills and a 100 psi air station is provided for the use of a pneumatic nailer.

HOW:

Preliminary Work:

1. Carpenter Sketcher prepares the Installation Sequence Sketch (Figure #1).

2. The Carpenter Sketcher then prepares the Fabrication Sketches for the Prefab Partitions (Figure #2).

3. Electrical Sketcher prepares the Electrical and Telephone Facilities Sketch (Figure #3) and the Fabrication Schedule for Switches, Receptacles, and Telephone Jacks (Figure #4).

4. Electrical branch circuits and junction boxes are installed throughout the building. The junction boxes are located to receive the circuits from the receptacles in the partitions after they are installed...

5. Locations of partitions are marked on the building floor. It is important to use double line marking to facilitate rapid positioning of prefabricated partitions from either side of the units. Mark the installation sequence number of each partition on the floor also.

6. Carpenter etches concrete floor in spots approximately 18" on centers along the layout markings and on each side of door openings... Etching improves the bond of the epoxy adhesive that is used to fasten the partitions to the floor and the concrete. A 30% muriatic acid solution is used for the etching. The acid is poured in spots on the floor and after bubbling stops the spots are washed. Clean substrates are very important in adhesive bonding to concrete.

Fabrication and Installation Work:

Fabrication:

1. All studs, plates, and other framing material are pre-cut in the carpenter shop, identified, stored on farm wagon, and delivered to fabrication assembly line.

EXHIBIT B-7
Shop fabricating, transporting, and installing wood stud and gypsum board partitions (*continued*).

2. In the electrical shop all telephone jacks are prefabricated, wire for receptacle circuits is cut to proper lengths, and delivered to the assembly line.

3. One man removes the pre-cut studs from farm wagon and places them on the framing table... The stud spacing is marked on the framing table, therefore, no measuring is required.

4. Two men nail the top and bottom plates on the studs...and then slide the framework to the next position...

5. Two electricians install the electrical receptacles with wiring and the telephone jacks with conduit and pull wire... Figure #22 [not reproduced here] shows the man using the templates for locating the back-to-back receptacles. Two men slide partition to the next position.

6. Two carpenters apply gypsum board to the upper side of the framework... The partition is then tilted up onto the transport rack and the gypsum board is applied to the reverse side...

7. Two electricians then complete wiring and hook-up of electrical receptacles...

8. When the transport rack is full (6 partitions...), it is moved to the field by fork-lift truck...

   An empty transport rack is then moved into the assembly area and fabrication continues. The partitions are fabricated in sequenced groups of six so that each section may be removed from the transport rack in the correct sequence for installation. As an example, partition numbers 1-6 are fabricated in the order 6-1, 7-12 are fabricated in the order 12-7, etc. The partitions will then be stored on the transport rack in the proper sequence for installation, i.e., 1-6, 7-12, etc.

Installation:

1. Screw the eye bolts into the top plate of the partition for securing the lifting bar... The holes for the eye bolts are pre-drilled during the fabrication of the partitions.

2. A hoist is hooked to the lifting bar and the partition moved into the building...

3. Apply the epoxy adhesive to the floor and adjoining wall using a wood spatula... Note in Figure #32 [not reproduced here] that wood blocks have been nailed to the concrete wall. These blocks are temporary plumb guides and braces for the partition.

4. Set the partition into place and brace the partition temporarily as needed until the adhesive gains its initial set... This procedure is repeated until all partitions have been installed.

5. The joints of the partitions are taped and spackled using conventional methods...

EXHIBIT B-7
Shop fabricating, transporting, and installing wood stud and gypsum board partitions (*continued*).

Inspection and Quality Control:

1. In fabrication, stud material should be inspected to assure that straight studs are used as the end studs of partition panels. This is important because in joining two panels end to end, the straighter the abutting studs, the better the joint.

2. Final alignment of the corridor partitions is very important to the appearance of the completed partition. The partitions should be aligned and properly braced before the epoxy adhesive gains its initial set; if not, it is impossible to move the partitions without causing damage.

EXHIBIT B-7
Shop fabricating, transporting, and installing wood stud and gypsum board partitions (*continued*).

Requisition and purchase

1. Specify shop-applied grounds

2. Fittings to have hubs at all openings

3. Establish firm delivery dates

4. Specify delivery on open truck

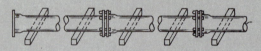

Receiving and storage

1. Deliver directly to point of installation with vendor's truck

2. When necessary to stockpile, place directly from vendor's truck to farm wagons in storage yard

Preparation

Procure and place dunnage approximately 5 feet from established centerline of trench

Preassemble

1. Place pipe directly from truck to dunnage

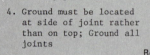

2. Install gland follower, gasket, and ground

3. Make up joint

4. Ground must be located at side of joint rather than on top; Ground all joints

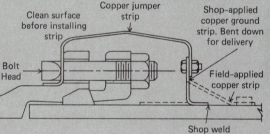

Clean surface before installing strip

Copper jumper strip

Shop-applied copper ground strip. Bent down for delivery

Bolt Head

Field-applied copper strip

Shop weld

Installation

1. Begin excavating with trenching machine

2. Follow trencher, lower pipe to ditch with power equipment; Generally, a yard crane and a 15,000 pound fork truck will be sufficient

3. Backfill and tamp between joints (to 12 inches above pipe) by hand, the balance with dozer (Caution: Do not run dozer parallel to pipe)

EXHIBIT B-8
Installing mechanical-joint, ductile-iron pipe.

Bridles

1. Bridle valves and bends as specified in project specifications and drawings.

2. Use purchased clamps and site fabricated bridle rods. (Eng. Std. F-4E).

Kicker blocks

1. Form and pour concrete kickers at all bends and tees on high pressure lines.

2. When pouring kicker blocks, be sure concrete is poured from back of pipe to undisturbed soil.

3. Only in isolated cases will bridles and kicker blocks be needed on the same line.

Hydrostatic test

1. Fabricate head for hydrostatic test.

2. Blank off end of pipe with test head. Vent pipe must be in uppermost position.

3. Fill line with water and vent all air from line.

4. Shut off vent valve, open pump line valve, and purge pump lines with water. Fill supply barrel.

5. Make sure all valves, except one from test pressure jump, are closed.

6. Pump line pressure up to specified test pressure (generally 1-1/2 times working pressure) and close pump line valve. (If test pressure is difficult to obtain, recheck to see that all air is removed from line.)

7. Test to stand for a minimum of 1 hour or as specified.

8. If pressure drop is excessive, tighten flange joints which show signs of leakage and re-test. (See Leakage Calculations [below].)

9. Complete inspection and have authorized personnel complete necessary test forms.

Electrical resistance test

1. Check E.R. of each joint while pipe is under hydrostatic test.

2. File a small spot on each side of the joint for good contact.

3. Check the resistance with a low range Ohm meter and reject any joint that measures more than 0.05 Ohm.

4. Correct any joint by tightening flange bolts.
   Caution: Do not tighten bolts with line under hydrostatic test.

EXHIBIT B-8
Installing mechanical-joint, ductile-iron pipe (*continued*).

5. Joints that still fail test must be bridged with 4/0 copper cable.

   Note: When it is impossible to test individual joints, a satisfactory, though less effective, test may be used to check up to 15 lengths of pipe. The test is identical to the one for individual joints except test leads should bridge the total group of joints, not exceeding 15.

   The resistance of each group must not be more than 0.025 Ohms times the number of joints being measured.

   Complete backfill on completion of all tests.

Leakage calculations

Leakage can be calculated by determining the amount of water required to refill line and using:

$$L = \frac{ND5P}{1850}$$

L = Leakage in gallons per hour.

N = Number of joints.

D = Nominal diameter of pipe in inches.

P = Average test pressure during test in psig.

Deflection of joints

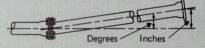

Maximum Deflection Full Length Pipe

Size of Pipe	Maximum Joint Deflection in Degrees	Deflection in Inches	Approximate Radius in Feet of Curve Produced by Succession of Joints
		18 ft.Length	18 ft.Length
4	5°	19	205
6	5°	19	205
8	5°	19	205
10	5°	19	205
12	5°	19	205
14	4°	15	260
16	4°	15	260
18	3°	11	345
20	3°	11	345
24	3°	11	345
30	3°	11	345
36	3°	11	345

EXHIBIT B-8
Installing mechanical-joint, ductile-iron pipe (*continued*).

526

SCOPE

Receive, unload, shake-out, erect and align structural steel. Bolting methods are described in CM 4.1-6 [not reproduced here].

ENGINEERED PLANNING

Information:

1. Steel drawings are marked, for area sequence of delivery, as soon as they are available. Three copies of marked drawings are prepared. One copy is retained by the project, one copy forwarded to vendor and one copy forwarded to the M and E Section.

    Arrangements are made by the site to secure an early copy of vendor detail drawings. These drawings are marked for sequence within areas (large jobs) and content of individual shipments. Base plate dimensions and anchor bolt drilling can also be obtained from these drawings.

2. Shipping instructions are clearly designated by the site after economic evaluation of alternate methods available. Truck-trailer shipment is usually more economical over-all for lots up to 100 tons. It is frequently more economical for larger jobs due to decreased site handling and more efficient sequencing than is possible with rail shipment.

3. Investigate economics of having steel fabricator supply pre-cut handrail (angle) members, stair treads, channel iron door frames and grating.

4. Provide marked vendor erection drawings to preassembly and/or prebolting crews to designate exactly what steel will be preassembled prior to actual erection placement. Preassembly of sections at ground level, which would otherwise be raised piece by piece, offers substantial savings potential.

5. Safety: There are many recognized hazards to avoid in erecting steel. These are usually prepared for. Some hazards not so frequently recognized are:

    a. Small objects left on beams which have to be "walked" by bolt-up crews.

    b. "Grandstanding" or "showing off"--the fearless also have "accidents".

    c. Wet spots or temporary slippery spots caused by liquid spillings, dew, etc.

    d. Proper ground storage--stability. Injuries to steel erectors happen quite often on the ground rather than in the air.

Tools:

1. Usual complement of ironworker's tools.

2. Floats and rigging for same.

EXHIBIT B-9
Erecting and aligning structural steel.

3. Guys for plumbing and alignment.

4. See CM 4.1-6 [not reproduced here] for bolting tools required.

5. Bolt and tool wagon for bolt-up crews.

Materials:

1. Specify that dunnage be placed under steel to facilitate unloading.

2. Request bolt and base plate shipment with first lot of steel. Bolts and washers are to be segregated and packaged by size.

3. Evaluate the desirability of having fabricator apply finish paint to steel as well as primer.

4. All steel shipments are to be made in either open top truck trailers or gondola rail cars.

Work Area:

1. Jobs involving 300 to 400 tons of steel, or more, usually require a shake-out yard removed from the actual erection area. Shake-out for smaller jobs takes place adjacent to erection area except where lack of space prohibits it.

2. The shake-out area is provided with adequate dunnage and particular attention is directed toward providing access to steel, in the sequence desirable, for erection.

3. Arrange adequate space for a preassembly and prebolting area adjacent to the erection area. Lifts and hauls of large preassembled units of structural steel are seldom practical.

HOW:

1. Steel is received and checked against vendor's fabrication and erection drawings by erection crew foreman. Shake-out and arrange storage area in pre-planned sequence to suit erection schedule. Information on shortages should be forwarded to appropriate supervision at once. Provide dunnage and arrange steel for easy pick-up for later transportation and/or erection.

2. Steel to be preassembled and/or prebolted before actual erection is segregated in an area reserved for this purpose.

3. Transport steel from storage area (where used) to erection area with a straddle carrier when available, or a flat bed trailer (float). Use adequate dunnage on flat bed.

   Small quantities of relatively light steel may be transported from storage areas with fork lifts or wagons. Special heavy frame wagons may be justified for large, multiple-building erection jobs.

4. Set shims for column base plates. Pre-grouted shims are usually not necessary for light construction and single-story structures. Base plates should be welded to columns for these structures.

EXHIBIT B-9
Erecting and aligning structural steel (*continued*).

5. Set and level base plates if supplied loose.

6. Position erection crane for maximum reach before moving is required. Set columns and girder beams according to preplanned erection sequence. See CM 4.1-6 [not reproduced here] for bolting methods.

7. Follow column and beam erection with cross and wind bracing.

8. Bolt-up crew plumbs and aligns steel using prefabricated guy lines with an eye in each end and ratchet type take-up. For one-story structures, plumb and level tolerance, 1:500, for multi-story buildings, 1:100 for exterior columns, elevator shafts and other critical components.

9. Bolt-up crew follows with final checking and bolting up of all connections.

Preassembly and Prebolting:

1. Preassembly crew, working from marked vendor erection drawings, begins assembly of designated sections on the ground... Preassembly area is kept as close as possible to erection area.

2. Preassembled sections are prebolted on the ground for installation as a completed unit... Weld on masonry wall anchors and ties.

3. Preassembly and prebolted section is hoisted to a vertical position preparatory to raising to second-level erection point... Note that erection crane is positioned so that no "walking" with the section is required.

4. Move section into final position and bolt to existing steel...

5. A preassembled and prebolted platform assembly being positioned for erection is shown in Figure 8 [not reproduced here]. Consideration should be given to installation of completely assembled stair towers including grate-platforms and stair treads...

Inspection and Quality Control:

1. Check plumb and level of steel carefully before final bolt-up. Be positive that they are within tolerance limits. Check elevation of critical equipment support steel.

2. See that all erection bolts are removed and permanent bolts are properly tightened. See CM 4.1-6 [not reproduced here] for bolt inspection procedure.

3. Check sag rods and windbracing for correct installation.

4. Check column splices for proper seating of members.

Prefabricate plywood wall form sections complete with reinforcing steel, asse ble into largest units practical and field install units on footings.

EXHIBIT B-9
Erecting and aligning structural steel (*continued*).

SCOPE

This method covers the procedures, tools and equipment to be used in
handling and erecting pre-insulated pipe, duct, and equipment to minimize
damage to the insulation and finish.

ENGINEERED PLANNING

Information:

1. The insulation of products in the shop presents a somewhat delicate
   materials handling problem. There are three things that must be
   considered before a shop insulated product can be rigged for handling.
   These items are:

   (a) The compressive strength of the insulation material--can the
       localized load of the product be supported by the insulation
       material without crushing it?

   (b) The durability of the insulation finish--can it stand considerable
       handling or must it be protected with some type of covering at
       the rigging point to prevent possible damage?

   (c) The rigging of the product--can it be rigged from a point where
       the insulation will not be involved as in the case of many cubes?...

2. Good rigging practices must be insisted upon. Rig straight runs of
   pipe correctly to eliminate any possibility of the pipe sliding out of
   the insulation. Lifting lugs welded directly to the pipe or friction
   collar clamps placed securely on the pipe should be considered...

   The wrapping of several tie wires around the pipe when the insulation
   is slightly larger than the pipe to assure a snug fit at the insulation
   section is permissable and helps prevent rotation of the covering while
   installing the remaining sections.

   Each site should ascertain that correct rigging and handling procedures
   are being used in the erection of pre-insulated pipe.

3. Only thoroughly trained and instructed crews should be used to handle
   pre-insulated items. Crews must be educated in proper handling techniques.

4. The pre-insulated item should be erected in its final location as soon
   as possible after it arrives in the area. Insulated pipe, duct, and
   equipment can easily be damaged if left stored in an area where other
   work is being performed.

5. Wagons and pallets should be loaded so that the product can be removed
   easily without the damage that usually results from shuffling...

6. Lifting lugs should be placed on all products where practical. This
   will eliminate contact with the insulation...

7. Equipment necessary for handling insulated products should be readily
   available in the field and shops areas...

EXHIBIT B-10
Handling and protecting preinsulated pipe and equipment.

530

Tools:

    1. Nylon Slings (2500 lb test)

    2. Rope Ladders

    3. Pipe Clamps

    4. Lifting Lugs Welded to Pipe

    5. Usual Complement of Rigging Tools

Materials:

    1. Foam rubber skins,* scraps of insulation, or any other materials that will afford protection by cushioning the load.

    2. Pipe hanger saddles.

Work Area:

    1. Allow adequate space in your shop and field areas to allow finished pieces to be stored temporarily without having to be stacked. The field lag areas should be near the points of erection. Pre-insulated products should be erected in their final locations as soon as possible after they arrive in the area. Field production crews should be trained and assigned to the installation.

    2. Provide adequate dunnage in the lag areas for keeping the insulated products off the ground.

    3. Provide storage for nylon slings, rope ladders, and padding materials in lag areas. These tools and materials should always be readily available in the areas where finished products are to be rigged.

HOW:

1. When a product is to be insulated, planning in the initial fabrication stages will guarantee that it can be handled and erected with a minimum of damage to the finish.

2. Pre-planning should begin before the item to be insulated enters the shop. In many cases large sections of pipe or duct can be insulated directly on a farm wagon or pallet and moved to the installation point without having to be re-handled prior to being rigged for erection...

3. The pipe, duct, or equipment rigging crews and material handlers should be conscientiously trained craftsmen with a desire to protect the materials they handle. As much consideration should be given in selecting these people for their jobs as is given in selecting people for any other critical job.

----

    *Foam rubber skins: trimmings from foam rubber manufacturing process may be purchased in sizes up to 5'x8'x1" from United Shredding & Salvage Company, Philadelphia, Pennsylvania. These "skins" provide economical cushioning for delicate finishes.

EXHIBIT B-10
Handling and protecting preinsulated pipe and equipment (*continued*).

4. All pre-insulated products should be loaded on wagons or pallets in a single distributed layer. Each unit can then be unloaded as needed with a minimum of shuffling and contact with other pieces...

5. Pre-insulated products should always be lifted from the wagon or pallet, and never dragged.

6. Where possible, lift the loaded wagon or pallet into the building. In this manner the finished products can be brought closer to their erection point before they have to be handled...

7. Have protective equipment readily available in all construction areas.

8. A nylon sling can be used on aluminum "Hypalon" and "Tedlar" finishes.

   NOTE: Nylon slings should never be used on hot pipe, duct, or equipment.

9. On heavier pipe with a very delicate insulation finish, simple protective measures such as several layers of asbestos cloth, burlap or sponge rubber provide an excellent cushioning for the finish...

10. A rope ladder on more rugged finishes is sufficient.

11. A combination of both asbestos cloth and rope ladders on "Tedlar" and other delicate finished products will prevent rigging damage...

12. Where possible, combine the finished pipe, duct, motors and vessels into assemblies or cubes. In this manner large quantities of finished materials can be rigged for handling with relatively few rigging points...

13. When lifting lugs or beams are required, they should be considered in the initial planning for the fabrication of the product... Locations for lifting lugs should be determined at the time of sketching.

INSPECTION AND QUALITY CONTROL

Close follow-up and retraining, where necessary, is essential to the success of this Method.

Before this Method is adopted, the following site policy must be established:

1. Pipe, duct, and equipment can be pre-insulated and pre-finished at the shop.

2. Pre-insulated pipe, duct, and equipment can be handled and installed without damage.

EXHIBIT B-10
Handling and protecting preinsulated pipe and equipment (*continued*).

EXAMPLES OF PRODUCTIVITY-IMPROVEMENT STUDIES

This Appendix offers eight examples of productivity-improvement studies drawn from the very large collection available to the authors. They report actual field studies covering a few of the more common operations in utility, carpentry, concrete, and industrial construction. They are intended to reinforce the concepts and procedures outlined in Chap. 8 and to demonstrate again that productivity-improvement efforts can pay large dividends.

Two reasons why practices such as these given here are all too common are:

1 There is a strong tendency to resist challenging the way things have been or are being done. The reasons are almost endless. As examples, layout of work areas is assumed to be efficient, current work methods and tools are accepted as satisfactory, and proposals for new approaches always involve risk.

2 When pushed to get going on the work, even able field supervisors fail to plan each task step by step and in detail.

If the examples given here are to be useful, those studying them must realize that they are not exceptions; they represent the way construction often gets done. Neither must they be taken as a criticism of construction people and their abilities. Construction workers are frequently under pressure in and must often contend with a tough environment, and the authors have too much admiration for them to nitpick about their capabilities.

EXAMPLE C-1

Crew-Balance Study from Time-Lapse Film of Pipe Installation In Trench

The original crew of four men and a tractor crane showed a large amount of nonproductive time. In the sequence shown in Fig. C-1-1 as existing method, the tractor crane swung a piece of pipe into the ditch, then held it in midair while the end was mortared. The crane then lowered the pipe while the man in the ditch guided it into place. A man stood by on the bank to hook onto the pipe, and an extra man in the trench held the pipe steady while it was being mortared. The time that the crane held the pipe for mortar was nonproductive, and the man on the bank was useful only about 30 percent of the time. The changed sequence is also charted in Fig. C-1-1 as proposed method. By moving the mortaring operations to the bank, the problem of transporting the mortar to the trench was eliminated, the cycle was speeded up (time reduced from 96 to 78 seconds), and one man less was required for this part of the operation. In both methods the production of the machine was worker-paced.

EXAMPLE C-2

Crew Balance in Concrete-Placing Operation

Figure C-2-1 is a crew-balance study of a concrete-placing operation on the foundation of a large industrial building. Concrete could be chuted directly from trucks into the forms. The first group of vertical bars in Fig. C-2-1 shows what each man of the seven-man crew did while a truck was being unloaded. The lower group of bars shows suggested work distribution if the crew were reduced to four men. Note that there was no difference in the time required to unload a truck. The right-hand pair of bars shows percentage of working and nonproductive time for the actual and proposed crews.

It would be incorrect to say that the data shown in Fig. C-2-1 are conclusive. Possibly there are good reasons why the larger crew was required. On the other hand, the proposed change should be given serious consideration because of the 31 percent decrease in nonproductive time. It would also be incorrect to say that the proposed solution is the best one. Some of the remaining uses of time, such as holding the chute and splash board, might not be essential or could be eliminated by a change in procedure. The real lesson in this example lies in the fact that a critical look at a routine procedure has suggested several avenues for cost reduction on an operation that in the minds of supervision had been entirely satisfactory.

EXAMPLE C-3

Process Chart Study of Conduit Installation[1]

Conduit of various diameters is an important element in any large industrial project. Some of it is of relatively large diameter and includes long straight runs (see Fig. C-3-1). Often changes in direction are involved (see Fig. C-3-2). On completion of the project,

---

[1]This is one of a variety of productivity-improvement efforts undertaken as a part of the Richmond Lubricating Oil Project described in Exhibit 11-1.

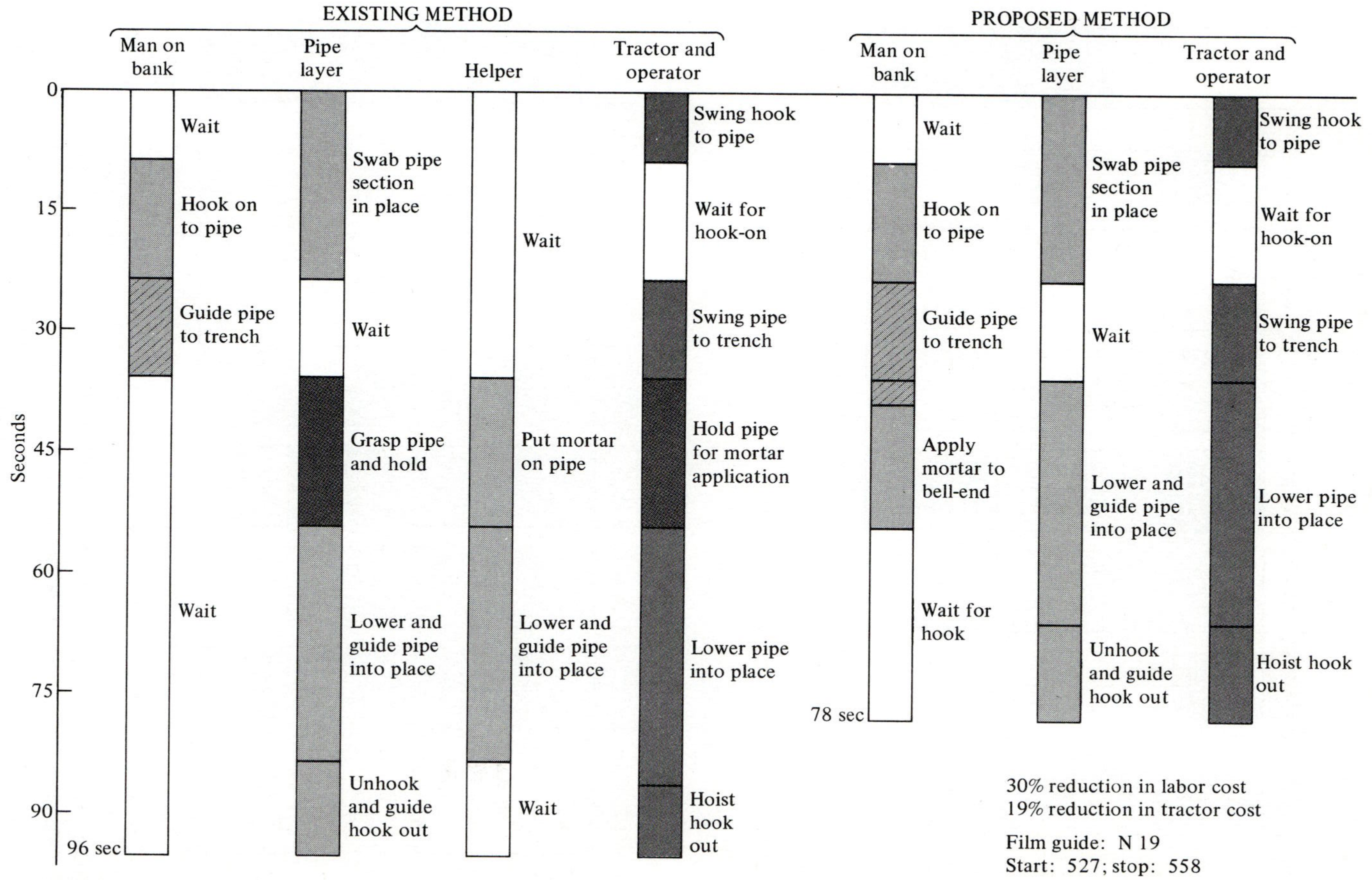
EXISTING METHOD
PROPOSED METHOD
Man on bank
Pipe layer
Helper
Tractor and operator
Seconds
0
15
30
45
60
75
90
96 sec
78 sec
Wait
Hook on to pipe
Guide pipe to trench
Wait
Swab pipe section in place
Wait
Grasp pipe and hold
Lower and guide pipe into place
Unhook and guide hook out
Wait
Put mortar on pipe
Lower and guide pipe into place
Wait
Swing hook to pipe
Wait for hook-on
Swing pipe to trench
Hold pipe for mortar application
Lower pipe into place
Hoist hook out
Wait
Hook on to pipe
Guide pipe to trench
Apply mortar to bell-end
Wait for hook
Swab pipe section in place
Wait
Lower and guide pipe into place
Unhook and guide hook out
Swing hook to pipe
Wait for hook-on
Swing pipe to trench
Lower pipe into place
Hoist hook out
30% reduction in labor cost
19% reduction in tractor cost
Film guide: N 19
Start: 527; stop: 558

FIGURE C-1-1
Crew-balance study of pipe installation in a trench.

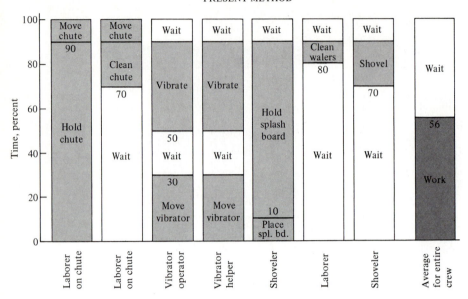

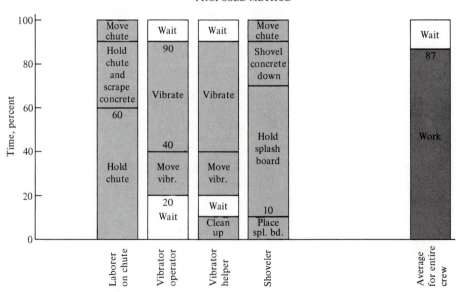

FIGURE C-2-1
Crew-balance study of concrete-placement gang.

FIGURE C-3-1
Photograph showing a straight run of large-diameter conduit at the Richmond Lube Oil Project.

conduit at some locations may be several layers deep. Installing it is labor-intensive and can be a major cost element. On this project total conduit length was 730,000 linear feet; the labor budget for it was 13 percent of the project total.

This example concerns placing and coupling individual lengths of conduit to form long straight runs such as those shown in Fig. C-3-1. Traditionally, local electricians have installed runs of conduit by positioning the first two 20-foot lengths, screwing them together, and pulling them forward with a tugger (see Fig. C-3-3). Several additional lengths were then added successively and all pulled into place with the tugger.

An early planning session of supervision and crew was held with the aim of setting a target rate (0.020 worker-hours per foot of conduit). At subsequent sessions the crew viewed time-lapse film and, through discussion, developed improved methods. As time progressed, the following modifications were introduced:

1 Arrangements were made to store enough conduit at the working level to satisfy a day's needs, thus releasing the crane to other crews.

2 Causes of delays between cycles, such as changing work areas, getting tools, sharing the tugger, and suffering interruptions in electric power were examined and means of improvement were developed.

3 A better shackle for hooking the tugger line to the forward end of the conduit string was developed, thereby making hookups easier and preventing damage to the threads on the conduit.

4 A cone-shaped fitting for an electric drill was developed by the foreman. It spun the next pipe length until it was almost tight, which significantly reduced connection and tightening time.

FIGURE C-3-2
Photograph showing the complexities of conduit installation at the Richmond Lube Oil Project.

FIGURE C-3-3
Procedure for pulling conduit into place.

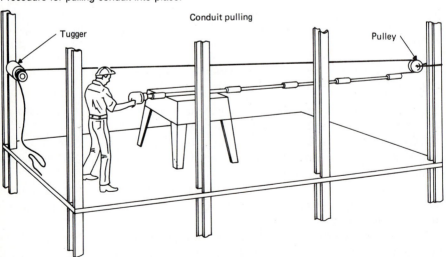

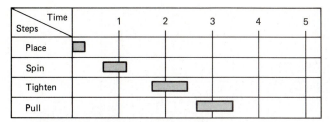

Crew size: 6 workers, average duration = 3.5 minutes per length
Labor: 6 x 3.5/60/20 = .0175 labor–hours per foot
Production = 60/3.5 x 20 = 342.5 feet per hour

FIGURE C-3-4a
Process chart—joint and pull cycle for one 20-foot length of conduit
(traditional method).

5 A sled was developed to fit on the front of the lengths of pipe so that it did not
hang up on the supports when pulled forward.

6 Rather than pulling each run of conduit forward separately, as shown in Fig. C-
3-3, five runs were connected and all of them were pulled forward at once. This required that a more elaborate connecting device and sled be fabricated.

Before and after times in the steps required to install the conduit, as recorded by time-
lapse filming, are shown in process-chart form in Figs. C-3-4a and C-3-4b. Comparisons
between them show not only the total magnitude of the improvements but also substantial
reduction in the delays between individual steps. Data recorded under the charts show a 42
percent reduction in worker-hours per foot of length, which is a substantial saving.

In this methods-improvement study, a crew size of six was taken as a given, and no
crew-balance study was undertaken. To do otherwise might have defeated the effort
before it got underway. Actually, during the progress of the work, the sixth member
was dropped from the crew without incident.

EXAMPLE C-4

Flow Diagram and Process Chart Analysis of a Fabrication
Shop for Cutting and Rolling Sheet-Metal Covers that Will Clad
Insulation for Piping

An existing sheet-metal fabrication shop was the original site for the sheet-metal cutting
and rolling operation. It was begun on a small scale and was crowded into other operations that were underway in the shop. Since output was to be increased, a new facility
was required, and a consultant was engaged to develop the layout. His responsibilities
were to be as follows:

1 Develop a new and more efficient layout for the operation.

2 Reduce the time lost by field crews looking for completed bundles of cladding that
were misplaced because of the present chaotic conditions in both workplace and storage areas.

Since the consultant was not asked to develop detailed work assignments for crew
members, the report did not include a crew-balance analysis.

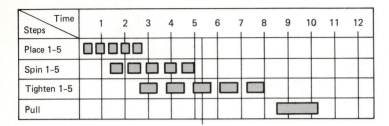

Time Steps	1	2	3	4	5	6	7	8	9	10	11	12
Place 1–5												
Spin 1–5												
Tighten 1–5												
Pull												

Crew size: 6 workers, average duration = 10.2 minutes per 5 lengths
Labor = 6 x 10.2/60/100 = 0.0102 labor hours per foot
Production = 60/10.2 x 100 = 588 feet per hour

FIGURE C-3-4b
Process chart—joint and pull cycle for five 20-foot lengths of conduit
(improved method).

The As Found Situation   Figure C-4-1 is a plan view of the area where the as-found operation was carried out by two workers. It was done in the following manner:

1 Rolls of uncut sheet metal 100 feet long and 3 feet wide were carried one at a time from the storage area in a nearby trailer to the corner of the shop.

2 Each roll was mounted in the roll rack which held it in a horizontal position.

3 The workers pulled the end of the roll forward across table *A* and through the cutter onto table *B*. The locations for successive cuts were marked off for the entire length of the 100-foot roll (lengths ranged from 3.5 to 4.0 feet or 25 to 30 pieces per 100-foot roll). The sheet was then rerolled.

FIGURE C-4-1
As found fabrication shop.

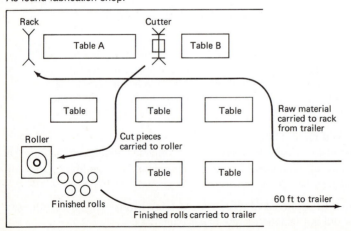

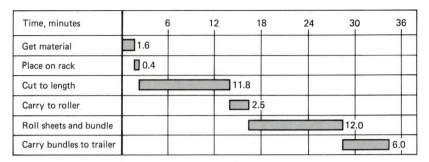

Average time per roll = 34.3 minutes (27 sheets)
Two worker crew = one 100 foot roll/68.6 worker minutes

FIGURE C-4-2
Cladding fabrication—as found process chart.

4 The sheet was successively pulled forward again, positioned for the next cut, and the cut was made.

5 Individual cut sheets were carried to the roller and run through it to form cylinders. The first rolled sheet of each 10 or so was tied with string, and about 9 others were coiled up inside it.

6 At intervals, finished rolls were carried to the trailer and piled randomly in it.

Figure C-4-2 is a process chart showing the times required for the two-person crew to carry out each of the steps described above. As noted in the figure, the average time to process a roll, as developed from time-lapse, was 34.3 minutes. Worker time was double that, or 68.6 minutes.

The Revised Plan   The layout for the new shop and a revised work method were developed by the consultant working with the foreman and crew. The revised layout appears in Fig. C-4-3. A process chart showing the steps and measured times used with the new layout and revised work method appears as Fig. C-4-4.

Changes introduced in the revised work method were:

1 Reduced transportation distances for raw and finished material and during the work process itself.

2 Reorganization of equipment placement and the cutting, rolling, and handling sequence.

Changes introduced in the revised process after the roll was positioned on the rack were:

1 With the desired length marked out on the table, it was possible to pull the sheet forward for the next cut, position it at the mark without making a separate measurement, and make the cut. Cutting was speeded up and could be done more easily and safely than before.

2 After cutting, individual sheets were carried from the table to the roller, shaped, and bundled as before.

3 Bundles were carried to storage in designated areas for pickup by the field crews.

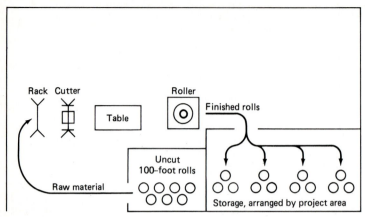

FIGURE C-4-3
Layout of new fabrication shop.

As noted in Fig. C-4-4, the change brought a 30.6 percent reduction in the cost of carrying out the operation. In addition, substantial unmeasured benefits resulted from the change. These included lessened disruption to other operations in the existing shop, a reduction in time losses to field crews since the specific material needed by each could be easily found, and better acceptance of the method by the crew, since its members had been involved in planning the operation.

EXAMPLE C-5

Flow-Diagram and Process-Chart Analysis of Electrical Cable-Cutting Operation

An addition to a major steel mill had been underway for 7 months when the electrical portion of the work began in earnest. At its peak, this phase would engage some 250

FIGURE C-4-4
Process chart of revised operation for cladding fabrication.

Time, minutes		6	12	18	24	30
Get material	1.1					
Place on rack	0.4					
Cut to length	4.5					
Carry to roller	1.8					
Roll sheets and bundle			12.0			
Carry bundles to pickup area				4.0		

Average time per roll = 23.8 minutes (27 sheets)
Two worker crew = one 100 foot roll/47.6 worker minutes
Comparison with earlier method: (68.6–47.6) 68.6 = 30.6% reduction

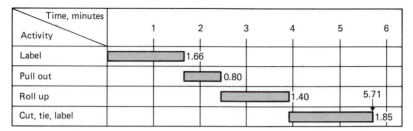

FIGURE C-5-1
Process chart for cable-cutting operation—original method.

workers. Among the operations to be carried out were installing cable trays and conduit, installing (pulling) cable, and completing motor installations and control terminations. At the time of this study of the cable-cutting operation, cable installation, an operation which would finally involve some 2,500,000 lineal feet, was just beginning.

Cable arrived on site on large spools. These were raised on jacks and mounted on a horizontal shaft supported at each end on stands. The cable could then be easily pulled from the spool, appropriate lengths could be measured (these were variable, but generally about 100 feet), rolled up by hand, cut off, and labeled a second time. It was then ready for pickup by the crews pulling the lengths into place. Tools and supplies employed by the crew were rudimentary, such as (1) a horizontal shaft, stands for it, and jacks for raising the spools onto the stands; (2) gummed labels and pens; (3) rolls of tape for tying the reeled cable segments; and (4) a hacksaw. This operation, as developed by a lead man and two workers, had been underway for a week when it was filmed by time-lapse and analyzed by a consultant.

The sequence of work was as follows:

1 The "Lead man" wrote the order number for the cable segment on a pregummed label and stuck it on the free end of the cable.

2 A worker, designated as "Blue" by the color of his clothing, pulled the cable off the reel by pacing the preordered distance. If several segments were to have the same length, he indicated the endpoint with a mark on the floor.

3 Blue then rolled the cable as he walked back to the starting point.

4 A second worker, designated as "Big brown," cut the cable with a hacksaw. Simultaneously, the Lead man tied the coiled segment and attached a second label.

A process chart showing the steps in this simple operation and the times each consumed is shown in Fig. C-5-1. A crew-balance chart, Fig. C-5-2, indicates the division

FIGURE C-5-2
Crew-balance chart for cable-cutting operation—original method.

Activity / Person	Label	Pull out	Roll up	Cut, tie, apply label	Legend
Leadman	/////////			/////////	///// Working
Big Brown				/////////	
Blue		/////////	/////////		□ Idle

of labor among the three parties. The percent working was 42. At $20 per hour for wages and benefits for each worker, the cost of cutting and rolling an average segment would be

$$\text{Cost per segment} = \frac{5.21 \text{ minutes}}{60 \text{ minutes per hour}} [(\$20 \text{ per hour})(3 \text{workers})] = \$5.21$$

Cost of preparing 25,000 segments by the original method = ($5.21)(25,000) = $130,000

The time-lapse film of the operation, Figs. C-5-1 and C-5-2, and the cost analysis were shown to the foreman, lead person, and crew. Their first response was embarrassed laughter. They then had to be convinced that they were seeing a true portrayal of their activities, including the amount of standing around and the high cost involved. After that, they were ready to work seriously with the consultant to answer his question, "How would you do it if you were going to do it right?"

The group proposed a complete restructuring of work assignments. One person would determine from the foremen of the installation crews their needs for cable the next day and prepare the necessary labels. The other two people would work independently, each applying the prepared first label, pulling out the cable, rolling it up, cutting and tying it, and applying the second label. Also, markings were placed on the floor so that measuring lengths for individual segments was not necessary. On seeing the plan, management approved it and provided the mountings for a second reel and two more suitable cable cutters. A process chart showing the revised operation, with two individuals working alone, is given in Fig. C-5-3.

Observations of this operation showed that the Lead man needed less than 3.55 minutes per cut segment to get the information from the foreman and write the two labels, including data on length. Any time savings for this are not considered in computing the cost of the revised operation. These are

$$\begin{matrix}\text{Cost per segment for} \\ \text{the revised method}\end{matrix} = \frac{3.55 \text{ minutes}}{60 \text{ minutes per hour}} [(\$20 \text{ per hour})(3 \text{ men})] = \$3.55$$

$$\begin{matrix}\text{Cost of preparing} \\ \text{25,000 segments}\end{matrix} = (\$3.55)(25,000) = \$88,700$$

The estimated savings by changing methods = $130,000 − $88,700 = $41,300, or 32 percent.

FIGURE C-5-3
Cable-cutting operation—revised method.

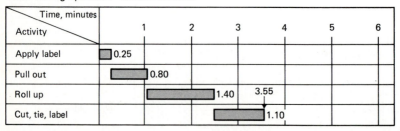

Activity / Time, minutes	1	2	3	4	5	6
Apply label	0.25					
Pull out	0.80					
Roll up		1.40	3.55			
Cut, tie, label			1.10			

Two among other possible but unmeasured benefits of the method change are:

1 Orders are picked up from the foremen at the crews' work locations, making it unnecessary for the foremen to leave their crews unsupervised when they turn in orders.

2 There are fewer errors and less waste in providing the proper lengths. With the previous method, cuts had been too generous to ensure that they were adequate.

This example is intended to illustrate useful techniques for improving the efficiency of crews. But if other work is not provided to utilize the time that is saved, the measured benefits will not be achieved. Management must then face the broader question, "How can it utilize the time saved for some other productive purpose?"

EXAMPLE C-6

Flow Diagram and Process Chart for Wooden Bridging

Figure C-6-1 is a very simple illustration of the process chart and flow diagram. It shows a typical lumber-cutting operation on an industrial building project. One carpenter working alone was making bridging pieces by cutting long 2 × 4's into 18-inch lengths having beveled ends. Two nails were started in each end of each cut piece. Then the pieces were bundled for handling. Flow diagrams for the present and proposed methods appear to the left and process charts to the right.

The symbols used in the process charts are standard ones employed by industrial engineers. Their meanings appear at the lower right in the figure. They provide a quick way for determining which kind of step is involved. Where process charts are employed on a routine basis, a template for forming these symbols is useful. A special process-chart paper is also available with all of the symbols printed side by side on each line. With this paper, the steps can be marked simply by drawing a zigzag line through the appropriate symbols.

A study of the figure shows that under the present method, ten operations and two transportations were required. Under the proposed method, the number of operations is reduced to seven and the transportations to one. Actually, all that is suggested is to move the nailing table over against the saw table. This avoids having to stack the material on the ground, pick it up again, and carry it 15 feet. This is not a spectacular change but it is well worthwhile if the quantity of work is at all large.

This particular example was adopted because it illustrates how a process chart can open up new lines of thought. Application of the suggestions listed earlier in Fig. 4-5, page 79, could readily evoke a number of questions, for example:

Why use bridging? Are there other types (metal, solid, etc.)?
Why cut the bridging on the job? Could it be more cheaply done at the mill?
Could the lumber pile be more conveniently located?
Could the sawing or nailing procedures be simplified by employing stops or jigs?
Could saw cuts be made on several pieces of material at once?
Why tie the material in bundles? Could lots be handled more easily on pallets or by other means?
Why store the completed bundles on the ground? Why not place them on a cheap trailer or cart that could be towed or wheeled to the point of use?

This example illustrates again that a review of an operation by some orderly process may lead to costs savings.

Flow diagrams   Process charts

PRESENT METHOD

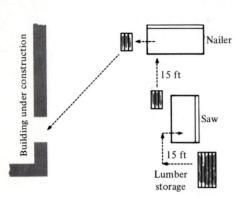

▼		2 × 4-in. lumber in storage
1	●	Picked up by carpenter
	→	Carried to saw—15 ft
2	●	Set on saw table
3	●	Cut to length
4	●	Stacked on ground
	◗	Accumulated on ground
5	●	Armload picked up by carpenter
	→	Carried to nailer's table—15 ft
6	●	Stacked on table
7	●	Four nails started
8	●	Stacked on ground
9	●	Placed on wire
10	●	Tied in bundles
▼		Stacked on ground until needed

PROPOSED METHOD

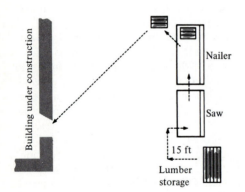

▼		2 × 4-in. lumber in storage
1	●	Picked up by carpenter
	→	Carried to saw—15 ft
2	●	Set on saw table
3	●	Cut to length
	◗	Accumulated on table
4	●	Placed on nailer's table
5	●	Four nails started
6	●	Placed on wire
7	●	Tied in bundles
▼		Stacked on ground until needed

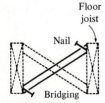

SUMMARY

Step	Old	New	Save
"Do" operation	10	7	3
Transportation	2	1	1
Delays	1	1	0
Storage	2	2	0
Distance	30	15	15

Symbols

●	"Do" operation
→	Transportation
◗	Delays
▼	Storage
■	Inspection

FIGURE C-6-1
Flow diagram and process chart for lumber-cutting operation.

546

Much more complex operations than the one just presented can be studied fruitfully by the process-chart method. In making such studies, the analyst must be consistent in following either worker, material, or machine. The analysis becomes jumbled if the chart follows first one and then another.

Process charts have proved particularly effective in studying precutting, prefabrication, and other repetitive operations.

EXAMPLE C-7

Crew-Balance Chart, Flow Diagram, and Process Chart for Sealing Wooden Roof Decking

This study was made on a timber sealing operation for a large school complex. Wooden tongue-and-grooved roof decking (3 inches × 6 inches) of various lengths were to be sealed on the edges and top with a water-repellent sealer. This operation was originally accomplished with a three-person crew: one painter utilizing a spray gun and two laborers lifting one beam at a time. This particular crew task was picked because it was highly repetitive (because of a large number of beams to be sprayed) and because it offered some excellent examples of the advantages to be gained by use of crew-balance analysis.

Original Method The original method of beam sealing is outlined in Fig. C-7-1. The general configuration of the work area and location of beam stacks and other equipment is shown in Fig. C-7-2. The production rate was 80 beams per hour utilizing this method, with the painter active 32 percent of the time and the laborers active 80 percent of the time. Several problems should be noted after a close analysis of Fig. C-7-1.

1 The painter is completely dependent upon the work of the laborers, and the laborers are completely dependent upon the work of the painter. Under the original method, the jobs of each are mutually dependent, thereby producing inactivity for both groups that could not be eliminated by working more productively.

2 There are essentially three handling operations done by the laborers for each beam: first, removing the beams from the original pile and stacking them on edge; second, spreading the beams on the horses after edges are painted; third, removing sprayed beams from the horses to the finished pile.

3 Finally, the level of activity for the laborers is relatively high, even though they are not very productive. The solution to finding a more productive operation involves developing a better method, not simply increasing the level of activity.

The flow diagram and process chart of the present method are presented in Fig. C-7-2.

The flow diagram shows that the work is linearly laid out, but the site is spread out, which requires excessive transportation. The process chart reveals that each board undergoes six operations and is transported two times for a total of 80 worker-feet.

Improvements Work improvements for this crew task involve the answers to two questions: Can the number of component tasks be reduced? Can the two groups—

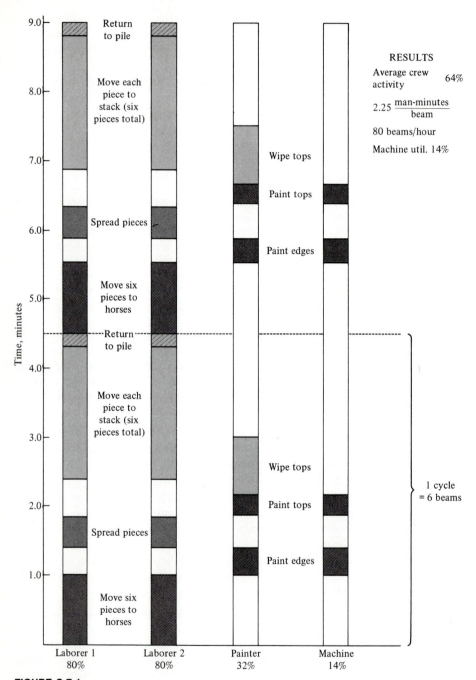

FIGURE C-7-1
Crew-balance chart of original method for sealing roof beams. Chart shows 12 beams being sealed in two cycles.

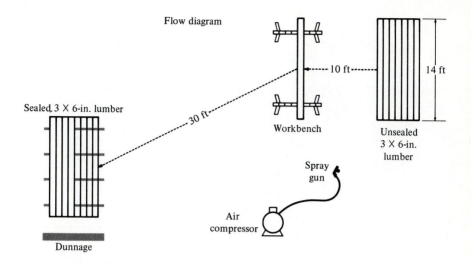

Flow diagram

Process chart

▼ 3 × 6-in. T & G lumber in storage
➡ Carried 10 ft by two laborers (one board at a time)
◗ Boards stacked six high on horses
● Painter seals one edge (sprayed six at a time)
● Painter seals other edge (sprayed six at a time)
● Boards spread face up by two laborers
● Painter sprays face (sprayed six at a time)
● Painter wipes face (one at a time)
➡ Carried 30 ft by two laborers (one board at a time)
● Stacked on ground
▼ Stored until needed

SUMMARY

Operations	6
Transportations	2
Distance	40 ft (80 man-ft)
Delays	1
Storage	2

FIGURE C-7-2
Flow diagram and process chart of present method of sealing roof beams.

painter and laborers—be made independent of each other so that neither interferes with the other or has to wait for the other to finish?

1 The number of component tasks was reduced by following a different painting procedure. A special sawhorse was designed, as illustrated in Fig. C-7-3. By utilizing this horse to hold six beams, the top and one side of each beam could be sprayed and wiped without intermediate positioning. The remaining side to be painted was sprayed initially by spraying the whole face of the unsprayed stock at one time. Hence the material-handling operations by the laborers was reduced to simply loading and unloading the sawhorses with six beams at a time.

2 The second problem, that of eliminating the dependencies of the painter on the

FIGURE C-7-3
Proposed sawhorse design to improve beam sealing
operation.

laborers, and vice versa, could be solved by establishing two work centers adjacent to
each other, each with its set of horses and stacks of painted and unpainted beams. With
this configuration, the painter could spray the face of one pile and the edge and top of
the six beams on one set of horses while the laborers simultaneously unloaded and then
reloaded the other set of horses. Since both these operations take about the same
amount of time, each group could switch to the opposite set of horses without waiting
for the other.

Proposed Method   The proposed method, presented in Fig. C-7-4, includes procedures discussed above. By this method, the beam production could be increased to
216 beams per hour, with the painter 100 percent active and the laborers 90 percent
active. While it is recognized that this level of activity may be somewhat unrealistic,
the 216 beams per hour figure is the maximum production possible. Any decrease in
production would depend upon a conscious slackening of activity by the workers rather
than an inherent inefficiency in the work method. In summary, management (in this
case, the painter was the actual subcontractor) could realize a reduction in job labor
costs of up to 270 percent by initiating the few improvements suggested above.

The proposed flow diagram and process chart are shown in Fig. C-7-5. Note that the
work area has been compressed so that the workers will be required to move a minimal

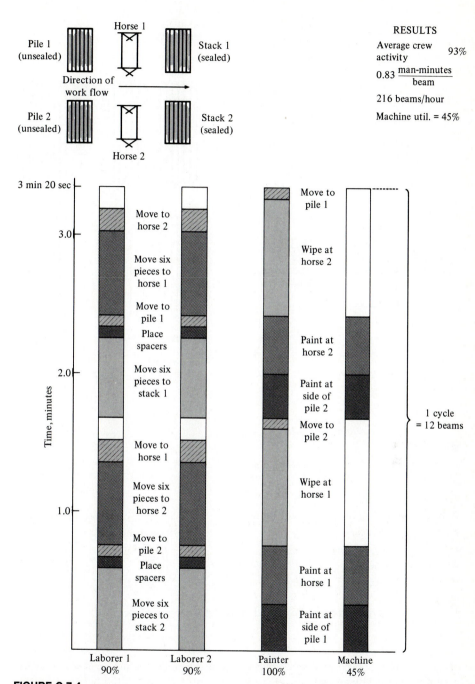

FIGURE C-7-4
Crew-balance chart of proposed method for sealing roof beams. Chart shows 12 beams being sealed in one cycle.

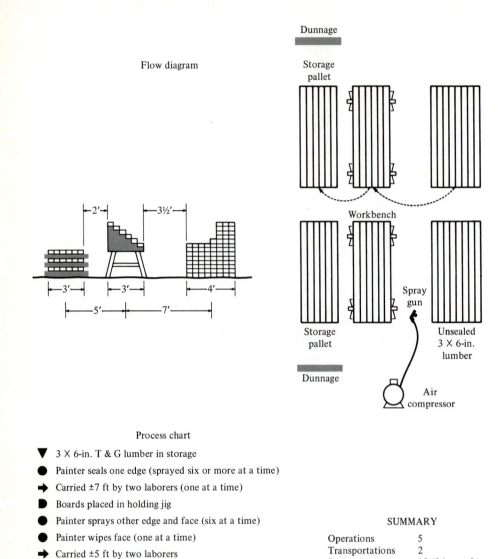

FIGURE C-7-5
Flow diagram and process chart of proposed method of sealing roof beams.

distance. The proposed process requires only five operations and transportation has been reduced to 24 man-feet.

By stacking the incoming materials in line, the horses can be advanced as the new material is depleted and the stored material increases, so the operation can move through the storage pile.

This proposed solution should not be considered the ultimate answer, but it does represent a large increase in potential with a very small capital outlay (the steps for the horses). The reader may be able to suggest other solutions.

EXAMPLE C-8

Gang Process Chart for Concrete Curb and Gutter Forms

Under certain circumstances, a gang process chart may have merit. In Fig. C-8-1 the participation of each worker and machine in each step of a process is shown on a single chart. It is common to include a cost breakdown for each operation directly on the chart so that the big-money items become readily apparent. Before the gang process chart can be prepared, the activities of each worker and each machine must be observed and timed, much as is done for other work-simplification methods.

The gang process chart of Fig. C-8-1 brought interesting results in one aspect of the process. The observer became concerned over the time required to set and stake the gutter board and to set and support the face board. A more detailed study indicated that much of the trouble was caused by the form boards splitting at the ends. The final outcome was to purchase a supply of patented clamps that largely eliminated the need for nailing. Lower labor costs and more reuses of the forms resulted. This example illustrates again one of the principal advantages of work-simplification studies: They force a critical examination of present methods.

Date **5/18/88**   Observer **J. Thomas**   Chart no. **18**   Job No. **172**

Subject **Forms for Concrete Curb & Gutter**   Cost code **42**

Work cycle **Move & set forms**   Quantity **206**   Unit **Lin Ft**

Equipment **Flat-bed truck – small tools**   Est. cost, pres. prop. _____

Superintendent **O'Brien**   Foreman **Martinez**   Act. cost, pres. prop. **$0.828**

Craft   No. in gang **8**

Distance, ft	Time, begin	Time, end	Time used, minutes	Foreman	Cem. Mason	Cem. Mason	Laborer	Laborer	Laborer	Laborer	Laborer	Truck #15 Driver	Cost of work during step or time interval	Cost of step	Step number	Description of step
800	8:00	8:15	15	①									$2.55	2.55	1	Inspect rough grade
1500	8:00	8:30	30		2→							2→	17.01	17.01	2	Bring up back board
1500	8:30	8:55	25		3→							3→	14.16	14.16	3	Bring up stakes
	12:30	1:15	45	④	④	④		④	④	④			37.44	37.44	4	Set & stake back board
	12:25	12:55	30							⑤			3.48	3.48	5	Distribute stakes
	12:55	1:15	20							⑥			2.31	8.10	6	Pull nails
1500	12:30	1:00	30		7→							7→	17.01	17.01	7	Bring up gutter board
1100	1:00	1:15	15		8→							8→	8.49	8.49	8	Bring up face board
	1:15	1:45	30	⑨	⑨	⑨		⑨	⑨	⑨	⑥		28.56	25.08	9	Set & stake gutter board
	1:45	2:05	20	⑩	⑩	⑩		⑩	⑩	⑩	⑥		19.05	16.74	10	Set & support face board
	2:05	2:45	40					⑪	⑪	⑪	⑪		18.81	18.81	11	Backfill forms
200	2:45	3:00	15	⑫									2.52	2.52	12	Final inspection
															13	
															14	

Total											17139	
Minutes	125	95	95	100	135	135	135	140	100	OK		
Hours	2.08	1.58	1.58	1.67	2.25	2.25	2.25	2.33	1.67			
$	48.76	18.94	14.94	11.64	15.69	15.69	18.69	16.26	45.00	179.76		

SUMMARY

Operations _____ **6**
Transports _____ **4**
Delays _____ **0**
Inspections _____ **2**
Quantity _____ **206 LF**
Unit cost _____ **$0.8.26**
Unit man hrs _____ **0.086**

FIGURE C-8-1
Gang process chart for form assembly and erection for concrete curb and gutter. (*Modification of form proposed by George E. Deatherage.*)

THE CAL/OSHA GUIDE FOR SAFETY IN CONSTRUCTION

This Appendix consists of abstracts from the *Cal/OSHA Guide for the Construction Industry,* which is distributed on request in a pocket-sized paperback. This guide is considered to be one of the best readily available safety checklists. Possibly one-third of the *Guide* has been selected for reproduction here; the sections chosen are those that are most pertinent to the more common forms of construction.

As noted at the beginning of the excerpt, the recommendations as such do not carry the force of law; their force lies in the clauses of the Administrative Code, which have been referenced. As a practical matter, however, failure to observe any of the recommendations could bring a citation that will, in most cases, be upheld by the courts.

This Appendix should be particularly useful to construction students or to others unfamiliar with construction operations in the field, since it provides a means of comparing actual safety practices with those considered proper by experts in the field. It can also be very useful to field supervisors, since it will remind them of rules that may, in the hustle and bustle of the job, be overlooked or forgotten.

In 1987, the governor of California abolished Cal/OSHA by vetoing funds for it, thus returning jurisdiction over safety to the federal government. His actions have been challenged in the courts. Regardless of the outcome of this legal action, the data that the *Guide* offers remain among the best available.

CAL/OSHA GUIDE FOR THE CONSTRUCTION INDUSTRY

This Guide is not meant to be a complete substitute for, or a legal interpretation of, the safety standards. Almost every sentence in this Guide is followed by a number. That number indicates the section in **TITLE 8** or the **CALIFORNIA LABOR CODE** from which information has been summarized. THE READER IS CAUTIONED TO REFER DIRECTLY TO TITLE 8 OR THE LABOR CODE FOR DETAILED AND EXACT INFORMATION, SPECIFICATIONS, OR EXCEPTIONS.

INTRODUCTION

Under the California Occupational Safety and Health Program (Cal/OSHA), the Department of Industrial Relations publishes a complete set of safety orders to provide safe working conditions in all places of employment in California. The *CALIFORNIA ADMINISTRATIVE CODE, TITLE 8—INDUSTRIAL RELATIONS* contains safety orders for specific industries, such as the *Construction Safety Orders,* as well as the *General Industry Safety Orders* which apply to all industries.

Employers must comply with all Cal/OSHA safety orders that apply to their type of business. All employers must meet reporting requirements, and employers with eleven or more employees must meet recordkeeping requirements, specified in the booklet *RECORDKEEPING AND REPORTING REQUIREMENTS UNDER THE CALIFORNIA OCCUPATIONAL SAFETY AND HEALTH ACT* available from the California Division of Labor Statistics and Research.

Employers are responsible for providing safe and healthful places of employment and working conditions for their employees. Employers must inform employees of their rights and obligations under the Cal/OSHA Program and post the Cal/OSHA poster *SAFETY AND HEALTH PROTECTION ON THE JOB* available from offices of the Division of Occupational Safety and Health.

The most effective way to prevent accidents and illness on the job is to have an accident and illness prevention program. *A WORKPLACE INJURY AND ILLNESS PREVENTION GUIDE* is available from the CAL/OSHA Consultation Service. *General Industry Safety Order 3203* and the *Construction Safety Orders* require employers to have an on-going accident prevention program which includes instruction to workers in safe work practices as well as scheduled periodic safety inspections of all work areas.

ACCIDENT PREVENTION PROGRAM REQUIREMENTS

The *Construction Safety Orders* require every employer to have an accident prevention program. The program must include:

● Frequent, regular inspections of each construction site, equipment, and vehicles by responsible supervisors.

● Immediate action to eliminate hazards.

● Monthly or more frequent meetings of all foremen (are recommended).

● Periodic (at least every 10 days) safety meetings of supervisor with employees.

● Adoption and use of a Code of Safe Practices.
1509

SAFETY PROGRAM RECOMMENDATIONS

An effective safety program should also include the following:

o Supervisors should be qualified in safety procedures and be held accountable for carrying them out.

o Accidents should be investigated to determine the cause.

o A plan of corrective action should be implemented following accidents or injuries.

o Safety effectiveness should be monitored by: the Insurance Experience Modification Rate and the Accident Incident Rate.

Safety Precautions

Every employer must require employees to use safety devices. The employer must use practices, means, methods, operations and processes which make the place of employment safe and healthful.
1511(a)

● Prior to starting work, the employer must survey the job site to determine the hazards and the safeguards necessary to ensure that work is performed safely. 1511(b)

● When a worker is first employed, he/she must be given instructions regarding job hazards, safety precautions and the employer's Code of Safe Practices. 1510(a)

● The employer shall permit only qualified or experienced employees to operate equipment or machinery. 1510(b)

ADMINISTRATIVE REQUIREMENTS UNDER CAL/OSHA

CAL/OSHA POSTER

The employer must display the Cal/OSHA Poster *Safety & Health Protection on the Job* in a conspicuous place at the jobsite. 8 CAC 340

CAL/OSHA RECORDKEEPING AND REPORTING

Every employer of 11 or more people must keep records and make reports as outlined in the booklet *Recordkeeping and Reporting Requirements Under the California Occupational Safety and Health Act,* available from the Division of Labor Statistics and Research. Every employer must report immediately by phone to a Division of Occupational Safety and Health (DOSH) District Office any work-related accident or illness causing serious injury or death to an employee. 8 CAC 342(a)

HOUSEKEEPING

Scrap lumber and debris must be kept reasonably cleared from work surfaces, passageways, and stairs. 1513(a)

Ground areas within 6 feet of buildings under construction must be reasonably free from irregularities. 1513(b)

Storage areas and walkways on construction sites must be maintained reasonably free of dangerous depressions, obstructions, and debris. 1513(c)

Piled or stacked material must be placed in stable stacks to prevent it from falling, slipping, or collapsing. 1549(a)

The face of a pile of cement bags (more than 5 feet high) must be tapered back. 1549(b)

Lumber piles must not exceed 16 feet in height if the lumber is handled manually, or 20 feet when handled with equipment. 1549(c)

PERSONAL PROTECTIVE EQUIPMENT
EYE AND FACE PROTECTION

Eye protection (or face shields) are required where there is an inherent probability of eye injury from flying particles, or light rays. 1516(a)

FOOT PROTECTION

Foot protection is required when workers are exposed to foot injury from hot, corrosive, or poisonous substances, or from falling objects.
1517(a)

HAND PROTECTION

Hand protection is required for workers exposed to cuts, burns, or corrosive, irritating substances.
1520(a)

BODY PROTECTION

Clothing appropriate for the work must be worn by workers. 1522(b)

Workers must not wear clothing saturated with flammable liquids, or corrosive or oxidizing agents.
1522(c)

HEARING PROTECTION

Noise levels above 90dBA are frequent during many construction operations. If noise exposures cannot be reduced to the allowable level, employees must wear acceptable hearing protection. 1521

HEAD PROTECTION

Employees exposed to flying or falling objects must wear approved head protection. 1515(a)

Note: Hard hats should be worn by everyone at a construction site.

RESPIRATORY PROTECTION

If engineering or operational controls are not practical to limit harmful exposure to airborne contaminants, a NIOSH-approved respirator must be worn by exposed employees. 1531(a)

Employees must wear supplied-air respirators (covering the head, neck and shoulders):
● During blasting, when dust may exceed limits specified in *GISO 5155.* 5151(b)(1)(B)
● During blasting with silica sand or where toxic material evolves. 5151(b)(1)(C)

Note: A dust filter respirator may be used for two hours, if concentration is less than ten times the limit in *GISO 5155.*

FIRST AID

First aid kits shall be provided on all job sites.
1512(c)

Emergency medical services shall be readily available. 1512(a)

Personnel trained and immediately available to provide first aid treatment must be provided.1512(b)

Note: Trained personnel must possess a current Red Cross (First Aid) certificate, or equivalent. 1512(a)(1)

SANITATION

An adequate supply of potable water must be provided at each job site. 1524(a)(1)

A toilet is required at each job site. 1526(b)

One toilet is required for each 20 employees or fraction thereof of each sex; urinals can be substituted for one half of the units. 1526(a)

Toilets must be kept clean and supplied with toilet paper. 1526(d)

Toilets are not required for mobile crews, provided transportation is available to toilets nearby. 1526(e)

Adequate washing facilities must be provided for employees engaged in operations involving harmful contaminants, paints, or coatings. 1527(a)

FIRE PROTECTION, PREVENTION and FIGHTING

The employer is responsible for establishing an effective fire prevention program. 1920(a)

SAFE LIFTING PROCEDURES

Safe manual lifting practices should be followed.
1510(a)

WORK SURFACES GUARDING
ACCESS

Safe, unobstructed access is required to all working and walking surfaces regardless of the conditions or duration of the job. The method of access may be by ladder, ramp, stairs or their equivalent and must be installed at all points where there is a break in elevation of 18 inches or more, at frequently traveled passageways, entries or exits. 1629(a)(3)

ACCESS PROHIBITIONS

—Gaining access by climbing structures or walking on undecked floor supports is prohibited. 1629
—At elevations over 15 feet high, walking on open beams or sliding down beams without safety belts is prohibited. Iron workers are not exempt.
1670
—Riding on loads, hooks, or slings of a derrick, hoist, or crane (except in unusual situations under controlled conditions) is prohibited. 1718(a)
—Riding concrete buckets is prohibited. 1720(c)(3)

STAIRWAYS

Stairs are a safe method of access to floors and working levels of buildings and scaffolds, and must be installed as follows:
● Stairs must be installed prior to raising studs for the next higher floor. 1629(b)(1)
● In steel frame buildings, stairs must be installed to each planked floor. 1629(b)(2)
● In concrete buildings, stairs must be installed to the floor which supports the vertical shoring system. 1629(b)(3)
● For buildings of three stories or less, at least 1 stairway is required. 1629(a)(4)
● For buildings of more than 3 stories, 2 or more stairways are required. 1629(a)(4)

Buildings 60 feet or more in height or 48 feet in depth require an elevator. 1630(a)

Stairs must be at least 24 inches wide and equipped with treads and handrails. 1629(a)(2)

Temporary stairs must have a 30-inch wide landing for every 12 feet of vertical rise. 1629(a)(2)

Stairs must be properly illuminated. 1626(c)

LADDERS

Ladders can be used for access to working surfaces above and below ground level on short-duration jobs prior to the installation of a permanent means of access.

Ladders may be used to provide access to:
° Building levels prior to installation of stairs
1629(a)(3)
° Roofs (steel buildings) 1629(b)(2)
° Scaffolds 1637(l)
° Levels in a vertical shoring system (concrete buildings) 1629(b)(3)

557

Ladder Specifications:

° Rungs must be at least 11½ inches long and spaced 12 inches apart vertically. 1675(h)

° Step ladders must not exceed 20 feet in length. 3278(d)(1)(A)

° Cleat ladders must not exceed 30 feet in length. Double cleat ladders are required for two-way traffic or when used by 25 or more employees. 1629(c)

° Extension ladders shall not exceed 44 feet in length. 1678(a)

° Overlapping section should not be less than 10 percent of the working length. 1678(b)

Job-built Ladders

Ladders constructed on the job must safely support the intended load, and:

Rungs must be made from clear, straight-grained lumber. 1676(c)

Cleats shall be uniformly spaced, 12 inches top-to-top. 1676(j)

Cleats must be nailed at each end with 3 ten-penny nails or equivalent. 1676(j)

Cleats must be blocked or notched into the side rails. 1676(j)

Width of single cleat ladders shall be 15 to 20 inches. 1676(f)

Rails must be made from select Douglas Fir without knots (or equivalent). 1676(b)

Rail splicing is permitted only if there is no loss of strength. 1676(b)

Ladder Safety:

—Face the ladder and keep hands free for climbing. 3278(c)(12)

—Don't stand on the top 3 rungs of rung ladders without handholds unless you are protected by a safety belt. 1675(k)

—Remove damaged ladders from use. 1675(b)

—Don't place ladders where they can be accidently struck or displaced. 1675(h)

—Secure ladders in use against displacement. 1675(j)

—Extend ladder side rails at least 3 feet above the landing, unless handholds are provided. 1675(i)

—Place ladders at approximately 75° pitch. 1675(g)

—Don't place planks on top (cap) of stepladders. 1675(f)

—Don't splice ladders together. 3278(e)(14)

—Do not use metal ladders for electrical work or near live electrical parts. 1675(l)

—Mark portable metal ladders:
CAUTION—DO NOT USE AROUND ELECTRICAL EQUIPMENT 3279(d)(12)

RAMPS AND RUNWAYS

Properly designed ramps provide safe means of access for foot or vehicle traffic.

Open sides and ends of ramps 7½ feet or more high must be guarded with standard railings. 1621(a)

Foot Ramps:

Ramps must be at least 20 inches wide and have a surface that does not spring. 1624(a)

If the ramp slope exceeds 2:10, beveled cleats should be placed 16 inches apart. 1624(b)

Wheelbarrow Ramps:

Ramps or runways must be firmly secured against displacement. 1623

Ramps over 3 feet high must be 2½ feet wide and planks must be firmly cleated together. 1623

Powered-buggy Ramps and Runways:

Ramps and runways must be at least 5 feet wide and designed with turn-outs for two-way traffic. 1625

Ramps must be designed to support at least 4 times the maximum load and curbs must be constructed (equivalent to 4" x 4" lumber.) 1625

Falsework ramps and runways must be designed for a load of 125 pounds per square foot. 1717(a)(2)

CONSTRUCTION ELEVATORS
Requirements:

o In addition to stairways, a construction elevator is required for structures or buildings 60 feet or more above ground or 48 feet below ground level. 1630(a)

o An elevator is required at demolition sites of 7 or more stories or 72 feet or more in height. 1735(s)

o Personnel hoists may be used at special construction sites, such as bridges and dams, if approved by a registered professional engineer. 1604.1(c)

o Use of endless belt-type (or wire rope guided) manlifts is prohibited. 1604.1(a)(3)

o Prior to use, construction elevators must be inspected and tested in the presence of a DOSH representative. A permit to operate is required. 1604.29(a)

o Ropes must be inspected at least once each 30 days, and records must be kept. 1604.25(j)

o A capacity plate must be posted inside the car. 1604.21(b)

o Statement—Compliance with American National Standards Institute (ANSI) A92.3-1973. 3638(b)(6)

Employees must receive operating instructions. 3638(c)

Platforms must have guardrails. 3642(a)(1)

Minimum platform width is 16 in. 3642(a)(4)

Powered equipment must be equipped with an emergency lowering means. 3642(c)

Powered units must have upper and lower controls, and be guarded against accidental operation. 3642(d)

Pinch and shear points must be guarded. 3643(a)

Operation—Refer to *GISO 3646* for operation guidelines.

EXCAVATIONS, TRENCHES, EARTHWORK

A permit by DOSH is required before starting work on excavations 5 feet deep or more which workers are required to enter. 341

Before Excavating:

● Determine the location of underground utilities and notify the owners. 1540(a)(1),(3)

● Inspect the area for hazards from moving ground. 1540(c)(1)

● Inspect the excavation after every rainstrom, earthquake, or other hazard-increasing occurrence. 1540(c)(2)

● Inspect the face, banks, and top daily when workers are exposed to falling or rolling material. 1546(a)

Hazards

Remove trees, poles, and boulders which may be hazardous. 1540(b)(4)

Shore, bench, slope, or use equivalent methods to protect workers in excavations 5 feet deep or more. 1540(d)

Locate spoil at least 2 feet from the edge of the excavation, or 1 foot from the edge when the excavation is less than 5 feet deep.   1540(e)(1)

In trenches 4 feet deep or more provide safe access within 25 feet of any work area.   1540(g)(2)

Install crossings with standard guardrails and toeboards when the excavation is more than 7½ feet deep.   1540(h)(2)

Do not excavate beneath the level of adjacent foundations, retaining walls, or other structures until a qualified person has determined that the work will not be hazardous. Support undermined sidewalks.   1540(j)(1),(2)

Shore, brace, or underpin structures when their stability is threatened. Inspect the structures daily.   1540(j)(3)

Erect barriers around excavations in remote work locations. Cover all wells, pits, shafts, and caissons. Backfill temporary wells, pits, and shafts when work is completed.   1540(m)(1)(A),(B)

Use additional bracing when vibration or external loads are a hazard.   1540(o)

Use standard shoring, sloping, and benching (See CSO 1541 and 1937 for specific requirements.), or shore according to plans prepared by a civil engineer registered in California.   1541

Shafts

Retain all wells and shafts over 5 feet deep which workers are required to enter, with lagging, spiling, or casing.   1542(a)(1)

Earthwork and Excavating

Install a bench or other method of working if the height and condition of the face pose a hazard.   1544(a)

Overburden

Do not allow a worker under a face or bank where stripping or other work constitutes a hazard. Use barriers or other devices to protect workers from rolling or sliding material.   1545(a)(b)

Face Inspection and Control

Prohibit overhanging banks except:
- When material is moved by mechanical means with controls at a safe distance
- When the bank is undercut by a stream and the monitor is located a safe distance from the bank.   1546(c)(1),(2)

When falling rock is a hazard, station a worker at the face to give warning.   1546(d)

Provide enough light for safe night work.   1546(e)

Post KEEP OUT signs at dangerous non-work areas, or barricade such areas.   1546(g)

Protection of Workers at the Face

Prohibit work above or below workers if it is a hazard.   1547(a)

On top of the bank:
- Fence with guardrails or ropes.
- Use a railed platform.
- Use safety belts and lieflines.
- Use portable staging.
- Use a boatswains chair or skips especially designed for faces. When using a boatswains chair, use a safety belt and a lifeline equipped with a descent control.   1547(b)(2)

At the foot of the bank:
- Remove loose rock.
- Maintain a ready exit.   1547(b)(5)

Larger shafts over 4 feet square must be guarded by a lagging system made from proportionally larger materials (determined by a registered safety engineer)   1542(c)

GUARDRAILS

The open sides of all work surfaces 7½ feet high or higher must be guarded by railings (or periphery cables, or workers must be protected by safety belts or nets.)   1621(a)

TOEBOARDS are required on scaffolds and interior building openings if workers work or pass below.

Exception: structural steel crafts   1621(b)

SAFETY BELTS AND SAFETY NETS

SAFETY BELTS

If not otherwise protected, workers must use safety belts and lifelines when:
- Work is performed from thrustouts or similar locations.   1669(a)
- Exposed to the hazard of falling more than 15 feet from buildings, bridges, structures, or construction members such as trusses, beams, purlins, or plates.   1669(a)

Exception: When the installation of safety belts causes a greater hazard, then adequate risk control must be maintained under constant and immediate supervision.   1669(c)

SCAFFOLDS

GENERAL REQUIREMENTS:

- Scaffolds must be provided for work that cannot be done safely while standing on solid construction at least 20 inches wide, or from ladders.   1637(a)

    Exception: short duration work less than 15 feet high and less than 20 inches wide.

- Scaffolds must conform to design standards, or be designed by a licensed engineer.   1637(i)
- Standards are based on stress grade lumber; metal or aluminum may be substituted if structural integrity is maintained.   1637(b)

Required safety factors:
— Hardware, beams and components—4   1637(f)
— Ropes—6   1658(e)
— Platforms must be capable of supporting the intended load.   1637(k)
— A permit from DOSH is required for the erection of scaffolds exceeding 3 stories or 36 feet in height.   341(a)(2)

ACCESS

Safe, unobstructed access must be provided to all platforms.   1637(l)

Ladders must conform to ladder standards, and rails must extend 3 feet above the platform or handholds must be provided.   1644(a)(7)

SECURING SCAFFOLDS

Scaffolds must be built solidly and tied-off with a double wrap of No. 12 wire (or equivalent).1640(a)(3)

Wood pole scaffolds must be tied every 20 feet horizontally and vertically for light trades. 1640(a)(3) and every 15 feet for heavy trades. 1641(d)(1)
Metal scaffolds must be tied every 26 feet vertically and every 30 feet horizontally 1644(a)(4)

PLANKING
Standard: 2 x 10 in. (nominal) 1637(e)
Span: (maximum)
o Light Trades @25 psf—10 ft.
o Medium Trades @50 psf—8 ft.
o Heavy Trades @75 psf—7 ft. 1637(e)
Overlap (ledger/support)—6 in. (minimum) 1640(a)
—18 in. (maximum) 1637(e)

RAILINGS
Standard guardrails must be installed on open sides and ends of scaffold platforms which are 7½ feet or higher. 1621(a)

Exceptions: 1644(a)(5)

— X-brace to substitute for midrail;
X-brace must intersect 20–36 inches above platform.
— X-brace to substitute for top rail;
X-brace must intersect 42–48 inches above platform and midrail must be 19–25 inches above platform. Maximum vertical distance allowed between centers of each X-brace is 48 inches at the uprights.

Toeboards are required where workers work or pass below. 1621(b)
Exceptions: suspended scaffolds, horse scaffolds, ladderjack scaffolds

HEIGHT LIMITS
Maximum height (unless designed by a licensed engineer):
o Wood (frame/post)— 60 ft. 1643
o Tube and coupler — 125 ft. 1644(b)(5)
o Tubular welded — 125 ft. 1644(c)(7)
o Horse (single) — 16 ft. 1647(e)
o Horse (tiered) — 10 ft. 1647(e)

PROHIBITED SCAFFOLDS:
o shore scaffolds o brick or blocks o lean to scaffolds
o nailed brackets o jack scaffolds o. loose tile
o stilts o unstable objects 1637(h)

SPECIALIZED SCAFFOLDS
The requirements listed below are unique to each specialized scaffold and replace or augment the general requirements.

Tube and Coupler Scaffolds
All posts, ledgers, ribbons and bracing must be constructed from 2 inch OD steel tubing. The top work level and all additional levels must also be planked. 1644(b)

Tube and Coupler Scaffold systems require:

— X-bracing, across the width of the ledger frame, every 3rd ledger horizontally (e.g. 1, 3, 6, etc.) on the first and every fourth tier (see Illustration A). 1644(b)(10)
— Longitudinal diagonal bracing (at approximately 45°) installed across the entire front and back faces of the scaffold system every 5th frame horizontally and in both directions (see Illustration A). 1644(b)(11)

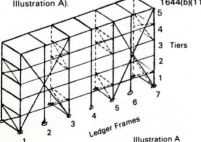

Illustration A

Tubular Welded Scaffolds
These units are commercially fabricated. Panels/frames must stack or nest on each other in true alignment.

Frame connections must be pinned. 1644(c)

TOOLS
Tools must be kept clean and in good repair. 1699(a)

Only trained or experienced employees may operate tools, machines, or equipment. 1510(b)

Control Switches (Powered Hand Tools)
The following tools must be equipped with a constant contact 'on-off' switch: 1707(a)(1)

* Drills
* Tappers
* Fastener Drivers
* Grinders
* Disc and Belt Sanders
* Reciprocating and Saber Saws
* Tools similar to those above

Circular (Skil) Saws
Chain Saws (electric, hydraulic, and pneumatic)

* An additional single motion lock-on control is also permitted.

Exceptions:
Concrete Vibrators
Concrete Breakers
Powered Tampers
Jack Hammers
Rock Drills
Tools similar to those above

Gasoline-powered chain saws must be equipped with a constant pressure throttle which will shut off the power to the chain drive when released. 1707(a)(1)

Hoisting or lowering electric tools by their cords is prohibited. 1707(b)(1)

POWER TOOLS
Power-operated tools should be grounded or be of the double-insulated type. They should be kept out of wet locations. 2395.45

PNEUMATIC TOOLS
Safety clips are required to prevent dies from being accidentally expelled from the barrel. 1695

Pneumatic Nailers and Staplers
When operating at more than 100 psi pressure, the tool must have a safety device to prevent operation (when the muzzle is not in contact with the surface). 1704(a)

When not in use, pneumatic tools must be disconnected from the air supply at the tool. 1704(b)

When the tool is in use, all personnel except tool operators may not come within 10 feet of the area. 1704(d)

Tools must have a pressure reduction safety device at the air pressure source on all air hoses larger than ½ inch. 1704(c)

Jack Hammers
Operators must wear hearing protective devices if

the noise level exceeds allowable exposures. 5099
Allowable exposure:

Sound Level (dbA)	Time Per Day (hrs.)
90	8
95	4
100	2
105	1
110	½	5097(a)

HEAVY CONSTRUCTION EQUIPMENT

GENERAL

- Repairs must not be made to powered equipment until workers are protected from movement of the equipment or its parts.   1595(a)
- Wherever mobile equipment operation encroaches upon a public thoroughfare, a system of traffic controls must be used.   1598(a)
- Flagmen (wearing orange vests) are required at all locations where barricades and warning signs cannot control the moving traffic.   1599(a)
- Jobsite vehicles must meet design requirements as follows:
   —Operable service, emergency, and parking brake.   1597(a)
   —Two operable headlights and taillights for night operation.   1597(b)
   —Windshield wipers and defogging equipment as required.   1597(d)
   —Seatbelts are required if the vehicle has Roll-Over Protection Structures (ROPS).   1597(g)
   —Fenders or mud flaps are required.   1597(i)
- Vehicles used to transport employees must have adequate seating.   1597(f)
- Vehicles (and systems) must be checked for proper operation at the beginning of each shift. 1597(j)

ROLL-OVER PROTECTION (ROPS) and SEAT BELTS

ROPS and seat belts are required for the following equipment with a brake horsepower rating above 20:

   crawler tractor
   bulldozer
   front end loader
   motor grader
   scraper
   tractor (except side boom pipe laying)
   water wagon prime mover
   rollers and compactors (over 5950 lbs.)   1596(a)

HAULAGE AND EARTH MOVING EQUIPMENT

GENERAL

Every vehicle (with a body capacity of 2.5 cubic yards or more) used to haul construction material must be equipped with an automatic back-up alarm which sounds immediately on backing.   1592(a)
All vehicles and graders must be equipped with a manually operated warning device.   1592(c)

CRANES

GENERAL

Each crane, derrick, and cableway exceeding 3 tons capacity must be certified annually by a qualified person.   1588.2(a)

All cranes must be equipped with an operable warning device controllable by the operator.
1582.17
No crane shall be operated with wheels or tracks off the ground unless properly bearing on outriggers.   1587.4(a)
A signalman shall be provided when the point of operation is not in full and direct view of the crane operator.   1587.12(a)

ROOFING OPERATIONS

(Hot operations only)
Do not carry hot up ladders.   1725(a)
Station an attendant within 100 feet of any kettle not equipped with a thermostat.   1725(d)
Do not locate LPG cylinders where the burner will increase the temperature of the cylinder.   1725(g)
A Class BC fire extinguisher shall be kept near each kettle in use as shown below:

   Less than 150 gallons (kettle capacity)   8:B.C.
   150 to 350 gallons ....   16:B.C.
   Larger than 350 gallons ....   20:B.C.

Coal Tar Operations

Use skin protection.   1728(a)
Washing or cleansing facilities must be available.
1728(c)
Use respirators and eye protection in confined spaces not adequately ventilated.   1728(b)

Bucket Capacity (Max.)

   Carry buckets .... 6 gal.
   Mop buckets .... 9½ gal.   1729(a)(2)(4)

Compressed-air-fueled kettles must have a relief valve set for a pressure not to exceed 60 psi.
1726(c)

Work on roofs over 20 feet high or on roofs with equipment which the operator must pull backwards requires the following protective devices:

— for monolithic roof coverings (e.g. built-up roofs) where the slope is 0:12 through 4:12:

   warning lines and headers or,
   safety belts and lines or,
   catch platforms with guardrails or,
   scaffold platforms or,
   eave barriers or,
   standard railings and toeboards   1730(b)

— for monolithic roof coverings where the slope exceeds 4:12:

   parapets, 24 inches or higher or,
   safety belts and lines or,
   catch platforms or,
   eave barriers or,
   standard railings and toeboards   1730(c)

Index